THE PRINCIPLES & PRACTICE OF BREWING

BEER & ALE

Second Edition

by

Walter J. Sykes

Wexford Press
2008

PREFACE.

AMONG the achievements of the latter half of the present century, few have been fraught with more beneficial consequences than the practical application of the deductions of science to manufacturing operations. Nearly every important industry has passed from the empirical to the scientific stage. This revolution has not been effected at a leap, but has been of relatively slow growth. Much prejudice had to be overcome, and for a long time the intervention of science was regarded with considerable mistrust. Perhaps in no other industry was this the case more than in that of brewing. Now all this is changed; the benefits which science is able to confer are generally recognised and appreciated, and where science was formerly repulsed, she is now welcomed with open arms. That this is true with regard to the fermentation industries, is shown by the number of Institutions which have arisen in different parts of the world, during the last ten years, with the object of investigating the laws which govern the processes and operations carried on in malting, brewing, distilling, &c. In our own country, which has been one of the most backward in this respect, we have seen the founding of the Institute of Brewing and other kindred Institutions, which, together with their conjoint Journal, pay considerable attention to the scientific aspect of brewing.

During this short period an enormous amount of experimental investigation has taken place, the records of which are scattered over a vast amount of literature. It is the object of the present work to bring together, within the limits of a volume of a handy size, the salient points of this mass of literature, and to present, as it were, a bird's-eye view of the science and art of brewing.

The two fundamental principles on which so much depends, viz., the action of diastase on starch, and the alcoholic fermentation caused by yeast, have been treated at considerable length, and as far as possible in historic sequence, so as to present these subjects to the reader in the most interesting form. In the Chemical portion of the book an attempt has been made to classify the substances used by brewers, and those appertaining to the science of Brewing, in a systematic form consonant with the arrangement generally adopted in modern works on organic chemistry. Chapters on the Physics of brewing, such as heat and density, are also included. Considerable space has been allotted to a description of elementary Biology, a subject of great importance to the brewer, but one which has been strangely neglected in most works on brewing. The general methods of bacteriological research are also given, and a chapter is devoted to instructions for the use of the microscope, without the aid of which instrument bacteriology and the culture of pure yeast would be an impossibility. The interpretation of the results of the analysis of the materials used in brewing has been given, but the analytical methods of testing them have been omitted, as they would have made the book too voluminous, and constitute in themselves a subject sufficiently large to form a separate treatise.

In the practical portion of the work, the general methods employed in Malting and Brewing are given, but it is evidently impossible to include all the minor details, since these differ in almost every brewery, and can only be learned by actual experience in the brewery itself. The author trusts that the work will be found not only useful to those who are commencing a brewing career, and for whom it is primarily intended, but also to older, practical men who may be desirous of increasing their knowledge of the scientific side of brewing.

With these objects in view, he has attempted to present each subject in as lucid and intelligible a manner as possible, often, it is to be feared, at the sacrifice of literary elegance. Numerous references are given throughout the work solely for the benefit of those who wish to enlarge their knowledge of any particular subject; in the great majority of cases these have been verified from the originals, and the remainder have invariably been taken from reliable sources.

The author would here express his indebtedness to many friends for much kind assistance in the compilation of the work, particularly to Mr. John Marriott and Mr. Harris S. Beaman, who have revised, respectively, the sections on Practical Malting and Practical Brewing; also to Mr. A. Ashe for kindly looking over the microscopical portion, and to Mr. Arthur R. Ling (Editor of the *Journal of the Federated Institutes of Brewing*), and Mr. Charles A. Mitchell for assistance in correcting the proofs. A few of the illustrations in the bacteriological portion are from blocks lent by Mr. C. Gerhardt, of the Chemical Instrument Repository, Bonn, and some of those of brewery utensils are supplied by Messrs. Llewelyn & James of Bristol.

WALTER J. SYKES, M.D.

CROYDON, *August,* 1897.

TABLE OF CONTENTS.

INTRODUCTION 1

PART I.

CHAPTER I.

PHYSICAL PRINCIPLES INVOLVED IN BREWING OPERATIONS.

	PAGE		PAGE
Heat	3	Specific Heat	10
Definition	3	Latent Heat	12
Conduction	3	Evaporation	14
Convection	4	Density and Specific Gravity .	15
Radiation	5	Density	15
Exhaustion	5	Specific Gravity . . .	15
The Thermometer . . .	6	Hydrometers	18
Heat Unit	10	Saccharometers . . .	23

CHAPTER II.

CHEMISTRY, WITH SPECIAL REFERENCE TO THE MATERIALS USED IN BREWING.

The Elements	25	Hydrazones	39
Combining Proportions . .	25	Osazones	41
Valency	26	Optical Stereoisomerides . .	41
Molecules	28	Constitution of the Sugar Group	45
Chemistry of the Carbon Compounds	29	Disaccharide Sugars . . .	49
		Sugar Estimation . . .	51
The Paraffins	30	Cupric Reducing Power . .	53
The Alcohols	31	Solution Factor . . .	53
The Aldehydes	32	Molecular Weight . . .	55
The Aldoses	33	The Polarimeter and Specific Rotatory Power . . .	57
The Ketones and Ketoses . .	34		
The Acetic Acid Series . .	33	Cellulose	62
The Compound Ethers . .	36	Starch	64
The Amides	37	Granulose	68
Phenylhydrazine . . .	38	Action of Iodine on Starch .	68

xii   CONTENTS.

	PAGE		PAGE
Gelatinisation of Starch	69	Investigations of O'Sullivan	126
Constitution of the Starch Molecule	70	Musculus and Gruber	132
		Brown and Heron	133
Soluble Starch	72	Brown and Morris	142
Amylodextrins	74	Maltodextrin	146
Maltose	76	Molecular Weight of Starch	149
Glucose	78	Brown and Morris's Second Hypothesis	150
Fructose or Lævulose	81		
Cane Sugar	83	Isomaltose	151
Invert Sugar	84	Dextrins	155
Raffinose	86	Investigations on Isomaltose by Ling and Baker	158
α and β Amylan	86		
Gum	87	Brown and Morris	160
Furfural	87	Ost	161
Pectin	88	Metamaltose	164
The Proteids or Albuminoids	89	Achroodextrin III.	164
The Proteids of Barley	92	Ling and Baker on the Dextrins	165
The Proteids of Maize	96	Action of Enzymes on Proteids	165
The Proteids of Malt	96	Nitrogenous Constituents of Malt	169
The Amides	99		
Asparagine	99	Hops	170
Glutamine, &c.	100	The Hop Plant	171
The Enzymes	101	Chemical Constituents of Hops	173
The Diastases	105		165
The Enzymes of Yeast	112	Essential Oil	173
The Chemical Composition of Yeast	114	Resins	174
		Tannins	175
Action of Diastase on Starch	122	Diastase of Hops	177

PART II.

CHAPTER III.

THE MICROSCOPE.

Microscope, General Description of the	179	Bacteriological Methods	205
		Sterilisation	205
Nose-Piece	183	Thermostat	207
Stage	183	Sterilising Apparatus	208
Diaphragms	184	Culture Media	209
Sub-stage	184	Cultivation Plate	211
Objectives	185	Moist Chamber	213
Eye-Pieces	188	Incubator	213
Illuminating Apparatus	188	Colonies	213
Choice of Microscope	189	Impression Preparations	215
Micro-Chemical Reagents	191	Petri Dishes	215
Microscopical Manipulation	192	Bacteriological Examination of Water	216
Examination of Yeast	198		
Hanging-Drop Method	199	Hansen's Method	218
Examination of Bacteria	200	Wichmann's Method	221
Microscopical Preparations	203	Bacteriological Examination of Air	222

CHAPTER IV.

VEGETABLE BIOLOGY.

	PAGE
The Living Cell	224
Protococcus pluvialis	224
Osmosis	229
The Yeast Plant	232
The Wild Yeasts	237
The Mycoderms	239
The Torulæ	239
The Bacteria	239
Fermentation and Putrefaction	243
Bacterium termo	245
Butyric Acid Bacteria	246
Bacillus viscosus	247
The Acetic Bacteria	248
The Sarcinæ	249
The Mould Fungi	250
Mucor mucedo	250
Mucor racemosus	254
Other Mucors	255
Simple Multicellular Organisms	256
Penicillium glaucum	256
Eurotium aspergillus glaucus	260
Botrytis cinerea	262
Aspergillus oryzæ	264
Chalara mycoderma	264
Oidium lactis	265
Oidium lupuli	265
Dematium pullulans	266
Monilia candida	268
Mould Fungi Dangerous on Brewing Premises	268
The Higher Plants	270
Germination of Barley	271
Structure of Barleycorn	274
Modification of Endosperm	275
Parasitic Nature of Germ	278

CHAPTER V.

FERMENTATION.

	PAGE
Fermentation, Antiquity of	280
Views on, during the Period of the Alchemists	281
during the Period of Iatro-Chemistry	282
during the Age of Phlogiston	284
during the Period of the Foundation of Modern Chemistry	285
Views of Liebig on	289
Views of Traube	291
The Physiological Theory of Fermentation	292
Doctrine of Spontaneous Evolution	294
Pasteur on Spontaneous Evolution	296
Tyndall's Experiments	299
Sterilisation of Organic Fluids	301
Competition amongst Micro-Organisms	303
Transformation of Organisms	304
Distribution of Atmospheric Germs	305
Miquel's Experiments on	310
Koch's and other Methods of Air Investigation	310
Hansen's Investigations on the Air of Breweries	313
Pasteur's Experiments on Fermentation	314
Mould Fungi as Ferments	317
Pasteur's Experiments on Yeast	318
Growth with and without Air	319
Pasteur's Theory of Fermentation	323
Meyer's Theory	327
Brefeld's	327
Schützenberger's	329
Nägeli's Moleculo-Physical Theory	330
Hoppe-Seyler's Theory	334

xiv   CONTENTS.

	PAGE		PAGE
Adrian Brown's Experiments on Fermentation.	335	Hansen's and other Pure Yeast Cultivation Apparatus	351
Investigations of Hansen	342	Advantages of Hansen's Pure Single-Cell Yeast	360
Pure Cultures from a Single Cell	343	Differences in the Action of the various Yeasts	362
Varieties of Yeast	345	Energy of Fermentation	365
Hansen's Six Species and Varieties.	345	Hayduck's Apparatus for Determining	365
Introduction of Pure Yeast Cultures into the Brewery	349		

PART III.

CHAPTER VI.

WATER.

Water, Chemical Composition of	367	Artificial Treatment of Waters	376
Occurrence	367	Kainit	378
General Constituents	370	Influence of Boiling	382
Statement of Results of Analysis of	370	Organic Constitution	382
Hardness	372	Effect of Filtration	383
Waters Suitable for the Production of Different Classes of Ale	373	Methods of Water Analysis	384
		Microscopic Examination of Water Sediments	392

CHAPTER VII.

BARLEY AND MALTING.

Barley	394	Saladin's	411
Choice of	395	Hemming's	411
Vitality	395	Drying Kiln	413
Age	397	Drying	414
Malting	400	Changes Effected in Drying	415
Steeping	401	Storage	416
Steep-Water	402	Chemical Examination of Barley	417
Germination of Barley	405	Malt Substitutes	418
Flooring	407	Quality of Malt	420
Sprinkling	407	Chemical Examination of Malt	422
Withering	409	Ready-formed Sugars	426
Pneumatic Malting	409	Maltol	427
Galland's System	409		

CONTENTS.

CHAPTER VIII.

BREWERY PLANT.

	PAGE		PAGE
Gravitation Brewery	428	Hop Back	442
Cold and Hot Liquor Backs	430	Coolers	442
Malt Mill	431	Refrigerators	442
Grist Case	434	Collecting Vessel	443
Mash Tun	434	Fermenting Vessels	444
Rakes	436	Burton Union System	448
Mashing Machines	436	Attemperators	448
Sparge	438	Parachutes	449
Coppers	439	Racking Squares	451
Fire	439	Vats and Casks	451
Steam	441		

CHAPTER IX.

BREWING.

	PAGE		PAGE
Estimation of Quantities for the Brew	453	Addition of Yeast	472
Amount of Liquor Required	454	Change of Yeast	475
Hardening Materials	455	Fermentation Temperatures	476
Mashing	456	Dressing	476
Use of Subsidiary Apparatus	459	Appearance of Heads	477
Black Beers	462	Cleansing System	477
Sparging	462	Stone Square System	479
Boiling	464	Settling and Racking	480
Action of Hop-tannin Bodies	467	Dry Hopping	481
Cooling	469	Secondary Fermentation	481
Refrigerating	470	Priming	482
Collection of Wort	470	Antiseptics	483
Extract Yielded	471	Fining	484
Fermentation	472	Bottled Ales and Bottling	487

CHAPTER X.

BEER AND ITS DISEASES.

	PAGE		PAGE
Flavour and Aroma	490	Brightness	493
Condition	491	Turbidity	493
Palate Fulness	491	Ropiness	495
Head	492		

	PAGE
BIBLIOGRAPHY	496
APPENDIX A.—SOLUTION WEIGHT AND SOLUTION FACTOR	497
APPENDIX B.—SPECIFIC ROTATORY POWER	498
APPENDIX C.—THE LAW OF DEFINITE RELATION	499
APPENDIX D.—ALCOHOLIC FERMENTATION WITHOUT YEAST-CELLS	499
APPENDIX E.—FERMENTATION IN A VACUUM	501
INDEX	503

LIST OF ILLUSTRATIONS.

FIG.		PAGE
1.	Specific Gravity Bottle	17
2.	Hydrometer	18
3.	Hydrometer	19
4.	Hydrometer	20
5.	Valency of Elements	27
6.	Laurent Polarimeter	58
6a.	Schmidt and Haensch Half-shadow Polarimeter	59
7.	Starch Cells from Barley, Wheat, Rice, Maize	66
7a.	Starch Cells from Potato and Sago	67
8, 9, 10.	Seed of the Hop	172
11.	Microscope	181
12.	Hot-air Oven	206
13.	Thermostat	207
14.	Koch's Steam-Steriliser	208
15.	Basket for Steam-Steriliser	208
16.	Autoclave	209
17.	Plate-Box	212
18.	Levelling Apparatus	212
19.	Moist Chamber	213
20.	Support for Cultures	213
21.	Incubating Chamber	214
22.	Petri-Dishes	215
23.	Wolfhügel's Counting Apparatus	217
24.	Protococcus pluvialis	225
25.	Osmotic Apparatus	229
26.	Budding Yeast	232
26a.	Yeast Spores (*Hansen*)	235
26b.	Germination of Yeast Spores (*Hansen*)	236
27.	Saccharomyces apiculatus (*Hansen*)	238
28.	Schizo-saccharomyces pombe (*Lindner*)	238
29.	Bacteria (*Baumgarten*)	240
29a.	Bacterium termo	245
30.	Butyric Acid Bacteria	247
31.	Bacterium aceti (*Hansen*)	248
32.	Bacterium Pastorianum (*Hansen*)	248
33.	Bacterium Kützingianum (*Hansen*)	248
34.	Mucor mucedo (*Brefeld and Kny*)	251
35.	Gemmæ of Mucor	254
36.	Mucor circinelloides (*Van Tieghem and Gayon*)	255
37.	Penicillium glaucum (*Brefeld and Zopf*)	257
38.	Eurotium aspergillus glaucus (*De Bary*)	260
39.	Botrytis cinerea (*De Bary*)	262
40.	Chalara mycoderma (*Hansen*)	265

xviii LIST OF ILLUSTRATIONS

FIG.		PAGE
41.	Oidium lactis (Hansen)	266
42.	Dematium pullulans (Loew)	267
43.	Monilia candida (Hansen)	268
44.	Sections from a Grain of Barley—	
	A. The Germ and its Appendages	272
	B. Section showing Secretory Layer of Epithelium	273
	C. Section showing Aleurone Layer	273
45.	Yeast Growing in a Fermenting Fluid	293
46.	Pasteur Flask	298
47.	Bottcher Moist Chamber (Hansen)	343
48.	Saccharomyces cerevisiæ I. (Hansen)	346
49.	Saccharomyces Pastorianus I. (Hansen)	346
50.	Saccharomyces Pastorianus II. (Hansen)	347
51.	Saccharomyces Pastorianus III. (Hansen)	347
52.	Saccharomyces ellipsoideus I. (Hansen)	348
53.	Saccharomyces ellipsoideus II. (Hansen)	348
54.	Carlsberg Flask (Hansen)	351
55.	Hansen's Pure Yeast Culture Apparatus	354
56.	General View of Hansen's Apparatus	355
57.	Yeast Propagating Apparatus of Bergh and Jörgensen	359
58.	Hayduck's Apparatus for Ascertaining the Fermentative Energy of Yeast	365
59.	Diagrammatic Section of the Strata Forming the London Basin	368
60.	Coldewe's Germinating Apparatus	396
61.	Vogel's Apparatus for Testing Barley for Transparency	398
62.	General View of a Pneumatic Malthouse at Chicago	410
63.	Section of Pneumatic Malting Drum	412
64.	Section of Pneumatic Malting Drum	413
65.	Schematic View of a Gravitation Brewery	429
66.	Mash-tun	435
67.	External Mashing Machine (Steel's improved)	437
68.	Sparge	438
69.	Fire Copper	440
69a.	Steam Copper	441
70.	Yorkshire Stone Square	446
71.	Burton Union	448
72.	Fermenting Round with Parachute	450

Plate illustrating the Arrangement of the Atoms of a Molecule in Space . . . . *to face page* 42

LIST OF ABBREVIATIONS.

Anleitung z. Quant. Chem. Anal.	Anleitung zur quantitativen chemischen Analyse. Fresenius.
Agricult. Chem.	Lehrbuch der Agriculturchemie. R. Sachsse.
Ann. Chim. et Phys.	Annales de Chimie et de Physique.
Ann. d. Chem. u. Pharm.	Annalen der Chemie and Pharmacie. (Liebig's.)
Beit. z. Kennt. d. Stärke.	Beiträge zur Kentniss der Stärkegruppe. Nägeli.
Berichte	Berichte der deutschen chemischen Gesellschaft.
Botan. Zeit.	Botanische Zeitung.
Bull. Soc. Chim.	Bulletin de la Société Chimique de Paris.
Central. f. Agriculturchem.	Biedermann's Centralblatt für Agriculturchemie.
Chem. Zeit.	Chemiker Zeitung.
Compt Rend.	Comptes Rendus de l'Académie des Sciences. Paris.
Jahresber. d. Chemie	Jahresbericht der Chemie. (Liebig and Kopp.)
Journ. Amer. Chem. Soc.	Journal of the American Chemical Society.
Journ. f. Landwirthsch.	Journal für Landwirthschaft.
Journ. f. prakt. Chem.	Journal für praktische Chemie.
Landwirth. Jahrbuch.	Landwirthschaftliche Jahrbücher.
Med. Carlsb. Lab.	Compt Rendu des Travaux Carlsberg Laboratorium.
Phys. Chem. d. Pflanzen.	Physiologische Chemie der Pflanzen. Ebermayer.
Sitz. d. bayer. Acad.	Sitzungsberichte der königlich-bayerische Akademie der Wissenschaften. München.
Trans. Lab. Club	Transactions of the Laboratory Club.
Woch. für Brau.	Wochenschrift für Brauerei.
Zeit. f. angew. Chemie.	Zeitschrift für angewandte Chemie.
Zeit. f. Biologie.	Zeitschrift für Biologie.
Zeit. f. Brau.	Zeitschrift für Brauerei.
Zeit. f. d. gesammt. Brau.	Zeitschrift für das gesammte Brauwesen.
Zeit. f. phys. Chemie	Zeitschrift für physiologische Chemie.
Zeit. f. Rübenzucker Ind.	Zeitschrift für Rübenzucker-industrie.

A
THEORETICAL AND PRACTICAL
TREATISE ON BREWING.

INTRODUCTION.

THE principles involved in the processes of malting and brewing fall under the three scientific heads of Physics, Chemistry, and Biology.

To the first of these categories belong all such operations as the heating, evaporation, and freezing of water, in which no chemical change is produced, but simply an alteration of its physical state; also all those operations which are of a purely mechanical nature, such as the raising of water by pumps, the sorting and cleansing of barley, &c.

Under the second head are included all those processes in which a chemical change is brought about in some of the materials employed. To these the operation of mashing belongs, in which the starch of the malt is so profoundly modified that it no longer exists as such, but is metamorphosed by the agency of the unorganised enzyme diastase into sugar and other bodies that differ considerably in chemical composition from the original starch, and, consequently, possess entirely different properties.

To the third category belong those processes in which chemical changes are effected through the agency of life, either in organisms or by their aid. Of such a nature is the process of malting, in which, during the germination of the barley, profound chemical changes are brought about in the constituents of the barleycorn by vital agency. So, also, in the fermentation of wort, which is effected through the instrumentality of the living organism "yeast," whereby the sugar of the wort is converted into a series of bodies, alcohol, carbon dioxide, &c., entirely different in their properties from the original sugar.

The broad lines on which the operations of malting and brewing are based had been fixed by practical experience long before anything was known of the scientific principles underlying the methods employed, and a number of rules gradually came to be formulated, by the observance of which the brewer was enabled, although he might possess no scientific knowledge, to carry on his operations with a greater or less degree of success. But the brewer who worked under such conditions was little better than an animated machine; he simply followed a certain routine, and knew nothing whatever of the why and wherefore of the various processes going on around him. So long as circumstances were favourable, and everything went on smoothly, he obtained good results; but when some trifling irregularity occurred in the manipulation of the highly sensitive materials he employed, then the whole brewing process became disorganised, often with the effect of entailing serious pecuniary losses. In such a case the nonscientific brewer is helpless. As Bersch says, he is just as unable to detect the cause of his trouble, with a view to its remedy, as an ordinary individual is to locate the fault in his watch when it has stopped from some hidden cause. His attempts to bring matters into their regular groove are merely the outcome of so much conjecture; he may by chance hit upon the right thing to do under the circumstances, or his endeavours may make matters still worse. Fortunately, this state of affairs is now almost a thing of the past; there is hardly to be found nowadays a brewer who does not possess more or less knowledge of the principles involved in the daily routine of his work.

It is the province of science to afford an adequate explanation of the principles involved in those chemical and biological changes which take place in the materials dealt with in the process of brewing, and just in proportion as these principles become understood and practically applied, so is precision and certainty introduced into actual working. The brewer who possesses an intimate acquaintance with the principles governing his operations is able to make allowances for variations in his materials and alterations in his working conditions; he is therefore in an infinitely better position to secure uniformly good results than one who is only able to work by rule of thumb.

But the theoretical explanation of the phenomena involved in some of the processes employed in the manufacture of malt and beer lead into the most recondite, and, at times, into quite unexplored territories of scientific inquiry; hence, where this is the case, an adequate scientific explanation of the observed phenomena is as yet unattainable; consequently, brewing still remains to some extent an empirical art. But as science, which is now making rapid strides, progresses, so will the remnant of empiricism which still remains disappear from the brewery, and real definite scientific knowledge, with its consequent advantages, take its place.

As brewing is a commercial undertaking, the brewer who wishes to be a master of his business must possess, in addition to much scientific knowledge, a commercial training. Like every other workman, he has to handle tools, and thorough efficiency in the use of those appliances in the brewery which in reality are his tools can only be obtained by actual experience in the brewery and malthouse, hence the value of an apprenticeship. It is only the harmonious blending of theory with practice that will enable a brewer to carry on his operations with that enhanced degree of certainty and with that economy which are so requisite in these days of keen competition.

PART I.

CHAPTER I.

PHYSICAL PRINCIPLES INVOLVED IN THE PROCESS OF BREWING.

HEAT.

MANY of the processes which are absolutely indispensable in the operation of brewing, such as boiling, evaporating, cooling, &c., are effected through the agency of heat.

Definition of Heat.—Heat may be defined as that agent which, when applied to the human body, produces the sensation of "warmth" or "cold." The tendency of all bodies, when in contact with one another, is to assume a state of equilibrium as regards heat, and when this state of equilibrium has been attained, they are said to be of the same temperature. Thus when we touch a body which is hotter than the temperature of our own bodies, it gives off heat to our hand, and consequently we feel the sensation of warmth; conversely, if the body touched be colder than ourselves, some of the heat of our hand passes to it, and we perceive the sensation of cold, which is really the feeling caused by heat passing away from our body. It must not be supposed that any actual substance is transferred in either of these cases, for heat is not a material substance, but a state of movement in the imponderable ether which pervades all substances and all space. It is, however, exceedingly convenient to speak of heat as if it were a material substance passing from one body to another, and one which conduces much to simplicity in describing the phenomena of heat.

Conduction of Heat.—If, while holding the end of a rod of copper, we place the other end in the flame of a spirit-lamp or other source of heat, we find that the whole rod speedily becomes so hot that we can no longer hold it. If, however, we try a similar experiment with a rod of glass, we find that the end which is opposite to that in the flame never becomes perceptibly hot, however long we may hold it. The reason for this is, that copper and the other metals belong to that class of bodies which allow heat to pass along them readily, and are hence called "good conductors;" whilst glass belongs to the opposite class called "bad conductors," since they only allow heat to pass along them with extreme difficulty. These two properties of bodies are very frequently made use of in the construction of brewing utensils; for instance, when the brewer wishes to prevent as much as possible the loss of heat from his steam-pipes, or from his mash-tun, if it is a metal one, he clothes these portions of his apparatus with badly conducting materials, such as asbestos, or a mixture of cork and clay, &c. When, on the other hand, he wants the heat to pass rapidly from his worts to his cooling liquor, he interposes between the two fluids in his refrigerator a thin sheet of copper, which is an excellent conductor of heat.

Convection of Heat.—Liquids as a rule are exceedingly bad conductors of heat. If, for instance, a lump of ice is put in the bottom of a test tube, a small coin placed over it to retain it in position, the test tube partially filled with water, and the flame of a lamp applied to the tube just beneath the surface of the fluid, its upper layer may be boiled for some considerable length of time without melting the ice. As is well known, the whole of the water or other fluid contained in a vessel may be readily heated, provided the flame or other source of heat is applied to the bottom of the vessel; but here another method of conveying heat—viz., convection—steps in. This method may be ocularly demonstrated by taking a large glass flask and nearly filling it with water, adding a few pinches of powdered cochineal or some other body of about the same specific gravity as water, and placing beneath the flask a lamp or other source of heat. It will be seen that a current is soon established, which ascends through the centre of the fluid, and which, on arriving at its surface, bends over in every direction like the foliage of a palm-tree, and forms a number of descending currents. These travel downwards in proximity with the sides of the flask until they reach its heated portion, when they again ascend in the hot central current; thus, in its turn, every portion of the water becomes exposed to the action of the heat. The reason why this occurs is that the water when heated expands and becomes specifically lighter than cold water; consequently it ascends, whilst the colder portions simultaneously descend to replace the portions which have ascended, and this action goes on continuously until the temperature is reached at which water boils. Hence, in the process of heating by convection, the force of gravity also comes into play.

In cooling, the reverse process takes place; the portion of water in contact with the air, or with some substance colder than itself, loses its heat, contracts, and becomes heavier; consequently it sinks, the warmer portions meanwhile ascend, and, after being cooled, descend; this goes on continuously until the whole of the fluid arrives at the same temperature as that of the cooling medium. Hence, any apparatus for cooling a mass of fluid, such as the attemperator in a fermenting vat, should be placed near the surface of the liquid, or the portion of fluid above it will not receive its due share of the cooling effect of the apparatus.

Gases are also bad conductors of heat; they also are heated by convection. A familiar instance of this is the draught produced by the fire in an ordinary chimney. The air heated by the fire expands, becomes specifically lighter, and consequently rises, forming a current which is rendered visible by the particles of carbon carried along with it. Currents of colder air continually descend from the ceiling and sides of the room to take the place of the heated air which has passed away. In order therefore to heat the air in a room equally, the source of heat should be placed as near the floor as possible. Conversely, when it is sought to cool the air of a room, the agency for the withdrawal of heat is placed as near the roof as possible; thus the refrigerating pipes occasionally employed for keeping down the temperature of a fermenting-room or storage-cellar are placed near the ceiling.

Radiation of Heat.—There is yet another method by which heat is propagated, viz., radiation. Every hot body, in addition to losing heat by convection and conduction, also loses heat by radiation. This may be shown by hanging a hot body in a vacuum, when it is found to rapidly grow colder, though the loss in this case cannot be due either to conduction or convection. Heat propagated by radiation moves in straight lines, precisely in the same way as light does; indeed, both light and heat are caused by vibrations of the ether, which are exactly similar in all respects except that of wave length. Thus a body, on being gradually heated, at first only emits heat rays; as it gradually grows hotter it begins to emit rays of light as well, and the hotter it becomes the greater the number of light rays it sends forth. A familiar illustration of this is seen in the glowing carbon filament of an incandescent electric lamp, which, being intensely heated by the current of electricity passing through it, emits a large number of light rays.

The rate at which heat passes away from a hot body by radiation is much influenced by the state of its surface; a bright polished surface radiates heat with comparative slowness, whilst a dead black surface radiates heat with great rapidity. Conversely, a body which radiates heat readily also absorbs it rapidly, and a body which parts with its heat by radiation slowly also absorbs it slowly. Hence the cover of a mash-tun, when made of copper, should be kept as bright as possible, in order to prevent loss of heat by radiation. Steam-pipes, and all the vessels in the brewery from which it is sought to prevent loss of heat by radiation, are covered with some non-conducting material.

Expansion by Heat.—Almost all bodies expand when heated. Of the various classes of bodies, gases expand the most, solids the least, and liquids occupy an intermediary position. Each body has a degree of expansibility under the influence of heat peculiar to itself; thus, amongst metals, zinc expands the most, and probably platinum the least. Of fluids, those which are most volatile as a rule expand most. Alcohol, for instance, is more expansible by heat than water, water more than mercury. All gases, on the other hand, have the same degree of expansibility for the same amount of heat, or so nearly so that the exceedingly small deviation which under certain circumstances they make from this rule need only be taken into account in the most delicate researches.

The Thermometer.—In many operations connected with brewing it is of the highest importance to have some convenient means for ascertaining correctly the temperature of the various substances we have to deal with, such as that of the malt on the kiln floor, or of malt on the drying kiln, the temperature of the worts at various stages of the process of brewing, that of the air in the fermenting-rooms, storing-cellars, &c. This is fulfilled in an eminent degree by the "thermometer," an instrument which is constructed in the following manner:—A glass tube of very fine and even bore is taken, and a cylindrical or spherical bulb is blown at one end of it. The bulb is then heated to drive out a portion of the air which it contains, the open end of the tube dipped into a vessel containing mercury, and the bulb allowed to cool. As the bulb cools, its contained air contracts, and a portion of mercury is drawn into it. This is then boiled, in order that its vapour may drive out and replace the remainder of the air and any moisture left in the bulb and tube. Whilst the whole is hot, and the mercurial vapour escaping freely from the open end of the tube, this end is placed under the surface of hot mercury, and the whole allowed to gradually cool down. When completely cold, the whole of the bulb and tube will be filled with mercury. The bulb is now gently heated, and a portion of the mercury driven out, the end of the tube being sealed up with the blow-pipe while the column of mercury still fills the tube. As the mercury contracts, a vacuum is left behind; if the tube were sealed while full of air, it would burst when the mercury again expanded. The tube ought then to be calibrated, in order to ascertain what differences there are in the diameter of its bore. For this purpose a small column of mercury about a third of an inch is detached from the main body of the fluid, moved up and down the tube, and its length noted in the various portions of the tube by the aid of a microscope. It is obvious that where the tube is narrower the detached column will be longer than in the wider portions; these differences are noted, and allowances made for them when the tube is subsequently graduated. The next point is to find two fixed points of temperature, and for this purpose the melting-point of ice and the boiling-point of water are chosen as being those most readily attainable. In order to find the position of the point in the mercurial column which corresponds to the freezing-point of water, the thermometer is placed in pounded ice in a melting state, care being taken that the water produced by the thawing of the ice has a ready means of passing away. When the column has assumed a stationary position, a small mark is made on the tube indicating its height. The next operation is to subject the thermometer to the action of steam evolved from water boiling at the pressure of the atmosphere, and this is effected by placing it in a long, upright, cylindrical vessel, in the bottom of which water is kept boiling. An aperture is provided at the side of the vessel for the free escape of steam, so that the pressure inside the vessel may not rise beyond that of the external atmosphere. When, after a little time, the end of the column has become stationary, another mark is made on the tube to indicate the temperature of boiling water. This is not an absolutely fixed point; water boils at different temperatures, according as the pressure to which it is subjected varies. Thus, when the air has a temperature of $32°$ F., and the barometer indicates a pressure equal to 29.315 inches of mercury, water boils at $211°$ F.; when it indicates a pressure of 30.444 inches, the air being at $32°$ F., water boils at $212.9°$ F. In order that this variation may not give rise to practical difficulties, the boiling-point of water ($212°$ F.) has been fixed by a Government Commission, who have determined that the temperature at which distilled water boils under a pressure of 29.905 inches of mercury, with air at $32°$ F., shall be indicated by $212°$ F. Consequently, when the position for this point is ascertained, with the air at any other pressure and temperature than this, a corresponding correction must be made. These two fixed points having been ascertained, the space between them is divided into degrees. In England it is customary to divide the space into 180 degrees, as proposed by Fahrenheit; in France it is divided into 100 degrees, on the plan introduced by Celsius; so that $0°$ indicates the freezing-point of water and $100°$ its boiling-point, hence this is called the "Centigrade scale." This latter is by far the most rational method of division, and consequently is the one almost exclusively used by scientific workers throughout the world. There is yet another method, that of Reaumur, which is much used in the breweries of Germany and Austria; in this the space is divided into 80 parts. By whatever method the thermometer is graduated, degrees of the same size as those between the two ascertained fixed points are marked off above and below these points, always taking into consideration the necessary corrections for the calibre of the tube. Other fluids besides mercury are sometimes used for filling thermometers, such as alcohol, carbon disulphide, &c. But though alcohol has the advantage of not freezing at low temperatures (mercury freezes at $-39°$ C., $-38°$ F.), it has the disadvantage of boiling at the comparatively low temperature of $78.39°$ C. ($173.1°$ F.), so that a thermometer filled with alcohol would be entirely unfitted for many purposes in the brewery. Another reason which leads to the choice of mercury for this purpose is that this fluid expands in equal increments for each degree of heat between the freezing and boiling points of water, a property not shared by most other liquids or solids.

Conversion of the Readings of the various Thermometric Scales into each other.—To reduce degrees Centigrade to degrees Fahrenheit the following formula may be employed :—

$$C.° \times \frac{180}{100} + 32 = F.°; \quad 60° \text{ C. would therefore be } 60 \times \frac{9}{5} + 32 = 140° \text{ F.}$$

To convert degrees Fahrenheit into degrees Centigrade :—

$$(F.° - 32) \times \frac{5}{9} = C.°; \quad 140° \text{ F. would be } (140 - 32) \times \frac{5}{9} = 60° \text{ C.}$$

Similarly, degrees Reaumur to degrees Fahrenheit and the converse are effected by the following formulæ :—

$$R.° \times \frac{9}{4} + 32 = F.° \qquad (F.° - 32) \times \frac{4}{9} = R.°$$

Sources of Error in the Thermometer.—The bulb of an ordinary glass thermometer gradually contracts to a slight extent for some time after it is made; consequently, if graduated immediately after it is filled, its readings after a time become too high. This error may amount to + 1° C. (1.8° F.), which is about the extreme limit of the change. Consequently, all thermometers should be allowed to rest for some considerable length of time after being filled before they are graduated. All thermometers used for delicate researches are now made of the so-called Jena glass, which contains, in addition to the ordinary constituents of glass, a small quantity of zinc. When a thermometer tube made of this glass is exposed to a temperature of 300° C. for thirty hours, after subsequent filling and graduation its bulb does not contract on use, or at most to an almost imperceptible extent. Consequently, it is advisable to have in every brewery a standard thermometer of this kind, so that the rest of the thermometers can be tested against it for correctness from time to time. A thermometer which has been graduated in an upright position should always be used in the same position, since the hydrostatic pressure of the column of mercury tends to enlarge to a slight extent the capacity of the bulb, and as the capacity of the bore is very small relatively to that of the bulb, small alterations in the latter are enormously magnified in the reading of the column. If a thermometer be suddenly heated and then cooled, as, for instance, if it be heated to the temperature of boiling water and then suddenly plunged into ice, its freezing-point will have altered slightly, and may read 0.1° C. higher, and only in ten days or a fortnight will it have returned to its original reading. In taking readings with a thermometer, the instrument should always be immersed in the liquid or other substance which it is desired to take the temperature of as far as the length of the column of mercury extends. This is always carefully done when the two fixed points of the

SOURCES OF ERROR IN THE THERMOMETER.

instrument are ascertained for the purpose of graduation. For instance, if a thermometer has its bulb only immersed in boiling water, the stem and the column of mercury it contains will only have the temperature of the air which surrounds them, and obviously the column will be shorter than if it were at the temperature of boiling water, hence the reading will be slightly too low. The tube being also cold, contraction will diminish its bore to a slight extent, and this, by lengthening the column of mercury, will tend to rectify the error to a small extent. But if an attempt be made to ascertain the temperature of any liquid approaching that of boiling water in this way, the error in the reading of the thermometer may be from 2° to 3° F. too low.

As many of the thermometers met with in commerce, owing to careless graduation, have the freezing-point several degrees out of position, all these instruments, before being taken into actual use, should be carefully tested against a normal thermometer; and this should be repeated every few months, in order to ascertain that their indications remain correct.

Thermometers are of various forms. For instance, the form used for ascertaining the temperature of the mash-tun is of considerable length, and is enclosed in a wooden or metallic tube, a sliding door or other suitable arrangement being provided to cover and protect the bulb when not in use, whilst the uppermost portion where the readings are taken is also protected by a small door.

In taking the temperature of a body with the thermometer, it is necessary that the instrument should come into as intimate contact with it as possible. When the body is gaseous this presents no difficulty. In taking the temperature of solid or liquid bodies, such as malt, water, wort, &c., care must be taken not only that the bulb, but also that the whole of the mercurial column is immersed in the substance, only so much of the column being left exposed as is just necessary to obtain a reading. In taking the temperature of a mash considerable time must be allowed to elapse, as the heat has to penetrate through a considerable amount of material before the tube and mercurial column reach the temperature of the mash. It is also advisable to take the temperature of the mash in several places, then to add all the various readings together, and divide their sum by the number of readings taken. In this way we obtain an average reading. When taking the temperature of malt on the drying kiln, care must be taken that its bulb does not come in contact with the floor, or too high a reading will be obtained. In order to obviate this, thermometers made for the drying kiln have a peculiarly shaped stand, which prevents the bulb coming in contact with the tiles.

An instrument called a "thermograph" has been devised for obtaining a continuous record of the temperature of the malt when on the floor of the drying kiln. It contains a cylinder, driven by clockwork, which revolves once in twenty-four hours, to which a sheet of ruled paper is attached. On this the record is marked by a pen attached to a peculiarly constructed thermometer.

Unit of Heat.—It must be remembered that the thermometer only measures the *intensity* of heat; it does not measure its *quantity*. In order to measure the quantity of heat, two other factors must be taken into account—the *specific heat* of the substance in question and its *weight*. It is a matter of great convenience to have some unit for heat, just as we have for weights and measures. The quantity of heat usually taken for this unit is that required to raise one kilogramme of water from 0° to 1° C., but any other quantity of water and degree of heat on any other scale may be taken; consequently, as a matter of convenience, we shall occasionally take for the unit the quantity of heat required to raise one pound of water from 32° to 33° Fahrenheit.

Specific Heat, or Heat Capacity.—Bodies, when indicating the same degree of sensible temperature, differ very considerably in the amount of heat which an equal weight of them contains. If two pieces of metal of the same size, shape, and weight—the one of silver and the other of copper—are suspended by means of pieces of thin wire in a vessel of boiling water for a few minutes, until they have attained the temperature of the water, and are then withdrawn and suspended in the air, their temperature will gradually fall; the piece of silver, however, will be found to lose its heat much more rapidly than the piece of copper. Hence we may conclude from this that an equal weight of copper at the temperature of boiling water contains much more heat than an equal weight of silver. As the copper parts with much more heat than the silver does in cooling, it must have required much more heat to raise it to the temperature of boiling water. Consequently, we say that the specific heat, or the heat capacity, of copper is greater than that of silver. Taking as the thermal unit the quantity of heat required to raise one kilogramme of water from 0° to 1° C., the specific heat of any other body is assumed to be the quantity of heat which is necessary to raise the temperature of one kilogramme of it from 0° to 1° C. It has been found by properly conducted experiments that, taking the specific heat of water as 1, that of copper is 0.095, and that of silver 0.057.[1]

---

[1] A curious relation appears to exist between the specific heats and the atomic weights of the elements, the product of the specific heat of an element multiplied by its atomic weight being nearly a constant quantity, thus:—

	Specific Heat.	Atomic Weight.	Product.
Copper	0.0951	63.5	6.0389
Silver	0.0570	108.0	6.9560
Magnesium	0.2499	24.0	5.9976

The numbers in the last column are known as the atomic heats of the respective elements. Compound bodies also have molecular heats, but these vary with the class of compounds; thus with the chlorides of the metals it is about 19, with their carbonates about 21 to 22.

SPECIFIC HEAT OF MALT.

Mercury has a much less specific heat than water; consequently, if a pound of mercury having a temperature of 100° C. is shaken up with a pound of water at 0° C., the resulting temperature of the mixture will not be the mean, 50° (as would have been the case if a pound of water at 100° had been mixed with the same weight of water at 0°), but will be found to be 3.2° C. The mercury has, therefore, lost 96.8° of heat to raise the temperature of an equal weight of water 3.2° only. Conversely, if a pound of water at 100° C. be shaken up with a pound of mercury at 0° C., the temperature of the mixture will be 96.8° C. Thus one unit of heat would raise a pound of mercury $\frac{96.8}{3.2} = 30.25°$ C., consequently $\frac{1}{30.25}$ (or 0.033) of a unit of heat would raise that quantity of mercury one degree; 0.033 is therefore the specific heat of mercury.

Specific Heat of Malt.—Malt has a specific heat rather less than half that of water, viz., 0.42, a point which it is highly necessary to take into consideration when calculating what the initial heat of a mash is to be. This is worked in the following manner:—Supposing an initial mash-heat of 150° F. is required, and the mash liquor is to be used at the rate of two barrels to each quarter of malt, which latter has a temperature of 50° F., what must be the temperature of the mash liquor? A quarter of malt weighs 336 lbs., and two barrels of water weigh 720 lbs.; that is in the proportion of 7 lbs. of malt to 15 lbs. of water. Taking as the unit of heat that quantity necessary to raise the temperature of 1 lb. of water from 32° to 33° F., the first question to be ascertained is how many of these units will be required to raise 7 lbs. of malt from 50° to 150°. Now, to raise 7 lbs. of water from 50° F. to 150° F. would require 700 units of heat; but as malt only requires $\frac{42}{100}$ as many units to raise it to a given temperature as water does, 7 lbs. of malt will require $700 \times \frac{42}{100}$ (or 700×0.42) = 294 units. Since 15 lbs. of water at 150° F. contains 2250 heat units, we must add on to this sum the 294 units necessary to raise the 7 lbs. malt from 50° to 150°, which will give 2250 + 294 = 2544, and this sum divided by 15 (15 lbs. water) will give the number of units of heat required to be contained in each lb. of mash liquor, or, what is the same thing, the temperature of the water, viz., $\frac{2544}{15}$ = 169.6° F. Or the method of calculation may be concisely stated as in the following formula:—

$$\frac{7 \times 100 \times 0.42 + 15 \times 150}{15}$$

The same principle has to be kept in mind in adding piece

liquor. Supposing that we have a mash of 40 quarters of malt and 80 barrels of water at 150° F., and that we wish to raise its temperature to 155° F. by the addition of piece liquor at the rate of half a barrel per quarter, what will be the required temperature of the piece liquor? The proportions of malt and water in the mash are, as before, 7 lbs. malt to 15 lbs. water; the piece liquor to be added will be in the proportion to these of 3.75 lbs., *i.e.* a fourth of 15 lbs. In order to raise each 7 lbs. of malt from 150° to 155° F. there will be required $7 \times 5 \times 0.42 = 14.7$ units of heat. The amount of heat required to raise the temperature of each 15 lbs. of water already in the mash will be $15 \times 5 = 75$ units. The number of units required to raise each 7 lbs. of malt and each 15 lbs. of water will be therefore $14.7 + 75 = 89.7$, and these will have to be contained in each 3.75 lbs. of piece liquor, over and above the $3.75 \times 155° = 581.25$ necessary to maintain its temperature at 155° F. after the admixture has taken place; this is $581.25 + 89.70 = 670.95$, and this sum divided by 3.75 will give the number of heat units which each pound of underlet must contain to effect this purpose, or, in other words, it must have that temperature. This is $\frac{670.95}{3.75} = 178.9°$ F. Expressed in a formula, the calculation appears as follows:—

$$\frac{7 \times 5 \times 0.42 + 15 \times 5 + 3.75 \times 155}{3.75}$$

Latent Heat of Solidification.—All solids, when they pass into the liquid state, absorb a certain amount of heat which can no longer be detected by the indications of the thermometer; in other words, a certain amount of heat has passed from the sensible into the latent state. If 1 lb. of snow or pounded ice be added to 1 lb. of water at a temperature of 79° C., when all the snow or ice is melted the resulting 2 lbs. of water will be found to have a temperature of 0° C.; whereas, if 1 lb. of water at 79° C. had been added to 1 lb. of water at 0° C., the result would have been 2 lbs. of water at 39.5°, or the mean between 79° and 0°. In melting, the 1 lb. of ice has absorbed 79° of heat, of which the thermometer now gives no indication; hence 79° have disappeared or, in other words, become latent; they have been used in converting the ice from the solid to the liquid state. All other solids behave in a similar manner in passing from the solid to the liquid state; they render latent a certain amount of heat, and this is retained so long as the substance remains liquid.

When a liquid passes from the liquid to the solid state, its latent heat is evolved and becomes again sensible. It is difficult to illustrate this with ice, but the phenomenon may be readily shown in another manner. If to water boiling in a flask sodium sulphate be added until no more is dissolved, the flask rapidly corked while steam is still issuing from it, and the whole allowed to remain undisturbed until it is quite cold, the whole of the salt remains in solution, and what is known as a super-saturated solution of sodium sulphate is obtained. This is a solution which contains much more of the salt than cold water under ordinary circumstances would dissolve; for boiling water is able to dissolve 42.5 per cent. of this salt, whilst water at freezing-point can only take up 5 per cent. If the flask be now carefully uncorked, and the surface of the solution touched with a rough solid body, or a crystal of sodium sulphate dropped in, the whole mass instantly solidifies. That this conversion of the liquid into the solid state is accompanied by the evolution of heat may be readily perceived on grasping the flask with the hand; or, if a thermometer be introduced into the mass while still liquid, as solidification takes place the temperature will rise 18° or 20° C.

Latent Heat of Vaporisation.—When a liquid is converted into vapour, much heat is also rendered latent. This is very markedly the case with water. If a vessel containing water at 0° C. is placed over a steady source of heat, and the length of time noted which is required to raise its temperature to 100° C., which, say, is ten minutes, it will be found that a period of time about five times this length, or about fifty minutes, will have elapsed when all the water has evaporated away. Since the water received 100° of heat during the first ten minutes, it must have approximately received 500° during the fifty minutes. But we know that the temperature of boiling water at the pressure of the atmosphere never rises beyond a certain point, and that a thermometer introduced into the steam from boiling water indicates the same temperature as the water itself. Consequently, water in passing from the liquid to the gaseous state renders latent about four times as much heat as is necessary to raise it from 0° to 100° C. This may be also shown in the converse manner. If 537 grammes of water be placed in a tared beaker, which is swathed in cotton wool or some other bad conducting material so as to prevent loss of heat by radiation, a thermometer inserted into the water, and steam allowed to bubble through the liquid. If the beaker and its contents be removed and weighed immediately the thermometer indicates a temperature of 100° C., the water will be found just to have increased in weight by 100 grammes. This shows that the amount of heat contained in the steam of 100 grammes of water is able to raise the temperature of 537 grammes of water from 0° to 100° C. In addition to the 100° of sensible heat, steam therefore contains 437° of latent heat.

Theoretically, if no loss of heat took place by radiation, &c., the steam evolved from 100 gallons of water in the steam boiler would be able to bring to the boiling-point 437 gallons of water in the hot-liquor back, and yield in this way, taking into account the 100 gallons of water produced by the condensed steam, 537 gallons of water at boiling-point.

PHYSICAL PRINCIPLES.

Evaporation.—By this is understood the process in which a liquid quietly and silently passes into the state of vapour. The rapidity of the process depends upon several factors. The higher the temperature of the liquid, the more rapidly it evaporates. The larger the amount of surface exposed to the air, the greater the amount of evaporation. Thus a cooler is so constructed that the wort on it shall present a large exposed surface to the atmosphere, to increase the evaporation as much as possible; and as a large amount of heat is absorbed by water in passing from the liquid to the gaseous state, rapid evaporation means rapid cooling. Evaporation goes on much more rapidly in dry air than in air which is considerably charged with moisture; hence the comparative slowness with which cooling proceeds on a foggy day, when the atmosphere is saturated with moisture. The constant removal of those portions of air in immediate contact with the surface of the liquid which have become highly charged with aqueous vapour, and the substitution of dry air for these, also considerably increases evaporation. Hence coolers are invariably placed in a room, the windows of which are provided with louvre-boards for the admission of a current of air; and the process is often accelerated by placing a fan driven by power in the centre of the cooler, which causes rapid currents of air to traverse the surface of the liquid.

Evaporation goes on with extreme rapidity in a vacuum, so quickly that it is possible to freeze water by its own evaporation, if an arrangement is made to remove the aqueous vapour as fast as it is formed. This is generally accomplished by taking a flat vessel containing sulphuric acid, which has a great attraction for water, and placing over it a small metallic capsule containing the water to be frozen. These are placed under the bell-glass of an air-pump and the air exhausted, when evaporation takes place so rapidly that the temperature of the water is speedily brought down to its freezing-point. This principle of reducing the heat of a liquid and of those bodies with which it is in contact by rapid evaporation receives practical application in the action of freezing machines. In these, liquids which have a much lower boiling-point than water are generally employed, and they, when caused to rapidly evaporate by the partial vacuum maintained in the machine by its pump, absorb an enormous amount of heat. The following table gives the boiling-points of some of the liquids commonly used in freezing machines :—

	Boiling-points in Degs. C.	In Degs. F.
Ether	+ 35.5	+ 96.0
Sulphurous anhydride	− 10.0	− 14.0
Ammonia	− 33.0	− 27.4
Carbonic anhydride	− 78.2	− 108.8

DENSITY AND SPECIFIC GRAVITY.

One of the most frequent operations of a purely physical nature performed in the brewery is the determination of the specific gravity of wort, beer, and other fluids. The following is a brief description of the principles upon which this simple operation is based.

Density.—If we compare the relative proportions in size of a pound of lead and a pound of cork, we see at once that equal weights of the two bodies occupy very different volumes or spaces. Similarly, if we have a piece of cork and a piece of lead of the same size, the two bodies will have very different weights. Expressing these facts in more scientific language, we say that the densities of these two substances differ considerably; that lead possesses a much higher density than the cork. In familiar conversation, we often express the same idea when we say that a substance, such as lead, is "heavy;" that another substance, such as cork, is "light." Similarly, liquids have very different densities; thus a litre of water weighs 1000 grammes, a litre of mercury 13,600 grammes, a litre of alcohol 790 grammes. In order to express the density of a body, the simplest method is to weigh a known volume of it, and to divide the weight thus obtained by the volume; in this way the weight per unit volume is arrived at. Density, therefore, is only a relative term; we may say that the density of alcohol is $\frac{790}{13,600}$ that of mercury, or that the density of water is $\frac{1000}{13,600}$ that of mercury. But if a list of the varying densities of bodies is to be compiled, it is obviously necessary that the weight of a volume of each of them should be compared with the weight of an equal volume of some other body which shall serve as the standard, and that this latter substance should also be one which is readily attainable. In the case of solid and liquid bodies, water is chosen for this purpose; in that of gases, air—since these two substances are always at hand.

Specific Gravity.—This term "specific gravity of a body" expresses the ratio of its density to that of the substance chosen as the standard, that is—in the case of solids and liquids—water; or, in other words, the ratio between the weight of a given volume of the body under investigation and that of a similar volume of water. As the densities of bodies vary considerably with their temperatures, when the specific gravity of a substance has to be determined, not only must its own temperature be taken into account, but also that of the body with which it is to be compared. In taking the specific gravities of liquid or solid bodies a temperature of $4°$ C. ($39.2°$ F.) is often chosen, this being the

16   PHYSICAL PRINCIPLES.

temperature at which water is at its maximum density; for water, unlike most other substances which contract between this temperature and 0° C. (32° F.), expands. Since at some seasons of the year it is exceedingly inconvenient to have to bring bodies to such a low temperature, in ordinary practice a more readily attainable temperature is chosen, and this is 15.5° C. (60° F.). The specific gravity of gases is generally determined at the temperature of the air at the time this operation is performed, and the volume it would occupy at any other temperature found by calculation.

Methods of obtaining the Specific Gravities of Solids.—In ascertaining the specific gravity of a solid it would be well-nigh impossible, in most cases, to obtain a bulk of water equal to that of the solid. Advantage is therefore taken of the principle, first enunciated by Archimedes and applied by him to this particular purpose, that when a solid is immersed in water it loses exactly as much weight as that of a volume of water equal to its own bulk. In actual practice, the specific gravity of a body is found by weighing the substance first in air, then in water of a given temperature. This is accomplished by having a scale-pan attached to the balance, with shorter attachments to the beam than the ordinary pan has. On the under side of the former a hook is fixed, to which the body to be weighed in water is attached by a piece of thin string or wire. The weight of the body is first taken in air, and then a vessel containing water is placed under the scale-pan, and the body again weighed immersed in water. These two weights furnish the required data. Thus, supposing a piece of lead weighs in air 100 grammes, and when weighed in water it weighs 91.111 grammes; by subtracting 91.111 from 100 we obtain 8.889, and this is the weight of a volume of water equal to that of the lead. A simple proportion sum then gives the specific gravity of the lead:—

$$8.889 : 100 :: 1 : 11.25$$

Weight of equal vol. of water.   Weight of substance in air.   Specific gravity of water.   Specific gravity of lead.

In the case where the specific gravity of a substance which is soluble in water has to be taken, as, for instance, that of a crystal of sugar, it is first weighed in air, then in some fluid in which it is insoluble, such as oil of turpentine. The specific gravity of the oil of turpentine, or other fluid used for this purpose, is also ascertained, when a simple calculation gives the specific gravity of the body in question.

Specific Gravity of Gases.—These are determined in a very simple manner. A glass globe provided with a stop-cock is weighed full of air; the air is then exhausted as completely as possible by an air-pump, and the weight of the empty globe taken. This will be less than the first weight, and this loss is the weight of the air which the globe contained at the temperature and the pressure of

THE SPECIFIC GRAVITIES OF LIQUIDS.

the atmosphere at the time the experiment was made. The globe is now filled with the gas to be examined, its weight again taken, due attention being paid to the temperature of the gas and the pressure to which it is subjected. If these latter are the same as when the globe filled with air was weighed, then the following simple proportion sum gives the specific gravity of the gas, taking air as the standard :—

Weight of air : weight of gas :: 1 : specific gravity of gas.

If the gas be weighed at a different temperature or at a different pressure, then these varying conditions will have to be taken into account and the requisite calculation made.

Methods of taking the Specific Gravities of Liquids.—This is by far the most important branch of the subject, so far as the brewer is concerned, since, in his numerous operations, he is very often guided by the varying specific gravities of the liquids with which he is constantly dealing. There are several methods of taking the specific gravity of a liquid, some of which leave nothing to be desired in the matter of exactitude.

The Specific Gravity Bottle.—Amongst the most accurate instruments for obtaining the specific gravity of a liquid is the specific gravity bottle (Fig. 1). This consists of a light glass flask, generally made to contain some even quantity of distilled water at 15.5° C. (60° F.), such as 50 or 100 grammes, or 500 or 1000 grains. An accurately ground stopper, perforated by a fine hole, fits into the neck of the little flask. It is so arranged that when the flask, including the perforation in the stopper, is filled with pure water at the temperature of 15.5° C. (60° F.), the quantity of water in the flask just amounts to one of the weights mentioned above. In using the apparatus the flask is completely filled with the liquid to be examined, and the stopper inserted, care being taken that the liquid has a temperature of 60° F. ; the exterior of the flask is then wiped perfectly dry and the whole weighed. The weight of the flask, which must have been previously ascertained (usually a brass counterpoise equal to the weight of the flask is supplied along with it, and this obviates the first part of the next calculation), is now deducted from the total weight; this gives the weight of a volume of the liquid equal to that of the water which the flask holds. The following calculation then gives the specific gravity of the liquid under examination :—

FIG. 1.—Specific Gravity Bottle.

Weight of : weight of :: specific gravity : specific gravity
water   liquid   of water   of liquid.

PHYSICAL PRINCIPLES.

Supposing the experiment to have been conducted with a flask holding 100 grammes of water, and the weight of the corresponding volume of the liquid under examination is found to be 100.42 grammes, then the calculation, taking the specific gravity of water as 1 (or, as it is commonly expressed, water = 1), is as follows:—

$$100 : 100.42 :: 1 : 1.0042.$$

If the specific gravity of water be taken as 100 (water = 100), then the specific gravity of the liquid expressed under these conditions would be 100.42.

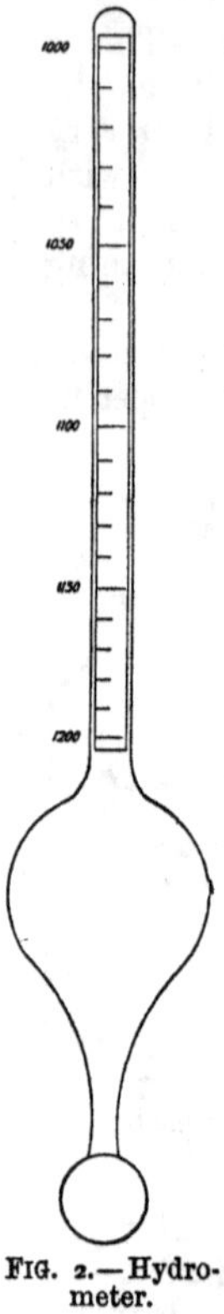

In the brewery the specific gravity of water is always taken as 1000 (water = 1000), and the specific gravity of the above liquid, expressed in this way, would be 1004.2.

It is obvious that a flask containing any quantity could be made to serve this purpose, only in such a case a calculation would have to be made each time, whereas, with a flask holding 1000 definite units of weight, the result can be obtained by mere inspection; or, with a flask which contains 50 or 500 units, the result needs simply multiplying by 20 or 2.

The Hydrometer.—This instrument is the one most frequently used for ascertaining the specific gravities of liquids in manufacturing operations, since it is by far the easiest and most convenient to manipulate. At the same time its indications are not absolutely accurate, but they are sufficiently so for most practical purposes. Its usual form is shown in the accompanying cut, Fig. 2. The instrument is generally made of glass, sometimes of metal, and consists of an elongated cylindrical bulb, having attached to its upper end a glass tube of larger or smaller diameter according to the delicacy required in the readings of the instrument, and enclosing a scale of paper or ivory. The lower portion is a small glass bulb, partially filled with small shot or mercury. This not only compels the hydrometer to float in an upright position, but also enables the maker to regulate its weight. The action of the instrument depends on the principle "that a body, when floating in a liquid, displaces a volume of the liquid equal to its own weight." This may be illustrated by taking a piece of perfectly cylindrical glass tube, Fig. 3, and closing its lower end. A mark is then made on the tube at some little distance from its top, and the length between this mark and the bottom of the tube divided into twelve equal parts by lines etched on the tube. To the lines figures as shown in the cut are appended. The tube is now loaded with shot

FIG. 2.—Hydrometer.

THE HYDROMETER.

or mercury until, when placed in water, it sinks to the point marked 10. The volume of water displaced will be equal to that of the tube from figure 10 to its lower extremity, and this volume of water will have the same weight as the tube and its contents have when weighed in air. For example, let the volume of water displaced by the tube when immersed to line 10 be 100 c.c. (cubic centimetres), and as 1 cubic centimetre of water weighs one gramme, the weight of the apparatus will be 100 grammes. If the tube be now placed in a liquid which has a higher specific gravity than water, it will not sink so far as the line 10. When placed in such a liquid, let the tube sink to the line marked 5; then it is obvious that the volume of liquid displaced in this case is only 50 c.c., or half the size of that of the water displaced in the previous experiment. But since this quantity of the liquid must have the same weight as the volume of water previously displaced, and as it is half its size, 50 c.c. of it must weigh 100 grammes, and a given volume of such a liquid will weigh twice as much as an equal volume of water, or in other words, it has a specific gravity twice that of water. If, on the other hand, the tube be placed in a liquid specifically lighter than water, it will sink farther than the mark 10. Let it be placed in a liquid in which it sinks to the mark 12. Obviously, then, twelve volumes of such a liquid weigh just as much as ten volumes of water, consequently the specific gravity of the fluid would be $\frac{10}{12}$ that of water. Every line on the scale corresponds to a certain specific gravity, and this can be found in each case by a sum in simple proportion, thus:—

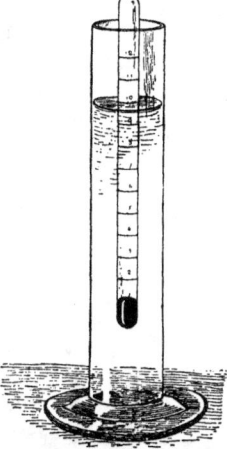

FIG. 3.—Experimental Hydrometer.

Vol. of liquid : vol. of water :: sp. gravity of : x
displaced   displaced   water

Supposing the tube to sink to the line marked 8, and that we take the specific gravity of water as 1000, then—

8 : 10 :: 1000 : 1250.

In using this apparatus, we make the converse observation to that which we make with the specific gravity bottle: with the hydrometer we ascertain the *relative volumes* of the *same weight* of liquids; with the specific gravity bottle we ascertain the *relative weights* of the *same volume* of liquids. Obviously, then, by means of the instrument constructed in the before-mentioned way we could roughly determine the specific gravity of many liquids. In order to be able to make the determinations of the instrument more exact, the form shown in Fig. 4 is given to it. The expansion of the stem does not in the least affect the principle of the action of the instrument, so long as we know the relation between the volume of the graduated portion to that of the whole apparatus. Supposing the instrument Fig. 4 weighs 100 grammes, and when immersed in water up to the mark 90 displaces 90 grammes of water,. and that, when immersed in water and left to itself, it sinks to the point 100 and then displaces 100 grammes of water, and that the distance from 90 to 100 is divided into ten equal parts, the specific gravity which corresponds to any of these divisions can be ascertained as in the former case. For instance, let the instrument sink in a liquid to the mark 95, then—

$$95 : 100 :: 1000 : 1052.863$$

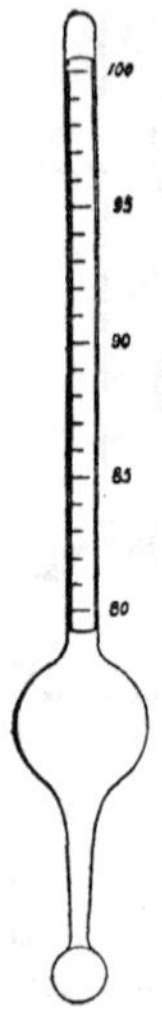

Volumeters.—Instruments made on the principle just mentioned are also termed "volumeters" because they express the volumes of liquids equal to a given weight, *i.e.*, the weight of the instrument. Of such a nature is the volumeter of Gay-Lussac. It is, however, exceedingly inconvenient to have either to make a calculation or to consult a table each time the instrument is used; consequently, instruments are generally graduated so that the reading of the scale gives directly the specific gravity of the liquid under examination; and this form is the one exclusively used by brewers in this country.

Instruments which directly Indicate Degrees of Gravity. — The principle underlying the graduation of all instruments which show degrees of specific gravity directly may be explained in the following manner. Suppose the perfectly cylindrical tube (Fig. 3) to be so weighted that, when floating freely in water, it sinks to the point marked 10; the volume of water displaced by the tube is equal to the volume of the tube from its lower end to 10; call this quantity 1 volume. Let this distance in the tube be divided into 100 equal spaces and numbered consecutively. From what has been said before, we know that if the tube when immersed in another liquid only sinks half the distance it did in the previous case, or to the point marked 50, then that fluid has twice the specific gravity of water. If, when placed in a third liquid, it only sinks a third, that is, to the point marked 33.3, then such a liquid has three times the specific gravity of water. Hence it follows that the specific gravities 1, 2, 3, are proportional to the volumes $\frac{1}{1}, \frac{1}{2}, \frac{1}{3}$; and this may be

FIG. 4.—Hydrometer.

HYDROMETERS.

extended to any degree of fineness by employing closer numbers. Thus specific gravities 1, 1.01, 1.02, 1.03, are proportional to volumes $\frac{1}{1}, \frac{1}{1.01}, \frac{1}{1.02}, \frac{1}{1.03}$. These latter fractions equal respectively the volumes 1, 0.99009, 0.98039, 0.97087; they are, in other words, the reciprocals[1] of the specific gravities; and since the specific gravities are in arithmetical progression, it follows that the figures for the corresponding volumes are in harmonic[2] progression. When, therefore, a hydrometer is so graduated as to show directly specific gravities, its graduations are the reciprocals of the specific gravities it is constructed to indicate, expressed as fractions of the volume of water which it displaces when allowed to float freely in that liquid. As these numbers are in harmonic progression, the intervals of such a scale will not be equal, but become gradually closer together at that portion of the scale which indicates the higher gravities. This may be readily seen on observing the scale of a hydrometer, Fig. 2 (page 18). This inequality of the intervals between the divisions of the scale becomes less and less apparent as the indications of the instrument increase in delicacy; but it is obvious that they must exist even in instruments constructed to show very minute differences in specific gravity.

Hydrometers which Indicate Pounds per Barrel.—These instruments in reality only indicate specific gravity in another way. Long's saccharometer, which may be taken as an illustration, is made of metal, and has a flat stem, on the opposite sides of which are two scales, the one reading from 0 (specific gravity of water) to 25 lbs. per barrel, the other commencing with 25 lbs. per barrel and extending up to 52 lbs. When the gravity of the liquid being examined exceeds 25 lbs. per barrel, a small weight is attached to the lower part of the instrument, and the readings are then taken from the second scale. As we shall recognise later on, the expression "pounds per barrel" is an exceedingly convenient one for use in the brewery.

Relation of Pounds per Barrel to Specific Gravity.—This (taking water at 1000) is as follows:—A barrel is equal to 36 gallons, and as a gallon of water weighs 10 lbs., a barrel of water will weigh 360 lbs. A barrel of another liquid, such as wort, which has a higher specific gravity than water, will weigh more than 360 lbs. On taking the gravity of such a liquid with Long's saccharometer, or a similar instrument indicating pounds per barrel,

---

[1] The reciprocal of a number is the quotient obtained by dividing unity by that number.

[2] Numbers are said to be in arithmetical progression when they increase or decrease by equal differences, as 0, 3, 6, 9, 12. When a series of numbers are in harmonic progression, the ratio between the first and third of any three contiguous numbers is as the difference between the first and second is to the difference between the second and third; thus the figures 2, 3, 6 are in harmonic progression, and consequently $2 : 6 :: 3-2 : 6-3$.

we ascertain how many more pounds a barrel or 36 gallons of such liquid weighs over and above 360 lbs. Supposing a sample of wort, when tested in this manner, shows 18 lbs. per barrel, a barrel of such a wort will weigh 360 + 18 = 378 lbs. Its specific gravity may be found by the following sum in proportion:—

$$\underset{\substack{\text{Weight of barrel}\\ \text{of water}}}{360} : \underset{\substack{\text{Weight of barrel}\\ \text{of wort}}}{378} :: \underset{\substack{\text{Weight of 1000 parts}\\ \text{of water}}}{1000} : \underset{\substack{\text{Weight of 1000 parts}\\ \text{of wort}}}{1050}$$

Conversely, the weight of a barrel of wort may be found from the specific gravity in the following manner:—

$$1000 : 1050 :: 360 : 378$$

and 378 — 360 equals 18 lbs.

Or the weight per barrel may also be found from the specific gravity in the following manner, where the weight in excess of the 1000 is employed:—

$$1000 : 360 :: 50 : 18.$$

As 360 ÷ 1000 = 0.36, the same result is obtained by multiplying the 50 by 0.36 thus, 1050 − 1000 × 0.36 = 18 lbs. per barrel. The converse operation, conversion of pounds per barrel to specific gravity, may be effected by dividing the number of pounds per barrel by 0.36, and adding the sum thus obtained to 1000, as 18 ÷ 0.36 = 50, and 50 + 1000 = 1050. The number 36, being the number of gallons contained in a barrel, is one easily remembered. Specific gravity, when expressed in the usual manner (*i.e.*, water = 1000), may be regarded as the weight in excess in pounds over and above the weight of a hundred gallons of water, that is, 1000 lbs., just as pounds per barrel are the excess in weight over and above 36 gallons (or 360 lbs.) of water.

Hydrometers which Indicate directly the Percentage of Substance in Solution.—Of this nature is the hydrometer of Balling, almost exclusively used in Continental breweries. Its indications show how many parts of cane-sugar are contained in 100 parts by weight of a solution at a temperature of 14° R. (63.5° F.). It may seem surprising that a hydrometer indicating percentages of cane-sugar should be used in brewing operations, instead of one which indicates percentages of brewers' extract. It has, however, proved so difficult to ascertain what the true percentages of brewers' extract corresponding to the various degrees of gravity are, that it is found advisable to take some readily accessible substance, a solution of which can be easily prepared of any required strength. Though the instrument does not express true amounts of brewers' extract, yet its indications are strictly comparable amongst themselves, and this is the question of real importance to the brewer. On the Continent, therefore, the gravity of a wort is usually expressed in percentages of extract according to Balling's scale, and the amount of gravity lost during fermentation in a similar manner.

Tralles' Alcoholimeter.—This instrument is also constructed on an empirical scale, and indicates directly the amount of alcohol by volume contained in mixtures of alcohol and water at a temperature of 60° F.

Bates' Saccharometer.—It is obvious that the length of the scale of any hydrometer of a convenient size is limited, if its divisions are to be so wide apart that they can be read with accuracy. Consequently, in order to deal with the various specific gravities of the fluids met with in the brewery, it is necessary to have several instruments with different scales. This, though of no importance when the hydrometers are kept in one room, becomes a positive disadvantage when they have to be carried about from place to place. A number of attempts have been made to obviate this inconvenience and to contrive a single instrument by means of which, with slight alterations, all the gravities met with in the brewery may be measured. The problem was satisfactorily solved by Bates, who constructed an instrument provided with five removable weights, or "poises," as they are termed, and capable of estimating the specific gravity of any liquid from 1000° to 1150° (water = 1000). This is the instrument universally employed by the Excise Authorities, and also very frequently by brewers.

Bates' saccharometer is constructed of brass, and has much the same form as an ordinary hydrometer, but the stem is square instead of cylindrical. In place of the lower bulb, which in the ordinary hydrometer is weighted with mercury or shot, there is a ring to which can be attached in turn, by means of a conical hole in its under side, a number of poises of different weights and bulks, each of which is provided with a conical peg. In this way a portable instrument, reading from 1000° to 1150° of specific gravity (water = 1000), is obtained. The scale has thirty principal divisions divided into halves, and, with the smallest poise attached, the instrument indicates degrees of gravity from 1000 to 1030; that is, the reading of the scale *plus* 1000. The other poises are marked respectively 30, 60, 90, 120, and, when any one of these is attached to the instrument, the number marked on the poise is added to the reading of the scale and their sum added to the 1000. Thus, if the instrument with the poise marked 60 attached, is immersed in a liquid and sinks to the mark 15 on its scale, then the specific gravity indicated will be 15 + 60 + 1000 = 1075°. The principles which underlie the construction of the instrument may be explained in the following manner. Suppose that an instrument has its stem so divided into thirty divisions that these indicate degrees of specific gravity from 1000 to 1030, and that the instrument weighs 103 grammes. It is evident that when immersed in water it will displace 103 c.c. (1 cubic centimetre of water weighs one gramme), and that when placed in a liquid of specific gravity 1030 it will displace 100 c.c. If to the instrument a poise be attached which has a bulk of 3 c.c. and a weight of 6.18 grammes, it will sink to its zero mark when immersed in a liquid of specific gravity 1030, because the total bulk of the instrument is now 106 c.c., and this multiplied by $1.03 = 109.18$ grammes. When immersed in a liquid of specific gravity 1060 it will sink to the point marked 30, since it then displaces 103 c.c. of liquid and $103 \times 1.06 = 109.18$ grammes. This may also be shown to be true of every other division on its scale. If another poise having a volume of 6 c.c. and weighing 12.54 grammes is attached, then the instrument registers from 1060 to 1090 of specific gravity, for either 109 c.c. of a liquid of specific gravity 1060, or 106 c.c. of a fluid of specific gravity 1090, weigh 115.54 grammes. It is obvious that this process can be extended, each successive poise having an additional volume of 3 c.c. and a constantly increasing weight. The following considerations will show how this weight is obtained :—Suppose that without increasing the original volume of the instrument (103 c.c.), we wish to make it sink to zero in a liquid of a specific gravity of 1030, we shall have to increase its weight by $(103 \times 1.03) - 103 = 3.09$ grammes. If we add on a volume of 3 c.c. this will displace $3 \times 1.03 = 3.09$ grammes, consequently this weight will have to be added to the instrument in addition to the former one to cause it to sink to zero, and $3.09 + 3.09 = 6.18$ grammes. The weights of the other poises may be found in a similar manner. It is obvious that whatever the original volume of the instrument may be, the respective volumes and weights of the poises may be obtained from the above data by simple proportion.

CHAPTER II.

CHEMISTRY, WITH SPECIAL REFERENCE TO THE MATERIALS USED IN BREWING.

BEFORE entering into a detailed description of the chemistry of the materials used in brewing, it will be well to devote a short space to the consideration of a few of the fundamental principles of chemistry.

The Elements.—The chemist is able to split up most of the bodies which occur in nature into two or more substances. This process is called "chemical analysis."

After continuing the splitting up for a time, a stage is reached in which a variety of bodies are obtained which resist all attempts at further division. Suppose the chemist takes a piece of chalk; he is readily able to split this body into calcium, carbon, and oxygen; but he can go no further than this, for, however much chemists have endeavoured to split up calcium, carbon, or oxygen into other bodies, their efforts have so far been unsuccessful. Substances of this nature, which it has been found impossible to further chemically subdivide, are called "elements."

Working in this way, chemists have been able to discover some 67 different elements. Comparatively few of these, however, concern the brewer, as only some 14 of them occur in the materials used in brewing.

Chemical Symbols.—In describing the elements, it is customary to indicate each of them by a symbol. This is either the first letter or the first two letters of the Latin name of the element. Thus H is used to denote hydrogen, O oxygen, Ca calcium, Cl chlorine, &c. In most instances the initial letters of both the English and Latin names of the elements are identical; in a few, as, for instance, K (kalium) potassium, Na (natrium) sodium, &c., they are not.

Atomic Combining Proportions.—When elements combine with each other, the combination always takes place in some definite relative weight or multiple of that weight. Thus when oxygen and calcium combine, it is always in the proportion of 16 parts by weight of oxygen and 40 parts of calcium. The number 16 is called the atomic weight of oxygen, and 40 that of calcium. The symbol is, however, used to express something beyond the mere name of the element; it not only denotes the element, but its atomic weight as well. Thus when we write FeO (ferrous oxide), we not only express a compound of iron and oxygen, but also at the same time signify that the compound consists of one atom or 56 parts by weight of iron, and one atom or 16 parts by weight of oxygen. If we have 72 ounces of ferrous oxide, we know at once from its formula that it contains 56 ounces of iron and 16 ounces of oxygen; or if it were 72 pounds, tons, or grains, the other figures would be respectively pounds, tons, or grains. In such a compound as water, where two atoms of hydrogen are united to one atom of oxygen, a small figure 2 is appended to the symbol for hydrogen, H_2; H_2O therefore denotes that water is composed of two atoms of hydrogen combined with one atom of oxygen. Similarly NH_3 (ammonia) stands for one atom of nitrogen combined with three atoms of hydrogen; CH_4 (methane), one atom carbon and four atoms hydrogen.

Table of the more Common Elements.—The following table contains a list of those elements which are principally met with in brewing chemistry, together with their symbols and atomic weights:—

Hydrogen,	H	1	Sulphur,	S	32
Carbon,	C	12	Chlorine,	Cl	35.5
Nitrogen,	N	14	Potassium (Kalium),	K	39
Oxygen,	O	16	Calcium,	Ca	40
Sodium (Natrium),	Na	23	Iron (Ferrum),	Fe	56
Magnesium,	Mg	24	Copper (Cuprum),	Cu	63
Aluminium,	Al	27.5	Silver (Argentum),	Ag	108
Silicon,	Si	28	Iodine,	I	127
Phosphorus,	P	31			

Valency.—Another characteristic of the atoms of the elements is that with some of them the atom is capable of uniting with one atom of another element, as hydrogen, which only forms one compound with chlorine, hydrochloric acid gas, HCl. Such elements are termed "univalent." The atom of others is able to unite with several atoms of another element, as we have just seen in the case of water, where one atom of oxygen unites with two atoms of hydrogen. Hence the element oxygen is termed "bivalent." "Trivalent" nitrogen is able to unite with three atoms of hydrogen to form NH_3 (ammonia), and "quadrivalent" carbon can unite with four hydrogen atoms to form methane, CH_4. This atom-fixing power is called "valency" or "equivalence," and the greatest number of hydrogen atoms, or their equivalent, which any particular element can fix is taken as its standard of valency. In the case of such an element as calcium, which does not combine directly with hydrogen, we are compelled to ascertain its valency indirectly. Calcium combines readily with oxygen to form calcium oxide, CaO, and since, so far as valency is concerned, one atom of oxygen is equivalent to two atoms of hydrogen, calcium must be bivalent.

Bonds.—It is exceedingly convenient to be able to give visible expression to the valencies or atom-fixing powers of elements, and for this purpose short lines drawn from the symbol of the element, technically called "bonds" or "bonds of affinity," are frequently

CHEMICAL UNION.

made use of. Thus one short line affixed to the symbol H— tells us that the hydrogen atom possesses one bond of affinity; three similar lines to nitrogen —N— that it has three, —C— that carbon has four, and so on with the rest of the elements.

Chemical Union.—When two atoms enter into chemical combination with each other, the bonds of the one are supposed to become linked with those of the other. Thus in the compound H—Cl, the bond of the H atom is linked to the bond of the Cl atom; in water H—O—H the two bonds of the oxygen atom are linked to the bonds of the two hydrogen atoms. Similarly,

$$\text{H}-\text{N}-\text{H} \quad \text{and} \quad \text{H}-\overset{\text{H}}{\underset{\text{H}}{\text{C}}}-\text{H}$$

In order to more vividly picture the valency of elements, and the way in which their bonds of affinity unite, small differently coloured balls with short pegs affixed to them are often employed. Thus hydrogen may be represented by a white ball with one peg, chlorine by a yellow ball with one peg, oxygen by a red ball with two pegs, nitrogen by a blue ball with three pegs, and carbon by a black ball with four pegs. The pegs are easily connected by small pieces of rubber tubing. The following compounds illustrated in this way appear thus:—

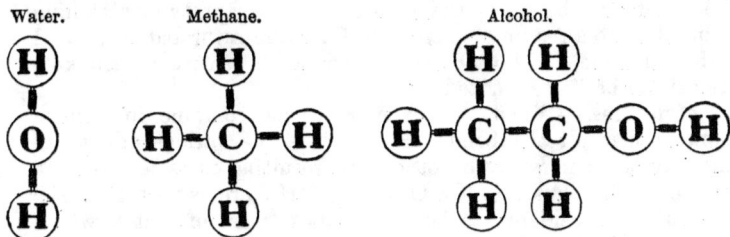

FIG. 5.—Valency of Elements.

Constitution.—Any plan of this kind which gives us an idea as to the position in which the various atoms of a compound are arranged with reference to one another is called a "structural" or "constitutional formula"; that is, it expresses the structure or constitution of the compound.

It must be distinctly borne in mind that all this doctrine of valency and linkage of atoms is purely hypothetical. It has, however, proved itself of such immense assistance in explaining the way in which compounds are built up or pulled to pieces, that it may fairly be considered that some such arrangement of the atoms really exists, though we are by no means certain as to what its exact nature may be.

Saturated Compounds.—When, as in the cases above mentioned, *all* the bonds of affinity of an element are linked with those of one or more other atoms, such an atom is said to be "satisfied." Thus a univalent atom requires one univalent atom to satisfy it, as in H—Cl (hydrochloric acid gas). A bivalent atom requires either two univalent atoms or one bivalent atom, as in calcium chloride $Ca=Cl_2$, or calcium oxide $Ca=O$. Similarly, a trivalent atom requires three univalent, and a quadrivalent atom four univalent or two bivalent atoms to satisfy it, as in—

$$\text{Methane} \quad C\begin{smallmatrix}H\\H\\H\\H\end{smallmatrix} \quad \text{or in carbon dioxide} \quad C\begin{smallmatrix}=O\\=O\end{smallmatrix}$$

Such compounds as the above, in which all the bonds of affinity of the atoms are satisfied, are known as "saturated compounds." They are possessed of considerable stability, and are incapable of taking up more atoms.

Unsaturated Compounds.—In such compounds as carbon monoxide, $=C=O$, two of the bonds of the carbon atom are unsatisfied, and as this latter pair of bonds are ready to attach themselves to other atoms in order to become satisfied, the resulting compound is unstable and prone to change.

Thus the simple ignition of carbon monoxide gas in air affords the unsatisfied bonds of the carbon the opportunity of attaching themselves to an atom of oxygen to form the saturated compound carbon dioxide, $O=C=O$. Such compounds as carbon monoxide are said to be "unsaturated."

Molecules.—When two or more atoms combine in such a manner that all the bonds of affinity of the various atoms are satisfied, the one with the other, the resulting mass is termed a "molecule." Thus, H—Cl, $O=H_2$, $C\equiv H_4$, are molecules. The molecule is considered to be the smallest mass of matter which can exist in a free state.

From this it is inferred that the elements do not exist in the state of atoms, but as molecules, two or more atoms being always combined, so that the bond or bonds of affinity of the one atom satisfy the bond or bonds of the other. Thus, hydrogen is supposed to exist as H—H, oxygen as $O=O$, and so on with the rest of the elements.

But many chemical reactions are known in which an interchange of atoms takes place. For instance, when the metal sodium is placed in contact with water, the following reaction ensues:—

$$\underset{\text{Sodium.}}{Na-Na} + \underset{\text{Water.}}{\begin{matrix}H-O-H\\H-O-H\end{matrix}} = \underset{\text{Sodium hydroxide.}}{\begin{matrix}Na-O-H\\Na-O-H\end{matrix}} + \underset{\text{Hydrogen.}}{H-H}$$

In this, one of the hydrogen and one of the oxygen atoms of each molecule of water unite with one of the sodium atoms to form a molecule of sodium hydroxide. The remaining hydrogen atoms of the molecule of water are liberated, and unite to form a molecule of hydrogen. The atom is considered to be the smallest particle of an element which is capable of taking part in a chemical reaction.

Nascent State.—When, as in the case just mentioned, an element is liberated from a compound in atoms, it is said to be in the nascent state, and in this condition it manifests a degree of activity which it does not exhibit under ordinary circumstances. For instance, hydrogen gas may be passed through a dilute solution of nitric acid (HNO_3) without in the slightest degree affecting the composition of the acid. But if hydrogen be generated in the nascent state in such a solution, the chemical composition of the acid becomes entirely changed. The hydrogen, under these circumstances, is able to combine with and remove the whole of the oxygen of the nitric acid as water, whilst at the same time two other atoms of hydrogen attach themselves to the nitrogen to form ammonia. In this way nitric acid is converted into ammonia according to the following equation:—

$$HNO_3 + 8H\text{*} = 6H_2O\text{*} + NH_3.$$

Other elements, such as oxygen, nitrogen, &c., behave in a similar manner.

Chemistry of the Carbon Compounds.—By far the greater number of bodies with which we have to deal in the operation of brewing have at one time entered into the structure of and been formed by living organisms. Hence they belong to that domain of chemistry which was at one time called organic. Formerly the opinion was held that the so-called organic compounds were of a nature essentially different from the inorganic. Though the chemist could readily build up the latter class of bodies from the elements, he had been, up to that time, unable to form any of the substances belonging to the former class in a similar manner.

This view received its death-blow in 1828, when Wöhler produced urea artificially. Since then it has been found possible to build up in the laboratory an enormous number of substances which it was formerly supposed could only be formed by living organisms. Even such complex bodies as the sugars have been built up from the elements in recent times, and it is now fully recognised that, chemically, there is no difference between organic and inorganic compounds. They are composed of the same elements, and these obey the same laws of chemical combination in either class of bodies.

Since the element carbon enters into the composition of all the

---

* A large numeral placed before a symbol or group of symbols means that the symbol or group is to be multiplied by that number. Thus, in this equation 8H stands for eight atoms of hydrogen, and $6H_2O$ for six molecules of water.

so-called organic bodies, that which was formerly termed "organic chemistry" is generally spoken of now as "the chemistry of the carbon compounds."

The elements which enter into the composition of organic substances are, as a rule, extremely limited in number, being generally confined to not more than five; yet the manner in which, by the operation of a few simple laws, these few elements are varied in their relations one with another, and the enormous number of different substances which are produced in this way, naturally or artificially, is truly remarkable. One of the most striking features of this department of chemistry is the numerous series of compounds which are constructed on a definite plan common to all its members, each of which merely differs from the member preceding it as we proceed down the series by some simple relation, such as containing one atom of carbon and two atoms of hydrogen more. Such series are called "homologous" (ὁμός, like, λόγος, description), because all the members of any particular series being constructed on the same plan, the same description applies to each.

The first of these series, the paraffins, is a particularly important one, for it forms the foundation of many other homologous series which contain bodies of special interest to the brewer. We shall find as we proceed that the paraffins are, as it were, the skeletons of such bodies as alcohol, the sugars, and many other substances.

The Paraffins.—The first member of this series is
$$\begin{array}{c} H \\ | \\ H-C-H \\ | \\ H \end{array}$$
methane, a body in which the four bonds of the carbon atoms are each linked with an atom of hydrogen, and as all the bonds of the carbon atom are satisfied in the above compound, a molecule of considerable stability, and one which is incapable of attaching further atoms to itself, results.

The bonds of the carbon atoms have the power of satisfying one another under certain conditions, and by virtue of this property a series of bodies of the following constitution is formed:—

$$\underset{\text{Ethane.}}{\begin{array}{c} H\ H \\ |\ \ | \\ H-C-C-H \\ |\ \ | \\ H\ H \end{array}} \qquad \underset{\text{Propane.}}{\begin{array}{c} H\ H\ H \\ |\ \ |\ \ | \\ H-C-C-C-H \\ |\ \ |\ \ | \\ H\ H\ H \end{array}} \qquad \underset{\text{Tetrane.}}{\begin{array}{c} H\ H\ H\ H \\ |\ \ |\ \ |\ \ | \\ H-C-C-C-C-H \\ |\ \ |\ \ |\ \ | \\ H\ H\ H\ H \end{array}}$$

and so forth, in which the carbon atoms are linked together in what is termed an "open chain."

In this way the second and remaining members of the paraffin series are progressively built up:—

Methane	.	.	.	CH_4	Butane	.	.	.	C_4H_{10}
Ethane	.	.	.	C_2H_6	Pentane	.	.	.	C_5H_{12}
Propane	.	.	.	C_3H_8	Hexane	.	.	.	C_6H_{14}

SUBSTITUTION.

Each member of the series as we descend contains one carbon and two hydrogen atoms (CH_2) more than the preceding one. The list does not stop at hexane; all the members of the series from methane CH_4 to hexadecane $C_{16}H_{34}$ are known.

Substitution.—The paraffins being fully saturated, show no inclination to combine directly with other bodies, but they exhibit in a marked degree the curious property—shared by most organic compounds—of allowing one or more of their hydrogen atoms to be replaced by a univalent element or group. This process is called "substitution." For instance, it is possible to replace successively the hydrogen atoms in methane by the univalent element chlorine, and obtain in this manner the following four bodies:—

$$\begin{array}{cccc}
\text{Cl} & \text{Cl} & \text{Cl} & \text{Cl} \\
| & | & | & | \\
\text{H—C—H} & \text{H—C—Cl} & \text{H—C—Cl} & \text{Cl—C—Cl} \\
| & | & | & | \\
\text{H} & \text{H} & \text{Cl} & \text{Cl} \\
\text{Monochloromethane.} & \text{Dichloromethane.} & \text{Trichloromethane} & \text{Tetrachloromethane} \\
 & & \text{(Chloroform.)} &
\end{array}$$

Unsaturated Groups.—Many unsaturated groups are met with, such as H—O— (hydroxyl), which body may be regarded as water H—O—H *minus* one atom of hydrogen. Such groups are incapable of independent existence, but often exist in pairs, the free bond or bonds of the one group satisfying those of the other; thus hydrogen peroxide (H—O)—(O—H) is a well-known compound.

But the property of these groups which chiefly concerns us is their power of replacing the elements, and thus forming what are termed "substitution compounds." The univalent group H—O— (hydroxyl) often takes the place of an atom of hydrogen, as, for instance, when it replaces one of the hydrogen atoms in methane,

and yields
$$\begin{array}{c} \text{H} \\ | \\ \text{H—C—O—H} \\ | \\ \text{H} \end{array}$$
or methylic alcohol.

Compound Radicles.—Unsaturated groups, which, so far as substitution is concerned, behave in this way are often called "compound radicles." The members of the paraffin series, for instance, when deprived of an atom of hydrogen, form such radicles. Of such a nature are methyl CH_3, that is methane CH_4 *minus* one atom of hydrogen; ethyl C_2H_5 which is similarly related to ethane C_2H_6, and so on. Obviously such bodies cannot exist in the free state.

The Alcohols.—When one of the hydrogen atoms of a paraffin is replaced by a hydroxyl group we obtain a monohydric alcohol. In this way the homologous series of alcohols, of which methylic alcohol forms the first member, is produced, each member differing from its predecessor by CH_2. The following are the first six members:—

Methylic alcohol	CH_3OH	Butylic alcohol	C_4H_9OH
Ethylic alcohol	C_2H_5OH	Pentylic alcohol	$C_5H_{11}OH$
Propylic alcohol	C_3H_7OH	Hexylic alcohol	$C_6H_{13}OH$

The Glycols or Dihydric Alcohols.—When two atoms of the hydrogen of a paraffin are replaced by hydroxyl groups, we obtain another homologous series of alcohols, known as the "glycols."

Thus ethane—

$$\begin{array}{cc} H & H \\ | & | \\ H-C-C-H \\ | & | \\ H & H \end{array} \text{ or } CH_3 \cdot CH_3 \text{ yields ethylene glycol } \begin{array}{cc} H & H \\ | & | \\ OH-C-C-OH \\ | & | \\ H & H \end{array}$$

or $CH_2(OH) \cdot CH_2(OH)$; and propylene $CH_3 \cdot CH_2 \cdot CH_3$ gives propylene glycol $CH_2(OH) \cdot CH_2 \cdot CH_2(OH)$.

The Trihydric Alcohols.—When three of the hydrogen atoms of a paraffin are replaced by three hydroxyl groups, another homologous series of alcohols is produced of which glycerol (glycerin) is the first member; thus propane

$$\begin{array}{ccc} H & H & H \\ | & | & | \\ H-C-C-C-H \\ | & | & | \\ H & H & H \end{array} \text{ or } CH_3 \cdot CH_2 \cdot CH_3 \text{ yields } \begin{array}{ccc} H & H & H \\ | & | & | \\ H-C-C-C-H \\ | & | & | \\ OH & OH & OH \end{array} \text{ or } CH_2OH \cdot CHOH \cdot CH_2OH$$

glycerol. This and amylic glycerol are the only two members of the series known.

Tetrahydric and **Pentahydric Alcohols** are also known.

Hexahydric Alcohols.—When in the paraffin hexane C_6H_{14}, six of the hydrogen atoms are replaced by hydroxyl groups, the following alcohol, mannitol, is obtained—

$$CH_2OH \cdot CHOH \cdot CHOH \cdot CHOH \cdot CHOH \cdot CH_2OH,$$

found in the drug manna, which is excreted by certain tropical trees. Sorbitol, though a different body, has the same constitution; it is what is termed an optical stereoisomeride (page 41).

The Aldehydes.—Most of the monohydric alcohols, when they are submitted to the action of certain oxidising agents (such as potassium dichromate and sulphuric acid), have two atoms of hydrogen removed from their molecule, and a body called an "aldehyde" (*alcohol dehydrogenatum*) results. Methylic alcohol when passed over a red-hot spiral of platinum wire yields methylic aldehyde or formaldehyde—

$$\begin{array}{c} H \\ | \\ H-C-OH + O = \\ | \\ H \end{array} \quad H-C\begin{array}{c} \diagup O \\ \diagdown H \end{array} \quad + H_2O$$

Methylic alcohol. Formaldehyde. Water.

ALDOSES.

Similarly, ordinary alcohol yields ethylic or ordinary aldehyde—

$$CH_3 \cdot CH_2OH + O = CH_3 \cdot COH + H_2O$$
$$\text{Alcohol.} \qquad \qquad \text{Aldehyde.} \quad \text{Water.}$$

Thus the aldehydes form another homologous series. They frequently behave like unsaturated bodies, as if one bond of the C and one of the O in the COH group were unsatisfied.[1] These unsatisfied bonds exhibit an affinity for oxygen, and in virtue of this are able to reduce many metallic oxides. They also form compounds with phenylhydrazine (see page 39).

The Aldoses.—These bodies are of an aldehydic nature; they contain the group COH, reduce alkaline metallic solutions, and yield characteristic crystalline compounds with phenylhydrazine. They are usually sweet to the taste, often optically active, and some of them are capable of undergoing the alcoholic fermentation. The names triose, tetrose, &c., have been proposed by E. Fischer for these bodies, according to the number of carbon atoms they respectively contain, whilst the general name of "aldose" for the whole group was suggested by Professor Armstrong. The following members are known :—

Triose .	. $CH_2OH \cdot CHOH \cdot COH$.	. Glycerose.
Tetrose .	. $CH_2OH \cdot 2(CHOH) \cdot COH$.	. Erythrose.
Pentose .	. $CH_2OH \cdot 3(CHOH) \cdot COH$.	. Arabinose, xylose.
Hexose .	. $CH_2OH \cdot 4(CHOH) \cdot COH$.	. Glucose, galactose, mannose.
Heptose .	. $CH_2OH \cdot 5(CHOH) \cdot COH$.	. Heptose.
Octose .	. $CH_2OH \cdot 6(CHOH) \cdot COH$.	. Octose.
Nonose .	. $CH_2OH \cdot 7(CHOH) \cdot COH$.	. Nonose.

Of these, the pentoses and hexoses are of particular interest to the brewer. The pentoses he meets with to a certain extent in his materials, and to the hexoses or their derivatives belong all the sugars which he subjects to the process of alcoholic fermentation. It is a remarkable fact that of the known aldoses only those which contain three carbon atoms, or a multiple of three, are fermentable by brewers' yeast—that is, the trioses, hexoses, and nonoses.[2] The pentose sugar arabinose, though unfermentable by yeast, is attacked and broken down by an organism, the *Bacillus ethaceticus*, yielding ethyl alcohol, acetic acid, carbon dioxide, hydrogen, and traces of succinic acid.

The Ketones.—This homologous series of bodies may be regarded as derivatives of the aldehydes, in which the hydrogen atom of the COH group is replaced by a compound radicle derived from a paraffin. Thus when the H atom in aldehyde is replaced by CH_3 (methyl), acetone is formed—

$$CH_3{-}CO{-}H \qquad \qquad CH_3{-}CO{-}CH_3$$
$$\text{Aldehyde.} \qquad \qquad \qquad \text{Acetone.}$$

---

[1] See formula, p. 34.

[2] And this property is only possessed by nono-mannose, nono-glucose being unfermentable.

The group CO (carbonyl) is the distinguishing characteristic both of the aldehydes and of the ketones. Mixed ketones also exist, such as methyl ethyl ketone, $CH_3 \cdot CO \cdot C_2H_5$.

Ketoses.—These bodies bear the same relation to the ketones that the aldoses do to the aldehydes. Fructose (lævulose), one of the constituents of malt and of invert sugar, is of this nature. It has the following constitution :—

$$CH_2OH \cdot CHOH \cdot CHOH \cdot CHOH \cdot CO \cdot CH_2OH$$

and may be considered as derived from butyl-methyl ketone, $CH_3 \cdot CH_2 \cdot CH_2 \cdot CH_2 \cdot CO \cdot CH_3$.

Cyanhydrins.—These bodies are produced by the direct union of hydrocyanic acid with an aldehyde or a ketone, thus :—

$$\underset{\text{Aldehyde.}}{CH_3 \cdot COH} + \underset{\text{Hydrocyanic acid.}}{HCN} = \underset{\text{Aldehyde cyanhydrin.}}{CH_3 \cdot CHOH \cdot CN}$$

$$\underset{\text{Acetone.}}{(CH_3)_2 \cdot CO} + \underset{\text{Hydrocyanic acid.}}{HCN} = \underset{\text{Acetone cyanhydrin.}}{(CH_3)_2 \cdot COH \cdot CN}$$

Similar compounds are formed by the aldoses and ketoses, thus :—

$$\underset{\text{Arabinose.}}{CH_2OH \cdot 3(CHOH) \cdot COH} + \underset{\text{Hydrocyanic acid.}}{HCN} = \underset{\text{Arabinose cyanhydrin.}}{CH_2OH \cdot 4(CHOH) \cdot CN}$$

When the cyanhydrins are acted upon by a mineral acid, the CN group resolves itself into the COOH group, ammonia being formed simultaneously, thus :—

$$\underset{\text{Arabinose cyanhydrin.}}{CH_2OH \cdot 4(CHOH) \cdot CN} + HCl + 2(H_2O) = \underset{\text{Mannonic acid.}}{CH_2OH \cdot 4(CHOH) \cdot COOH} + NH_4Cl$$

This is an exceedingly interesting reaction, since it enables us to pass from a lower to a higher compound. In the instance just given we pass from arabinose to mannose.

Acetic Acid Series.—When the two free bonds of an aldehyde derived from a paraffin are satisfied with an atom of oxygen, we obtain a series of acids of which formic acid is the first member, thus :—

$$\underset{\text{Formaldehyde.}}{\overset{|}{\underset{|}{H-C-O}}} + O = \underset{\text{Formic acid.}}{\overset{|}{\underset{|}{H-C\!=\!O}}}$$
$$HOH$$

Ordinary aldehyde similarly yields acetic acid :—

$$CH_3 \cdot COH + O = CH_3 \cdot COOH$$

CONVERSION OF ACIDS TO ALDEHYDES.

The aldoses behave in a similar manner, yielding the corresponding acids of the series. Thus glucose yields gluconic acid:—

$$CH_2OH \cdot 4(CHOH) \cdot COH + O = CH_2OH \cdot 4(CHOH) \cdot COOH$$
Glucose.   Gluconic acid.

Conversion of Acids into Aldehydes and Alcohols.—It will be seen that it is possible by a process of gradual oxidation to convert an alcohol first into an aldehyde, and then by further oxidation to convert the aldehyde into an acid. By operating in the reverse way, that is to say, if, instead of using an oxidising, we use a deoxidising agent, or what is commonly called a "reducing" agent, we can, so far as the acids corresponding to the aldoses are concerned, reverse the operation. Thus, if gluconic acid be acted upon at a low temperature by nascent hydrogen, which may be readily generated from sodium amalgam,[1] the acid is first converted into the aldose in the following manner:—

$$CH_2OH \cdot 4(CHOH) \cdot COOH + 2H = CH_2OH \cdot 4(CHOH) \cdot COH + H_2O$$
Gluconic acid.   Glucose.

By continuing the process two atoms of hydrogen are added on to the COH group, and the alcohol, which contains the CH_2OH group, then results.

Lactones.—Here may be mentioned a peculiar tendency possessed by certain acids, amongst which are those derived from the sugars, of losing a molecule of water and becoming converted into a lactone when evaporated to a strong aqueous solution. Thus gluconic acid, $CH_2OH \cdot 4(CHOH) \cdot COOH$, becomes gluconic lactone, $CH_2OH \cdot CHOH \cdot CH \cdot 2(CHOH) \cdot CO$, with an O bridge. These bodies have a neutral reaction.

Lactic Acid.—The series of acids to which this substance, which is frequently met with in fermentations, belongs is closely allied to the acetic series. When, in the third member of that series, propionic acid, $CH_3 \cdot CH_2 \cdot COOH$, one of the hydrogen atoms of the middle carbon atom, is replaced by hydroxyl, lactic acid, $CH_3 \cdot CH(OH) \cdot COOH$, is formed.

Oxalic Acid is derived from the first dihydric alcohol or glycol by oxidation of the two CH_2OH groups to COOH, thus:—

$$\begin{matrix}CH_2OH \\ CH_2OH\end{matrix} + 4O = \begin{matrix}COOH \\ COOH\end{matrix} + 2H_2O$$
Glycol.   Oxalic acid.   Water.

It is one of the acids occurring in malt, and is often seen in the form of calcium oxalate, which forms very characteristic octahedral crystals, in specimens of yeast when examined under the microscope.

---

[1] Made by warming 100 grammes of mercury in a porcelain dish, covering the surface with a layer of solid paraffin melted, and plunging into the mercury (in pieces about the size of a pea, one at a time) about 5 grammes of metallic sodium.

Calcium oxalate, together with a small amount of organic matter, frequently deposits on the sides and on the attemperating coils of the fermentation vessels.

Succinic Acid.—This acid, which is a constant product of the alcoholic fermentation, is derived from the third member of the glycol series: butylene glycol, $CH_2OH \cdot CH_2 \cdot CH_2 \cdot CH_2OH$, which, when oxidised, yields succinic acid, $COOH \cdot CH_2 \cdot CH_2 \cdot COOH$.

Salts of Organic Acids.—All the organic acids share with the mineral acids the property of combining with bases to form salts. Thus when sodium hydroxide and acetic acid are brought together, the following reaction ensues:—

$$CH_3 \cdot COOH + NaHO = CH_3 \cdot COONa + H_2O$$
$$\text{Acetic acid.} \quad \text{Sodium hydroxide.} \quad \text{Sodium acetate.} \quad \text{Water.}$$

In other words, the hydrogen of the COOH (carboxyl) group can be replaced by a metal.

Bibasic Acids.—Acids, such as oxalic and succinic, which possess two COOH groups, are able to combine with two atoms of a univalent or with one atom of a bivalent base. Thus we have $COONa \cdot COONa$, sodium oxalate; and $\begin{smallmatrix}COO\\COO\end{smallmatrix}\!\!>\!\!Ca$, calcium oxalate. They also form acid salts in which the hydrogen of one of the COOH groups only is replaced by a base, as in acid oxalate of potassium, $COOH \cdot COOK$.

Compound Ethers[1] or Esters.—These bodies are analogous in constitution to the salts just mentioned. In their formation the alcohols behave in the same manner as the metallic hydroxides. Ethylic alcohol may be regarded as the hydroxide of ethyl, $C_2H_5(OH)$, just as caustic soda is the hydroxide of sodium, $Na(OH)$, and both can be made to combine with an acid in a similar manner, thus:—

$$CH_3 \cdot COOH + C_2H_5(OH) = CH_3 \cdot COO(C_2H_5) + H_2O$$
$$\text{Acetic acid.} \quad \text{Alcohol.} \quad \text{Ethyl acetate.} \quad \text{Water.}$$

$$CH_3 \cdot COOH + K(OH) = CH_3 \cdot COOK + H_2O$$
$$\text{Acetic acid.} \quad \text{Potassium hydroxide.} \quad \text{Potassium acetate.} \quad \text{Water.}$$

The reaction does not take place nearly so readily as with a metallic base, consequently the process which has to be adopted to cause such a combination is much more complicated.

---

[1] The esters or compound ethers must not be confounded with the ordinary ethers. These bear the same relation to the metallic oxides as the alcohols do to the hydroxides. Thus ordinary ethylic ether has the constitution $\begin{smallmatrix}C_2H_5\\C_2H_5\end{smallmatrix}\!O$ just as sodium oxide is $\begin{smallmatrix}Na\\Na\end{smallmatrix}\!O$. In order to avoid confusion between the two bodies the compound ethers are now often called esters.

THE AMIDES.

Many of the esters are extremely fragrant bodies, and on their presence depends the flavour of many fruits, the bouquet of wines, and to some extent the flavour of beer. Vinegar chiefly derives its fragrant odour from the presence of a small amount of ethylic acetate. The pear flavour of the confectioner is amylic acetate; pine-apple flavour is ethylic butyrate.

When an ester is boiled with an alcoholic solution of a metallic hydroxide, such as caustic soda, the following decomposition ensues :—

$$CH_3 \cdot COO(C_2H_5) \; + \; Na(HO) \; = \; CH_3 \cdot COONa \; + \; C_2H_5(OH)$$
Ethyl acetate.   Sodium hydroxide.   Sodium acetate.   Alcohol.

This reaction is used for the quantitative determination of the esters, since it is comparatively easy to ascertain the amount of alkali neutralised by the acid liberated in the decomposition.

The Amides.—Another univalent group, NH_2 (amidogen), often replaces a hydrogen atom, or the hydroxyl group, in organic compounds, and the body so formed is called an "amide." When the body so acted upon is an organic acid, the compound thus formed may be an amido-acid, or an acid amide, according as the position of the NH_2 group differs. As an example of the former may be cited amido-acetic acid,

$$H-\underset{NH_2}{\overset{H}{C}}-COOH,$$

in which one of the H atoms of the CH_3 group of acetic acid is replaced by amidogen.[1]

When in an organic acid the hydroxyl group of the CO(OH) group is replaced by NH_2, an acid amide is formed. Acetamide, $CH_3 \cdot CO(NH_2)$, is such a body.

Asparagine.—This is the acid amide of amido-succinic acid. The relations of the three bodies are shown in the following formulæ:[2]—

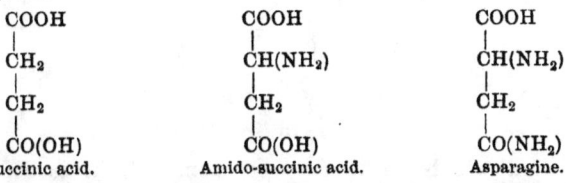

Succinic acid.   Amido-succinic acid.   Asparagine.

---

[1] The amido-acids are destitute of acid taste, do not redden litmus, and can be crystallised from alcohol in the presence of ammonia. These phenomena strongly point to the fact that the COOH group of the acid no longer exists as such. Consequently, it is considered that the formula

$$\begin{array}{c} H_2C-NH_3 \\ || \\ O=C-O \end{array}$$

in which nitrogen is quinquivalent, expresses their constitution more accurately.

[2] If, however, the formula, as suggested in the preceding note, is the

Asparagine was first discovered in asparagus, hence its name; it is found in most germinating plants. It forms large white transparent crystals, and is a highly diffusible substance. Young seedlings of the *Lupinus luteus*, when grown in the dark to a length of four or five inches, contain one fifth of their dry weight of asparagine. This is to us an interesting substance, since it is found in malt rootlets, and can serve as a nitrogenous food for yeast. On being fermented with yeast in the presence of albuminoids, it is converted into ammonium succinate.

Glutamine.—Another amide found in plants is closely allied to asparagine, from which it differs by having a hydrogen atom replaced by methyl, thus:—

$$\begin{array}{c} COOH \\ CH(NH_2) \\ CH(CH_3) \\ CO(NH_2) \end{array}$$

It is the acid amide of glutamic acid.

Phenylhydrazine.—This substance, which has played such an important part in recent times in the elucidation of the constitution of the members of the sugar group, and which has proved itself of inestimable service in the identification of the various sugars, is a derivative of benzene C_6H_6. The composition of benzene will strike the reader as being somewhat extraordinary, especially when he is told that it behaves, under most circumstances, like a fully saturated body; indeed, it is difficult to conceive how such a body can be built up with carbon which has four bonds of affinity. The theory first proposed by Kekulé to explain this apparent anomaly assumes that benzene has the following constitution:—

in which it will be seen that those bonds of the carbon atoms which are not attached to hydrogen are attached to one another.

---

more correct one, then probably amido-succinic acid has the constitution

$$\begin{array}{cc} COOH & CO(NH_2) \\ | & | \\ H-C-H, \text{ and asparagine} & H-C-H \\ | & | \\ H-C-NH_3 & H-C-NH_3 \\ | \; | & | \; | \\ O=C-O & O=C-O \end{array}$$

HYDRAZONES.

Several other slightly differing hypotheses have also been propounded to explain the constitution of this compound. When the carbon atoms are arranged in a compound in this way, they are said to form a "closed chain."

If one of the hydrogen atoms in benzene be replaced by the univalent group $-NH-NH_2$ (which is hydrazine, H_2N-NH_2, *minus* an atom of hydrogen), the compound phenylhydrazine, $C_6H_5-NH-NH_2$, is obtained.

Hydrazones.—When one molecule of phenylhydrazine combines with one molecule of an aldehyde, ketone, aldose, or ketose, a body termed a "phenylhydrazone," or, more shortly, a "hydrazone," results, thus:—

$$CH_3-CH\;O\;+\;H_2\;N-HN-C_6H_5\;=\;CH_3-CHN-HN-C_6H_5\;+\;H_2O$$
Aldehyde.   Phenylhydrazine.   Aldehyde-hydrazone.   Water.

$$(CH_3)_2-C\;O\;+\;H_2\;N-HN-C_6H_5\;=\;(CH_3)_2-CN-HN-C_6H_5\;+\;H_2O$$
Acetone.   Phenylhydrazine.   Acetone-hydrazone.   Water.

Similarly, an aldose yields an aldose-hydrazone, and a ketose a ketose-hydrazone:—

$$CH_2OH-4(CHOH)-CHO\;+\;H_2N-HN-C_6H_5\;=$$
Glucose.   Phenylhydrazine.

$$CH_2OH-4(CHOH)-CHN-HN-C_6H_5\;+\;H_2O$$
Glucose-hydrazone.   Water.

$$CH_2OH-3(CHOH)-\underset{\underset{CH_2OH}{|}}{CO}\;+\;H_2N-HN-C_6H_5\;=$$
Fructose.   Phenylhydrazine.

$$CH_2OH-3(CHOH)-\underset{\underset{CH_2OH}{|}}{CN}-HN-C_6H_5\;+\;H_2O$$
Fructose-hydrazone.   Water.

The hydrazones are for the most part colourless compounds, generally soluble in water, and are only occasionally of use in determining the nature of a sugar. Under the influence of strong hydrochloric acid, the hydrazones split up into phenylhydrazine (which combines with the acid) and the original body from which they were derived.

Osazones.—When an aldose or ketose in aqueous solution is heated for several hours on the water-bath with phenylhydrazine acetate, in the proportions of one molecule of the former to three molecules of the latter, a somewhat complicated reaction ensues, the result of which is the combination of one molecule of the sugar with two molecules of phenylhydrazine, a molecule of aniline and another of ammonia being simultaneously formed. The reaction between this reagent and an aldose is illustrated in the following equation:[1]—

$$\begin{array}{ll} \text{CH}_2\text{OH} & \text{CH}_2\text{OH} \\ 3(\text{CHOH}) & 3(\text{CHOH}) \\ (a)\text{CHH} \; \vdots \text{O} + \text{H}_2 \vdots \text{N—HN—C}_6\text{H}_5 = \text{CN—HN—C}_6\text{H}_5 + \text{H}_2 + \text{H}_2\text{O} \\ \\ \text{CH} \vdots \text{O} + \text{H}_2 \vdots \text{N—HN—C}_6\text{H}_5 \quad \text{CHN—HN—C}_6\text{H}_5 + \text{H}_2\text{O} \\ \quad \text{Glucose.} & \quad \text{Glucosazone.} \end{array}$$

The two hydrogen atoms which are liberated from the CHOH group immediately adjoining the COH group (marked (a)) combine with the third molecule of phenylhydrazine to form aniline and ammonia, thus:—

$$\underset{\text{Phenylhydrazine.}}{\text{H}_2\text{N—HN—C}_6\text{H}_5} + 2\text{H} = \underset{\text{Aniline.}}{\text{H}_2\text{N—C}_6\text{H}_5} + \underset{\text{Ammonia.}}{\text{NH}_3}$$

A somewhat analogous reaction ensues when a ketose is similarly treated; the first molecule of phenylhydrazine combines with the CO group, the second molecule then combines with the lowest CH_2OH group, from which two atoms of hydrogen are liberated, and these reacting on the third molecule of the reagent, as in the previous case form aniline and ammonia. The following equation illustrates the reaction:—

$$\begin{array}{ll} \text{CH}_2\text{OH} & \text{CH}_2\text{OH} \\ 3(\text{CHOH}) & 3(\text{CHOH}) \\ (a)\text{C} \vdots \text{O} + \text{H}_2 \vdots \text{N—HN—C}_6\text{H}_5 = \text{CN—HN—C}_6\text{H}_5 + \text{H}_2\text{O} \\ \\ \text{CH}_2\text{H} \vdots \text{O} + \text{H}_2 \vdots \text{N—HN—C}_6\text{H}_5 \quad \text{CHN—HN—C}_6\text{H}_5 + 2\text{H} + \text{H}_2\text{O} \\ \text{Fructose.} \quad \text{Phenylhydrazine.} & \quad \text{Glucosazone.} \end{array}$$

It may be observed that whether we start with the aldose glucose or the ketose fructose, the same body, glucosazone, is obtained. This is a proof of the intimate relation between the aldose and ketose, of which we shall have to speak further on; and it further shows that the opposite rotatory properties of these two related bodies is dependent on the asymmetric carbon atom marked (a) in each of the formulæ, since this is the only asymmetric atom which suffers change in the foregoing reactions.

The osazones are mostly crystalline bodies sparingly soluble in water, which differ from one another in solubility, melting-point, and optical behaviour, and on account of these properties they have proved of the utmost value in the identification of sugars. Along with the hydrazones they have been found particularly valuable in the discovery of new sugars, several of which have been already found by their aid.

[1] For the sake of simplicity the acetic acid is left out of the formulæ.

Osones.—When an osazone is heated with strong hydrochloric acid, the phenylhydrazine residues combine with the acid, and a body of the following composition, $CH_2OH \cdot 3(CHOH) \cdot CO \cdot COH$, called an "osone," is left. When an osone is subjected to the action of nascent hydrogen, evolved from finely divided zinc and acetic acid, two atoms of hydrogen become attached to the COH group. Thus when glucosone, $CH_2OH \cdot 3(CHOH) \cdot CO \cdot COH$, is treated in this way, fructose is obtained.

This reaction is extremely interesting, for by it we are able to transform an aldose into its corresponding ketose. This, if optically active, always polarises in the opposite direction to that of the aldose from which it was derived.[1]

The reverse process, the conversion of a ketose into its corresponding aldose, can also be effected in the following manner. Fructose (lævulose), on being treated with sodium amalgam, is partially converted into mannitol, just as mannose is when subjected to the same treatment; and mannitol, on treatment with dilute nitric acid, yields mannose. Consequently we have here a method of passing from the ketose sugar fructose (lævulose) to the aldose mannose.

Optical Stereoisomerides.—When two bodies have the same structural formula, and are possessed of similar properties, except that of their action on polarised light, and other physical differences, they are called "optical stereoisomerides." The peculiar action of bodies optically active to polarised light is explained by a theory enunciated by Le Bel and Van't Hoff, who consider this action to depend on the presence of one or more asymmetric carbon atoms in such compounds, and all experience so far confirms the validity of this theory.

Asymmetric Carbon Atoms.—A carbon atom is called "asymmetric" when its four bonds are linked to four different elements or four different groups. Thus, taking the four bodies—

H	H	CH_3	CH_3
|	|	|	|
H—C—H	H—C—H	H—C—H	H—C—COOH
|	|	|	|
H	OH	OH	OH
Methane.	Methylic alcohol.	Ethylic alcohol.	Lactic acid.

the first three of these bodies possess no optical activity, because two or more bonds of the carbon atom are linked to the same element. But when, as in the last body, lactic acid, they are all different, then the body is optically active.

Cause of the Opposite Opticities of Stereoisomerides.—One of the simplest explanations of this phenomenon is afforded by the two lactic acids, one of which polarises to the right, the other to

---

[1] Another interesting method of converting an aldose into its corresponding ketose is by reducing its osazone with zinc dust and glacial acetic acid. Taking glucosazone, for example, the following body is produced: $CH_2OH \cdot 3(CHOH) \cdot CO \cdot CH_2NH_2$, and this, on treatment with nitrous acid in the cold, yields fructose.

the left. The various groups may be arranged about the carbon atom in two distinct ways, thus:—

$$CH_3-\underset{\underset{OH}{|}}{\overset{\overset{H}{|}}{C}}-COOH \qquad COOH-\underset{\underset{OH}{|}}{\overset{\overset{H}{|}}{C}}-CH_3$$

If these formulæ be written on slips of thin paper and superimposed, it will be found that, in whatever way they are turned about, the groups can never be made to coincide. The second of these formulæ has exactly the same configuration as that of the reflection of the first in a mirror. But the arrangement of the atoms and molecules which causes stereoisomerism cannot always be represented in such a simple manner as in the foregoing example, nor portrayed on a flat surface like a sheet of paper. If, for instance, the bonds of affinity of the carbon atom were all in the same plane, then such

bodies as $H-\underset{\underset{Cl}{|}}{\overset{\overset{Cl}{|}}{C}}-H$ and $H-\underset{\underset{H}{|}}{\overset{\overset{Cl}{|}}{C}}-Cl$ ought to be different, but they

are not. To explain this and similar phenomena, it is assumed that the bonds of the carbon atom are arranged around it at four equidistant points. Thus, for the purpose of illustration, we may represent the carbon atom as a round ball on the surface of which four points are marked out at equal distances from one another. These will represent the four centres of attraction. If at each of these four points holes are bored into the ball in a direction towards its centre and pegs inserted into them, these projections will roughly represent the position of the bonds of the carbon atom. Obviously all the points marked on the surface of the sphere are equidistant from its centre and from another: hence they are of equal value. The first figure on the opposite page illustrates this.

This and several of the following diagrams are constructed so as to represent what may be termed a right and left eye view of the objects. If, in looking at them, the eyes are slightly converged, or if, in other words, they are viewed with a slight squint, the two pictures will become visually superimposed, and upon continuing to gaze at them for a few seconds a stereoscopic effect is produced, and the diagram stands out in bold relief. Should any difficulty be found in obtaining this slight divergence of the eyes, the point of the index finger may be placed midway between the two diagrams, and, whilst the attention is kept steadily fixed on it, gradually moved towards the eyes. As the finger moves, the diagrams appear to approach each other, until at last they coincide. The finger is then moved out of sight, and the attention concentrated on the figure, when it shortly stands out in relief, and it is then possible to appreciate the correct position in space of the bonds and their attached molecules or groups.

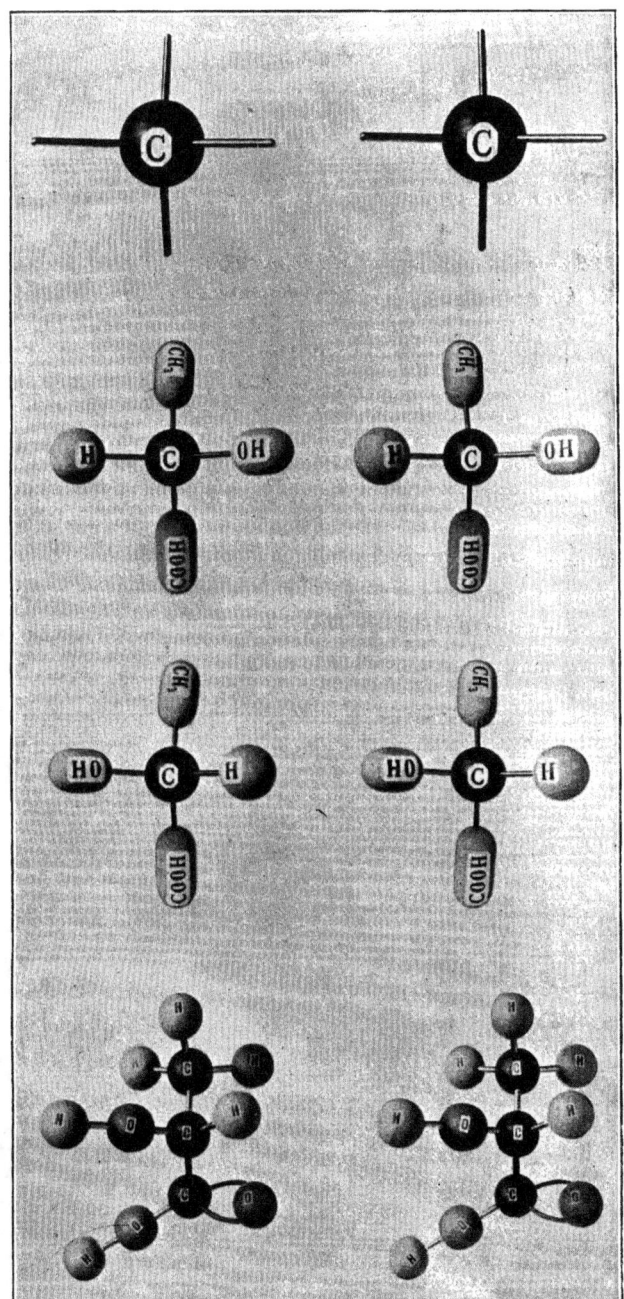

To face p. 42.]

ASYMMETRIC CARBON ATOMS.

The next series of diagrams represents in a similar manner the stereoisometrical construction of the two lactic acids, the attached groups being represented for the sake of simplicity by cylindroid balls. In the last diagram the atoms entering into the construction of the various groups are shown.

Bodies like lactic acid, which contain one asymmetric carbon atom, have three modifications—dextro, lævo, and inactive. The last modification is obtained by dissolving equal quantities of the right and left bodies, and evaporating the solution. Combination between the two active bodies ensues, the result being the formation of the inactive one. The right and left modifications exhibit, with the exception of slight differences in their crystalline form, the same physical properties, such as the same melting-point, solubility, &c. The inactive form, or "racemic modification," as it is termed, often, on the contrary, exhibits considerable differences in its physical properties; it can be separated into its active components by appropriate treatment.

Substances, the Molecule of which contains two or more Asymmetric Carbon Atoms.—As the number of the asymmetric carbon atoms in the molecule of a substance increases, so does the number of the stereoisomeric modifications in which it is capable of existing, but the latter increase at a much higher rate than the former. Thus four bodies having the structural formula of tartaric acid are known, the dextro-, lævo-, inactive tartaric acids and racemic acid, all of which contain two asymmetric carbon atoms. The following formulæ show the arrangement of the dextro- and lævo-acids:—

$$\begin{array}{cc} \text{COOH} & \text{COOH} \\ | & | \\ \text{H—C—OH} \;+ & \text{OH—C—H} \;- \\ | & | \\ \text{OH—C—H} \;+ & \text{H—C—OH} \;- \\ | & | \\ \text{COOH} & \text{COOH} \end{array}$$

In the former both atoms are positive, namely, rotate to the right; in the latter both are negative. The third modification, inactive or meso-tartaric acid, has the following arrangement:—

$$\begin{array}{c} \text{COOH} \\ | \\ \text{H—C—OH} \;+ \\ | \\ \text{H—C—OH} \;- \\ | \\ \text{COOH} \end{array}$$

One atom is positive and the other negative; consequently they neutralise or compensate one another, and the result is an inactive body. The compensation in such cases as this is said to be "internal." The fourth modification, racemic acid, is a natural product, but may be prepared artificially by dissolving equal quantities of the dextro- and lævo-tartaric acids, and evaporating the solution.

As the two asymmetric atoms of the one acid counterbalance those of the other, the compensation in this case is said to be "external." Racemic acid was the first body in which this peculiar phenomenon was observed, hence this form is often called the racemic modification. Substances which exist in this modification can often be separated into their optically active components. It is a curious fact that the vast majority of the optically active substances which the chemist is able to form synthetically are obtained in the racemic modification.

The following Table, taken from Tollen's *Handbuch der Kohlehydraten*,[1] shows the possible number of the optical stereo-isomerides of the hex-aldoses, which contain four asymmetric carbon atoms. $\overset{H}{\underset{OH}{C}}$ denotes a plus (+) atom, $\overset{OH}{\underset{H}{C}}$ a negative (−) one. Of these sixteen possible members, R-glucose is the ordinary glucose; eleven others have been synthesised by E. Fischer, and four are still unknown:—

1.	CH$_2$OH	$\overset{H}{\underset{OH}{C}}$	$\overset{H}{\underset{OH}{C}}$	$\overset{OH}{\underset{H}{C}}$	$\overset{H}{\underset{OH}{C}}$	COH	R-Glucose (Grape-sugar).
2.	CH$_2$OH	$\overset{OH}{\underset{H}{C}}$	$\overset{OH}{\underset{H}{C}}$	$\overset{H}{\underset{OH}{C}}$	$\overset{OH}{\underset{H}{C}}$	COH	L-Glucose.
3.	CH$_2$OH	$\overset{H}{\underset{OH}{C}}$	$\overset{H}{\underset{OH}{C}}$	$\overset{OH}{\underset{H}{C}}$	$\overset{OH}{\underset{H}{C}}$	COH	R-Mannose.
4.	CH$_2$OH	$\overset{OH}{\underset{H}{C}}$	$\overset{OH}{\underset{H}{C}}$	$\overset{H}{\underset{OH}{C}}$	$\overset{H}{\underset{OH}{C}}$	COH	L-Mannose.
5.	CH$_2$OH	$\overset{OH}{\underset{H}{C}}$	$\overset{H}{\underset{OH}{C}}$	$\overset{OH}{\underset{H}{C}}$	$\overset{OH}{\underset{H}{C}}$	COH	R-Gulose.
6.	CH$_2$OH	$\overset{H}{\underset{OH}{C}}$	$\overset{OH}{\underset{H}{C}}$	$\overset{H}{\underset{OH}{C}}$	$\overset{H}{\underset{OH}{C}}$	COH	L-Gulose.
7.	CH$_2$OH	$\overset{H}{\underset{OH}{C}}$	$\overset{OH}{\underset{H}{C}}$	$\overset{OH}{\underset{H}{C}}$	$\overset{H}{\underset{OH}{C}}$	COH	R-Galactose.
8.	CH$_2$OH	$\overset{OH}{\underset{H}{C}}$	$\overset{H}{\underset{OH}{C}}$	$\overset{H}{\underset{OH}{C}}$	$\overset{OH}{\underset{H}{C}}$	COH	L-Galactose.
9.	CH$_2$OH	$\overset{H}{\underset{OH}{C}}$	$\overset{OH}{\underset{H}{C}}$	$\overset{OH}{\underset{H}{C}}$	$\overset{OH}{\underset{H}{C}}$	COH	R-Talose.
10.	CH$_2$OH	$\overset{OH}{\underset{H}{C}}$	$\overset{H}{\underset{OH}{C}}$	$\overset{H}{\underset{OH}{C}}$	$\overset{H}{\underset{OH}{C}}$	COH	L-Talose.
11.	CH$_2$OH	$\overset{OH}{\underset{H}{C}}$	$\overset{H}{\underset{OH}{C}}$	$\overset{OH}{\underset{H}{C}}$	$\overset{H}{\underset{OH}{C}}$	COH	R-Idose.

[1] Vol. ii. p. 14.

CONSTITUTION OF THE SUGAR GROUP.

12. $CH_2OH \quad \overset{H}{\underset{OH}{C}} \quad \overset{OH}{\underset{H}{C}} \quad \overset{H}{\underset{OH}{C}} \quad \overset{OH}{\underset{H}{C}} \quad COH \quad$ L-Idose.

13. $CH_2OH \quad \overset{H}{\underset{OH}{C}} \quad \overset{H}{\underset{OH}{C}} \quad \overset{H}{\underset{OH}{C}} \quad \overset{OH}{\underset{H}{C}} \quad COH \quad$ Unknown.

14. $CH_2OH \quad \overset{OH}{\underset{H}{C}} \quad \overset{OH}{\underset{H}{C}} \quad \overset{OH}{\underset{H}{C}} \quad \overset{H}{\underset{OH}{C}} \quad COH \quad$ Unknown.

15. $CH_2OH \quad \overset{H}{\underset{OH}{C}} \quad \overset{H}{\underset{OH}{C}} \quad \overset{H}{\underset{OH}{C}} \quad \overset{H}{\underset{OH}{C}} \quad COH \quad$ Unknown.

16. $CH_2OH \quad \overset{OH}{\underset{H}{C}} \quad \overset{OH}{\underset{H}{C}} \quad \overset{OH}{\underset{H}{C}} \quad \overset{OH}{\underset{H}{C}} \quad COH \quad$ Unknown.

The number of stereoisomeric forms in which a compound is capable of existing is found by the formula 2^n, n representing the number of asymmetric atoms contained in the molecule. Thus with bodies having the structural formula of glucose, which contains four asymmetric atoms, $n = 4$, consequently 2 when raised to its fourth power gives $2 \times 2 \times 2 \times 2 = 16$, the possible number of these stereoisomerides. Similarly, the other aldoses have the following numbers:—

Trioses, tetroses, pentoses, hexoses, heptoses, octoses, nonoses, &c.
  2   4   8   16   32   64   128

Constitution of the Sugar Group.—This was for many years a puzzle to chemists, and the views put forward were matters of almost pure conjecture until Killiani in 1885 [1] furnished evidence to show that lævulose (fructose) was a ketonic body which had the following constitution: $CH_2OH \cdot 3(CHOH) \cdot CO \cdot CH_2OH$. Killiani obtained by treating lævulose with hydrocyanic acid (HCN) its cyanhydrin, $CH_2OH \cdot 3(CHOH) \cdot \underset{CN}{COH} \cdot CH_2OH$ (page 34). This body, when hydrolysed with strong hydrochloric acid, yielded the following acid: $CH_2OH \cdot 3(CHOH) \cdot \underset{COOH}{COH} \cdot CH_2OH$. On treating this acid with hydriodic acid the following body was produced: $CH_3 \cdot 3(CH_2) \cdot CH \cdot \underset{COOH}{CH}$, which is methyl-butyl acetic acid. This indicates that the CO group is situated in the place indicated by the above formula for lævulose, for the added COOH group must undoubtedly attach itself to the CO group. Had it become attached to any other of the groups, an entirely different acid would have resulted.

Turning his attention to glucose, he concluded that the constitution of this body was as follows: $CH_2OH \cdot 4(CHOH) \cdot COH$; for on treating this sugar in an exactly similar manner to that in which he had treated lævulose (fructose), the following acid

[1] *Berichte*, xviii. 3066.

was finally obtained: $CH_3 \cdot 5(CH_2) \cdot COOH$, which is normal heptylic acid; and this distinctly shows that the COH group, to which the COOH group had attached itself, is at the end of the chain.

The most remarkable work, however, on the constitution of the sugar group is that of Emil Fischer, to be found in volumes xviii. to xxiii. of the *Berichte* of the German Chemical Society, of which the following is a short résumé:—

Fischer found that the hexahydric alcohol mannitol, $CH_2OH \cdot 4(CHOH) \cdot CH_2OH$, on treatment with dilute nitric acid, yielded the aldose $CH_2OH \cdot 4(CHOH) \cdot COH$, mannose, which in many respects resembles glucose. It ferments in contact with beer yeast, reduces Fehling's solution, and rotates the polarised ray to the right, but not so powerfully as glucose. When acted on by bromine it is further oxidised to mannonic acid, $CH_2OH \cdot 4(CHOH) \cdot COOH$. When a solution of this latter acid is evaporated to a syrup, a molecule of water is disengaged, and mannonic acid lactone,
$$CH_2OH \cdot CHOH \cdot CH \cdot 2(CHOH) \cdot CO$$
$$\underbrace{\qquad O \qquad}$$
is formed.

Some years before this Killiani had obtained by boiling certain gums with dilute sulphuric acid the pentose sugar arabinose, $CH_2OH \cdot 3(CHOH) \cdot COH$. On treating this with hydrocyanic acid its cyanhydrin was obtained, and this, on hydrolysis, yielded an acid isomeric with the mannonic acid mentioned above. Fischer, on evaporating the acid obtained from arabinose to a syrup, obtained a lactone similar in every other respect to that of mannonic acid, with the single exception that it was lævo-rotatory, its opticity being $[a]_D - 54.8$, whilst the lactone of mannonic acid was dextro-rotatory, and had an opticity of $[a]_D + 53.81$. Making allowance for experimental error, it may be considered that these two bodies possess equal and opposite opticities, that they are, in fact, what is known as stereoisomerides, and this was the first instance in which such a relation had been observed amongst the members of the sugar group. When an equal weight of the dextro-rotatory lactone and an equal weight of the lævo-rotatory one were dissolved in water, combination ensued and an inactive lactone resulted, which could only be resolved into its optically opposed components by special treatment.

Mannose Group.—This includes the following bodies, which were all that had been produced up to that time:—

Nature.	Members.	R. Series.	I. Series.	L. Series.
Lactone	Mannonic acid lactone	+	o	−
Acid	Mannonic acid	+	o	−
Aldose	Mannose	+	o	−
Alcohol	Mannitol	+	o	−
Ketose	Lævulose (Fructose)	−	o	+

Here then arose a difficulty with the names of the ketoses, for one of them polarised to the right, another to the left, and the other was inactive. To avoid this difficulty Fischer proposed the name "fructose" for this class of sugars. Also, to signify the series to which any set of members belonged, he suggested appending the letters R. I. L., for right, inactive, and left, according to the direction in which the aldose member of the series polarised. Ordinary lævulose would then be described as "R. fructose," for though it polarises strongly to the left, the aldose of the series to which it belongs polarises to the right.

We have seen that by treatment with a reducing agent (p. 41) it is possible to pass down the series and obtain in turn any member as far as mannitol—that is, it is possible to obtain the ketose from the aldose group through the osazone and osone (p. 41); that by using oxidising agents it is also possible to proceed in the opposite direction, and obtain any member of the series from mannitol upwards. Consequently, if it were possible to obtain any one member of either the R., I., or L. series, all the members belonging to that particular series could be produced from it; and if any member of the inactive series could be separated into its optically opposed components, we should then be in a position to obtain any member whatever of the three series. This has been found possible.

Resolution of the I. into the R. and L. Series.—This was effected by adding strychnine[1] to a solution of I. mannonic acid, and thus forming I. strychnine mannonate. This salt when boiled in alcohol is resolved into the salts of the R. and L. acids; that of the L. strychnine mannonate, being difficultly soluble in alcohol, falls out of solution. This salt, after the removal of the strychnine from it, yields pure L. mannonic acid, from which all the members of the L. series can be formed. The whole of the L. strychnine mannonate is not, however, precipitated from the alcoholic solution, and to obtain the R. salt in a pure condition, the strychnine is first removed from the solution of the mixed acids and morphine added. From this solution, on long-standing, crystals of pure R. morphine mannonate separate. After removing the morphine from these, pure R. mannonic acid is left, from which all the members of the R. series can be formed.

Glucose.—As observed before, this sugar very much resembles mannose. It possesses the same structural formula, forms with phenylhydrazine the same osazone, which in turn yields the same osone and ketose (fructose or lævulose). Upon reduction, however, with sodium amalgam, glucose does not yield the same alcohol as mannose, but an isomeric one; mannose yields mannitol, glucose yields sorbitol. Their opticities also differ, that of glucose being much higher than that of mannose; their hydrazones also

---

[1] Strychnine is a poisonous alkaloid found in the beans and other parts of the *Strychnos nux vomica*. Morphine is similarly an alkaloid prepared from the juice of the white poppy (*Papaver albus*).

differ; gluco-hydrazone is freely soluble in water, manno-hydrazone is comparatively insoluble. Moreover, gluconic acid forms with calcium a distinct salt, differing in form from the corresponding calcium mannonate. All these facts point to mannose and glucose being stereoisomerides. Fischer considers that the difference in the opticities of mannose and glucose is due to the arrangement of the groups contiguous to the asymmetric CHOH group, which adjoins the COH group in these two sugars. After various futile attempts to convert mannose into glucose, this was at last effected by heating R. mannonic acid with quinoline[1] to 140° C., when a part of the mannonic acid was converted into gluconic acid. After the removal of the quinoline the mixed acids were converted into their respective brucine[2] salts. The brucine mannonate, being insoluble, was precipitated, and the brucine gluconate remained in solution, from which, after removing the brucine, gluconic acid was left. This, on reduction by sodium amalgam, yielded glucose in all respects identical with the ordinary sugar. L. or I. mannonic acid, by similar treatment, yields L. or I. glucose.

Synthetic Sugars.—It has been shown that from any one of the members of the mannitol series the whole of the other members can be formed; hence, if we could build up any single member of a series from the elements, it would be possible to produce the whole of the series from it.

So long since as 1861, Butlerow observed that an aqueous solution of formaldehyde, under the influence of certain bases, yielded by condensation a sugar-like body to which he gave the name "methylenitan." Löw, using as the base lead oxide, formed the same or a similar substance, and called it "formose." Fischer obtained, by acting on acrolein bromide with barium hydrate,[3] and afterwards from glycerose, $CH_2OH·CHOH·COH$, a sugar-like body which he named "a-acrose." He repeated Löw's process, and found formose to be a mixture of various aldehyde and ketone sugars, for, on treating it with excess of phenylhydrazine acetate, a mixture of osazones was formed, all of which were, with one exception, soluble in alcohol or ether. The insoluble one when treated with hydrochloric acid yielded an osone, which, on reduction with zinc and acetic acid, yielded a-acrose. This body (formed in either of the three ways) afforded, on reduction with sodium amalgam, a body to which he gave the name of "a-acritol," and this resembled mannitol in all other properties but that of opticity, it was optically inactive. It struck him that a-acritol might be I. mannitol, and a-acrose I. fructose; and

---

[1] Quinoline, C_9H_7N, is a mobile strongly-refracting liquid, first obtained by distilling cinchonine, an alkaloid contained in Peruvian bark, with caustic potash. It is now obtained synthetically by heating together aniline, glycerol, nitro-benzene, and sulphuric acid.

[2] Brucine is another poisonous alkaloid, obtained from the *Strychnos nux vomica*.

[3] $2C_3H_4Br_2O + 2Ba(OH)_2 = C_6H_{12}O_6 + 2BaBr.$

FORMATION OF SUGAR IN LIVING PLANTS.

upon further investigation, this turned out to be the case. Having now obtained two of the members of the mannitol series, it became possible to obtain any other member, and since methylic alcohol, which under proper treatment yields formaldehyde, can be built up from the elements, it follows that the whole of the mannitol series can be similarly synthetised.

Formation of Sugar in Living Plants.—The formation of a sugar from the condensation of formaldehyde is of great interest, since it has been assumed by several chemists and physiologists that this is really the process by which sugar, which is now known to be the first product of assimilation in plants,[1] is formed. By the agency of the chlorophyll cells of the plant, carbonic acid, H_2CO_3, is supposed to lose two atoms of oxygen and become formaldehyde, HCOH. Six molecules of this body uniting together could form either a molecule of glucose or fructose, and then, by the further loss of a molecule of water, become starch.

Disaccharide Sugars, or $C_{12}H_{22}O_{11}$ Type.—Very interesting is the manner in which Fischer has elucidated the constitution of two of these sugars, lactose (milk-sugar) and maltose. Milk-sugar was the first body operated on, and the discovery of its constitution served to settle that of maltose.

When milk-sugar is treated with phenylhydrazine acetate, its osazone (lactosazone) is precipitated, and this, when acted upon by hydrochloric acid, yields the osone. If this be boiled with hydrochloric acid, inversion takes place, and a mixture of equal molecules of galactose[2] and glucosone is obtained. If the mixture be again treated with phenylhydrazine acetate, glucosazone is precipitated almost in the cold (since an osone combines much more readily with phenylhydrazine than an aldehyde), and after heating for some time, galactososazone is also precipitated. This shows that the molecule of milk-sugar is built up from a molecule of galactose and a molecule of glucose, and that the COH group is attached to the glucose molecule, since it is to this the phenylhydrazine first attaches itself. The COH group of the galactose is evidently modified in the union of the two molecules, and this arrangement may be probably expressed by the following formula:—

$$CH_2OH \cdot 4(CHOH) \cdot CH$$
$$CH_2O \cdot CHOH \cdot CHOH \cdot CHO \cdot CHOH \cdot COH$$

Thus the two sugar molecules in combining lose a molecule of water:—
$$2(C_6H_{12}O_6) = C_{12}H_{22}O_{11} + H_2O.$$

---

[1] Brown and Morris, *Journal of the Chemical Society*, 1893, p. 624.
[2] A sugar having the same structural formula as glucose, in fact a stereo-isomeride of it.

Maltose when treated in the same manner gives evidence of a similar structure, only in this case the molecules from which it is derived are both glucose ones. The formula for maltose is the same as that shown above for milk-sugar, the two sugars being stereoisomerides.

Lactobionic and Maltobionic Acids.—Another proof that the constitution of lactose and maltose as given above is correct is derived from the consideration of acids which these sugars respectively yield when treated with bromine.

Lactose gives lactobionic acid :—

When this is inverted by boiling with dilute hydrochloric acid, galactose and gluconic acid result :—

$C_{12}H_{22}O_{12} + H_2O = CH_2OH \cdot 4(CHOH) \cdot COH + CH_2OH \cdot 4(CHOH) \cdot COOH$
Lactobionic acid.   Galactose.   Gluconic acid.

When maltose is similarly treated with bromine, maltobionic acid is formed, and this, on treatment with hydrochloric acid, yields a molecule of glucose and one of gluconic acid.

To the disaccharides also belongs cane sugar, $C_{12}H_{22}O_{11}$, which on hydrolysis is converted into equal molecules of glucose and fructose :—

$C_{12}H_{22}O_{11} + H_2O = C_6H_{12}O_6 + C_6H_{12}O_6$
Cane sugar.   Water.   Glucose.   Fructose.

Cane sugar does not reduce Fehling's solution, nor does it form a compound with phenylhydrazine. Its constitution is not known with absolute certainty. E. Fischer has suggested the following probable structural formula for this sugar :—

$$O\!\!<\!\!\begin{array}{l}CH\\CHOH\\CHOH\\CH\\CHOH\\CH_2OH\end{array}\!\!\!-\!\!O\!\!-\!\!\!\begin{array}{l}CH_2OH\\C\\CHOH\\CHOH\\CH\\CH_2OH\end{array}\!\!>\!\!O$$

Manno-heptose.—This sugar, which has the following structural composition, $CH_2OH \cdot 5(CHOH) \cdot COH$, has been obtained synthetically by E. Fischer by treating mannose with hydrocyanic acid, when its cyanhydrin is formed, which, on being hydrolysed, yields heptonic acid. The acid having been obtained, all the members of the heptose series can be formed from it. Mannooctose and manno-nonose have been obtained in a similar manner; the former from heptose, the latter from octose.

Estimation of Sugar by Fehling's Solution.—Most of the sugars, in common with other bodies of an aldehydic nature, exhibit (under certain circumstances) a strong affinity for oxygen. In virtue of this property, they are able to reduce the oxides of many metals (such as silver, gold, platinum, mercury, and bismuth, when present in alkaline solution) to the metallic form. The metal is precipitated in most cases as a fine powder, in others in the form of a metallic mirror clinging to the sides of the vessel. When an alkaline solution of a cupric salt is acted upon by a reducing sugar, the action does not proceed so far as to withdraw the whole of the oxygen from the cupric oxide; only half of it is removed, the cupric oxide, $2CuO$, being converted into cuprous oxide, Cu_2O. When a solution of grape-sugar is added to a dilute solution of cupric sulphate, and the mixture is made strongly alkaline with sodium hydrate, there separates out, at the ordinary temperature slowly, but on heating almost immediately, a yellowish red precipitate of cuprous hydroxide, $Cu_2(HO)_2$, which gradually loses its molecule of combined water and passes into cuprous oxide, which has a bright red colour. This reaction was first used by Trommer as a qualitative test for sugar. It was shown by Barreswill, in 1846, that a solution containing cupric sulphate, potassium or sodium tartrate, and a caustic alkali, will, if these bodies are mixed in proper proportions, remain unaltered on boiling, but that if to the boiling solution a trace of a reducing sugar be added, then an immediate precipitation of cuprous oxide takes place. He also pointed out that this reaction could be utilised for the quantitative estimation of sugar. Some time afterwards Fehling devoted much attention to the subject of sugar analysis, and essentially improved the solution originally proposed by Barreswill. This, which is known as "Fehling's solution," is in common use at the present day; it has the following composition:—Pure crystallised sulphate of copper, 34.639 grammes; sodium and potassium tartrate (Rochelle salt), 173 grammes; caustic soda, 60 grammes. These salts are dissolved separately in distilled water, their solutions mixed, and the volume of the mixture made up to 1000 c.c. The solution was used at first volumetrically, 10 c.c. of the Fehling's solution being added to 40 c.c. of water, the whole brought to the boil, and, while boiling, a dilute solution containing between 0.5 and 1.0 of sugar per 100 c.c. slowly added from a burette until the blue colour of the mixture was entirely discharged. The quantity of grape-sugar required to decolorise 10 c.c. of Fehling's solution, when used in this way, is 0.05 gramme, consequently the amount of this sugar contained in any solution can be found by a simple calculation.

It was shown by Neubauer[1] that one equivalent of grape-sugar

[1] *Archiv der Pharm.*, ii. 72, 278.

(180) reduced ten equivalents of cupric oxide (397); but Soxhlet, who subsequently made an exhaustive examination of the subject, found that this was only true when the Fehling's solution was used in the way specified above, and that any change in the dilution of the Fehling's solution, or in the strength of the sugar solution, altered the proportions between the amount of cupric oxide reduced and the quantity of sugar as given by Neubauer. The result of Soxhlet's investigations was an improvement in the volumetric method of sugar estimation which leaves little to be desired in the way of exactitude.

But Fehling's solution can be used in an entirely different manner. In this the dilute solution of sugar is added to an excess of boiling Fehling's solution, the precipitated cuprous oxide filtered off, washed, and weighed, either as such, or after conversion into cupric oxide, or after reduction to metallic copper. The first of these (weighing as cuprous oxide) has been all but abandoned, as the oxide is very liable to absorb oxygen during drying, and increase in weight. The second has been much used, and when performed under certain fixed conditions, such as those laid down by O'Sullivan,[1] is said to yield results of considerable accuracy. But certain disadvantages attend this and all similar processes which necessitate the employment of a filter-paper. Cuprous oxide comes down in a state of such exceedingly fine division that it is extremely difficult to obtain a filter-paper which will retain the precipitate completely; consequently, a small portion almost invariably passes through the filter and is lost. Filter-paper has also a great attraction for copper salts, which cling so obstinately to it that they cannot be removed perfectly by washing, and this apparently increases the weight of the precipitate obtained.

In order to avoid these sources of error, it was proposed by Soxhlet to use an asbestos filter, and, in order to avoid errors arising from incomplete oxidation of the cuprous oxide, to reduce it to metallic copper. The asbestos layer is placed in a filter-tube which is so constructed as to permit the precipitate to be readily reduced in a current of hydrogen afterwards.

Tables for Sugar Determinations.—As has been stated before, the concentration of the solutions used has a marked influence on the relation between the amount of sugar and the amount of cuprous oxide reduced. In order to obviate this source of error, it has been proposed to always use the same quantity of Fehling's solution (60 c.c. undiluted), and to invariably use the same quantity (25 c.c.) of a sugar solution containing not more than 1 per cent. of sugar, then to ascertain from a table the amount of sugar which corresponds to the quantity of copper obtained. Such a table was constructed by Allihn,[2] showing the amount of glucose equivalent to any amount of copper from 0.01 to 0.463 gramme obtained under

[1] *Journal of the Chemical Society*, 1876, ii. 131.
[2] Fresenius, *Quant. Analysis*, ii. 507.

specified conditions. In making an estimation, 60 c.c. of undiluted Fehling's solution are placed in a beaker of about 300 c.c. capacity, brought to the boil, 25 c.c. of the sugar solution (which must not contain more than 1 per cent. of sugar) added from a pipette, and the whole allowed to boil up for four minutes. The precipitate is collected, reduced, and weighed in the Soxhlet filter-tube described. The quantity of glucose contained in the 25 c.c. of sugar solution is found by inspecting the table; for instance, if the reduced copper weighed 0.364 gramme, the amount of glucose would be 0.1923 gramme. Meissl has given a similar table for invert sugar, and Wein another for maltose.[1] These have been collated in book form by Dr. Frew.[2]

Cupric Oxide Reducing Power of a Sugar.—It is a matter of great convenience to have a standard to which the reducing powers of the various sugars or mixtures of several of them can be compared. For this standard O'Sullivan proposed [3] to take the amount of cupric oxide reduced by one gramme of glucose, this being the sugar the reducing power of which was first determined, and call this 100. The quantity of cupric oxide reduced by one gramme of glucose (using O'Sullivan's process) was found to be 2.205 grammes; this he called 100. Later on he suggested that the letter K should stand for the cupric reducing power of glucose; the K of glucose is therefore 100. If a gramme of another sugar is found to reduce only 1.345 grammes of cupric oxide, its K or cupric reducing power is found by the following sum in simple proportion:—

$$2.205 : 1.345 :: 100 : x, \text{ and } x = 61.$$

The K of such a sugar is therefore 61. Since in the case of the products of the conversion of starch this plan is somewhat cumbersome, it is now more usual to state these reducing powers in terms of maltose. Thus a solution of these bodies which had a reducing power three-quarters that of maltose would be expressed as R_{75}, instead of $K_{45.75}$.

Solution Factor.—In the course of investigations connected with the carbohydrates it is constantly necessary to ascertain the amount of substance present in a solution. In ordinary chemical practice this is usually effected by evaporating a known quantity of the solution to dryness and then weighing the residue. With many of the bodies met with in brewing chemistry, owing partly to their powerful attraction for moisture and partly to their liability to decompose on long-continued heating, it is difficult to obtain a correct result in this way. Consequently, it has become

---

[1] It is unfortunate that in these methods the original Fehling's solution is not employed. Each author makes some slight difference in its composition, and this influences the results obtained.

[2] Tables for the Quantitative Estimation of the Sugars (Spon).

[3] *Journal of the Chemical Society*, 1876, ii. 930.

customary to calculate the amount of the carbohydrates present in a solution from its specific gravity. O'Sullivan[1] found that when 10 grammes of pure maltose or pure dextrin were dissolved in so much water that the solution measured exactly 100 c.c. at a temperature of 60° F. (15.5° C.), the specific gravity of the solution (10 per cent.) was 1038.5, water being taken at 1000. Assuming that the strengths of such solutions were strictly proportional to their specific gravities, a 1 per cent. solution would have a specific gravity of 1003.85, and solutions containing intermediate quantities would have gravities expressed by intermediate values. Consequently, if each per cent. of either of these bodies raised the specific gravity of a solution by 3.85, it would be a simple matter to ascertain the amount present in solution; it would be only necessary to subtract 1000 from the specific gravity of the solution and divide the figure so obtained by 3.85. Thus 100 c.c. of a solution of maltose of a specific gravity of 1055 would contain $\frac{1055-1000}{3.85} = 14.285$ grammes of that substance. Hence, according to O'Sullivan, the solution factor for maltose or dextrin is 3.85. In course of time it was found that this factor was not quite correct. Brown and Heron[2] came to the conclusion that 3.85 was too low for maltose, the correct divisor being 3.9314, and O'Sullivan afterwards gave the fresh divisor 3.95 for starch conversion products. It was afterwards found that with solutions of the various carbohydrates the specific gravity of the solution was not strictly proportional to the amount of substance contained in solution, and Brown and Heron[3] proposed the use of 3.86 in all cases as a solution divisor; but this is only correct for a 10 per cent. solution of cane-sugar, which has a specific gravity of 1038.6 at 60° F. Although this factor is not absolutely correct for the products of starch conversion, it was considered that to have one uniform divisor when determining the opticities and reducing powers of substances in solution would be extremely convenient.[4]

In the majority of cases where bodies connected with brewing have to be examined for cupric reducing power, their amount is most frequently deduced from the specific gravity of their solutions by means of a solution factor. Consequently, it is usual to affix in small characters the solution factor which has been used in any particular case, such as $K_{3.86}$ or $K_{3.85}$. This means that the amount of solid matter in the first instance was estimated by the 3.86 factor, in the second by the 3.85. The absolute reducing power may be readily obtained from the statement of these on the 3.86 factor, when the true solution factor is known. Thus the reducing power of maltose is

---

[1] *Journal of the Chemical Society*, 1876, p. 129.
[2] Ibid., 1879, p. 618.
[3] Ibid., p. 602.
[4] See Appendix A.

$K_{3.86} = 61$. If, as O'Sullivan assumed, the true solution factor for maltose is 3.9314; then its real reducing power is

$$3.86 : 3.9314 :: 61 : 62.12.$$

Elementary Composition.—In investigating the chemical composition of a carbohydrate, such, for instance, as glucose, a weighed portion of the substance is placed in a tube packed with cupric oxide, and burned in a current of dry oxygen. The carbon of the sugar is oxidised to form carbon dioxide, and its hydrogen to form water. These are collected and weighed; the weight of the carbon dioxide thus formed corresponds to the weight of carbon present in the sugar, and that of the water to the weight of the hydrogen. The amount of these two constituents having been determined, the remainder, which is known to be oxygen, is arrived at by difference. For instance, if 300 milligrammes of a substance burnt in this way yielded 440 milligrammes of carbon dioxide and 180 milligrammes of water, from this we should know that it contained 120 milligrammes of carbon and 20 milligrammes of hydrogen; and as these add up to 140, the remaining 160 milligrammes must be oxygen. From these data a simple proportion sum gives the percentage composition of the substance; in this case it will be carbon 40 per cent., hydrogen 6.67 per cent., oxygen 53.33 per cent. The relative number of the atoms in the molecule is found by dividing these figures by the atomic weights of the respective elements, thus:—

$$\text{Carbon} \quad 40 \div 12 = 3.33 \div 3.33 = 1$$
$$\text{Hydrogen} \quad 6.67 \div 1 = 6.67 \div 3.33 = 2$$
$$\text{Oxygen} \quad 53.33 \div 16 = 3.33 \div 3.33 = 1$$

That is, for each atom of carbon present there are two atoms of hydrogen and one of oxygen. But the question now arises: What is the relative weight of the molecule? If it can be shown to be 30, then the formula of the molecule is CH_2O (formaldehyde); if 60, then its formula is $C_2H_4O_2$ (acetic acid); if 90, then its formula is $C_3H_6O_3$ (lactic acid); if 180, its formula would be $C_6H_{12}O_6$ (glucose).

Determination of Molecular Weight.—The molecular weight of a body may be determined in various ways. In the case of a volatile liquid, such as alcohol, it can be found by comparing the weight of a given volume of its vapour with the weight of a similar volume of hydrogen. If the substance be an acid, its molecular weight can be generally ascertained by the amount of silver contained in its silver salt. Many organic bodies form compounds with the metals, and their molecular weights can be deduced from the amounts of metal they combine with.

Raoult's Method.—But the method chiefly employed in brewing chemistry, where the determinations generally refer to the molecular weights of the carbohydrates, is the cryoscopic or freezing method, first adopted by Raoult. It depends upon the following considerations: All bodies when dissolved in a given solvent lower the freezing-point of the latter, and with the same substance this is proportional to the amount of the substance present in the solution. For instance, if 5 grammes of cane sugar are dissolved in 100 grammes of water, and the freezing-point of the resulting solution is lowered 0.3° C., then a solution of 10 grammes of the same sugar in a similar weight of water will have its freezing-point lowered 0.6° C.

But if equal weights of two substances which differ considerably in their molecular weights, as, for instance, glucose (molecular weight 180) and cane sugar (molecular weight 342) are dissolved in equal weights of water, each solution introduced into a test-tube, and the tubes placed in a freezing mixture, the temperature required by the glucose solution for solidification to take place will be found lower than that required by the cane-sugar solution. Hence in this case the depression of the freezing-point is not equivalent to the actual weights of the substances present in solution. It has, however, been found to be definitely related to the molecular weights of these substances.[1]

If it were possible to dissolve a molecular weight of any substance in grammes, as, for instance, 180 grammes of glucose in 100 grammes of water, the freezing-point of such a solution would have a temperature 10.55° C. lower than the freezing-point of water. As it is impossible to do this, and for other reasons which we cannot enter into here it is only practicable to employ somewhat dilute solutions, the only available method is to obtain the actual molecular weight by a calculation. This is effected in the following manner when water is the solvent employed:—

$$\text{Molecular weight} = \frac{19 \times \text{grms. of substance dissolved in 100 grms. of water}}{\text{Observed depression of the freezing-point}}$$

The figure 19 is what is termed the constant for water.

For instance, 10 grammes of glucose were dissolved in 100 grammes of water, and the depression of the freezing-point of the solution found to be 1.07° C.; then—

$$\frac{19 \times 10}{1.07} = 177.5.$$

This is very close to the real molecular weight of glucose, which is 180. It may be here mentioned that the figures obtained, owing to disturbing influences, are only approximate, and never absolutely exact.

---

[1] This law does not hold good for inorganic salts, which behave as though they contained a larger number of molecules than they actually do. This, as explained by the law of Arrhenius, is supposed to depend on what is termed the molecular disassociation into the *ions* by the action of the solvent which is assumed to take place in dilute solutions of such salts.

Five grammes of cane sugar dissolved in 100 grammes of water were found to give a depression of 0.285° C.; this, by the above formula, gives a molecular weight of 333, which closely approximates to the true value, 342.

Similarly, Brown and Morris[1] found that 8.6114 grammes of maltodextrin dissolved in 94.71 grammes of water gave a depression of 0.180° C. This, when calculated out, gives a molecular weight of 960, and from this and other considerations they considered that the real molecular weight of this substance was 990.

The Polarimeter and Specific Rotatory Power.—When a beam of polarised light[2] passes through certain substances, it suffers a certain amount of twisting on its axis, or, in other words, it is rotated. The solutions of many substances, such as dextrin, maltose, glucose, &c., possess the power of rotating the polarised ray, and are consequently said to be optically active. Some bodies twist the ray round to the right; these are termed "dextro-rotatory": others twist it to the left; they are "lævo-rotatory." All solutions of an optically active substance of the same strength, when traversed by the polarised ray in layers of the same thickness, rotate the ray to the same extent, *i.e.*, each substance has a definite rotatory power. When this is calculated upon a certain definite strength of solution, and upon a layer of such solution of a definite thickness, an expression is found which is termed the "specific rotatory power" or "the opticity" of the substance. The function of the polarimeter is to determine the amount of rotation which the polarised ray suffers in passing through a layer of the solution of an optically active substance. From this, by taking certain other factors into account, the specific rotatory power of the substance can be found. When the specific rotatory power of a substance present in a solution is known, the polarimeter enables us to estimate its amount.

Laurent Polarimeter.—There are several varieties of polarimeters, or "polariscopes," as they are often incorrectly called, but the two most commonly used forms are the Laurent and the Schmidt and Haensch half-shadow. The former, which is shown in Fig. 6, is in form something like a telescope. The source of light is placed opposite the end of the instrument, J, and at a distance of a few inches from it. In the tube beneath H, is a prism of Iceland spar, the "polariser," which converts one-half the ordinary light passing through it into polarised light; the other half being reflected to the side and absorbed. At G is a thin plate of quartz, covering half the field. Its function is to increase the delicacy of the indications of the instrument. Next follows the trough F, with

---

[1] *Journal of the Chemical Society*, 1889, p. 465.
[2] The description of the nature and properties of polarised light is one of the most intricate branches of the science of optics, hence it is impossible to enter into any details on this subject here.

light-tight lid, in which the tube filled with the solution under observation is placed. Next comes the tube E, which contains another prism of Iceland spar, the "analyser," and the eyepiece A. The tube E can be rotated on its axis by means of the milled head seen underneath; it carries with it the index C, which indicates the observed rotation on the scale of the dial-plate B. The light employed with this instrument is monochromatic, and is obtained from a Bunsen burner, in which a pellet of sodium chloride is supported. The instrument is used in a partly darkened room. The observation tubes are generally made of glass, their ends being ground perfectly true. A brass screw attached to each end enables the tube to be closed water-tight with small circular glass plates, which are retained in position by screw-caps. The lengths of the tubes usually employed are 1, 2, and 2.2 decimetres.[1]

In using the instrument, the lamp is lighted and placed in

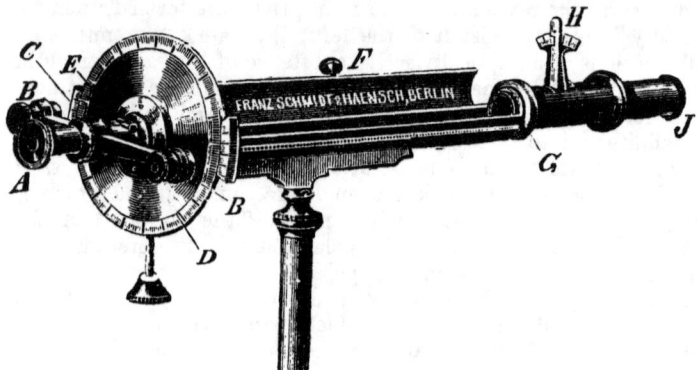

FIG. 6.—Laurent Polarimeter.

position, and an observation tube filled with water placed in the trough. On looking through the eyepiece with the index at zero, a circle of light will be seen, divided into two halves by a black line; the eyepiece is then adjusted until this line is perfectly sharp and distinct. If the instrument is in perfect adjustment, each half of the field will be equally illuminated. If this is not the case, the instrument must be adjusted. If now the observation tube is removed and a two-decimetre tube filled with a 10 per cent. solution of cane sugar inserted, the two halves of the field will be found to be unequally illuminated, but, by turning the index by means of the milled head in the same direction as that in which the hands of a watch go,[2] a position will be reached when both the fields are

---

[1] One decimetre = 3.937 inches.
[2] When the index has to be turned in this direction the angle read is plus (+), when in the opposite direction it is minus (−).

HALF-SHADOW POLARIMETER.

again of the same brightness. When this position is reached, the reading of the scale will indicate exactly the amount of twisting or rotation which the polarised beam has undergone in passing through a layer of 10 per cent. cane sugar solution 2 decimetres thick; and from this, as will afterwards be shown, can be deduced the specific rotatory power or opticity of cane sugar. The scale on the dial-plate is divided into 360 degrees, and, by means of the vernier attached to the index, can be read to the sixtieth of a degree.

Schmidt and Haensch Half-Shadow Polarimeter.—In this instrument, which is illustrated in Fig. 6a, the appearance of the circle of light is the same as in the Laurent; the light is, however, obtained from an ordinary gas or paraffin lamp flame, placed at a

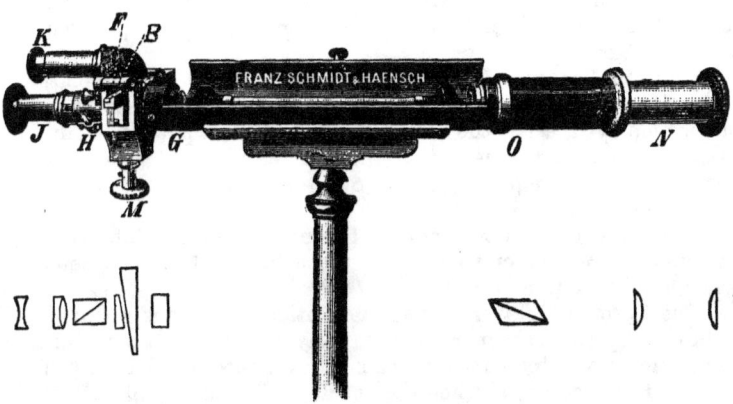

FIG. 6a.—Schmidt and Haensch Half-Shadow Polarimeter.

little distance from the end of the instrument, N. The adjustment of the index is made by the milled head M, and the scale is read through the eyepiece K. The scale is entirely different from that of the Laurent. As the instrument was constructed specially for use in sugar manufactories and refineries, its scale is constructed so as to show percentages of cane sugar directly. It has consequently 100 divisions, which by means of the vernier can be read to tenths of a division. When the two-decimetre tube is filled with a solution, 100 c.c. of which contain 26.048 grammes of *pure* cane sugar, the instrument indicates 100 scale divisions. If an impure sugar is employed for making the solution, and with this the scale reading is 76.5 divisions, then such a sample contains only 76.5 per cent. of cane sugar. Owing to this difference in the scales of the two instruments, it is necessary in order to convert Schmidt and Haensch degrees into Laurent degrees to multiply by a factor. This, according to Rimbach, is 0.344, and it has been shown recently by Rolfe and Defren to be correct for starch conversion products. The Schmidt and Haensch instrument only gives absolutely accurate indications with cane sugar and substances whose solutions are of exactly the same dispersive power as quartz; it is fairly accurate also for the carbohydrates, such as glucose, maltose, and dextrin, but with substances such as tartaric acid its readings would be quite inaccurate. The instrument is more convenient to use than the Laurent, because ordinary white light can be employed, and as the field of view is more strongly illuminated, it is easier to deal with coloured solutions; but in cases where absolute accuracy is requisite, the Laurent instrument should always be employed.

Sometimes the specific rotatory powers of bodies are found expressed in degrees $[a]j$. These relate to values obtained by older forms of the polarimeter. Twenty-four degrees $[a]j$ are equivalent to 21.67 degrees $[a]_D$.[1]

Solution Factor.—In stating observations made by the polarimeter, it is usual to append the solution factor in those cases where one is employed, as in the statement of reducing powers. Thus $[a]_{D3\cdot86} + 52.8$ would mean that this specific rotatory power had been obtained by employing the 3.86 factor to ascertain the amount of substance present in solution.

Temperature.—As the opticity of many substances is influenced by temperature, the temperature of the solution should be noted when an observation is made.

The Specific Rotatory Power or Opticity.—This is the angle indicated by the polarimeter when a layer of the substance one decimetre (3.937 ins.) in thickness is examined in that instrument. For instance, if when the one-decimetre tube is filled with English oil of turpentine, and placed in the Laurent polarimeter, the reading is $+21.5$, this number divided by the specific gravity of the turpentine gives directly the "opticity or specific rotatory power" of that particular sample of turpentine.

If, on filling a tube of the same length with French oil of turpentine, the angle observed is $-40.3°$, then the opticity of that particular sample may be obtained in a similar manner.

When, however, the opticity of a solid substance, such, for example, as cane sugar, has to be determined, the matter is not quite so simple, as obviously it is impossible to fill the tube with a solid substance like sugar. In such cases a solution of known strength of the body is made with a fluid which is without action on the polarised ray, and the opticity of the solution ascertained. For instance, when a 10 per cent. solution of pure cane sugar is examined in a tube 10 decimetres long, the angle observed is $+66.5°$. This angle is assumed to express the specific rotatory power or the opticity of cane sugar, and is expressed $[a]_D + 66.5°$.

[1] See Appendix B.

SPECIFIC ROTATORY POWER.

The opticity of a substance may therefore be described, as O'Sullivan has proposed, as the angle given by a layer of a 10 per cent. solution of that substance 10 decimetres thick. The use of a tube 10 decimetres long, owing to its unwieldiness, would be impracticable, and, moreover, solutions are rarely sufficiently colourless to permit of the use of a tube of such a length. Consequently, shorter tubes of 1 and 2 decimetres length are almost invariably employed, and the specific rotatory power obtained by a calculation.[1]

If, for instance, a 10 per cent. solution of pure cane sugar be examined in a one-decimetre tube, the angle observed will be 6.65°, and the multiplication of this figure by 10 will give the opticity of cane sugar. (As 1 decimetre is to 10 decimetres, so is 6.65° to 66.5°.)

If, instead of a 10 per cent. solution, one of 20 per cent. is employed, the angle observed will be twice as large, and will, consequently, have to be halved before being multiplied by 10. Should the two-decimetre tube be employed, then the reading will have to be halved and multiplied by 5.

The opticity is obtained in any case by the following equation:

$$[a]_D = \frac{R}{L \times \frac{C}{100}},$$

in which R is the reading of the polarimeter, L the length of the tube employed in decimetres, and C the number of grammes of substance contained in each 100 c.c. of the solution.

For instance, suppose we have a solution containing in every 100 c.c. 12.5 grammes of a body the opticity of which we wish to determine, and upon examining it in the two-decimetre tube, we obtain a reading of $+ 41.0°$; then—

$$[a]_D = \frac{+41.0}{2 \times \frac{12.5}{100}} = + 164°.$$

The opticity of specific rotatory power of such a substance is therefore $[a]_D = + 164°$.

When the specific rotatory power of a substance is known, the quantity of that substance in solution is readily determined by the polarimeter by the aid of the following formula:—

$$\frac{C}{100} = \frac{R}{L \times [a]_D}$$

---

[1] When the opticity of a substance is found in this way, it is probably never its real opticity, but only its *apparent* opticity, since it has been found that solvents, though themselves optically inactive, often have an appreciable effect on the opticity of the substance dissolved in them. This method, though obviously imperfect, is the only one possible, and even then it only indicates what would be the rotatory power of one decimetre of a 100 per cent. solution of cane sugar (if such a thing were possible).

C is the number of grammes of the substance in 100 c.c. of solution, R the observed angle, and $[a]_D$ the opticity of the substance, L being the length of the observation tube in decimetres. For instance, a solution of cane sugar of unknown strength is found to give a reading of $+17.29°$; then—

$$\frac{C}{100} = \frac{17.29}{2 \times 66.5}, \quad \text{and} \quad C = 13 \text{ grammes.}$$

Every 100 c.c. of the solution, therefore, contain 13 grammes of cane sugar.

When the weight and opticities of two bodies in solution are known, the relative proportions of each can easily be found. Suppose that a solution contains 16 grammes per 100 c.c. of a mixture of maltose (opticity $[a]_D + 135.4°$) and dextrin ($[a]_D + 195°$), and that, on examining the solution in the two-decimetre tube, the angle observed is $+52.842°$, i.e., the solid matter in solution has an opticity of $[a]_D + 165.2$, the relative proportions of maltose and dextrin in each gramme of the solid matter are found in the following manner:—

From the number found for the opticity of the mixed substances subtract the opticity of that which has the less rotatory power—in this case it will be 135.4—and divide the result by the difference of the rotatory powers of the two bodies, which here is 59.6.[1] This will be $165.2 - 135.4 = 29.8$, and $\frac{29.8}{59.6} = 0.5$ gramme dextrin. Each gramme of substance in solution consists therefore of 0.5 gramme of dextrin and 0.5 gramme of maltose, consequently the 16 grammes contain 8 grammes of maltose and the same weight of dextrin.

Cellulose.—This substance forms the groundwork of the tissues of every plant from the highest tree to the lowliest fungus. Its chemical composition is expressed by some multiple of the group $C_6H_{10}O_5$, but we are still in ignorance as to what that number is. Although one of the most abundant constituents of the vegetable kingdom, it is only met with in the pure state in very young cells, since, as these grow older, other substances are deposited in the cell-walls and in the interior of the cells. This is readily observable in the husk of barley, the cell-walls of which at first consist of pure cellulose, but which subsequently become incrusted with the tougher and more resistant lignin, in order that they may form a more effectual protection to the more delicate contents of the barley-seed. Cotton-wool or Swedish filter-paper may be cited as examples of nearly pure cellulose. In this condition cellulose is a

---

[1] Expressed algebraically, this would be as follows:—

$$195x \quad + \quad 135.4(1-x) \quad = \quad 165.2$$
Opticity of dextrin. Opticity of maltose. Opticity of mixture.

CELLULOSE.

tasteless and odourless body, insoluble in the ordinary solvents, such as water, alcohol, ether, &c., or by prolonged boiling with dilute solutions of the acids or alkalies. Cellulose is, however, soluble in a solution of cuprammonium oxide,[1] from which it is precipitated in an amorphous condition by acids, or by passing a current of carbon dioxide gas through the solution. It is also soluble in a strong solution of zinc chloride in hydrochloric acid.

Cellulose is coloured yellow or brown by a solution of iodine; but if previously treated either with a strong solution of sulphuric acid or of zinc chloride, it is then coloured blue by iodine.

When cellulose membranes have attained a certain age they exhibit the phenomenon of double refraction. Some of the celluloses, when first treated with somewhat concentrated sulphuric acid, yield glucose on prolonged boiling with a dilute solution of the same acid. These may, therefore, be regarded as polymerised anhydrides of this sugar. Other varieties, when similarly treated, yield mannose, and this is particularly characteristic of what are termed the "reserve celluloses," such, for instance, as those which are contained in large quantity in the ivory nut or the seed of the date. Here cellulose is deposited as a reserve material, just as starch is in other seeds, such, for instance, as in those of the barley plant. For this form of cellulose (which is somewhat more readily soluble in strong alkaline solutions than the former) the name of "hemi-cellulose" has been proposed. Often the cell-walls of plants consist of a mixture of these two forms of cellulose, which exist most probably in a state of chemical combination. It has been shown by Cross and Bevan that there is still another modification of cellulose present in many vegetable tissues, such as those of the barley plant; and these, since they contain a larger percentage of oxygen than the ordinary cellulose, they have named "oxy-celluloses." They are extremely resistant to the action of dilute alkalies. Like the pentose sugars, the oxy-celluloses, when distilled with hydrochloric acid, yield furfural; and, upon hydrolysis with dilute acids (1 per cent. sulphuric acid), are partially converted into a fermentable sugar that has probably the following constitution: $O_3H_8C_5\langle{}^O_O\rangle{}CH_2$. They point to the possibility,[2] by first boiling spent brewers' grains for two hours with 1 per cent. sulphuric acid solution, and then neutralising the acid with chalk, of obtaining a further portion of fermentable sugar, which might be added to the wort previous to fermentation. At all events, the sugar obtained in this way should be utilisable by the distiller and vinegar-maker.

The cellulose portion of the starch granules appears to be of

[1] Made by placing copper turnings in a vessel partially filled with strong solution of ammonia and permitting free access of air. The liquid gradually assumes a deep blue colour.

[2] *Journal of the Federated Institutes of Brewing*, 1897, p. 15.

a different nature to ordinary cellulose, since it is converted into soluble carbohydrates by the action of diastase.

The weight of the husk of barley which consists of lignified cellulose ranges from 7.2 to 14.9 per cent. It may be determined by soaking the barley, and placing it while still under water in the receiver of an air-pump, and exhausting the air. This removes the air contained in the barley; and, on again admitting air, the pressure of the atmosphere forces water into the spaces originally occupied with air, the effect being to so loosen the husks that they can be easily removed. They are then dried and weighed.

Starch or Amylum.—To the brewer this substance is the most important constituent of the barleycorn, of which it forms 60–66 per cent.; the sugar, which upon fermentation is to yield alcohol and carbon dioxide gas, is for the most part derived from starch.

Next to cellulose, starch, which is found in all green plants, is the most abundant material met with in the plant world. It is also stored up in considerable quantities in the seeds of many plants, such as the cereals, as a reserve material from which the seedling derives its nourishment during germination.

Characteristics of Starch.—Starch in a state of purity forms a white, tasteless, and inodorous powder consisting of small granules. The starch grains from different plants, when examined under the microscope, are found to differ so much in form and size that it is often possible to determine, by a simple microscopic examination, the plant from which the starch has been derived. The illustrations on pp. 66 and 67 show this. Though all these starches differ in size and shape, still they exhibit certain similarities in structure. Each has a dark point, central in some, eccentric in others, known as the "hilum." Round this are seen a series of concentric lines, an appearance caused by the peculiar structure of starch granules, which are built up of layers containing varying amounts of water. The hilum is always rich in water, and each layer alternately contains more and less water, the outside layer being always the poorest in water and richest in substance. As a consequence of the proportion of water increasing from the outside of the granule inwards to the hilum, fissures radiating from the hilum towards the periphery often arise as the granule becomes dry. By soaking starch granules in alcohol, which entirely deprives them of water, all appearance of stratification disappears, but the lines reappear if the granules are moistened with water. If they are treated with dilute alkalies or acids, or with a weak solution of chromic acid, the appearance of stratification is rendered much more distinct.

Starch is insoluble in the ordinary solvents, such as alcohol, ether, &c., and it is insoluble in a solution of cuprammonium oxide. The intact grains are completely insoluble in water, and hence starch is readily obtained from seeds, tubers, and other starch-containing organs. These are crushed or ground and washed with water on a sieve; the starch granules pass through the sieve along with the water, leaving the coarser portions of the tissues behind. On standing, the starch deposits on the bottom of the vessel placed to receive it and the liquid which passes through the sieve. It is then washed several times with water by decantation, and finally dried at a low temperature. Starch is generally made from rice in England, from potatoes and wheat in Germany, and from maize in Southern Europe.

The average specific gravity of air-dried starch is 1.503 to 1.504. Air-dried starch contains from 15 to 20 per cent. of hygroscopic water, the last traces of which it retains with great pertinacity; consequently, it is almost impossible to remove the whole of the moisture by heat alone without at the same time causing a chemical alteration in the starch substance itself. To avoid this it has been proposed by Dafert[1] to remove the hygroscopic water by drying starch in a vacuum at 100° C. (212° F.). Absolutely dry starch attracts moisture with such avidity, that when moistened with water a perceptible rise of temperature takes place. This shows that the last portions of water are in a state of chemical combination.

Formation of Starch.—The starch granules are deposited in certain cells in the plant tissue known as "starch-formers" or "amyloplasts." According to the theory propounded a long time since by Nägeli, starch granules do not grow like a crystal from within outwards (by apposition), but from without to within (by intersusception). A layer of starch is first deposited round the inner surface of the starch-forming cell, which is filled with a solution of the substance (probably sugar) from which the starch is derived. Another layer is deposited inside this, and so on until the cell is completely filled. According to this view, the outside layer of starch is the oldest and poorest in water; the portions about the hilum are the newest and richest in water. Attempts have been made by Schimper and Meyer[2] to disprove this theory, and to show that the starch granules grow by apposition; but the older theory (intersusception) of Nägeli appears to explain the whole of the phenomena observed in the more satisfactory manner, and is the one which has found the most general acceptance.

Microscopic Characters of Starch Granules.—In microscopic appearance barley-starch (Fig. 7, a) consists chiefly of two sets of granules, one very much larger than the other, also a few of an intermediate size. The larger granules are circular in shape, and when viewed from their sides have the appearance of bi-convex lenses; the smaller granules are either spherical or polygonal. The hilum is concentric, but is seldom well marked. The concentric markings are very indistinct, unless brought out by some reagent,

[1] *Journ. f. prakt. Chem.*, lxxiii. 51.
[2] *Botan. Zeit.*, 1881, pp. 185 and 841.

such as chromic acid. The smaller granules are entirely unstratified. The large grains vary in diameter from 10.8 to 32.8μ, the majority measuring 20.3μ. The lesser ones are from 1.6 to 6.4μ in width, the most frequent size met with being 4.6μ.[1]

Wheat-starch (Fig. 7, *b*) very much resembles barley-starch in appearance; the small granules of wheat are, however, smaller than those of barley, and the large granules show the concentric markings even less distinctly. The large granules vary in diameter from 11.1 to 41μ, and the smaller from 2.2 to 8.2μ.

Rice-starch (Fig. 7, *c*) consists of two kinds of granules, simple and compound. Both are polygonal in shape. A large polygonal or a star-shaped hollow takes the place of the hilum. The compound grains vary in diameter from 18 to 36μ, the simple grains from 3 to 7μ.

Maize-starch (Fig. 7, *d*) also consists of simple and compound granules. In the exterior horny portion of the corn the granules are polyhedral and compound, but the interior and more mealy

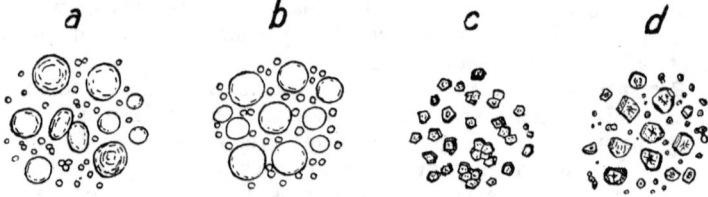

FIG. 7.—Starch Granules from—(*a*) Barley; (*b*) Wheat; (*c*) Rice; (*d*) Maize.

portions of the grain consist almost exclusively of simple granules, only a few really compound granules being met with in this position. The simple granules are spherical or elliptical; the compound granules, and also the small grains contained within them, are polygonal. The majority of the granules possess a hilum, or in its place a star-shaped hollow. The compound granules have a diameter of about 47μ, the simple vary from 7.2 to 32.5μ in diameter, the average being 20μ.

In *potato-starch* (Fig. 7, *e*, *f*) the large and mature granules are egg-shaped, the smaller and more immature spherical. The larger ones have an excentric hilum, around which the excentric markings are very distinct; the small granules exhibit no signs of stratification. The granules of this starch vary in diameter from 60 to 100μ, their most frequent size being 70μ. Potato-starch is known commercially as "farina." Of this there are two kinds, *prima*

[1] The micromillimetre, which equals the one-thousandth part of a millimetre, is an exceedingly convenient unit for microscopic measurements, and is denoted by the Greek letter μ. It is equal to the $\frac{1}{25,000}$ of an inch.

and *secunda*, distinguished by the size of the respective granules. The *prima* consists of large granules varying from 21 to 33μ in diameter; the *secunda* of granules from 12.5 to 21μ in diameter.

Sago-starch (Fig. 7, *g*) is somewhat elliptical in form, the hilum often appearing as a fissure situated in the long diameter of the granule. The cells have a tendency to split in two through their short diameter, thus producing muller-shaped granules. As met with in commerce, these characters are much modified by the heat and moisture used in the process which this starch undergoes in being made into grains.

Composition of Starch.—Starch chiefly consists of two bodies, granulose and starch cellulose. It also contains minute quantities of fat, nitrogenous bodies, and ash. The granulose of the starch grains is easily dissolved by several reagents which do not affect

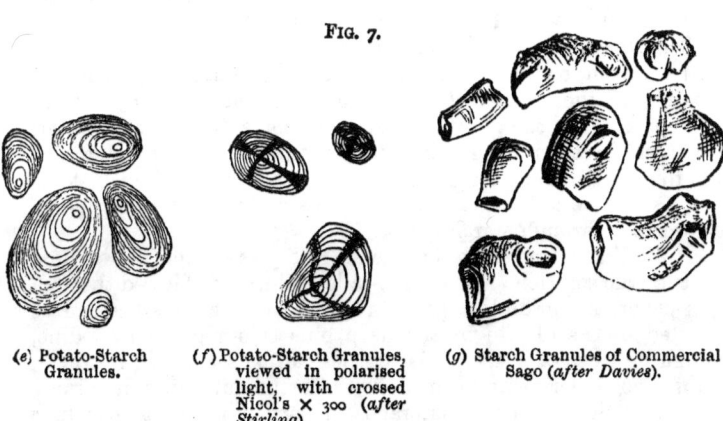

FIG. 7.

(*e*) Potato-Starch Granules.   (*f*) Potato-Starch Granules, viewed in polarised light, with crossed Nicol's × 300 (*after Stirling*).   (*g*) Starch Granules of Commercial Sago (*after Davies*).

the cellulose, consequently a separation of the two can be readily effected. If starch grains are digested with a little fresh saliva at 35° to 55° C. (95° to 131° F.), or treated at 60° C. (140° F.) with a saturated solution of common salt containing 1 per cent. of hydrochloric acid, the granulose gradually dissolves away, and a perfect skeleton of the cellulose is left behind which exactly resembles in form the original starch cell. The substance thus obtained is coloured from yellow to brown by iodine; if previously treated with sulphuric acid or chloride of zinc, iodine then gives a blue coloration. This is the general reaction given by cellulose (p. 63). Starch cellulose is soluble in cuprammonium hydrate.

Dragendorff[1] found, by treating various starches with 1 per cent. hydrochloric acid at 60° C. (140° F.), amounts of starch cellulose varying from 2 to 6 per cent. in the different starches;

[1] *Journ. f. Landwirthsch.*, 1862, p. 211.

and similar results were obtained by Nägeli.[1] If the action of the acid be continued too long, some of the cellulose dissolves, since it is not altogether unattacked by dilute acids. The outside thick envelopes are the most resistant.

According to the investigations of Sostegni,[2] starch cellulose consists of two substances, one of which is a fat. This portion, which constitutes about one-fifth of the whole substance, can be extracted by ether. It is a pure white crystalline body, which melts at 47° to 48° C. (117° to 119° F.). The remaining portion of the starch cellulose is soluble in a 2 per cent. solution of caustic potash, and from this solution it is precipitated as an amorphous substance on the addition of an acid. If heated for some time with a dilute mineral acid, it dissolves and yields glucose.

Starch cellulose is also dissolved by the enzyme cytase discovered in the seeds of the cereals by Brown and Morris.[3]

Granulose.—This substance forms by far the greater portion of the starch cell; it is coloured blue by iodine. When the structure of the starch cell has been destroyed, either by heat or mechanical force, the granulose can be dissolved out by cold water. Berzelius,[4] by triturating starch in a mortar with cold water, was able to dissolve everything but the cellulose portion. Nägeli states that he has seen the granulose dissolve under the microscope in sections cut through starch grains.

Notwithstanding these statements, it is questionable if granulose forms a true solution. When a so-called solution is filtered through filter-paper, the granulose passes through, and its presence can be detected in the filtrate by the deep blue colour given by iodine, though no trace of suspended particles can be seen. If, however, a diffusion experiment be made with parchment paper, no granulose passes through the membrane. Hence it is fair to conclude that the granulose is not in a state of true solution, but is merely divided into exceedingly fine particles, which are diffused through the liquid. These particles, though sufficiently small to be able to pass through filter-paper, are not small enough to pass through a diffusion membrane.

Granulose in its so-called aqueous solution has an opticity of about $[a]j + 200°$ and no reducing power; it is precipitated from such solutions by alcohol and tannin.

Behaviour of Starch with Iodine.—Starch grains or solutions of granulose are coloured an intense blue on the addition of a dilute solution of iodine. This reaction, which was discovered by Gaultier de Clanbry in 1814, is extremely delicate, and is frequently used as a test for the recognition of minute

---

[1] *Beit. z. Kennt. d. Stärke*, 1874.
[2] *Central. f. Agriculturchem.*, xv. 638.
[3] *Journal of the Chemical Society*, 1890, p. 468.
[4] *Jahresber. f. Chemie*, 1831, p. 201.

traces of starch or of iodine. It is much influenced by temperature; the lower the temperature the more sensitive is the reaction; at or near the boiling-point of water the coloration does not appear at all. It does not show itself so readily, nor is the colour so pure, in the presence of certain other bodies, such as tannin, malt extract, beer yeast, &c. Whether the blue substance formed by iodine and granulose is a true chemical compound or not has not been definitely settled.

The colour produced by iodine and starch grains is not always a pure blue, but varies with the nature of the starch. Thus, potato-starch gives a deep blue, whilst with wheat-starch the colour produced inclines somewhat to violet. This is accounted for by a larger amount of the cellulose, which is coloured brown by iodine, being contained in some of the starches.

Two starches have been met with by Dafert and Kreusler,[1] the one in a species of rice, the other in a kind of millet, which are coloured from red to brown by iodine.

Imbibition.—Starch is able to imbibe a certain amount of cold water, which simply swells the granules out to a slight extent without destroying their organised structure.

Gelatinisation of Starch.—If starch be mixed with cold water and the mixture be gradually heated, when a certain temperature (which varies for each different starch) is reached, the granules begin to swell, and when a certain point is reached, their cellulose envelopes burst. As the temperature rises, more and more of the contents of the granules escape, until at last a gelatinous mass is produced, thinner or thicker in consistence according to the amount of starch present; this is known as "starch-paste." The granules, when examined under the microscope, are now found to have lost all structural organisation, and this is never afterwards regained. If the paste be dried, a horny mass is left, which, when heated with water, merely swells up, but does not dissolve. When starch-paste is frozen, the mass afterwards thawed and the water pressed away, the starch after drying is left as a loose spongy white mass, which upon being boiled with water merely swells up, and exhibits no tendency to form a gelatinous paste. With gelatinisation the starch granules cease to be doubly refractive. The viscosity of starch-paste varies with the temperature, the same solution being much less viscous when hot than cold. The temperature at which the starch has been dried has a marked effect upon the viscosity of the paste derived from it. Brown and Heron[2] found that the viscosity of starch-paste made from starch dried at 30° C. (86° F.) was only a fourth of that made from the same starch when dried at 100° C. (212° F.).

If thin warm starch-paste be filtered, the portion passing through the filter is quite clear, and gives a deep blue colour with

---

[1] *Landwirth. Jahrbuch,* xiii. 767; xiv. 831.
[2] *Journal of the Chemical Society,* 1879, p. 614.

iodine; no solid matter can be detected in the liquid by a microscopic examination. Märker expresses doubt as to whether the granulose is really in a state of true solution; Brown and Heron hold an opposite opinion, and think that the granulose does go into solution, and that the viscosity of starch-paste is to be attributed to the swollen cellulose. A later view, taken by Moritz and Morris,[1] is that on boiling starch the granulose is converted into soluble starch.

Whether in a state of true solution or not, starch-paste is quite non-diffusible, and incapable of passing through a porous membrane.

Temperatures at which Gelatinisation Occurs.—The temperature at which gelatinisation commences varies with the different starches, and a higher temperature than this is always required for its completion.

The following table[2] gives the gelatinisation temperatures of several of the commoner varieties of starch:—

	C.	F
Potato-starch	65°	149°
Barley-starch	80°	176°
Green-malt-starch	85°	185°
Kilned-malt-starch	80°	176°
Oat-starch	85°	185°
Rye-starch	80°	176°
Wheat-starch	80°	176°
Rice-starch	80°	176°
Maize-starch	75°	167°

Starch entirely free from alkalies or acids may be heated with water under pressure to 150° C. (302° F.) without undergoing hydrolysis. It is, however, converted into that modification known as "soluble starch." This property is now frequently made use of in the analysis of substances which contain starch.

Action of Acids on Starch.—Starch is gelatinised and dissolved by fuming hydrochloric acid in the cold to a clear solution, which principally contains soluble starch. Dilute acids split up, by hydrolysis, the complex molecule of starch into simpler molecules —dextrins, maltose—the final product being glucose. Hydrochloric acid has the most intense action in this direction, sulphuric acid acts less energetically, and the organic acids, such as oxalic, are much slower and more feeble in their action. Elevation of temperature essentially hastens the hydrolytic action. When starch is allowed to stand for seven or eight days under a 7.5 per cent. solution of hydrochloric acid, it is converted into soluble starch.

Constitution of the Starch Molecule.[3]—Much doubt exists even as to the size of the starch molecule. Musculus and Gruber

---

[1] *Science of Brewing*, p. 95.
[2] Lintner, *Brauer u. Mälzer Kalendar*, 1880.
[3] The reading of this paragraph had better be postponed until the chapter on the action of diastase (p. 122) on starch has been perused.

considered that it contained the group $(C_{12}H_{20}O_{10})$ five or six times; Brown and Heron assumed it to be $10(C_{12}H_{20}O_{10})$; and Brown and Morris $20(C_{12}H_{20}O_{10})_5$.

With reference to its constitution we are as yet entirely in the dark. An important step towards the elucidation of this enigma was made by Scheibler and Mittelmeier,[1] who showed that the dextrins possess free carbonyl (CO) groups, and hence reduce Fehling's solution. They look upon starch granulose as consisting of an unknown number of glucose residues, united by a number of monocarbonyl groups and one dicarbonyl group, since, like cane sugar, it has no reducing power. Of the many conceivable formulæ the following two are put forward as the most probable :—

   I. R<R<R<R<R......R<R<>R>R......>R>R>R>R
  II. R<R<R<R<........R<R<R<>R>R

R indicates a glucose residue, < a monocarbonyl and <> a dicarbonyl union.

As the formulæ show, when splitting up takes place at a dicarbonyl union, the resulting molecules have each a free COH group, and each therefore possesses reducing powers. If it be assumed that, when starch is hydrolysed, dextrin and maltose are formed simultaneously, then the first phase of the action may be explained according to one or other of the following formulæ, in which ⋖ denotes a free COH group :—

   I. R<R<R<...............R<R<>R>R + H₂O
        Starch   Water.

 =R<R<R<...............R<R⋖ + ⋗R>R
   Dextrin.   Maltose.

or

  II. R<R<R<...............R<R<>R>R + H₂O
       Starch.   Water.

 =R<R⋖ + R<R<.......R<R<R<>R>R
  Maltose.   Dextrin.

In the first case, besides maltose, a dextrin with a free COH group, and consequently possessing reducing power, would be obtained; in the second, maltose and a dextrin without reducing power.

In a third paper the authors [2] consider that they had not only effectually disposed of the hypothesis of Brown and Morris of 1889, but had also placed the question of the chemical constitution of starch in an entirely new light. Taking into account the views of E. Fischer on the structural formulæ of the disaccharides maltose and lactose, they regard that of cane sugar as $C_6H_{11}O_5 \cdot O \cdot C_6H_{11}O_5$. Since they consider starch to have a similar constitution, it would be expressed by the formula $C_6H_{11}O_5 \cdot O \cdot (R)_x \cdot C_6H_{11}O_5$,

---

[1] *Berichte*, xxiii. 3060.   [2] *Ibid.*, 1893, p. 2930.

in which R_x represents an unknown number of glucose residues. As the hydrolysis of cane sugar may be represented by the equation—

$$C_6H_{11}O_5 \cdot O \cdot C_6H_{11}O_5 + H_2O = 2C_6H_{12}O_6$$

so may starch by—

$$C_6H_{11}O_5 \cdot O \cdot (R)_x \cdot C_6H_{11}O_5 + (x-1)H_2O = (C_6H_{12}O_6)_x$$

Applying their symbols to the various linkings, the formula for cane sugar becomes $C_6H_{11}O_5 <O> C_6H_{11}O_5$, and that of starch $C_6H_{11}O_5 > O \cdot (R)_x > C_6H_{11}O_5$, and maltose $C_6H_{11}O_5 < O \cdot C_6H_{11}O_5 <$. As they had already shown that the dextrins possessed all the qualitative properties of maltose, their general constitution may be expressed by the following formula :—$C_6H_{11}O_5 < O \cdot (R)_x \cdot C_6H_{11}O_5 <$, the number of the glucose residues R, varying as the dextrins themselves vary.

Scheibler and Mittelmeier do not consider that the breaking down of starch by hydrolysis takes place according to Musculus's theory, but believe it to occur in a manner similar to that in which raffinose $C_6H_{11}O_5 \overset{*}{<} O \cdot C_6H_{10}O_4 < O > C_6H_{11}O_5$ does.

This sugar breaks down in two distinct phases; in the first the monocarbonyl link (marked with an asterisk) gives way, a molecule of fructose being split off, leaving a residue of melibiose; in the second phase the dicarbonyl link gives way, and the melibiose breaks down into glucose and galactose.

Soluble Starch.—This body seems to be the very first product of the action of either diastase, acids, or heat on starch.

Preparation.—O'Sullivan[1] prepared this substance by dissolving starch at 73° to 74° C. (163° to 165° F.), with the least possible quantity of cold water extract of malt, which had been heated to the same temperature. The solution was boiled, filtered, and concentrated by evaporation, when, as it cooled, soluble starch fell out as a brilliant white precipitate. It was further purified by being dissolved in water and re-precipitated therefrom several times. When dried it seemed to suffer some slight modification; a small portion was rendered insoluble, and another small portion (1 to 5 per cent.) converted into maltose and dextrin. The purest preparation he was able to make had an opticity of $[a]j + 220°$ and a K of 0.78. Probably the reducing power was due to an impurity. Musculus and Gruber had previously described a similar body to which they assigned a K of 6.

In 1886 C. J. Lintner[2] stated that a soluble modification of starch could be readily formed by allowing ungelatinised potato-starch to remain in contact with 7.5 per cent. hydrochloric acid

---

[1] *Journal of the Chemical Society,* 1879, p. 772.
[2] *Journ. f. prakt. Chem.,* xxxiv. 378.

for seven or eight days. Starch when thus treated, and freed from acid by washing, dissolves in hot water without forming a viscid paste. Brown and Morris [1] carefully examined the product of this reaction, and fully confirmed Lintner's observations. They also found that the same result could be attained in a much shorter time by using stronger acid. When 12 per cent. hydrochloric acid was used, the action was completed in twenty-four hours. The change in property is unaccompanied by any structural modification; the starch granules, when examined under the microscope, show no alteration in form; their behaviour towards polarised light is precisely similar to that of the original starch grains. They also found that this body was absolutely identical with the soluble starch of O'Sullivan just described.

Ost obtained, by heating starch with water under a pressure of two or three atmospheres, a nearly clear solution of a body which had a rotatory power of $[a]_D = +196° - 197°$, but no reducing power; it gave, with aqueous solution of iodine in excess, a greenish blue colour, without any tendency to violet. The body separated from its hot aqueous solution on cooling, or could be precipitated from such a solution by alcohol, as a white powder, which, when recently precipitated, dissolved readily in cold water, but, after being dried over sulphuric acid, would only dissolve in hot water, and then not completely. Treatment of starch under higher pressures always led to the formation of dextrin. Ost considers that the term soluble starch will have to be restricted to starch which has passed into solution without the size of its molecule having been altered; in fact, he regards it as the same substance which passes into solution when starch is rubbed up with sand and water.

Soluble starch can be prepared, according to Zulkowski, by heating starch with glycerin to 190° C. (374° F.). It can also be prepared by submitting starch and water to a temperature of from 125° to 150° C. (257° to 302° F.) in a closed vessel.

Properties.—Soluble starch, when prepared by any of the above methods, is readily soluble in water of 60° to 70° C. (140° to 158° F.). When a strong solution (10 per cent.) is cooled down, the soluble starch separates out of solution as a white pasty mass (the same also happens when alcohol is added to a solution of soluble starch), which, on washing with alcohol and subsequent drying, becomes very friable. By submitting a dilute solution of soluble starch to a low temperature, it is possible to obtain the substance in the form of sphæro-crystals (*Lintner*). Soluble starch after once being dried is almost completely insoluble in cold water.

Soluble starch and its solutions are coloured an intense deep blue by iodine. Its opticity is given as $[a]_{j_{3.86}} + 216°$ by Brown and Morris, and as $[a]_D + 200°$ by Lintner. Brown, Morris, and Millar [2] state that the opticity of soluble starch in 2.5 to 4.5 per

[1] *Journal of the Chemical Society*, 1889, p. 450.
[2] *Ibid.*, 1896, p. 243.

cent. solutions is, at $15.5°$ C., $[a]_D = +202°$. It has no reducing action on Fehling's solution.

When acted on by diastase in the cold, it is degraded in ten minutes to the No. 8 equation (see page 141). When, however, the action of the diastase is allowed to proceed for ten or twelve days in the cold, soluble starch is entirely converted into maltose.[1]

Its molecular weight, as determined by Raoult's method, gave, in the hands of Brown and Morris[2] a molecular weight of from 20,000 to 30,000. According to the same observers, and on the assumption that the stable dextrin of the No. 8 equation constitutes one-fifth of the molecule of soluble starch, its molecular weight is 32,400, and its formula $5(C_{12}H_{20}O_{10})20$.

Amylodextrins.—Nägeli, in his work on the starches, published in 1874, describes two amylodextrins which he had obtained by allowing 12 per cent. hydrochloric acid to act on starch for from two to four months, at the end of which time the starch grains had lost their power of giving a blue colour with iodine.

Amylodextrin No. I. was obtained by decanting the supernatant acid from the altered starch cells, and freeing them from the last traces of acidity by repeated decantation with water and alcohol. The grains, after being dried between sheets of blotting-paper, were dissolved in hot water, the solution filtered, and then frozen. The amylodextrin separated out as a crystalloid substance, which, when examined under the microscope, was found to consist of round discs, each of which was composed of a mass of needle-shaped forms, grouped radially round the centre of the disc. When examined by polarised light, they exhibited the phenomenon of double refraction.

Amylodextrin II. was obtained from the acid fluid in which the starch had been immersed, while this still gave a blue coloration with iodine, by adding to it after filtration four or five times its volume of 93 per cent. alcohol. The white precipitate was collected on a filter and washed with alcohol, dissolved in hot water, and the solution frozen; the amylodextrin then separated out in discs exactly similar to those of Amylodextrin I. The only difference between the two was their behaviour with iodine; a crystal of this substance communicated a blue-violet colour to No. I. and a red-violet to No. II. It was thought that they were not individual bodies, but mixtures of two bodies, one of which is coloured violet by iodine, the other red, and Nägeli succeeded in separating two bodies which fulfilled these conditions. Amylodextrin I. apparently contains more of the modification which gives a violet colour, whilst in No. II. the red-colouring modification preponderates.

Brown and Morris[3] examined this substance very carefully, but were unable to fractionate it with alcohol. They found,

---

[1] Lintner, *Journ. f. prakt. Chem.*, 1887.
[2] *Journal of the Chemical Society*, 1889, p. 465.
[3] *Ibid.*, 1889, p. 455.

contrary to Nägeli's assertion, that it was slowly but sensibly diffusible. They concluded that it was a perfectly distinct and definite compound, having an opticity of $[a]j + 204.6$ and a κ of 10.12, corresponding to a composition of maltose 16.6 per cent. and dextrin 83.4 per cent. They regarded it as a compound of six dextrin groups and one maltose group, which would require an opticity of $[a]j + 206.11$ and a κ of 9.08. Experiments made with Raoult's freezing method also point strongly to its molecular weight being 2286, a number which corresponds closely with the composition indicated. When treated with fresh malt-extract at any temperature under 60° C. (140° F.), it was completely hydrolysed to maltose, thus showing that it did not, like soluble starch, contain a dextrin residue. This curious fact led Brown and Morris to make a series of experiments to ascertain when the dextrin residue disappeared. The results of these are shown in the following table, in which potato-starch immersed in an 11 per cent. solution of hydrochloric acid was tested from time to time by treatment with fresh malt-extract:—

Time of Treatment with HCl.	Iodine Reaction.	Apparent Composition.	
		Maltose.	Dextrin.
		Per Cent.	Per Cent.
After 48 hours	Pure blue	81.41	18.59
After 12 days	Reddish-purple	87.70	12.30
After 21 days	Light reddish-brown	85.77	14.23
After 33 days	Light reddish-brown	90.10	9.90
After 66 days	Light reddish-brown	92.05	7.95
After 100 days	Light reddish-brown	92.24	7.76
After 8½ years	Light reddish-brown	97.00	3.00

This last preparation, when dissolved in water and re-precipitated by alcohol, gave pure amylodextrin, which, on treatment with fresh malt-extract, yielded maltose only.

The solution of hydrochloric acid used in the experiment was found to contain glucose equal in weight to 40 per cent. of the starch originally taken.

Maltose.—This sugar is produced whenever starch is submitted to the action of certain enzymes, such as diastase, ptyalin (from saliva), or trypsin (from the pancreas), under proper conditions. It is also one of the primary products of the action of acids upon starch.

Maltose was discovered by De Saussure in 1819, and again described by Dubrunfant in 1847. These facts seem to have been forgotten until O'Sullivan in 1872 re-discovered the sugar and brought it prominently into notice (see p. 126).

Preparation.—This sugar has been usually prepared by making a starch conversion with cold-water malt-extract, under those conditions which yield the largest proportion of maltose, the conversion being afterwards freed from dextrin by alcohol, which precipitates this body and dissolves the maltose. The impure maltose obtained by evaporating the alcoholic solution is then further purified by several re-solutions in alcohol and subsequent evaporations, after which it is obtained in a crystalline state (see p. 126).

It can be prepared much more readily by allowing diastase to act on a solution of gelatinised starch for a considerable length of time in the cold. For this purpose 100 grammes of potato-starch are intimately mixed with two litres of cold water, and the whole raised to a temperature sufficient to thoroughly gelatinise the starch. The mixture is then cooled down to 30° C. (86° F.), 0.5 gramme diastase dissolved in water, and 5 cubic centimetres of chloroform added, and the whole well shaken up. The object of the addition of chloroform is to prevent the growth of bacteria. The whole is well corked, and allowed to stand from ten to twelve days at the ordinary temperature. The solution, which should then give no turbidity with alcohol (absence of dextrin), is evaporated to a syrup, and this, on standing for a few days, solidifies to a crystalline mass. The mass is dissolved in boiling 80 per cent. alcohol, and this solution on evaporation yields crystals of pure maltose.

Characteristics.—Maltose crystallises from its aqueous solution in the form of fine needles, which consist of sharply pointed prisms containing one molecule of water of crystallisation, $C_{12}H_{22}O_{11} + H_2O$, which can be driven off by exposing the crystals to a temperature of 100° C. (212° F.) in a vacuum, when anhydrous maltose remains. Anhydrous maltose can also be obtained in crystals by slowly evaporating a strong alcoholic solution of the sugar.

Maltose is readily soluble in water; its solution has a sweet taste. It is only slightly soluble in alcohol, much more so in alcohol containing water, the more dilute the alcohol the greater being its solubility.

Opticity.—The opticity of maltose has been variously stated by different observers. O'Sullivan,[1] using as solution divisor 3.85, gives $[\alpha]j + 150°$, and later on, using the more correct divisor 3.95, $[\alpha]j + 154°$. Brown and Heron,[2] using 3.86 as solution divisor, give $[\alpha]j + 150°$. When the requisite calculation is made, it will be seen that these last two numbers are almost identical. The absolute opticity of maltose as given by Brown and Heron (*l.c.*), who found that the correct solution factor for a 5 per cent. solution was 3.9314, is $[\alpha]j + 153.1$, which equals $[\alpha]_D + 137.4$. Tollens has recently given the opticity as $[\alpha]_D + 137$. Brown, Morris, and Millar[3] give $[\alpha]_D = 137.93°$ as the absolute opticity of maltose according to their latest investigations. According to Meissl, the

[1] *Journal of the Chemical Society*, 1876, p. 479.
[2] *Ibid.*, 1879, p. 619. [3] *Ibid.*, 1896, p. 243.

opticity of a solution of maltose is little influenced by concentration, but considerably by temperature. He gives the following formula for making corrections for concentration and temperature :—

$$[\alpha] = 140.375 - 0.01837P - 0.095T$$

in which P equals the percentage of maltose, and T the temperature of the solution. Thus a 10 per cent. solution at a temperature of 20° C. (68° F.) would have an opticity of $[\alpha]_D + 138.3°$. A freshly prepared solution of maltose shows an opticity of some 18° less than its normal opticity; this, however, rises to the normal after the solution has stood for twenty-four hours, or it may be brought about instantaneously by the addition of a small quantity of solution of ammonia.

Cupric Reducing Power.—Maltose has a much lower reducing action on Fehling's solution than glucose has, only about two-thirds of the quantity of cuprous oxide being yielded by maltose which an equal weight of glucose would yield. O'Sullivan, in 1876, using the 3.85 solution factor, gave K65 as its reducing power; Brown and Morris[1] consider κ61 with the 3.86 solution factor to be more correct; and O'Sullivan, in 1884,[2] using the 3.95 factor, gives 62.5, which, when calculated out to the 3.86 factor, gives 61.07. According to Soxhlet, with an excess of undiluted Fehling's solution and an approximately 1 per cent. solution of maltose, 100 parts of maltose invariably yield 127.3 parts of cupric oxide, or 113 parts of copper. A series of tables for the estimation of maltose have been prepared by Kjeldahl.[3] Maltose has no reducing action upon Barfoed's solution (see p. 81).

Action of Hydrolysts on Maltose.—Diastase has no hydrolysing action upon maltose, consequently this sugar is the lowest stage to which starch can be degraded by the action of diastase. Brown and Heron found that the secretion of the Peyerian glands of the small intestine contained an enzyme which readily converted maltose into glucose. Dilute mineral acids, assisted by heat, hydrolyse maltose to glucose thus :—

$$\underset{\text{Maltose.}}{C_{12}H_{22}O_{11}} + \underset{\text{Water.}}{H_2O} = \underset{\text{Glucose.}}{2(C_6H_{12}O_6)}$$

a reaction which is frequently made use of in the estimation of maltose in mixtures. Maltose is not nearly so easily hydrolysed by acids as cane sugar. According to Meissl, it offers five times as much resistance to hydrolysis as cane sugar.

Action of Yeast.—Maltose is easily and completely fermentable by yeast. It was supposed to ferment directly, that is, without previous hydrolysis to glucose, but it has been shown recently that this is not so. Yeast contains an enzyme which converts maltose into glucose.

[1] *Journal of the Chemical Society,* 1879, p. 619.   [2] *Ibid.,* 1884.
[3] *Med. Carlsb. Lab.,* 1895, p. 1; *Analyst,* xx. 227.

Osazone.—When maltose is treated with phenylhydrazine acetate, phenylmaltosazone, $C_{24}H_{32}N_4O_9$, is formed, which precipitates, as the solution cools, in the form of yellow crystalline needles, soluble in hot water but insoluble in cold. The osazone melts at 190°–200° C. with decomposition.

When heated with nitric acid, maltose yields saccharic acid, $COOH.4(CHOH).COOH$; when acted upon by bromine it yields maltobionic acid:—

$$CH_2OH \cdot 4(CHOH) \cdot CH$$
$$CH_2O \cdot CHOH \cdot CHOH \cdot CHO \cdot CHOH \cdot COOH$$

Constitution.—Maltose has been shown by Emil Fischer to be a disaccharide formed by the union of two glucose residues. These are probably united in the following manner:—

$$CH_2OH \cdot 4(CHOH) \cdot CH$$
$$CH_2O \cdot CHOH \cdot CHOH \cdot CHO \cdot CHOH \cdot COH$$

consequently the molecule has only one free COH group.

Glucose, Grape Sugar, or Dextrose.—This is one of the sugars found in great abundance in the vegetable kingdom. Ripe sweet fruits, such as grapes, plums, figs, &c., contain, in addition to fructose (lævulose) and small quantities of cane sugar, large quantities of glucose. It is met with in various other parts of plants, as the stems of the cereals, and in the flowers of many plants from which bees derive honey, a substance consisting largely of glucose. It is found in small quantities in the seeds of the cereals, such as barley. Compounds of glucose with numerous and very different substances are frequently met with in the vegetable kingdom. These are called glucosides, which, under the hydrolysing action of dilute acids or enzymes, readily split up into glucose and their other constituents.

Preparation.—Glucose may be prepared in several ways, as, for instance, by the inversion of cane sugar, or by the hydrolysis of starch, cellulose, dextrin, &c.

To secure glucose in a pure state, the former of these methods is generally used. Soxhlet gives the following directions for obtaining crystals of pure glucose. A mixture, made with 12 litres of 90 per cent. alcohol and 480 c.c. of fuming hydrochloric acid, is heated to 45° C. (113° F.), and into this is stirred 2 kilogrammes of powdered cane sugar, in small portions at a time, care being taken that the heat of the mixture never exceeds 50° C. (122° F.). In two hours the cane sugar is completely inverted—that is, converted into a mixture of glucose and lævulose:—

$$\underset{\text{Cane sugar.}}{C_{12}H_{22}O_{11}} + \underset{\text{Water.}}{H_2O} = \underset{\text{Glucose.}}{C_6H_{12}O_6} + \underset{\text{Lævulose.}}{C_6H_{12}O_6}$$

PREPARATION.

When the solution has become cold, a small quantity of crystallised glucose is stirred in, in order to start crystallisation. The whole is allowed to remain for twenty-four hours, at the end of which time the crystals are removed, washed with 90 per cent. alcohol until free from all traces of hydrochloric acid, washed with absolute alcohol, and recrystallised from pure methylic alcohol. In this way crystalline crusts of anhydrous glucose are obtained. If these crusts are melted with 12 per cent. of water on the water-bath, the glucose crystallises out on cooling. The crystals, after removal, are then dried over sulphuric acid; in this way hydrated glucose which crystallises with one molecule of water is obtained, $C_6H_{12}O_6 + H_2O$.

On the manufacturing scale, glucose is almost invariably made by submitting starch to the hydrolysing action of dilute sulphuric acid assisted by heat. The process is conducted sometimes in open vessels, at other times in closed vessels under pressure; in either case, the apparatus is provided with some means of heating its contents, such as a steam-coil, or a steam-pipe perforated with small holes. The quantity of acid used per cwt. of starch varies with different manufacturers, but is, as a rule, from six to eight pounds, and this is mixed with from forty to fifty gallons of water. Half the quantity of the dilute sulphuric acid required for the conversion is introduced into the converting vessel, and steam turned on until its contents boil. The starch, intimately mixed with the remaining half of the dilute acid, is now admitted in a small stream, and the admission so regulated that the liquid in the converter never ceases boiling, by which means an equal distribution of the starch is ensured. The whole is boiled for about five hours, and during the latter part of this time the solution is frequently tested, first with iodine, and, when this ceases to give a reaction, with alcohol. When the addition of two volumes of alcohol to one volume of the cooled solution no longer causes turbidity, the boiling is continued for another half-hour, and this portion of the operation is then terminated. The solution of glucose is then run into another vessel, and the free sulphuric acid it contains neutralised by the addition of whiting mixed with water. This latter is added in small portions at a time to prevent the frothing, due to the escape of carbon dioxide gas, becoming excessive, or there would be a loss of liquid by overflowing. When, on being tested with litmus paper, the liquid shows only a faint acid reaction, it is allowed to remain quiescent until the calcium sulphate has subsided. The clear liquid is drawn off from the precipitate, and that portion which adheres to the latter, removed by filtration. The solution is then evaporated by being passed over an apparatus constructed after the manner of an upright refrigerator with steam circulating inside its tubes, until a concentration of about 58 to 60 per cent. of sugar is reached. During this process some calcium sulphate separates out, which is removed partly by subsidence and partly by filtration, after this the sugar solution is filtered through animal charcoal to remove colour and the more or less unpleasant smell and odour which it possesses. The solution is again subjected to further evaporation in a vacuum pan, until it has a concentration of about 80 per cent., when it is run into shallow vessels, where, under constant stirring, it is allowed to cool. The cooled semi-solid mass is then introduced into conical moulds holding from 30 to 60 lbs., in which it is left to solidify, afterwards it is cut into irregular-sized lumps by a machine in construction very much like a hay-chopper.

Glucose obtained in this manner is a somewhat hard, opaque, sweet substance, varying in colour from pure white to brown, according to the amount of care expended in decolorising it by animal charcoal. It has frequently a bitter and harsh flavour. F. Birneisel gives the following average composition for commercial glucoses:—

	Per Cent.
Glucose	68.0
Dextrin	12.0
Ash	0.2
Water	20.0

Soxhlet has shown that by increasing the amount of liquid to 9 parts of half per cent. sulphuric acid for each part of starch, and heating the mixture for half-an-hour under a pressure of one atmosphere, a much better yield of glucose can be obtained than by the process as usually carried out. In this manner from 95 to 96 per cent. of the theoretical amount of glucose can be obtained. If the solution be evaporated in a vacuum to such a degree of concentration that it has a specific gravity of 1370 to 1380 (water = 1000) at 90° C., or, better still, a gravity of 1400 to 1420, the syrup slowly cooled down to 30° to 35° C., and kept at this temperature until complete solidification has taken place, it is possible to obtain the glucose as the hydrate, $C_6H_{12}O_6 + H_2O$, in a solid transparent form and with a distinct crystalline structure.

Characteristics.—Glucose crystallises from concentrated aqueous solutions, or from methylic or ethylic alcohol, in the anhydrous state, $C_6H_{12}O_6$. According as the process has taken place quickly or slowly, the sugar crystallises in fine needles or well-formed prisms, which belong to the triclinic system. Glucose also combines with water to form the compound $C_6H_{12}O_6 + H_2O$; the combination takes place slowly in the cold, but quickly if the solution be boiled. This sugar is very soluble in water, easily soluble in dilute alcohol, but difficultly soluble in absolute alcohol. Solutions of glucose rotate the polarised ray to the right, and concentrated solutions have proportionally a higher opticity than dilute ones. Tollens has given the two following formulæ, the one for hydrated, the other for anhydrous glucose, by which the true opticity of a

FRUCTOSE.

solution of any degree of concentration may be found. For $C_6H_{12}O_6 + H_2O$—

$$[\alpha]_D = 47.73 + 0.015534P + 0.0003883P^2$$

and for $C_6H_{12}O_6$—

$$[\alpha]_D = 52.50 + 0.018796P + 0.00051683P^2$$

in which P denotes the parts by weight of glucose in 100 parts by weight of the solution.

Solutions of glucose show in a marked degree the phenomenon of birotation. In a freshly prepared solution of glucose the opticity is nearly twice as great as when the solution has stood for some length of time. The birotation is very quickly removed by boiling the solution, two minutes' boiling being sufficient to effect this object, or by the addition of a small quantity of ammonia.

Glucose readily combines with oxygen, and hence is able to reduce the oxides of several metals to the metallic form; whilst in the case of some other metals, such as copper, the higher oxide is reduced to the lower, that is, CuO to Cu_2O. On this property the methods for determining glucose quantitatively are founded (see p. 51). Glucose reduces Barfoed's solution in the cold.[1]

Anhydrous glucose melts without decomposition at a temperature of 146° C. (295° F.); at a higher temperature it suffers decomposition, simultaneously losing water, and glucosan, a body having the formula $C_6H_{10}O_5$, is formed; when submitted to a still higher temperature, the mass gradually darkens in colour, and a dark brown body, caramel, is obtained. Oxidised with nitric acid, glucose yields saccharic and oxalic acids.

Treated with phenylhydrazine acetate, glucose yields an osazone which crystallises in beautiful yellow needles, which begin to separate out of the solution even while it is on the water-bath. Glucosazone melts at 204° to 205° C., and is much more insoluble in either hot or cold water, or in alcohol, than maltosazone.

Glucose readily undergoes the vinous fermentation, yielding in practice 48.67 per cent. of alcohol.

Constitution.—Glucose is an aldehyde sugar belonging to the mannitol group. Its constitution is expressed by the following formula—$CH_2OH \cdot 4(CHOH) \cdot COH$. When acted on by nascent hydrogen, it is converted into sorbitol; when subjected to limited oxidation with bromine, it yields gluconic acid, $CH_2OH \cdot 4(CHOH) \cdot COOH$.

Fructose, Lævulose, or Fruit Sugar.—This sugar, associated with glucose in invert sugar, is abundantly met with in nature.

Preparation.—Fructose can be prepared from inulin, a body which seems to take the place of starch as a reserve material in

---

[1] Barfoed's solution is prepared by dissolving one part of neutral crystallised cupric acetate in fifteen parts of water; to every 200 c.c. of this solution 5 c.c. of acetic acid, containing 38 per cent. of anhydrous acetic acid, are added.

the tubers of certain plants, such as the dahlia, Jerusalem artichoke, &c. This body, on being hydrolysed with a dilute acid, yields fructose as starch, under similar circumstances, yields glucose. It has been shown that there is no series of intermediate bodies between inulin and fructose as there is between starch and glucose. As shown by A. Wohl,[1] inulin, like cane sugar, can be inverted with very minute traces of hydrochloric acid (0.01 per cent. of the weight of ash-free inulin). For the preparation of crystalline fructose, Wohl directs 200 grammes of inulin to be introduced into a 500 c.c. flask with 50 c.c. of water. Hydrochloric acid is then added, in the case of a sample of inulin which contains 0.2 of ash, in the proportion of about one-half the percentage of the ash; when the ash amounts to from 0.2 to 0.4, then in a proportion of two-fifths of that percentage. The whole is heated in boiling water-bath for half-an-hour, reckoned from the time the mass begins to soften; the free acid is then neutralised either by calcium carbonate or sodium carbonate, and the resulting syrup poured into a litre of warm (commercial) absolute alcohol, a little blood charcoal being added to remove colour, and the whole allowed to stand for twelve hours, at the end of which time the clear solution is decanted off from a small quantity of syrup which forms. The clear liquid, after filtration, is evaporated at a gentle heat under reduced pressure to a thick syrup, and this is mixed with three or four times its bulk of absolute alcohol, allowed to stand for twelve hours, and the solution decanted. A crystal of pure fructose is then introduced into the clear solution to start crystallisation, which is further promoted by stirring.

At the end of twenty-four hours pure anhydrous fructose, equal in weight to one-third of the syrup, is deposited. After three days a second crop forms, and by concentrating the solution by evaporation further quantities may be obtained.

Fructose may be also obtained from invert sugar by taking advantage of the fact that it forms a comparatively insoluble compound with calcium hydrate, whilst glucose forms a soluble one.

Properties.—Fructose, crystallised from an alcoholic solution, forms a mass of fine brilliant silky crystals, which sometimes reach a length of 10 millimetres. Their melting-point is 95° C. Fructose rotates the polarised ray strongly to the left, hence its original name, lævulose; but as it has been found that sugars of a similar constitution exist, which are either dextro-rotatory or inactive, it has been proposed by Emil Fischer to discontinue the name "lævulose," and use "fructose" instead.[2] The opticity of fructose is in a high degree dependent on the temperature of its solution; for a temperature of 20° C. with a 10 per cent. solution,

---

[1] *Berichte*, xxiii. 2107.

[2] According to Wiley (*Jour. Amer. Chem. Soc.*, 1896, p. 81), the opticity of fructose at 0° C. (32° F.) is $[a]_D = -108°$, whilst at 88° C. (190° F.) it is equal to that of glucose, the alteration in the rotatory power being strictly

Parcus and Tollens[1] give $[a]_D = -92°$ to $-92.5°$. Its reducing power is somewhat less than that of glucose. According to Soxhlet, 0.05 gramme of fructose in a 1 per cent. solution reduces 9.3 c.c. of Fehling's solution diluted with four volumes of water, whilst, under similar conditions, the same quantity of glucose reduces 10.1 c.c. of that solution. Fructose is completely fermentable by yeast, but does not ferment so rapidly as glucose. Treated with phenylhydrazine acetate, it forms glucosazone identical with that obtained from glucose. Fructose is a ketone sugar, and its constitution may be expressed by the formula, $CH_2OH \cdot 3(CHOH) \cdot CO \cdot CH_2OH$. When acted on by nascent hydrogen it yields mannitol and sorbitol. Fructose is very easily decomposed, mere boiling with water effecting its decomposition, as is evidenced by its solution turning brown.

Cane Sugar or Saccharose.—This, the oldest and best known of the sugars, was originally derived from the sugar-cane (*Saccharum officinarum*), a native of the East Indies, the juice of which contains from 16 to 18 per cent. of sugar. Much sugar is now obtained from the root of the beet (*Beta vulgaris*), the juice of which contains 10 to 12 per cent. of sugar, and, by careful cultivation, this may be increased to 16 per cent. and upwards. It also occurs in the sap of many other plants, such as the sugar millet (10 to 11 per cent.), the stem of the maize plant (7 to 8 per cent.), and is probably one of the first products of assimilation in many plants. Cane sugar is found along with invert sugar in the juice of fruits and in many other parts of plants; also, as shown by Kuhnemann, and later by O'Sullivan, it is one of the sugars of barley and malt.

Properties.—Cane sugar crystallises in colourless anhydrous crystals, which belong to the monoclinic system, and which have a specific gravity of 1.58. It is very soluble in water; a saturated solution at 15° C. (60° F.) contains 66.3 per cent. of sugar, whilst a similar solution at 100° C. (212° F.) contains 82.5 per cent. It is very slightly soluble in absolute alcohol, more so in dilute alcohol, the solubility increasing with the dilution of the alcohol. It is insoluble in ether. The opticity of cane sugar in a 10 per cent. solution is $[a]_D + 66.51$; but is higher than this in solutions more dilute, lower in solutions more concentrated than this. For solutions between 18 and 69 per cent. the following formula may be used: $[a]_D = +66.386 + 0.015035P - 0.0003986P$, in which P equals the percentage of sugar in solution. It has no reducing action on Fehling's solution. Cane sugar melts at 160° C. (320° F.), and, on cooling, solidifies to an amorphous mass, which, after some time, becomes crystalline. When heated for some time to 170° C. (338° F.) it is split up into glucose and fructose (lævulose). When exposed to a higher temperature, 180° to 200° C. (356° to 392° F.), the

proportional to the degree of temperature. This property furnishes a valuable method for estimating fructose in a mixture of other sugars.

[1] *Berichte*, xxiv. 2000.

mass becomes first yellow, then brown; in this latter condition it is known as caramel, a substance much used for colouring purposes. If heated to a still higher temperature, decomposition takes place, with evolution of combustible gases and acid vapours, a light porous mass of carbon being left behind (sugar-charcoal).

When cane sugar is exposed to the action of dilute mineral acids, it suffers inversion, and is converted into the mixture of equal molecules of glucose and fructose known as invert sugar. This process is much accelerated by heat. Invertase, an enzyme found in yeast, has likewise the power of inverting cane sugar; so also has the inverting enzyme contained in malt.

Cane sugar forms compounds (saccharates) with several bases, such as potassium, strontium, lead, &c., all of which are decomposed by carbonic acid. It does not form a compound with phenylhydrazine, but when heated in a solution of the acetate of that base, it first suffers inversion into glucose and fructose, and both these bodies unite with the phenylhydrazine to form glucosazone.

Cane sugar is completely fermentable by yeast, but the fermentation is not a direct one, the sugar being first split up into glucose and fructose by the invertase of the yeast. Many bacteria, when introduced into cane sugar solutions, are able to induce peculiar fermentations, in which such bodies as lactic acid, butyric acid, mannitol, &c., are produced.

Difference between Beet and Sugar-Cane Sugar.—Saccharose derived from the sugar-cane is, especially after inversion, frequently used in the brewery as a partial substitute for malt. There are two varieties of cane sugar met with in commerce; one of these is derived from the sugar-cane, the other from the beet-root; the former of these is the one exclusively used for brewing purposes, because cane sugar, so employed, is invariably used in its raw state; the cost of refined sugar would be prohibitive. The impurities of beetroot sugar are of a nauseous character, whilst those of sugar-cane sugar have an agreeable, full, and luscious flavour; and so much is this the case, that the impure cane sugars are more valuable for brewing purposes than the refined, since the raw varieties yield to beer their luscious flavour. If, however, the cost of refined beet-sugar[1] would permit of its use in the brewery, there could be no objection to its employment there.

Invert Sugar.—This sugar, which may be regarded as a mixture of glucose and fructose, is found very abundantly in nature; forms the chief constituent of honey; it is also contained in the juice of the grape and in many ripe fruits. It arises from the inversion of cane sugar by means of an acid or an enzyme, thus:—

$$C_{12}H_{22}O_{11} + H_2O = C_6O_{12}O_6 + C_6H_{12}O_6$$
$$\text{Cane sugar.} \qquad \qquad \text{Glucose.} \qquad \text{Fructose.}$$

---

[1] A patent has been taken out by Dr. Wohl, of Berlin, for removing the objectionable constituents of raw beetroot sugar by passing a current of air and steam through this substance when in solution.

Preparation.—For this purpose the mineral acids are the most efficient, hydrochloric acid being the most powerful. As pointed out by Wohl,[1] a minute trace of this latter acid is, under certain conditions, able to effect the complete inversion of cane sugar. Thus, when to an 80 per cent. solution of the sugar 0.004 per cent. ($\frac{1}{200}$ of a per cent. on the sugar) of hydrochloric acid is added, and the whole kept at the temperature of boiling water for an hour, with constant stirring, complete inversion takes place, and a pure colourless solution of invert sugar is obtained.

Carbonic acid is able to effect the inversion of cane sugar, for if a solution of this sugar be saturated with carbonic acid, inversion slowly takes place; but if the action is assisted by heat and pressure, complete inversion results in from twenty to thirty minutes.

Properties.—Invert sugar is met with as a thick syrup, from which, after a time, crystals of glucose separate. As invert sugar contains half its weight of fructose, which, at ordinary temperatures, is more powerfully lævo-rotatory than an equal quantity of glucose is dextro-rotatory, it also is lævo-rotatory. The opticity of a solution of invert sugar is much affected by heat; when such a solution is gradually heated its opticity gradually diminishes, until, when a temperature of 87.2° C. is reached, the opticity is 0°; if the solution is heated still further, the opticity becomes right-handed. Tuchschmid has given the following formula for ascertaining the opticity of a solution of invert sugar of any temperature:—

$$[\alpha]_D = -(27.9 - 0.32t)$$

Its opticity is also influenced by the concentration of the solution, and for the correction of this O. Gubbe has given the following formula:—

$$[\alpha]_D = -23.305 + 0.01612 \cdot q + 0.00022391 \cdot q$$

in which q denotes the amount of water in the solution, therefore $100 - q$ denotes the amount of invert sugar.

The glucose constituent of freshly made invert sugar, when the latter is prepared by invertase, exhibits the phenomenon of birotation.

The reducing action of invert sugar on Fehling's solution is not so great as that of glucose; when that solution is diluted with four times its volume of water, 0.05 gramme of invert sugar in a 1 per cent. solution reduces 9.7 c.c. of Fehling's solution. Its κ is given as 96.6.

Heated with phenylhydrazine acetate it yields glucosazone.

Invert sugar (when pure) is completely and directly fermentable by yeast. Its glucose portion, however, ferments much more

[1] *Berichte*, xxiii. 2084.

rapidly than the fructose one, as may be seen from the following table:—

Time.	Glucose.	Fructose.	Difference in Milligrammes.
	Milligrammes per 100 c.c. Fermented.	Milligrammes per 100 c.c. Fermented.	
After 20 hours	557	279	278
After 40 hours	968	532	436
After 64 hours	1196	721	475
After 86 hours	1392	942	450
After 108 hours	1616	1194	422
After 130 hours	1796	1423	373
After 158 hours	1970	1767	203

This difference in the fermentability of the two sugars has been attributed by Bourquelot to the difference in their diffusive powers.

Raffinose.—This sugar ($C_{18}H_{32}O_{16} + 5H_2O$) is found in a very small quantity in beet-roots, and also, as shown by O'Sullivan, in barley. It crystallises in small needles or prisms, is easily soluble in water, slightly so in alcohol, and only possesses a faint sweet taste. Its opticity is $[\alpha]_D + 104.5$, and, like cane sugar, it has no reducing action on Fehling's solution. When heated for a short time with a dilute acid, it is split up into equal molecules of fructose and of a dissaccharide isomeric with milk-sugar called "melibiose." This latter sugar, on prolonged treatment with a dilute acid, splits up into galactose and glucose in equal molecules.[1] These reactions show that raffinose is a trisaccharide, that is, its molecule contains three C_6 groups; the glucose and galactose residues being united in a similar manner to that in which they combine to form milk-sugar, whilst the fructose residue is probably attached in some similar manner to that in which it is united with glucose in cane sugar. On fermentation raffinose behaves differently with the various yeasts; some are able to hydrolyse and ferment it completely, others only to partially invert it to melibiose and fructose, the latter sugar being alone fermented.

α- **and** β-**Amylan.**—These bodies were found by O'Sullivan in barley, wheat, and rye. Barley contains about 2 per cent. of α-amylan and about 0.3 per cent. of β-amylan. They were obtained by extracting barley, first with alcohol, then with water. The aqueous solution was concentrated by evaporation, and strong alcohol added, which precipitated the two bodies. The precipitate was then treated with cold water; this dissolved out the β-amylan and left the α-amylan. The latter was afterwards dissolved in dilute hydrochloric acid, and precipitated therefrom

---

[1] Scheibler and Mittelmeier. *Berichte*, xxii. p. 1678 and 3118.

by alcohol. Both are lævo-rotatory, and have the following opticities :—

$$\alpha\text{-Amylan} \quad \ldots \quad [\alpha]j = -24°$$
$$\beta\text{-Amylan} \quad \ldots \quad [\alpha]j = -73°$$

On being hydrolysed with dilute acids they are said to be converted into glucose.

Gum.—Lintner[1] has isolated from beer, and also from barley and malt, a gum very similar to that which can be extracted from wood by a dilute solution of caustic alkali, and which is named "xylan." It is also found in straw, bran, and brewers' grains. The gum was isolated from beer by treatment with dilute caustic alkali and cupric sulphate, the latter combining with the gum to form an insoluble precipitate. When obtained in the pure state, the gum forms a loose white powder, which is slowly soluble in cold water, but quickly soluble in hot. It does not, however, form a true solution, but the particles are merely finely suspended in the water. Even dilute solutions are very viscous, and filter with great difficulty. The gum is dextro-rotatory, does not reduce Fehling's solution, and is precipitated by lead acetate. Like many other gums, it gives a cherry-red colour with phloroglucin and hydrochloric acid, also a bluish-green colour with orcin and hydrochloric acid; its presence in beer may be demonstrated by these reagents. As shown by Lintner and Düll,[2] it may be regarded as galactoxylan; for when boiled with a dilute acid it is resolved into galactose and xylose. Evidently it is formed by the union of a molecule of galactose with a molecule of the pentaglucose sugar xylose, with the elimination of a molecule of water, thus :—

$$\underset{\text{Galactose.}}{C_6H_{12}O_6} + \underset{\text{Xylose.}}{C_5H_{10}O_5} - H_2O = \underset{\text{Galactoxylan.}}{C_{11}H_{20}O_{10}}$$

Furfural (*Furfur, Bran*).—One of the peculiar properties of the pentaglucoses is that, when boiled with a dilute acid, they yield a body having a peculiar odour, called "furfural."[3] Consequently, when galactoxylan is boiled with dilute hydrochloric acid, furfural is invariably formed. It is also formed to a slight

---

[1] *Brauer und Maltzer Kalendar*, 1890–91, ii. p. 121.
[2] *Zeit. angew. Chemie*, 1891, p. 538.
[3] Furfural is an oily liquid of an aldehydic nature, which boils at 160° C.
and has the constitution
$$\begin{array}{c} HC-HC \\ \| \quad \| \\ HC \quad HC-CHO \\ \diagdown O \diagup \end{array}$$
from which the respective alcohol and acid may be prepared. It is best obtained by distilling a mixture of bran with an equal weight of sulphuric acid diluted with three volumes of water. Furfural gives a beautiful red colour with aniline acetate, and when treated with phenylhydrazine acetate yields furfural phenylhydrazone.

extent during the process of mashing. The free acids of the mash first invert the galactoxylan, and then act on the xylose to form furfural. Furfural can be detected in beer and in distilled spirits, notably in brandy.

Pectin.—This is a substance nearly allied to the gums; it occurs in ripe juicy fruits, such as apples, pears, &c., also in roots, such as turnips, carrots, &c. If the juice of any of these substances be boiled to remove the coagulable albuminoids, filtered, and alcohol and hydrochloric acid added, a jelly-like precipitate of pectin falls. The formula of this body is $C_{32}H_{48}O_{32}$, as given by Fremy, who looks upon it as a transformation product of another body contained in unripe fruits named "pectose." An enzyme, pectase, is said to effect this transformation. Ullick has isolated substances of this nature from beer by rendering it strongly alkaline with caustic soda solution. A precipitate forms, which may be separated into two portions by treatment with dilute hydrochloric acid. The portion soluble in hydrochloric acid is precipitated by the addition of a large quantity of absolute alcohol to the acid solution, the precipitate washed with alcohol, and purified by being dissolved several times in water and reprecipitated therefrom by alcohol. In this way a white body is obtained, soluble in cold water, and having an opticity of $[a]_D = 113°$. The portion difficultly soluble in hydrochloric acid is washed with cold water and dissolved in boiling water. A strongly coloured solution is obtained in this way, which, on being decolorised with sulphurous acid, yields a precipitate, which, when again dissolved in water, has an opticity of $[a]_D = +288°$. Another pectinous body with an opticity of $[a]_D = +36°$ still remains in the sulphurous acid solution. Similar substances have been found in apples, sugar beet roots, and ordinary beet roots. These bodies possess in a high degree the property of yielding viscous solutions, and, as the palate-fulness of a beer is dependent up to a certain point on viscosity, the pectinous bodies must contribute in some degree to its palate-fulness.

The Fatty Constituents of Barley.—Many seeds, such as linseed, rape-seed, castor-oil seeds, &c., contain as a reserve material large quantities of oil, and in no seed is oil or fat entirely absent. Barley contains about 2.5 per cent., rye about 2.0 per cent., whilst linseed contains 37 per cent. and rape-seed 45 per cent. Most fats are compounds of glyceryl and several of the fatty acids, being, in fact, esters or compound ethers. Thus stearic acid combined with glyceryl forms stearin; palmitic acid, palmitin; and oleic acid, olein. When any one of these bodies is subjected to the action of a caustic alkali, this combines with the fatty acid and glyceryl is liberated, which combines with the HO groups of the caustic alkali to form glycerol; thus:—

$$(C_{17}H_{35}COO)_3 \cdot C_3H_5 + 3HONa = 3C_{17}H_{35}COONa + C_3H_5HO$$

Stearin. Caustic soda. Sodium stearate. Glycerol.

THE PROTEIDS OR ALBUMINOIDS.

Such a process is called "saponification," and on it is based the manufacture of soap.

The fat of barley can be readily obtained from the ground corn by extraction with ether. On removal of the ether by evaporation, a yellowish-brown limpid oil is left behind, which possesses an agreeable odour; but this is soon lost and replaced by an offensive one, if the oil is allowed to remain in contact with the air. On standing, the barley-oil separates into two nearly equal portions; one of these is fluid at the ordinary temperature, whilst the other solidifies to a mass of crystals. The composition of barley-fat has been investigated by a number of observers. König and Kieson found it to contain olein, stearin, and palmitin, and also oleic, stearic, and palmitic acids. Lintner, senior, has shown that it contains cholesterol, a monatomic alcohol of the composition $C_{26}H_{43}(OH)$, which is a normal constituent of animal bile. It is an unsaponifiable body, in no way allied to the ordinary fats. It is fatty in appearance, insoluble in water, tasteless and odourless. It dissolves readily in boiling alcohol, in ether or chloroform, and is deposited from boiling alcohol, on cooling, as beautiful white scales, which have a lustre like mother-of-pearl. Cholesterol melts at 137° C., and sublimes at 200° C. A. Stellwag has detected the presence in barley-oil of lecithin, a body containing phosphorus, and which has the following composition, $C_{42}H_{84}NPO_9$. According to this observer, the composition of barley-oil is as follows:—

	Per Cent.
Free fatty acids	13.62
Neutral fat	77.78
Lecithin	4.24
Cholesterin	6.08

F. Beckmann found that on distilling brewers' grains with dilute sulphuric acid a volatile fatty acid passes over with the steam, and is found as a white substance floating on the water in the receiver. He named it "hordeic acid." It is insoluble in water, but readily dissolves in alcohol or ether; it crystallises from this latter solvent in irregular scales, which are unaltered on exposure to the air.

THE PROTEIDS OR ALBUMINOIDS.

A large number of substances occurring in the animal and vegetable kingdom, which are eminently associated with living matter, are included under the term "Proteids," or, as they are often called, the "Albuminoids." These bodies contain the elements nitrogen and sulphur, in addition to carbon, hydrogen, and oxygen. The different proteids have often been submitted to elementary analysis but the results obtained by different observers, when analysing what was presumably the same substance, have not, until recently, been very concordant. This has arisen from the difficulty which existed of obtaining the different members of the proteid group in a state of purity; it is only within comparatively recent times that this has been accomplished with any degree of success.

Reactions of the Proteids.—These bodies give the following characteristic reactions, which may be readily obtained with a solution of egg-albumin, made by mixing a small quantity of white of egg, which contains about 12 per cent. of albumin, with water and filtering the solution.

Precipitation by Nitric Acid.—When strong nitric acid is added to the aqueous solution of any of the proteids, a white precipitate forms, which turns yellow on heating the liquid. The addition of ammonia to the mixture after it has become cold causes the yellow precipitate to become orange-coloured.

Biuret Reaction.—On adding a few drops of a very dilute solution of cupric sulphate to the aqueous solution of a proteid, and afterwards a few drops of a strong solution of caustic soda, a colour is developed which varies with the different classes of proteid. The albumins give a violet colour; the proteoses, a reddish-violet; the peptones, a rose-red.

Millon's Reagent.—This, which is a solution of mercuric nitrate in nitric acid, yields, with proteids, a white precipitate which becomes reddish on boiling the fluid.

Evolution of Ammonia.—The proteids when strongly heated evolve ammonia, produced under the destructive action of heat. The ammonia may be recognised by its changing the colour of a piece of moist litmus paper from red to blue. A smell resembling that of burnt hair is also given off during the heating process.

Formation of Cyanides.—The proteids, when heated with metallic sodium, yield sodium cyanide. The presence of this compound can be detected by extracting the mass with water, adding a few drops of a solution of ferrous sulphate containing a little ferric sulphate, and digesting for a short time. On the addition of hydrochloric acid a blue or bluish-green colour is produced if proteids are present.

Molecular Constitution of the Proteids.—As to the nature of the molecular constitution of the proteids we are still in the dark. Undoubtedly the proteid molecule is a very large one, and, although some insight has been obtained as to the nature of the different molecular groups which combine to form the large complex molecule of the proteid, yet we know comparatively little as to how these groups are actually combined with one another. In the determination of the size of the proteid molecule, as the cryoscopic method is inapplicable, advantage has been taken of certain metallic compounds which the proteids form. For instance, in the compound of egg-albumin with copper it has

EFFECTS OF HYDROLYSIS ON THE PROTEIDS.

been found that $C_{204}H_{322}N_{52}O_{66}S_2$ combine with one atom of copper. From the magnesium, sodium, and calcium compounds, which have been obtained with the crystalloid globulins obtained from various seeds, the formula $C_{292}H_{481}N_{90}O_{83}S_2$ has been deduced. Whether these values are true or only approximative, they serve to show that the proteid molecule is a very large one.

Effects of Hydrolysis on the Proteids.—When proteids are submitted to the action of certain hydrolysing agents, such as superheated steam, dilute mineral acids, caustic alkalies, or the enzyme trypsin, the large proteid molecule becomes finally split up into much smaller and less complex molecules; and it is from the study of these fragments, as it were, that we have obtained some little knowledge as to what is the nature of the molecular groups which enter into the composition of the proteids. By the action of boiling hydrochloric acid, to which a little stannous chloride has been added, which combination possesses an extremely powerful hydrolysing action, the proteid molecule can be broken down into comparatively small decomposition products, such as leucine, tyrosine, aspartic acid, glutamic acid, gluco-protein, lysin, and lysatine. When barium hydroxide is used as the hydrolysing agent, in addition to the formation of the before-mentioned products, much ammonia and carbon dioxide are evolved; and these gases are given off in the proportion of one molecule of water and one of urea, $CO\genfrac{}{}{0pt}{}{NH_2}{NH_2} + H_2O = CO_2 + 2NH_3$. When proteids are acted upon by certain enzymes, the splitting-up process does not extend nearly so far as this. The molecules of the bodies produced are much nearer in size to that of the original proteid molecule. Of such a nature are the proteoses and peptones.

The Various Members of the Proteid Groups.—It has been found possible to isolate the proteids from the substances with which they are associated, and to effect a fairly complete separation of the various members of the proteid groups from one another, by taking advantage of the varying solvent powers which different saline solutions exercise upon them, and also of the property possessed by certain salts of throwing out of solution certain members of the group when the solution is saturated with the particular salt. This latter process is technically known as "salting out," and by the employment of this method we are able to divide the various plant proteids into the following groups:—

Albumins.—Soluble in water, coagulated by heat.

Globulins.—(*a*.) **Vitellins.**—Insoluble in pure water, soluble in dilute saline solutions, coagulable in greater part by heat. (*b*.) **Myosins.**—Insoluble in pure water, soluble in dilute saline solutions, precipitated on saturation of the solution with sodium chloride, coagulable by heat.

Gliadin and Hordein.—Slightly soluble in water, readily soluble in 70 per cent. alcohol.

Glutenin.—Slightly soluble in hot water and hot alcohol, soluble in 0.1 per cent. solution of caustic potash, and in 0.2 per cent. hydrochloric acid, insoluble in saline solutions.

All these bodies, with the exception of the globulins, are obtained in an amorphous condition, in which state they probably exist in the plant. Many of the globulins have been obtained in a crystalloid form, and to some extent at least they exist in this condition in some seeds. Crystalloid proteids can be readily observed in the Brazil-nut and in the seed of the castor-oil plant. Cube-shaped proteid crystalloids occur in the potato, and may readily be seen by placing under the microscope a portion of the layer which is situated immediately below the skin of a boiled potato. If the starch cells are treated with iodine, the proteid crystals, since they are only stained yellow, can be readily distinguished from the blue starch grains.

The Proteoses.—These bodies are derived from the proteids by hydrolysis, and are also characterised by their behaviour when their aqueous solutions are saturated with certain salts. Thus a sharp distinction can be drawn in this way between the proteoses and the peptones, for if an aqueous solution of these two bodies is saturated with ammonium sulphate, the former are precipitated, the latter remains in solution. By taking advantage of similar properties, the group of the proteoses, as will be seen further on (page 166), can be separated into three members, proto-proteose, deutero-proteose, hetero-proteose, &c.

The Proteids of Barley.—Hitherto these bodies have received but scant attention on the part of chemists, the proteids of wheat having much more frequently formed the subject of chemical investigation. Mulder[1] found that barley contained 6 per cent. of albumin and plant gelatin; he obtained the latter by extracting barley-meal with hot alcohol. Von Bibra[2] considered that the proteid constituents of barley were albumin, plant gelatin, and casein, but gave no particulars concerning these substances. Kreusler found that the aqueous extract of ground barley contained an albumin which coagulated on boiling, and that hot 75 per cent. alcohol dissolved a substance which could be subsequently separated into three proteids, gluten-casein, gluten-fibrin, and mucedin, which were supposed to be identical with the bodies, having the same names, that, with another body, gliadin, Ritthausen[3] had isolated from wheat.

Isolation of the Proteids of Barley.—An elaborate series of investigations on the proteids of various grains and seeds have been made during the last few years by T. B. Osborne,[4] and we are indebted to him for much valuable information as to the

---

[1] *Physiol. Chem.*, i. 306.
[2] *Die Getreidearten, &c.*, p. 204.
[3] *Die Eiweisskörper.*
[4] Report of the Connecticut Agricultural Experiment Station, 1894.

nature and properties of the proteids occurring in these substances. The general methods of procedure employed will be briefly described, taking for this purpose his investigations into the proteids of the barleycorn.

Separation of the Albumin or Leucosin.—A large quantity of barley, say six or seven pounds, is ground into fine meal and extracted with 10 per cent. solution of sodium chloride. This dissolves out the albumin, globulin, and the proteose. If an attempt is made to dissolve out the albumin with water and leave the globulin behind, it does not succeed, for there are always a certain amount of soluble salts present in barley or other cereals, which dissolve in the water employed, and yield a dilute saline solution capable of dissolving the globulin. The salt solution thus obtained is next saturated with ammonium sulphate ; this precipitates all the proteids in solution. These, after being filtered off, are again dissolved in 10 per cent. salt (NaCl), the solution filtered and dialysed for several days, when the whole of the globulin is precipitated. Since this body is insoluble in pure water, it is precipitated when the salt has been removed from the solution by dialysis. The precipitated globulin is then collected on a filter, and the clear filtrate gradually heated to 65° C. (149° F.), which has the effect of coagulating and precipitating the albumin. The precipitated albumin is now filtered out, washed with warm water, alcohol, and ether, and dried over sulphuric acid.

Separation of the Globulin or Edestin.—The separation of the globulin is somewhat difficult, since the salt solution extracts a large amount of gum along with it ; and this difficulty is increased by the readiness with which globulin passes into an insoluble condition. It was prepared in as pure a state as possible by dissolving the globulin, obtained in the above-mentioned manner, in a 10 per cent. salt solution, and again dialysing ; the operations of solution and dialysation being repeated a second time. The proteid then separated in the form of small spheroids, which were filtered off, washed with water, alcohol, and ether, and dried over sulphuric acid. The purified substance, when dissolved in a 10 per cent. salt solution and gradually heated, became turbid at 90° C. (194° F.), but no coagulum actually formed until the solution actually boiled, and even then only a small portion of the substance was precipitated.

Separation of the Hordein.—In the first experiments on the isolation of this substance barley-meal was employed, but as this was found to yield a large quantity of colouring matter, ground pearl-barley—that is, barley from which the husk has been removed—was employed. From a quantity of this the bodies removable by a 10 per cent. solution were first extracted, and then sufficient alcohol added to the ground pearl-barley, while still wet with salt solution, to form with the water retained by the meal alcohol of

75 per cent. After this extract had been removed from the meal, another extraction was made with 75 per cent. alcohol. The two extracts were united, filtered, and concentrated to about a third of their bulk by evaporation on the water-bath. The proteid, which separated out as a plastic mass, was, after decanting off the supernatant fluid, macerated with distilled water, this poured off, and the hordein dissolved in 75 per cent. alcohol. The solution of the proteid thus obtained was poured into a quantity of distilled water, when the substance in solution separated. After removal of the water, the precipitated proteid was again dissolved in 75 per cent. alcohol, and the solution poured into a large quantity of absolute alcohol. The proteid did not separate, even after the addition of ether in considerable quantity, owing to the entire absence of mineral salts, these having been completely removed by the previous treatment. The addition of a little sodium chloride solution caused its immediate precipitation, the alcohol and ether retaining any fat which had been extracted along with the proteid. The precipitated hordein was then treated with successive portions of absolute alcohol and dried over sulphuric acid. Obtained in this way, it is a snow-white granular substance, which, when completely free from alcohol, is almost insoluble in cold water, slightly soluble in hot. Hot aqueous solutions do not yield a precipitate on cooling, nor do they coagulate on boiling, though they yield precipitates, not inconsiderable in amount, on the addition of salt. When hordein is dissolved in strong hydrochloric acid, it gives a beautiful crimson solution. When dissolved in a mixture of water and sulphuric acid in equal volumes, its solution has a red colour; a similar solution of wheat gliadin has a purple-red colour.

Proteid Insoluble in Water, in Saline Solution, and in Alcohol.—The quantity of this substance contained in barley-meal was ascertained by estimating the amount of nitrogen the meal contained after having been extracted with salt solution and alcohol; this quantity was then multiplied by the factor 5.9, on the assumption that the proteid substance contained 17 per cent. of nitrogen. Attempts to dissolve out the proteid with a weak solution of caustic potash resulted in failure. The previous treatment to which the barley-flour had been subjected apparently had rendered it almost insoluble, and the small amount which did dissolve was so much contaminated with gum that its further purification was impossible. Barley contains these various proteid bodies in about the following proportions:—

	Per Cent.
Leucosin (albumin)	0.30
Proteose	} 1.95
Edestin (globulin)	
Hordein	4.00
Insoluble proteid	4.50
Total	10.75

Average Percentage Composition of the Proteids of Barley.
—These are the means of a large number of analyses :—

	Carbon.	Hydrogen.	Nitrogen.	Sulphur.	Oxygen.
Leucosin . .	52.81	6.78	16.62	1.47	22.32
Edestin . . .	50.88	6.65	18.10	24.37	
Hordein . . .	54.29	6.80	17.21	0.83	20.87
Insoluble proteid .	Unknown.					

Gluten Group.—It has long been known that, on kneading the dough of wheat-flour under a stream of water, a tough, coherent, elastic mass is left behind, which can be pulled out into strings. This is the gluten, or the body which gives the coherent character to the dough made from wheat-flour, and the presence of which confers on that flour the property of yielding a light porous bread. Gluten cannot be obtained in this way from the meal of any other grain. It has been the subject of much investigation. Einhof[1] in 1805 discovered that a certain portion of the proteid matter of wheat could be extracted with hot alcohol, and this he considered to be identical with gluten. Various other observers worked in the same direction, notably Ritthausen,[2] who was able by fractionation to differentiate the substances extracted from wheat-flour with alcohol into three bodies—gluten-fibrin, gliadin, and mucedin, and a fourth, gluten-casein, which was insoluble in alcohol, but soluble in dilute alkaline solutions. He considered that these four bodies constituted the gluten of wheat, and that it was the gliadin which formed the binding material. This body was found to be absent in the flour of those grains which left no gluten behind on washing.

Weyl and Bischoff[3] have been led to the supposition that these bodies did not exist as such in the seed, but that the proteids of the wheat and other grains consisted chiefly of a globulin substance, which, when the flour was moistened for the purpose of extraction, was converted into the gluten bodies through the agency of an enzyme. Osborne and Vorhees,[4] in an exhaustive examination of the proteids of wheat, have come to the conclusion that this is not the case, but that the gluten bodies exist ready formed, though their number is not so large as had been hitherto supposed, there being but two, gliadin and glutenin. Ritthausen (*l.c.*) considered that there were three bodies belonging to the gluten

[1] *Journ. d. Chemie von Gehlen*, v. 131.
[2] *Die Eiweisskörper.*   [3] *Berichte*, xiii. 367.
[4] *American Chemical Journal*, xv. 392.

group in barley, viz., gluten-fibrin, gluten-casein, and mucedin. Osborne, as we have previously mentioned, finds that there are only two: hordein, which is apparently identical with the substance called mucedin by Ritthausen, but which has almost the same physical and chemical properties as the gliadin obtained from wheat, though it differs from it in composition; the second is the insoluble proteid which it was found impossible to isolate.

The Proteids of Maize.—This corn is now so extensively used for brewing purposes that a knowledge of its proteid constituents is a matter of interest and importance. The maize corn was subjected to an exhaustive examination by Chittenden and Osborne in 1891 and 1892, who found that it contained three globulins, one or more albumins, and a proteid soluble in alcohol. Direct extraction with water dissolved out a myosin-like globulin, but owing to the presence of soluble salts in the maize, the solvent was in reality a dilute saline solution. Further extraction with a 10 per cent. salt solution extracted a vitellin-like globulin. The third globulin was characterised by its extreme solubility in very dilute solutions of various salts, and especially of phosphates and sulphates. The first of these globulins, when dissolved in a 10 per cent. sodium chloride solution, coagulated at about 60° C. (140° F.); the second, when dissolved in a similar solution, hardly coagulated at all under the action of heat, but was precipitated on the addition of acetic acid; the third globulin coagulated in a 10 per cent. salt solution at about 62° C. (144° F.). By prolonged contact with water these globulins were gradually converted into insoluble modifications, apparently albuminates. Water also extracts from maize two albumin-like bodies, the first of which, when its solution is gradually heated, separates at between 60° and 70° C. (140° and 158° F.); a portion of the second separates when the fluid boils, but the whole of this proteid could not be removed in this way, probably because it became transformed into albumose. The only representative of the gluten group found in maize was a body termed "maize-fibrin" or "zein." This substance is extremely soluble in 75 per cent. alcohol, but is insoluble in either water, saline solutions, or dilute alkaline solutions.

The Proteids of Malt.—The proteids of barley undergo considerable modification during the process of malting; a large portion which are insoluble in water become soluble after the barley has germinated. This arises chiefly from the breaking down of the proteids into proteoses by the action of some proteolytic enzyme to some extent analogous to pepsin, since it seems to act best in a slightly acid solution, but resembling trypsin in its ability to degrade the proteids to a further stage than pepsin can.

The proteids of malt have been recently submitted to an exhaustive investigation by Osborne and Campbell.[1] By employing

---
[1] Report of the Connecticut Experimental Station, 1896, p. 239.

processes similar to those which they had used in the examination of the proteid matters of barley, they obtained the following bodies :—

Bynedestin.—A globulin, soluble in very dilute saline solutions, and therefore passing into the aqueous extract of malt, which is in reality a weak solution of the salts contained in the malt. It appears to replace the original edestin of the barley, from which it differs somewhat in composition, since bynedestin contains about 2 per cent. more carbon and 3 per cent. less nitrogen. It has the following percentage composition:—Carbon, 53.19; hydrogen, 6.69; nitrogen, 15.68; sulphur, 1.25; oxygen, 23.19.

Leucosin.—Identical in composition and properties with the albumin of the same name contained in barley.

Protoproteose I. has the same composition as leucosin, from which it is impossible to effect a complete separation. The proteose is precipitated from its aqueous solution by adding an equal weight of alcohol.

Protoproteose II.—Less readily precipitable than No. I. by alcohol. It has the following composition:—Carbon, 50.63; hydrogen, 6.67; nitrogen, 16.69; sulphur and oxygen, 26.01. It differs, therefore, considerably in composition from No. I.

Deuteroproteose.—A body which could not be separated from non-proteid impurities.

Heteroproteose.—A substance found in extremely small amount.

Bynin.—Insoluble in water or saline solutions, but readily soluble in dilute alcohol. It has the following composition:—Carbon, 55.03; hydrogen, 6.67; nitrogen, 16.26; sulphur, 0.84; oxygen, 21.20.

Insoluble Proteid.—This, which amounted to 3.80 per cent. of the total proteid matters, was insoluble in water, saline solutions, or alcohol; consequently it was impossible to study its composition or to determine its properties.

A sample of malt which contained altogether 7.84 per cent. of proteid matters yielded the following quantities of these substances, so far as they could be separated :—

	Per Cent.
Proteid, insoluble in salt solution or in alcohol	3.80
Bynin, soluble in dilute alcohol	1.25
Bynedestin, leucosin, and proteoses, soluble in water and salt solution — Coagulable	1.50
Uncoagulable	1.29
Total proteids	7.84

These results show "that in germination the proteids of barley undergo extensive changes before acquiring the properties of proteoses; that hordein disappears, and an alcohol-soluble body of entirely different composition takes its place; that edestin also disappears, and a new globulin is formed very different both in composition and properties. The albumin, on the other hand, appears to be unchanged in its character, but its quantity is increased. It is to be noted also that hordein and edestin are both replaced by proteids much richer in carbon and poorer in nitrogen."

Coagulation Temperatures of the Proteids of Malt.—Solutions of bynedestin (globulin) become turbid when heated to 65° C. (149° F.); flocks begin to form when the temperature has risen to 84° C. (183° F.); and these gradually increase in amount as the temperature rises to the boiling-point. The whole of the globulin is not, however, completely thrown out of solution even at this temperature, since the addition of dilute hydrochloric acid still produces an abundant precipitate.

The albumin is completely precipitated at a temperature of 59° C. (136° F.); consequently, this substance will be coagulated and left in the mash-tun. The bynin and insoluble proteid, being both insoluble in water or weak saline solutions, are of no moment to the brewer, though it is probable that, during their stay in the mash-tun, they may be converted to some slight extent into proteoses by the proteolytic enzymes of the malt.

At present we are unable to make use of the indications afforded by the relative proportion of the nitrogenous matter of malt contained in these groups, because so far no observations have been carried out to show how malts, in which these relative proportions differ from the normal, behave in the production of beer. But it is fair to presume that if this subject were thoroughly investigated in the direction indicated, important criteria would be established for the valuation of samples of malt.

According to the experiments of Hantke,[1] the following percentages of the total amount of the nitrogenous matter of a malt which he examined passed into solution at the following temperatures:—

	Per Cent.
16 hours' mash at 72° F.	31.78
1 hour's mash at 100° F.	52.16
1 hour's mash at 122° F.	37.51
1 hour's mash at 145° F.	32.75
1 hour's mash at 167° F.	32.80

Consequently the largest quantity is extracted at 100° F.

Of the total nitrogenous matter taken into solution by mashing at the following temperatures, the following percentages were precipitated on boiling the wort for an hour:—

	Per Cent.
16 hours' mash at 72° F.	21.12
1 hour's mash at 100° F.	15.24
1 hour's mash at 122° F.	14.18
1 hour's mash at 145° F.	22.80

---

[1] *Brewer and Maltster*, 1895, p. 1148.

THE AMIDE CONSTITUENTS OF PLANTS.

From this it appears that at a mashing temperature of 122° the smallest amount of nitrogenous matter coagulable on boiling is extracted; apparently, therefore, this temperature is the most favourable one for peptonisation.

THE AMIDE CONSTITUENTS OF PLANTS.

In addition to the proteids all vegetables contain another set of nitrogenous bodies, the amides, which, being of a crystalline nature, are eminently diffusible. We have already mentioned some of these bodies as being amongst the products of the digestion or breaking down of the proteids, but it is extremely probable that the amides in the living plant form a portion of the materials from which the proteids are constructed. The amides themselves are probably produced by a combination of the deoxidisation products of the chlorophyll cells with nitrogen, derived from the nitrates and ammonia contained in the fluid contents of the cells.

Asparagine.—One of the most frequently occurring of these amide bodies, and the one which has been longest known, is asparagine, so called because it was first found in asparagus sprouts. It forms large colourless crystals, readily soluble in water, but insoluble in alcohol. It has the following composition: $C_4H_8N_2O_3 + H_2O$. An aqueous solution of the ordinary asparagine rotates the polarised ray to the left, its opticity being $[\alpha]_D = -6.23°$, but a dextro-rotatory asparagine which rotates to the same angle in the opposite direction has been found in tares by Piutti. The dextro-rotatory body has a distinctly sweet taste; the lævo-rotatory is almost tasteless. When boiled with dilute acids, asparagine is readily transformed into aspartic acid and ammonia, thus: $C_4H_8N_2O_3 + H_2O = C_4H_7NO_4 + NH_3$. The same reaction ensues when it is boiled with a solution of a caustic alkali; only in this case the aspartic acid itself becomes partially hydrolysed to malic acid and ammonia, and even in simple aqueous solution asparagine suffers slight hydrolysis on prolonged boiling. Asparagine, when acted on by nitrous acid, suffers decomposition, and yields half its nitrogen in the gaseous state, according to the following equation:—

$$C_4H_8N_2O_3 + HNO_2 = C_4H_6O_5 + NH_3 + 2N$$
Asparagine.   Nitrous acid.   Malic acid.   Ammonia.   Nitrogen.

Sachsse[1] has based a method for the estimation of asparagine, and the amides generally, on this reaction. Asparagine is precipitated by mercuric nitrate, and the compound so obtained may be

---
[1] *Agricult. Chemie*, p. 390.

decomposed into asparagine and mercuric sulphide by the action of hydrogen sulphide.

Aspartic Acid.—This is the crystalline body, slightly soluble in water, which rotates the polarised ray to the right or left, according as the asparagine from which it was obtained was dextro- or lævo-rotatory. Treated with nitrous acid, aspartic acid yields up the whole of its nitrogen, thus :—

$$C_4H_7NO_4 + HNO_2 = C_4H_6O_5 + H_2O + 2N$$
Aspartic acid. Nitrous acid. Malic acid. Water. Nitrogen.

Glutamine.—This is a crystalline body having the following composition, $C_5H_{10}N_2O_3$. It readily suffers decomposition, being readily split up by the action of alkalies or acids, or by prolonged boiling of its aqueous solution, into glutaric acid and ammonia. In consequence of its unstable nature, glutamine is seldom met with in plants, though the frequent presence of glutaric acid leads to the presumption that this latter body is a decomposition product of glutamine. When treated with nitrous acid, glutamine yields up the whole of its nitrogen, thus:—

$$C_5H_{10}N_2O_3 + 2(HNO_2) = C_5H_8O_5 + 4N + 2(H_2O)$$
Glutamine. Nitrous acid. Oxyglutaric acid. Nitrogen. Water.

Glutaric Acid is a crystalline substance, slightly soluble in water, its solution being dextro-rotatory. A lævo-rotatory acid is also known. The former acid is found amongst the products of the hydrolysis of the proteids.

Tyrosine.—This body, which is hydroxyphenylamido-propionic acid, $C_6H_4{<}{OH}{>}CH(NH_2)COOH$, crystallises in fine brilliant silky needles, soluble with difficulty in cold water, slightly more so in hot water, but insoluble in alcohol. Its aqueous solution rotates the polarised ray to the left; a dextro-rotatory modification is also known. Treated with nitrous acid, it yields up nearly the whole of its nitrogen. It has long been known to be one of the products of proteolysis, and is frequently met with in germinating seeds.

Leucine or Amido-Caproic Acid.—An amide present in the products of proteolysis, and which is also found in germinating seeds. It consists of brilliant silvery-white plates, soluble in about twenty-seven parts of cold water, and slightly soluble in alcohol. Its aqueous solution is dextro-rotatory. It has been prepared synthetically. When treated with nitrous acid, it yields up the whole of its nitrogen, thus :—

$$C_6H_3NO_2 + HNO_2 = C_6H_{12}O_3 + H_2O + 2N$$
Leucine. Nitrous acid. Hydroxycaproic acid. Water. Nitrogen

THE ENZYMES.

Vernine.—An amide occurring in plants and having the composition $C_{16}H_{20}N_8O_8 + 3H_2O$. It has been isolated by Schulze and Bosshard. It crystallises in brilliant silky needles, dissolves with difficulty in cold water, readily in hot, but is insoluble in alcohol. Vernine, when heated with hydrochloric acid, yields guanine $C_5H_5N_5O$.

Betaine.—Another body of an amide nature, but which perhaps ought rather to be classed with the alkaloids. It has been isolated by Schulze from beet-roots. Its formula is:—

$$\begin{array}{c} CH_2\text{---}CO \\ | \qquad\qquad \rangle \\ N(CH_3)_3\cdot O \end{array}$$

Choline.—An alkaloidal substance which possesses a bitter taste, is of an exceedingly hygroscopic nature, and has an alkaline reaction. Its composition is $HO\cdot CH_2\cdot CH_2\cdot N(CH_3)_3\cdot OH$. It readily combines with carbon dioxide, and can be formed artificially from trimethylamine and glycol. It has been found in hops, in wort, and in beer.

The Amides Occurring in Malt.—Ullik was able to detect hypoxanthine, guanine, and vernine in wort and beer, also probably leucine and tyrosine, but he was unable to find either asparagine or glutamine. Amthor found vernine and another substance which appears to have a composition intermediate between xanthine and guanine. Düll[1] found the amine body choline in barley and malt. Asparagine appears to be contained in malt rootlets, but not in the malt itself.

THE ENZYMES OR HYDROLYSTS.

Enzymes.—There are found in all living organisms, be they animal or vegetable, certain remarkable nitrogenous bodies which are either actually albuminoids or are very closely allied to them. They are soluble in water and are coagulated by heat. Each one possesses the power of splitting up the complicated compound molecule of some particular organic body, such as starch, albumin, &c., when the enzyme and the substance are brought into intimate contact with each other under certain conditions, an absolutely necessary one of these being the presence of water. These nitrogenous bodies are often spoken of as the "unorganised ferments" or "enzymes," and the transformations they effect termed "fermentative processes." Since, in the changes which are brought about through their agency, water is almost invariably added to the molecule of one or both of the newly-formed substances, the process has

[1] *Chem. Zeit.*, 1893, p. 67.

been called "hydrolysis," and the agents concerned in the action "hydrolysts."

The selection of the name "ferment" for these bodies, and that of "fermentation" for the processes they induce, was an unhappy one, since these terms had been associated for a very long period with an entirely different set of bodies and processes, *i.e.*, the organised ferments, such as yeast, bacteria, &c. So much is this the case, that when fermentation is spoken of, one is extremely apt to connect this term with the older process denoted by that name, and hence confusion arises. When an organised ferment, such as yeast, acts upon sugar in solution, the appearances of boiling are more or less present; and since the word "fermentation" is derived from the Latin word *fervere* (to boil), it would seem much the more reasonable course to reserve this word for the designation of those processes with which it has been associated the longest, and which are more in accordance with its original signification. The word "ferment" would always then mean a living organism, and "enzyme" or "hydrolyst" an unorganised body. Of the two names for bodies belonging to the latter class, "hydrolyst"[1] is undoubtedly the better and more scientific one, since it denotes the action; and by combining with the word "hydrolyst" the name of the substance on which each particular enzyme acts, we obtain a distinctive name for each class. Thus the enzymes which act upon starch are called amylo-hydrolysts; those that act on proteids, proteo-hydrolysts, &c.

Action of Enzymes Similar to that of Acids.—Actions similar to those caused by the vegetable hydrolysts can be induced by several inorganic agents. Sulphuric, hydrochloric, oxalic, and many other acids, apparently act upon starch in a precisely analogous manner to that in which the amylo-hydrolyst diastase acts. During the process the acid neither suffers loss nor decomposition, and can be entirely recovered at the end of the experiment in an absolutely unaltered condition. Several metallic oxides, such as those of copper, cobalt, &c., behave in a similar manner with certain compounds which contain oxygen, such as hydrogen peroxide and the hypochlorites. They decompose these bodies when in solution, with the liberation of oxygen, the metallic oxides themselves suffering no change whatever in the process.

Purposes for which Enzymes are Produced.—The important work in nature assigned to the hydrolysts (enzymes) is the conversion of insoluble bodies into soluble and diffusible ones, or to render diffusible those bodies of a non-diffusible nature which already exist in solution. If these necessary changes did not take place, many substances could not be made use of by animal or vegetable organisms. For instance, when germination commences in the grain of barley, it is necessary that the starch

---

[1] Suggested by Professor Armstrong (*Journal of the Chemical Society*, 1890, p. 528).

stored up in its endosperm should be able to travel to the growing germ. This is first effected by an enzyme or amylo-hydrolyst (the translocation diastase) already present in the seed, afterwards by another hydrolyst (ordinary diastase), which is formed in considerable quantity during germination. The insoluble proteid substances in the grain are also rendered soluble and diffusible in a similar manner by another set of enzymes, the proteo-hydrolysts. But enzymic action is not confined to the process of germination alone, for enzymes are to be found in every part of the living plant, leaves, buds, tubers, stem, &c.

One of the most striking facts connected with the enzymes is that an exceedingly minute quantity of any one of them is able to transform an enormous amount of the substance on which it is capable of acting. The action of the hydrolysts has not yet received an absolutely certain explanation; the most probable conjecture is, that they are bodies whose molecules are in a high state of vibration, and that they are able to transmit these vibrations to other bodies in their immediate neighbourhood with such force as to overcome the stability of the molecules of the latter bodies. These are, as it were, shaken into bodies which possess a simpler and more stable molecular condition, upon which the vibrations of the enzyme are no longer capable of exercising a disruptive action.

Enzymes Derived from the Albuminoids.—If the hydrolysts are not themselves albuminoids, they are undoubtedly near derivatives of them, being most probably oxidation products. It is, however, somewhat doubtful whether they are really independent entities, or merely albuminoids which in a certain phase of their development temporarily assume this power. Amylo-hydrolysts are found in some organs of the plant which never contain starch, and where, in such cases, their presence is totally inexplicable unless they are to be made use of afterwards in the nutrition of the plant. It can hardly be imagined that the energy and material necessary for the production of these bodies would be expended for a totally useless purpose. The change from albuminoid to enzyme does not seem to be of a very profound nature. Baranetzky found that an aqueous extract of the resting tubers of the potato did not possess any amylolytic power when freshly prepared, but that, after remaining for a few days exposed to the air, the solution acquired this property, which it, however, again lost on further keeping. That this alteration could not be attributed to the entrance of bacteria is shown by the fact that the solution remained tasteless and odourless all the way through the experiment, and at the end of it possessed the same slight degree of acidity which it did at the commencement. According to Sachsse,[1] the simplest view to take of the hydrolysts, if we do not regard them as albuminoids, is to

---

[1] *Agricult. Chemic*, p. 364.

look upon them as near derivatives of these bodies, readily formed from them by the action of oxygen, and which just as readily resume their original condition. Viewed in this light, they are not exclusively formed for the specific purpose of inducing hydrolytic changes, but are general and normal products of the albuminoids produced by respiration in the plant. That oxidation plays an important part in the formation of the hydrolysts in the seed is shown by the well-known fact that seeds, when caused to germinate in a thick layer to which the oxygen of the atmosphere can only obtain but limited access, produce much less diastase than similar seeds do when they are spread out in a thinner layer and more freely exposed to the air. The maximum quantity of the enzyme is formed in seeds sown under natural conditions in the earth, since in this case each seed is isolated and secures an abundant supply of oxygen.

It has been shown by Reychler that when bodies belonging to the gluten-casein group are subjected to the action of dilute acids, a body possessed of amylolytic properties is formed, to which he gives the name of "artificial diastase." Lintner and Eckhardt, who have also investigated this subject, corroborate the results of Reychler, and find that this is especially the case with mucedin. When four grammes of mucedin were dissolved in 100 c.c. of 0.1 per cent. acetic acid, and the solution immediately tested for diastatic power, this was found to be $35.7°$. On standing at a temperature of $35°$ C. ($95°$ F.), the amylolytic power of the solution gradually increased, until at the end of twenty-one days it had risen to $83.3°$. A similar and equal effect was produced on the mucedin in ten days when dissolved in a 0.1 per cent. solution of hydrochloric acid, or in a 0.2 per cent. solution of acid potassium phosphate (H_2KPO_4). This effect, they suggest, may be caused by the action of the acid on an unknown "enzymogen"[1] present in the seed.

Brown and Heron have shown that when an aqueous extract of barley is submitted to the action of yeast for a few hours at a temperature of $30°$ C. ($86°$ F.) its power of transforming starch is much increased. This they attribute to some change brought about in the soluble albuminoids by the yeast.

In a state of complete dryness the enzymes will bear exposure to a temperature of $100°$ to $125°$ C. without losing the whole of their characteristic properties.

The Various Groups of the Enzymes.—Although the general action of the whole of the enzymes may be broadly defined as one of hydrolysis, yet each particular enzyme can only act upon a certain class of substances. Thus, diastase readily hydrolyses starch, but has no action on proteids; similarly pepsin, which hydrolyses proteids, is unable to affect starch. The enzymes

---

[1] Enzymogen is a body which under the action of an acid becomes converted into an enzyme.

THE ENZYMES.

may be divided into the following seven groups, in which they are arranged according to their respective specific actions :—

GROUP I......Diastatic Enzymes :—
    Diastase of secretion (malt)
    Translocation diastase (barley) } Convert starch into maltose and dextrin.
    Ptyalin (saliva)
    Glucase (maize) } Converts starch finally into glucose.

GROUP II.....Cyto-hydrolytic enzymes :—
    Cytase (malt)
    Enzymes of seeds in which the reserve material is cellulose } Transform cellulose into sugars, such as mannose, xylose, &c.

GROUP III...Pectin enzymes :—
    Enzymes which convert pectinous substances into vegetable jelly.

GROUP IV....Inverting enzymes :—
    Invertase (yeast)
    Invertase (malt) } Convert cane sugar into invert sugar.
    Glucase (yeast)
    Enzyme of the small intestine } Transform maltose into glucose.
    Enzymes of yeast which degrade the intermediate dextrins into maltose; these are especially present in the wild yeasts.
    Enzyme of kephir . . . Inverts milk-sugar.
    Probable enzyme of germinating barley } Converts maltose into cane sugar.

GROUP V.....Proteolytic enzymes :—
    Enzymes of malt and other vegetables (sometimes called peptase), but which have not yet been isolated } Convert proteids into proteoses, peptones, and amides.
    Trypsin (pancreas)
    Pepsin (stomach) } Converts proteids into proteoses and peptones, but not into amides.
    Peptonising ferments secreted by many bacteria.

GROUP VI....Glucosidal enzymes :—
    Emulsin (bitter almond) } Splits up amygdalin into oil of bitter almonds, hydrocyanic acid, and water.
    Many other enzymes which have the power of hydrolysing glucosides.

GROUP VII...Zymase of yeast :—
    Splits up sugar into alcohol and carbon dioxide.

The Diastases or the Amylolytic Enzymes.—The substance of this nature, which has been known for the longest time, is the diastase found in germinated barley; it was also the first enzyme isolated, though it was only obtained in an exceedingly impure condition. Since then it has been discovered that enzymes of this class are widely disseminated throughout the vegetable kingdom; scarcely a plant, or part of a plant, has been examined without detecting their presence. As has been shown by Lintner and Eckhardt, and by Brown and Morris, two distinct forms of diastase exist; one of these is present in considerable quantity in ungerminated barley, the other, which is secreted during the germination of barley, is largely present in malt.

Isolation of Diastase.—Numerous attempts have been made from time to time to obtain malt diastase in a state of purity, the first investigations on record in this direction being those of Payen and Persoz. Their method was as follows:—A strong cold-water[1] extract of malt was heated for half-an-hour to a temperature of 70° C. (158° F.), in order to coagulate certain albuminous matters; the extract so treated was then filtered off, and a large quantity of absolute alcohol added to the filtrate, which precipitated the diastase. The precipitate thus obtained was again dissolved in water, re-precipitated by alcohol, and dried at a temperature of 40° to 50° C. (104° to 122° F.) A white tasteless and odourless powder was obtained in this way, which, upon solution in water, was found to possess amylolytic properties. The diastase produced in this way is very weak in its action, since it suffers great deterioration at the high temperature to which the malt-extract is subjected. Various other attempts have been made from time to time to improve the methods of isolating diastase, such as that of extraction with glycerin, followed by precipitation by alcohol; precipitation by common salt, by acetate of lead, &c., but all these methods have proved unsatisfactory. A better process, which is in reality a modification of the method of C. O'Sullivan published in 1884, was proposed by C. J. Lintner. Lintner directs one part of powdered barley-malt or of sieved green malt to be digested for twenty-four hours or longer with two parts of 20 per cent. alcohol.[2] The solution is then filtered off, and twice, or at most two and a half times its volume of absolute alcohol added, with constant stirring. More than this quantity of alcohol should not be added, or slimy matters, possessed of little diastatic power, are simultaneously precipitated. The precipitated diastase quickly settles in yellowish-white flakes, from which the supernatant liquid is poured off, the precipitate brought on to a filter, and the alcoholic fluid removed as quickly as possible. The residue is removed from the filter-paper, placed in a mortar and well triturated with absolute alcohol, again placed on the filter,

---

[1] By cold-water extract of malt is meant a solution of the soluble portions of malt in cold water. It is prepared by grinding malt to a fine powder in a mill and mixing it with cold water in a beaker or other vessel. The mixture, after standing for a certain length of time, during which it is occasionally stirred, is filtered off. Malt-extract contains in solution the diastase and other soluble nitrogenous constituents, the gum, and the ready-formed sugars.

[2] Syilágyi employs 30 per cent. alcohol, so as to dissolve less inert proteid matter.

the alcohol removed and the precipitate washed with absolute alcohol. The precipitate is brought once more into the mortar, triturated with anhydrous ether, placed on the filter, and the ether removed as far as possible by the suction-pump, after which the precipitate is dried in a vacuum over sulphuric acid. The above-mentioned dehydration of the diastase by alcohol and ether is absolutely necessary in order to obtain the substance in the form of a loose powder, easy of solution, and possessing the maximum amount of diastatic power. If the dehydration is incomplete, the preparation dries to a tough horny mass, which is only partially soluble, and is possessed of little diastatic power. By repeated solution in water and re-precipitation by alcohol the enzyme may be entirely freed from carbohydrate matters.[1] The mineral constituents, which in the freshly precipitated preparation amount to 16 per cent., are exceedingly difficult to remove, but can be reduced by dialysis to a little less than 5 per cent. The most powerful diastase which Lintner was able to prepare in this manner, and presumably the purest, had the following percentage composition:—C. 46.66, H. 7.35, N. 10.41, S. 1.12, O. 34.46. It differed, therefore, considerably from the ordinary albuminoids in composition, being much richer in oxygen and poorer in nitrogen.

A most important series of investigations on diastase have recently been carried out by J. B. Osborne,[2] which have considerably advanced our knowledge of this substance. The diastase was prepared in the following manner:—Ten kilogrammes of ground malt were extracted with water, and the malt-extract thus obtained saturated with ammonium sulphate. This had the effect of precipitating the whole of the proteids in solution. The proteids were then suspended in four litres of water, and dialysed until much of the ammonium sulphate had been removed. The dialysed solution, after being filtered off from a residue consisting chiefly of globulin which had become insoluble, was again saturated with ammonium sulphate. The precipitate so obtained was suspended in 1500 c.c. of water, submitted to dialysis, and filtered off. Most of the globulin was converted by this treatment into the insoluble modification and remained behind. The solution of the malt proteids, *minus* nearly the whole of the globulin, was now dialysed into alcohol of specific gravity 0.84. In forty-eight hours the precipitate which had formed was removed; it was called "Precipitate No. 1." The solution was again dialysed into alcohol of the same strength for another forty-eight hours, when another precipitate formed; this was removed and called "Precipitate No. 2." The filtrate from this was dialysed into a rather larger quantity of somewhat stronger alcohol, when "Precipitate No. 3" formed and

---

[1] As shown by its solution, after being boiled for some time with a dilute acid, not being able to reduce Fehling's solution.

[2] Report of the Connecticut Agricultural Experimental Station, 1894, p. 192; also *Jour. Amer. Chem. Soc.*, 1895, p. 587.

was filtered off. The filtrate from this was similarly treated when "Precipitate No. 4" was obtained. To the filtrate from this a large quantity of alcohol was added, which threw down "Precipitate No. 5." Precipitate 1 was contaminated with much colouring matter, No. 2 less so; the others were nearly colourless. Their weights were approximately: No. 1, thirteen grammes; No. 2, eight grammes; No. 3, six grammes; No. 4, five grammes; No. 5, three grammes—altogether, thirty-five grammes.

Precipitate No. 1 was further purified by solution in water, when a large amount of insoluble matter was left behind; this was removed by filtration. The filtrate from this was first dialysed in water for several days to remove salts, and then dialysed in strong alcohol until nearly the whole of the proteids were thrown down. The precipitate thus obtained, after drying, weighed 2.11 grammes, and was almost entirely soluble in water. It had a diastatic power of 30° on Lintner's scale; its solution, when slowly heated, became turbid at 65° C., and deposited flocks at 70° C. After this coagulum had been filtered off, the fluid gave a strong biuret reaction, showing that the preparation consisted largely of proteoses.

Precipitate No. 2, after similar purification, yielded a powder almost completely soluble in water. Its solution, when heated, became turbid at 60°, and deposited flocks at 66°. Its diastatic power was 75.

Precipitate No. 3, after subsequent purification, yielded 3.0 grammes of a substance almost completely soluble in water. Its solution, when heated, became turbid at 55° C., and flocculent at 60° C. The filtrate from this gave a strong biuret reaction, indicating the presence of much proteose. The diastatic power of the preparation was 222.

Precipitate No. 4 was dissolved in water, filtered off, and dialysed for several days in water, afterwards in alcohol. Absolute alcohol was then added to the contents of the dialyser, when a precipitate weighing 4 grammes was obtained. Its solution, upon heating, became turbid at 50° C., and gave a large coagulum at 56° C. The filtrate from this gave a biuret reaction, showing that proteoses were present.

This preparation had a diastatic power of 600, and was the most powerful diastase which had ever been prepared. It was found to contain 0.66 per cent. of ash, and when subjected to analysis yielded the following results, which are corrected for ash. They show that it is a true proteid:—

	Ash-free.
Carbon	52.50
Hydrogen	6.72
Nitrogen	16.10
Sulphur	1.90
Oxygen	22.78
	100.00

The amount of sulphur is somewhat large for a proteid, but this was thought to be due to the retention of a trace of ammonium sulphate in the preparation.

Precipitate No. 5 dissolved completely in water, and, after being washed with absolute alcohol, weighed 2.87 grammes. Its aqueous solution, when heated, became turbid at 50° C., and yielded a small coagulum at 58° C. The filtrate from this gave a strong biuret reaction, showing that it contained much proteose. Its diastatic power was 15.

Nature of Diastase.—The purified precipitate from No. 4 was now closely investigated. Its coagulating point agrees exactly with that of the albumin (leucosin) obtained from wheat, rye, and barley, and it has also the same chemical composition. The aqueous extracts of all these grains have a powerful diastatic action. The amount of coagulable albumin in the purified preparation amounted to 53.2 per cent. These facts strongly point to the albumin as being the active substance, yet it was not found possible to establish any numerical relation between the amount of albumin contained in a substance and its diastatic power, though in no case was a diastatically active substance ever found which was entirely free from albumin. If the activity of the preparation were solely due to albumin, then the malt ought to have shown a much higher diastatic power than it did, judging from the amount of coagulable albumin which it contained. These facts led Osborne to surmise that diastase is a compound of albumin with some other substance, presumably a proteose. Kühne, in his attempts to isolate trypsin, had been led to a similar conclusion with reference to the nature of that body. These are, however, merely suppositions, for which as yet there is no direct evidence.

Properties of Osborne's Preparation of Diastase.—Its diastatic power in comparison with other so-called pure enzymes is very great; it is able within one hour to yield with soluble starch, at the temperature of 20° C. (68° F.), 2000 times its weight of maltose, together with an undetermined amount of dextrin. After being dried over sulphuric acid and kept for six months, it lost half its diastatic power; but even then it was able in seventeen hours to produce, at 20° C., 10,000 times its weight of maltose, besides an unknown quantity of dextrin. At 45° C. (113° F.) it produced a similar quantity of maltose in one hour, but at 50° C. (122° F.) much less maltose was formed, and at 55° C. (131° F.) very little.

In a solution of the preparation, which was compared with a malt-extract that had the same diastatic power, the liquefying power was found to be less; 5 c.c. of the malt-extract added to 10 c.c. of 2 per cent. starch-paste dissolved the starch completely in eight minutes, whilst 5 c.c. of the solution of the preparation required thirty-seven minutes. The malt-extract was also found to be more energetic in converting starch into bodies which gave no colour with iodine. For instance, 5 c.c. of the malt-extract added to a 10 c.c. solution of soluble starch caused the disappearance of the blue reaction with iodine in thirteen minutes, while it took thirty-eight minutes to arrive at the same stage with 5 c.c. of the dissolved preparation. To two portions of a starch-paste, each measuring 10 c.c., there was added to the one 5 c.c. malt-extract, and to the other the same amount of malt-extract which had been previously boiled and cooled, and to which a portion of the prepared diastase had been added, equivalent in action to that contained in the malt-extract. In the former case the starch was liquefied in seven minutes, in the latter in fourteen minutes, whilst thirty-seven minutes were required to produce the same result with a simple aqueous solution of the preparation. The solution of the diastase in boiled malt-extract was found just as active in converting starch into bodies which gave no colour with iodine as unboiled malt-extract. Evidently, then, the difference first observed was due to the conditions of the experiment (presence of amides, mineral salts, &c.), and not to a difference in the power of the enzyme. It has been shown by Effront that one of the amides, asparagine, has the power of considerably increasing the activity of diastase.

Test for Diastase.—Diastase, according to Lintner, responds to all the tests characteristic of the albuminoids, with the exception of that known as the biuret reaction. It gives, however, a deep blue colour with an alcoholic solution of gum guaiacum and hydrogen peroxide, a reaction which is given by no other proteid bodies but the enzymes. The best way of performing this test is to prepare a 1 per cent. solution of gum guaiacum in absolute alcohol, and this should always be prepared immediately before use. Ten c.c. of this solution are placed in a test-tube, a few drops of ordinary commercial hydrogen peroxide solution added, and, if this addition causes cloudiness, alcohol is added until the fluid again becomes clear. A single drop of a 0.05 per cent. solution of diastase shaken up with the mixture gives almost instantaneously a deep blue colour. So delicate is the test that the presence of diastase in the steep water of the barley cistern can be readily detected by its agency.

The Multiple Nature of Malt Diastase.—It is highly probable that the diastase of malt is not one single individual body, but a mixture of several diastases, for it has been shown that, when portions of the same solution of diastase are heated to different temperatures, and then allowed to act upon starch-paste, the products of their action are very different. Brown and Heron found that when the malt-extract, obtained by treating 200 grammes malt with 250 c.c. water, was slowly heated, a turbidity, coincident with coagulation, first appears in the fluid when a temperature of 46° C. (115° F.) was reached, and that if it was kept at this temperature for fifteen to twenty minutes, the maximum amount of coagulation for that particular temperature took place. Upon filtering off the precipitated matters and again heating the clear liquid, a fresh coagulation, reaching its maximum in fifteen to twenty minutes, commenced when a temperature slightly above 46° C. (115° F.) is reached. On repeating the process, it was found that with each successive rise in temperature a fresh coagulation took place, until, when a temperature of 95° C. (203° F.) was reached, all matter capable of coagulation by heat had been thrown down. The following table shows the quantities of coagulable matter which separated at different temperatures from 100 c.c. of malt-extract of the strength given above:—

Temperature.	Grammes of Proteid Matter Coagulated.	Percentage of Total Amount.
50° C. (122° F.)	0.044	19.1
60° C. (140° F.)	0.123	34.4
66° C. (151° F.)	0.155	13.9
66° C. (169° F.)	0.186	13.4
100° C. (212° F.)	0.230	19.2
		100.0

They state that every stage in the coagulation of malt-extract by heat is attended with a distinct modification of its amylolytic power, and, conversely, that they had never been able to discover any modification in starch-transforming power which was not attended with distinct coagulation.

Since at 80° to 89° the diastatic power of malt-extract is completely destroyed, and this is the temperature at which nearly the whole of the coagulable proteids were thrown down, they concluded that the diastatic power was a function of the soluble proteids, and was not due to a distinct agent. They found that, when malt-extract is filtered under slight pressure through a porous earthenware battery cell once or twice, all the proteids were retained within the cell, and that the filtered liquid was as completely deprived of its diastatic power as if it had been boiled.

Barley Diastase.—The fact that ungerminated grain contains an amylolytic enzyme has been known for a long time; its presence in the gluten of wheat had been detected as early as 1812 by Kirchoff. This fact had been lost sight of until quite recent times, when experiments were made on ungerminated grain, and its diastatic power ascertained by Lintner's method, in which soluble starch is used instead of starch-paste. It was then found that the diastatic power of ungerminated grain often equalled, and sometimes surpassed, that of ordinary malt. The diastase of barley has not as yet been isolated, consequently our knowledge concerning it is limited to its action upon starch. In this barley diastase differs widely from malt diastase; its power of liquefying gelatinised starch is very much inferior to that of malt diastase; the latter exerts as much liquefying action on a solution of gelatinised starch in eight minutes as an equivalent quantity of barley diastase does in twelve hours. It is also unable to corrode starch granules; this and the power of liquefying gelatinised starch appear, according to Brown and Morris, to go hand in hand. Lintner and Eckhardt[1] found considerable differences in the action of the two diastases at various temperatures. Thus the optimum temperature of barley diastase was from 40° to 50° C., that of malt from 45° to 55° C.; the former was as active at 4° C. (39° F.) as the latter was at 14.5° C. (58° F.).

Glucase.—An enzyme which is unable to act upon gelatinised starch, but which slowly converts soluble starch, more readily dextrin or maltose, into glucose. It has been said to be present in steeped barley, and also in malt, by several observers, who state that it exists there partly in a soluble and partly in an insoluble condition. Other authorities have been unable to detect the presence of this enzyme in either barley or malt. Glucase appears, from the researches of Géduld,[2] to be a normal constituent of maize. He prepared the enzyme by extracting ground maize with water, and precipitating the enzyme from the extract thus obtained with alcohol. Glucase gives the blue reaction with guaiacum and hydrogen peroxide. The action of this enzyme on dextrin and maltose is most energetic at temperatures between 50° and 60° C. (122° to 140° F.); at 60° the action becomes slower, and it is completely arrested at 70° C. (158° F.).

Laccase.—An enzyme, found by Bertrand[3] in the tree which yields lac, which possesses the power of absorbing oxygen from the air, and of evolving carbon dioxide. This action so closely resembles respiration that it has been thought to stand in some definite relation to this process. It is thought possible that an enzyme of this nature may exist in malt, which by its agency effects the transformation of a portion of the starch into water and carbon dioxide. A curious fact in connection with this enzyme is that it is not destroyed by boiling in water.

THE ENZYMES OF YEAST.

The cultivated yeasts contain several enzymes; the longest known of these is invertase, which possesses the power of inverting cane sugar. In order to obtain invertase, A. Meyer[4] allowed pressed yeast to stand for some time under alcohol, to kill the cells. The yeast was then freed from alcohol and triturated with sand and water to rupture the cells; the aqueous extract thus obtained was filtered off and alcohol added. The enzyme

---

[1] *Zeitsch. für Brau.*, 1886, p. 261.
[2] *Woch. für Brau.*, 1891, p. 545.
[3] *Compt. Rend.*, cxx. 266.
[4] Meyer, *Die Lehre v. d. Fermenten*, p. 64.

was precipitated in an impure condition, and subsequently dried over sulphuric acid in a vacuum. C. O'Sullivan allows pressed yeast to stand in a cool place until the mass liquefies; the fluid portion is then filtered off, and the enzyme precipitated with alcohol and dried, as in the foregoing process.

The most favourable temperature at which invertase exerts its action on cane sugar is, according to Kjeldahl,[1] for top fermentation yeast 55° to 57° C. (131° to 135° F.). According to J. O'Sullivan,[2] invertase cannot pass out of the yeast cell by exosmosis; consequently, the inversion of sugar, under normal conditions, takes place within the yeast cell.

Yeast Glucase.—In 1866 Borquelot[3] came to the conclusion that maltose suffered inversion before it was fermented by yeast, and noticed that when yeast was allowed to remain for several days in a solution of maltose saturated with chloroform, the opticity of the solution was diminished, presumably through the formation of glucose. Lintner,[4] in studying the fermentation of isomaltose, surmised that both this sugar and also maltose were converted into glucose by the invertase of the yeast before being fermented. E. Fischer[5] prepared, by extracting 1 part of dried yeast with 15 parts of water at 30° to 35.3° C. (86° to 95° F.), a solution which had the power of inverting maltose; but he was unable to obtain the same result with ordinary commercial invertase. Lintner[6] expressed the view that the maltose inverting ferment was distinct from invertase. He obtained the most powerful action on maltose by adding powdered yeast, which had been dried at a temperature of 40° C. (104° F.), to a solution of this sugar. The aqueous extract from the same powdered yeast exhibited much less activity, and the substance obtained by precipitation with alcohol still less. E. Fischer[7] showed that the enzyme was already present in the yeast cell, and was not formed during the process of drying, for on allowing ordinary yeast to act upon solutions of maltose which contained either 2 per cent. of thymol or 3 per cent. of toluene (added to arrest alcoholic fermentation), considerable quantities of glucose were formed. So long, therefore, as yeast is moist and fresh, the invertive action takes place within the cells.

Melibiase.—It has long been known that the top fermentation yeasts were unable to ferment more than a third part of melitriose (raffinose), but that the bottom fermentation yeasts were able to ferment this sugar completely. From this it would appear that the low yeasts contain an enzyme which is capable of inverting melibiase (see p. 86) which the top ones do not possess. That this is actually the case was demonstrated experimentally by

---

[1] *Med. Carlsb. Lab.*, i. 332. [2] *Proceedings of the Chemical Society*, 1892, p. 124.
[3] *Journ. d. l'Anatomie et Phys.*, 1866, p. 180. [4] *Zeit. f. Brau.*, 1892, p. 107.
[5] *Berichte*, 1894, p. 2986. [6] *Zeit. f. Brau.*, 1894, p. 415.
[7] *Berichte*, 1895, p. 1429.

Bau,[1] who called the enzyme "melibiase." Melibiase, like yeast glucase, is a comparatively insoluble body.

Other Inverting Enzymes.—It is almost certain that other enzymes are contained in such yeasts as the Frohberg varieties or the Schizo-sacc. Pombe, since these are able to invert some of the higher dextrins. It has been shown by E. Fischer that *Monilia candida*, which was supposed to be able to ferment cane sugar without previous inversion, contains in its protoplasm an insoluble enzyme which can effect the inversion of cane sugar.

THE CHEMICAL COMPOSITION OF YEAST.

Since yeast, under the older theories, was regarded as a putrefying substance which merely communicated a portion of its own internal motion to the fermentable substance, and neither gave off anything from itself to the fermenting fluid nor took up anything from the substances dissolved in that fluid, a detailed knowledge of its composition was comparatively a matter of indifference; consequently, the older analyses of yeast were mostly confined to a determination of its elementary composition. But when yeast was found to be a living organism, that not only derived its nutriment from the bodies contained in the fermenting fluid, but which also added its own excretions to that fluid, then it became necessary to extend this simple form of analysis, and to isolate and determine as far as possible the nature and composition of the various groups of substances which entered into its composition. In this way chemists endeavoured to study the relations subsisting between the structural compounds of yeast and of those substances which form its nutriment, much in the same way as an agricultural chemist studies the relation between the constituents of plants and those of the soil. Numerous investigations have been made in this direction, but the subject is one of considerable intricacy, for the examination of yeast, in addition to the inherent difficulties of this kind of analysis, presents some obstacles which are peculiarly its own. Yeast is always more or less contaminated with foreign impurities, and before it can be analysed with any degree of exactitude, these must be removed. It is generally sought to effect this object by washing with water, but in so doing some of the soluble constituents of the yeast are invariably dissolved and removed by the wash-water, whilst at the same time water is absorbed by the yeast, the result being an apparent lowering of the amount and alteration in the proportions of its various constituents. Moreover, yeast is a body whose chemical composition is in a perpetual state of change, and, consequently, analyses of the same yeast made at different times yield dissimilar results. The varieties of yeast which have invariably been chosen for investigation in this way are the high and low fermentation forms of the

[1] *Woch. f. Brau.*, 1895, p. 24.

Saccharomyces cerevisiæ. Little is known about the composition of other varieties of yeast, such as the *Sacc. ellipsoideus, apiculatus,* &c., or of the bacteria.

The Elementary Composition of Yeast.—The yeast cell consists of water and solid matter, the former constituting about 80 per cent. of its weight. The solid substance consists of carbon, hydrogen, oxygen, nitrogen, and sulphur, together with a small quantity of mineral substances (ash). The results of a number of elementary analyses of the dry substance of yeast, which was in each case freed from foreign substances by washing, are appended:—

	Mitscherlich.	Schlossberger.		Dumas.	Wagner.		Mulder.
	High Yeast.	High Yeast.	Low Yeast.		High Yeast.	Low Yeast.	High Yeast.
	Per Cent.	Per Cent.	Per Cent.	Per Cent.	Per Cent.	Per Cent.	Per Cent.
Carbon	47.0	49.9	48.0	50.6	49.8	44.4	50.8
Hydrogen	6.6	6.6	6.5	7.3	6.8	6.0	7.2
Nitrogen	10.0	12.1	9.8	15.0	9.2	9.2	11.1
Oxygen	35.8	31.4	35.7	37.1	34.2	35.8	?
Sulphur	0.6	?	?				

These analyses, as might be expected, exhibit considerable differences, but the most striking feature of the composition of yeast is the large amount of nitrogen which it contains when compared with that found in other plants,[1] and even in those belonging to the same genus. The common mushroom (*Agaricus campestris*), which has been found to contain from 7.26 per cent. to 7.33 per cent. of nitrogen, approaches it most nearly in this respect.

According to the analyses of yeast by Hayduck, the amount of nitrogen it contains may vary from 4 to 10 per cent., and this quantity may be altered at will by growing the yeast in a fluid poor or rich in assimilable nitrogenous food.

Wijsmann[2] found that the amount of nitrogen contained in yeast is in a constant state of variation. When 10 grammes of yeast, which contained 7.09 per cent. of nitrogen (reckoned on the dry substance), were added to a litre of malt wort, the nitrogen increased in the course of an hour to 9.9 per cent., in two hours it had slightly decreased to 9.6 per cent., and in three hours it was 9.55 per cent. In another experiment, where yeast was added to ordinary service-water containing in solution 0.7 per cent. of peptone, the 6.36 per cent. of nitrogen originally present in the yeast increased in ten minutes to 7 per cent., in fifty minutes to 7.08 per cent., and in eighty-five to 7.22 per cent. In further experiments yeast was added, in the one case to service-water containing in

---

[1] The nitrogen of the dry substance of ordinary plants ranges from 0.3 per cent. to 4.5 per cent. Ebermayer, *Phys. Chem. d. Pflanzen.*, i. 47.
[2] *Woch. f. Brau.*, 1891, p. 1887.

solution 0.25 per cent. of asparagine and 0.1 per cent. of potassium sulphate ; in the other, to a similar solution in which 3 per cent. of cane sugar had been dissolved. The amount of nitrogen of the original yeast (6.88 per cent.) varied in the following manner :—

Hours.	Minutes.	Without Sugar. Per Cent.	With Sugar. Per Cent.
2	21	8.10	9.27
3	5	7.91	9.24
3	40	7.43	9.55
5	0	7.45	9.24

Curiously, when yeast was immersed in a solution of acid ammonium phosphate, it quickly absorbed its maximum quantity of nitrogen, and this did not again diminish, as was the case when yeast was placed in malt wort. Wijsmann conjectures that yeast, at the commencement of fermentation and before it exercises its reproductive powers to the fullest extent, stores up a large quantity of nitrogenous material. It may be that this material is either assimilated by the protoplasm in preparation for budding, and, as budding proceeds, gradually returns to its original quantity, or that the nitrogenous bodies are at first simply absorbed in an unaltered condition, and are afterwards metamorphosed by the protoplasm. He concludes, from other experiments, that acid ammonium phosphate is not absorbed unchanged. In those cases where yeast was compelled to obtain its nitrogen from peptone, asparagine, or acid ammonium phosphate, the increase in its nitrogen was never so large as when it was grown in malt wort.

The Mineral Constituents of Yeast.—The amount of ash found in yeast by different observers varies from 2.5 per cent. to 9.0 per cent. The following table contains a number of analyses of the ash of yeast by different chemists :—

	Mitscherlich.		Bull.	Liebig.	Béchamp.
	High Yeast.	Low Yeast.	High Yeast from Weissbier.		
	Per Cent.	Per Cent.	Per Cent.	Per Cent.	Per Cent.
Phosphoric acid .	53.9	59.4	54.7	48.5	53.7
Potassium .	39.8	28.3	35.2	30.6	28.6
Soda .	...	...	0.5	...	1.4
Magnesia .	6.0	8.1	4.1	4.2	5.2
Lime .	1.0	4.3	4.5	2.1	2.4
Iron oxide .	...	...	0.6	...	5.0
Sulphuric acid [1] .	...	...	...	...	5.7
Hydrochloric acid	...	...	0.1	...	..

[1] Phosphoric acid acts under certain circumstances, as, for instance, in the case of the ignition of a yeast ash, as a more powerful acid than sulphuric acid. The latter may consequently be partially or completely driven off, and this is the reason why its presence has not been detected in a large number of analyses. Béchamp took special precautions to avoid loss from this cause.

Though the composition of the ash of yeast, as portrayed by the above analyses, varies considerably, yet these analyses all show that the most abundant of its constituents are phosphoric acid and potash, substances which have been proved to be indispensable in the nutrition of all the higher plants. The next body which occurs in fairly large quantity is magnesia, and this is also an indispensable substance. Lime and iron appear in smaller quantities (the large amount of iron shown in the analyses by Béchamp was probably due to contamination), and it is doubtful if the latter substance is an absolutely necessary constituent of yeast. Nägeli[1] considers that the potassium salts are dissolved in the cell-juice, whilst those of the alkaline earths are contained in the more solid tissues. In all the lower fungi it is supposed that the potassium bases are united with phosphoric or an organic acid, the potassium being in combination with phosphoric acid in the cell-juice as hydrogen-dipotassium phosphate (HK_2PO_4), a portion of the alkaline earths are assumed to exist as phosphates in the plasma, whilst other portions are deposited as compounds of organic acids (oxalic, &c.) in the cell-membrane.

Differentiation of the Different Groups of Compounds Occurring in Yeast.—The solid matter of yeast may be divided into two principal groups, the nitrogenous and the non-nitrogenous, each group being subdivided into several members.

Yeast Cellulose.—By the almost universal consent of all observers, the enveloping membrane of the yeast cell, like that of all other vegetable cells, consists of a variety of cellulose. Its amount has been estimated as forming from 18.5 per cent. (*Pasteur*) to 32 per cent. (*Payen*) of the dry yeast substance, the wide difference in the amounts given by various observers being undoubtedly due to the extreme ease with which cellulose undergoes alteration. It is very readily converted into sugar, an action which appears to take place naturally sometimes, as in the spontaneous fermentation which occurs in yeast when it is deprived of nutriment. Schlossberger, by treating yeast alternately with solution of caustic potash and acetic acid until nothing further was dissolved, and washing with water, obtained a residue of the following composition :—

	Per Cent.
Carbon	44.9
Hydrogen	6.7
Nitrogen	0.5
Ash	1.1

This corresponds with the composition of an impure cellulose.

Pasteur, on treating yeast with dilute sulphuric acid, found an amount of sugar produced which corresponded to 18.5 per cent. of cellulose. It is thought by some observers that the cellulose of

---

[1] *Sitz. d. bayer. Acad.*, 1889, Part iii.

yeast and of the rest of the fungi differs markedly in some of its properties from that of other plants, since it is insoluble in cuprammonium hydrate solution, and is readily attacked by dilute acids. Dreifuss,[1] on treating several of the fungi with dilute acid, alkali, alcohol, and ether, and then with concentrated caustic alkali at a temperature of 180° C., obtained a residue which, on boiling with a dilute acid, yielded glucose. This substance, which he considered to be the ordinary " plant cellulose," was found in the common mushroom, *Bacillus subtilis* and *Aspergillus glaucus*. Winterstein[2] obtained from certain fungi, on prolonged treatment by Schultz's method, a cellulose which differed considerably from ordinary plant cellulose. It only gave the blue or violet reaction with iodine and strong sulphuric acid after some time, and was but partially soluble in cuprammonium hydroxide. It contained from 1.64 to 3.94 per cent. of nitrogen, which could not have been derived from proteids. On hydrolysis with 70 per cent. sulphuric acid it yielded 65.95 per cent. of glucose, there being formed at the same time 12.52 per cent. of acetic acid. Gibson[3] was unable to obtain the crystalline variety of cellulose from fungi which he had obtained from other plants. Liebermann and V. Bitto[4] obtained yeast cellulose in a pure state by digesting yeast for several hours on the water-bath with a mixture of equal parts of hydrochloric acid and water, to which was added a small quantity of potassium chlorate. The residue was well washed, boiled for half-an-hour with 1.25 per cent. solution of acetic acid, and finally boiled with a 1.25 per cent. solution of potash. It gave the ordinary reaction of cellulose with iodine and zinc chloride solution.

Yeast Mucilage.—Nägeli obtained in the analysis of yeast 37 per cent. of cellulose and mucilage. This latter substance is one which occurs in great abundance in the seeds of certain plants, such as linseed, quince, &c. It shows its near relationship to cellulose in being coloured violet by iodine and sulphuric acid. On hydrolysis with dilute acids it yields a reducing sugar which is probably glucose. The mucilaginous coating discovered by Hansen, which, under certain conditions, forms around yeast cells, is probably of the same nature.

Yeast Gum.—Hessenland[5] obtained a gum on boiling yeast with lime, having the same formula as Scheibler's dextran, $C_6H_{10}O_5$, but which does not seem to be identical with it. On hydrolysis with dilute sulphuric acid this gum yielded chiefly D-mannose and a little glucose.

Salkowski[6] obtained, by acting on yeast with 3 per cent.

---

[1] *Zeit. f. phys. Chemie*, xviii. 358.
[2] *Berichte*, 1893, ii. 441.
[3] *La Cellule*, 1893.
[4] *Centralblatt f. Physiologie*, vii. 857.
[5] *Zeit. Rübenzucker Ind.*, 1892, p. 671.
[6] *Berichte*, xxvii. 495.

solution of caustic potash, and heating the resulting liquid with Fehling's solution, a copper compound of a gum. This was dissolved in hydrochloric acid, precipitated by alcohol, washed with alcohol and ether, and dried. In this way a fine white powder was obtained, which in some respects resembled gum arabic. It had a composition of $C_{12}H_{22}O_{11}$, and an opticity of $[\alpha]_D = +90.1°$. Acids converted this gum, of which the dry substance of yeast contained about 7 per cent., into a feebly dextro-rotatory sugar capable of undergoing fermentation. Salkowski considers that the gum prepared by Hessenland's method is a mixture of several substances.

The Fat of Yeast.—Yeast contains a certain amount of fat, which, when the cells are examined under the microscope, is seen in the form of bright, strongly refractive spots. Nägeli estimates its quantity at about 5 per cent., but considers that its amount varies, young cells being poor, older ones rich in fat.

Kulish[1] carefully investigated the composition of yeast-fat, and found it to consist chiefly of phytostearin and a glyceride of myristic acid; the glyceride of another acid, the nature of which he was unable to determine, also occurred in small quantity. The formation of fat in yeast takes place most rapidly when yeast reproduction proceeds most actively, and when yeast is supplied with an abundance of oxygen (*Nägeli*).

Glycogen.—This carbohydrate, which has been long known as one of the normal constituents of the animal liver, is found in yeast. In a pure condition it is a white powder, having the composition $C_6H_{10}O_5$, readily soluble in hot water, with which it forms an opalescent solution which does not reduce Fehling's solution. Glycogen is readily converted into glucose by boiling with a dilute acid; the enzymes ptyalin and diastase both convert it into a reducing sugar.

An aqueous solution of glycogen is coloured reddish-brown by iodine solution, a reaction by which the presence of glycogen may be readily detected in yeast cells. Laurent,[2] who studied the formation of this substance in yeast, found that it was produced as a reserve material, and that it might sometimes amount to 32.6 per cent. of the dry yeast substance. It could be formed by yeast from the salts of either lactic or succinic acid, from ammonium succinate, glycerin, glucose or cane sugar, dextrin, asparagine, &c. According to Max Cremer,[3] fresh yeast cells, when treated with a dilute solution of iodine in potassium iodide, exhibit the glycogen reaction in a marked degree, but yeast-cells which have been starved, such as those which have undergone spontaneous fermentation, are only coloured pale yellow. If the starved cells be placed in a solution of glucose, fructose, or cane sugar, the

[1] *Weinbau u. Weinhandel*, 1891, p. 250.
[2] *Botanische Zeit.*, xlviii. 719.
[3] *Zeit. f. Biologie*, xxxi. Part ii.

glycogen reaction can be again detected. He was unable to observe the formation of glycogen in similarly starved cells when these were introduced into a solution of one of the pentose sugars, or of glycerin, or of milk-sugar, all of which substances are unfermentable by ordinary yeast.

The Nitrogenous Constituents of Yeast.—As with the nitrogenous bodies of all other organisms, much uncertainty prevails with regard to the nature and composition of those of yeast, and though many attempts have been made from time to time to increase our knowledge in this direction, very much yet remains to be done.

Schlossberger[1] attempted to isolate the nitrogenous compounds of yeast by treating it with a dilute solution of caustic alkali. and neutralising the filtered liquid thus obtained with an acid. A substance came down in the form of a flocculent precipitate; it contained no ash, and had the following composition:—

	Per Cent.
Carbon	55.5
Hydrogen	7.5
Nitrogen	13.9
Sulphur	0.0

It is to be assumed that the remainder was oxygen. This substance, from the small amount of nitrogen which it contains, does not appear to be a normal albuminoid; it corresponds closely in composition with a body found by Schützenberger[2] amongst the products of the hydrolysis of albumin by dilute sulphuric acid, and called by him "hemi-protein," but which is now known as "anti-albumid." Mulder, by treating yeast with acetic acid, filtering off the resulting liquid and precipitation with ammonium carbonate, obtained a body more nearly resembling albumin in composition:—

	Per Cent.
Carbon	53.3
Hydrogen	7.0
Nitrogen	16.0

According to the investigations of Nägeli and Löw,[3] the plasma of a young yeast (which contained nearly 12 per cent. of nitrogen) yielded about 75 per cent. of proteids and about 2 per cent. of peptone. They state that during the gradual series of changes which yeast cells undergo previous to dying, the whole or nearly the whole of the proteids are transformed into peptones.[4] They give

---

[1] *Ann. d. Chem. u. Pharm.*, xi. 199.
[2] *Bull. Soc. Chim.*, xxiii. and xxiv.
[3] *Sitz. d. bayer. Acad.*, 1878, Part ii.
[4] By peptones is meant in this case substances precipitated by lead acetate.

the following analysis of a sample of yeast which contained 8 per cent. of nitrogen :—

	Per Cent.
Cellulose and mucilage	37
Proteids { Albumin	36
{ Substance resembling gluten-casein	9
Peptone	2
Fat	5
Ash	7
Extractive matters	4

The 4 per cent. of extractive matters were not precipitable by lead acetate; they contained a peptone-like body, small quantities of invertase, leucine, glucose, and still smaller quantities of glycerin, succinic acid, guanine, xanthine, sarkine, alcohol, and probably traces of inosite.

Nuclein.—This proteid, which seems to be a normal constituent of the nuclei of cells, has been isolated from yeast by treating it for a short time with a 5 per cent. solution of caustic potash. The resulting solution is filtered into dilute hydrochloric acid, the precipitate which forms is washed first with dilute hydrochloric acid, then with alcohol, boiled several times with alcohol, and finally dried over sulphuric acid. Thus obtained, it forms a white amorphous powder, insoluble in water and dilute acids, and readily soluble in dilute alkalies. When boiled for a considerable length of time with water, the greater portion of it is transformed into phosphoric acid and xanthine bodies. When fused with sodium carbonate and nitrate, it yields a mass containing phosphoric acid in large amount. Its probable composition, according to Miescher, is $C_{29}H_{49}N_9P_3O_{22}$. Nuclein is distinguished from the rest of the proteids by the large amount of phosphorus which it contains (9.59 per cent.), and by the extreme resistance which it offers to the hydrolysing action of pepsin. It gives an indistinct xantho-proteic reaction, but gives no reaction with Millon's reagent. According to recent views, nuclein is a compound of nucleic acid (a compound of albumin and phosphoric acid) with one or more of the xanthine bodies.

Specific Gravity of Yeast.—Guichard[1] found the specific gravity of yeast to be from 1.180 to 1.183 at a temperature of 16° C. (61° F.).

Products of the Putrefaction of Yeast.—These are of practical interest to the brewer, since putrefaction of yeast occasionally takes place in beer, the composition and properties of which it may affect if, as is sometimes the case, the beer is allowed to remain for a long time in contact with putrefying yeast. Muller[2] obtained, on adding calcium hydroxide to liquid putrefied yeast, a precipitate which, when distilled with dilute sulphuric acid, yielded

[1] *Bull. Soc. Chim.*, 1894, xi. 230.
[2] *Journ. f. prakt. Chemie*, lxx. 65.

a number of volatile acids. These were purified by conversion into their barium, sodium, or silver salts. Acetic acid was found in large, butyric acid in moderate quantity, also in smaller amount caprylic acid, together with traces of the higher acids of the same series. On distilling the liquid from which the precipitate had been obtained and collecting the distillate in dilute hydrochloric acid, and purification of the salts thus obtained, he was able to identify ammonia, trimethylamine, ethylamine (probably), amylamine, and caprylamine. This probably explains the formation of the methylamine sometimes found in wine. Lactic acid was also present, and this is not surprising, as yeast is often contaminated with lactic acid bacteria. Tyrosine crystallised out in chalk-like masses, and leucine was also found; both of these bodies are found amongst the products of the hydrolysis of albumin by dilute sulphuric acid. Bodies of a fusel-oil-like nature were also found, the principal one being amylic alcohol, and this lends some probability to the view held by some that fusel-oil is a body produced by the yeast when in a moribund condition.

THE ACTION OF DIASTASE ON STARCH.

This is a subject of the most profound importance to the brewer, for the products of the action of diastase upon starch differ widely, according as the circumstances vary under which these two bodies are allowed to interact, and the products formed in this way have an immense influence on the character of the finished beer.

The subject is one which has attracted an enormous amount of attention for a long series of years, and which has led to many differences of opinion amongst observers. Even at the present time these are far from being reconciled; and although our knowledge in recent years has made considerable progress, much yet remains to be explained and established with certainty.

In the opinion of the earliest observers, the transformation which took place when diastase acted upon gelatinised starch was that the starch molecule simply united with a molecule of water, and formed glucose according to the following equation:—

$$\underset{\text{Starch.}}{C_6H_{10}O_5} + \underset{\text{Water.}}{H_2O} = \underset{\text{Glucose.}}{C_6H_{12}O_6}$$

Later on it was found that another body, dextrin, was invariably produced in all starch transformations; and for a long time it was considered that the substance first produced was dextrin, this becoming afterwards converted into glucose. According to this view, dextrin is a body intermediary between starch and glucose.

ACTION OF DIASTASE ON STARCH.

In 1860 Musculus[1] brought forward evidence to show that this was not the case, but that in all starch transformations, be they effected either through the agency of diastase or of dilute acids, dextrin and glucose were produced simultaneously.

This view, though it was strongly controverted at the time it was brought forward, is the one which has survived, and which, in an extended form, has been found by the majority of workers on the subject to afford the best explanation of the observed phenomena.

Musculus was led to believe that the conversion invariably took place according to the following equation :—

$$3C_6H_{10}O_5 + H_2O = C_6H_{12}O_6 + 2C_6H_{10}O_5$$
$$\text{Starch.} \quad \text{Water.} \quad \text{Glucose.} \quad \text{Dextrin.}$$

that is, from every three molecules of starch two molecules of dextrin[2] and one molecule of glucose were formed.

Other observers, however (*Payen, Guerrin-Varry, Balling, Dubrunfaut*), in their experiments, obtained the two bodies dextrin and glucose in proportions widely differing from those which are required by the above equation.

In 1870 Schwartzer[3] pointed out that these discrepancies were caused by the observers not having taken into account the different temperatures at which the various starch conversions had been effected. He found that if the action were allowed to proceed at any temperature between 0° to 30° C. (32° to 86° F.), the products of the transformation were in the ratio of one molecule of dextrin to one molecule of glucose, according to the following equation :—

$$2C_6H_{10}O_5 + H_2O = C_6H_{10}O_5 + C_6H_{12}O_6$$
$$\text{Starch.} \quad \text{Water.} \quad \text{Dextrin.} \quad \text{Glucose.}$$

But if the process was conducted at a temperature above 60° C. (140° F.), two molecules of dextrin to one molecule of sugar were produced, thus :—

$$3C_6H_{10}O_5 + H_2O = 2C_6H_{10}O_5 + C_6H_{12}O_6$$
$$\text{Starch.} \quad \text{Water.} \quad \text{Dextrin.} \quad \text{Glucose.}$$

So far it had been considered that in all starch transformations but one dextrin was produced, but in 1871 Griessmayer[4] concluded from his experiments that two dextrins were formed; these he designated "Dextrin I. and II." The following year

---

[1] *Ann. Chim. Phys.* (3), lv. 203.
[2] The size of the starch and dextrin molecules at that time were regarded as $C_6H_{10}O_5$, which was, of course, an erroneous assumption.
[3] *Journ. f. prakt. Chemie* (2), i. 212.
[4] *Annalen*, clx. 40.

O'Sullivan[1] confirmed this, and called the two dextrins "a" and "β." In this same year Brücke[2] gave them the names "erythrodextrin" (ἐρυθρός, red) and "achroodextrin" (ἄχροος, without colour), since the first was coloured red by iodine, and the latter gave no colour with that reagent. This nomenclature has been retained to the present day. O'Sullivan carefully examined these two dextrins in the following manner:—

Starch conversions were made in some cases by the agency of diastase, in others by that of dilute sulphuric acid; the action being stopped sometimes when the conversion gave a deep brownish-red colour with iodine, at other times when this reagent had ceased to give a colour. In those cases in which the conversion had been effected with diastase, the further action of the enzyme was arrested by simply boiling the solution; in those in which sulphuric acid had been employed, by removing the acid by barium hydrate.

After the conversion the solutions were in each case evaporated to a small bulk, and strong alcohol (sp. gr. 0.830)[3] added in considerable quantity. The precipitate of impure dextrin thus obtained was further purified by being dissolved in water, and again precipitated with alcohol, the re-solution and re-precipitation being repeated in some cases as many as thirty times. In this way four preparations of dextrin were obtained, having the following opticities and reducing powers:—

$a.\ [a]j = +204°\quad K = 8.8 \quad|\quad c.\ [a]j = +204°\quad K = 9.0$
$b.\ [a]j = +205°\quad K = 8.5 \quad|\quad d.\ [a]j = +198°\quad K = 9.0$

a was obtained by the action of malt extract; b by that of dilute sulphuric acid; c by that of oxalic acid; d was prepared with dilute sulphuric acid, the action in this case being allowed to continue over eight hours. It was evident that in all these cases a considerable quantity of some substance having a reducing action upon Fehling's solution was present, and in order, if possible, to still further remove this, solutions of each of the above preparations were dissolved in water, and submitted to fermentation with yeast. When the fermentation had ceased, the liquid was filtered off, reduced to small bulk, and alcohol added. The dextrin thus obtained was several times dissolved in water, and re-precipitated by alcohol. The last precipitate, after being washed twice or thrice with alcohol, was pressed between folds of filter-paper, and finally dried in a vacuum over sulphuric acid. The dextrins obtained in this way were white, semi-crystalline, highly hygroscopic powders, not perceptibly soluble in cold alcohol of specific gravity 0.82, but readily soluble in water; and when one of these dehydrated dextrins was dissolved in water, its solution was accom-

---

[1] *Journal of the Chemical Society*, xxv. 579.
[2] *Wien. Acad. Bericht* (3), lxv. 126.
[3] This would contain 87 per cent. of alcohol and 13 per cent. of water by weight.

panied with evolution of heat. The following table exhibits the opticities and reducing powers of eight preparations made in this way:—

$e.$ $[a]j$	$= +213.0°$	$K = 2.03$		$i.$ $[a]j$	$= +212.7°$	$K = 1.40$
$f.$ $[a]j$	$= +212.0°$	$K = 2.20$		$j.$ $[a]j$	$= +213.0°$	$K = 1.15$
$g.$ $[a]j$	$= +212.7°$	$K = 1.24$		$k.$ $[a]j$	$= +213.5°$	$K = 0.80$
$h.$ $[a]j$	$= +213.1°$	$K = 1.20$		$l.$ $[a]j$	$= +214.0°$	$K = 1.03$

$e, f,$ and g are from the product of the action of diastase on starch-paste, the transformation being stopped when a portion of the solution, after being allowed to become cold, gave a reddish-brown colour with iodine (erythro-dextrin). h and i were prepared in the same way, the action being arrested when iodine ceased to give any coloration (achroo-dextrin). j was obtained from a transformation effected by dilute sulphuric acid, the acid being removed when iodine gave a reddish-brown colour (erythro-dextrin). k was produced similarly to j, with the exception that the acid was removed when iodine ceased to give any colour (achroo-dextrin). l was obtained by the action of oxalic acid, the action being stopped when the solution gave a reddish-brown colour with iodine (erythro-dextrin). Oxalic acid was found to be very much slower in its action than sulphuric acid.

O'Sullivan concluded from these results that all the eight preparations, if they could have been obtained in an absolutely pure condition, would have had no reducing action on Fehling's solution.

Elementary analyses were made with both the dextrins (erythro- and achroo-) after they had been perfectly freed from moisture in a current of dry air at 100° C. (212° F.); the results of which proved in a fairly conclusive manner that the composition of both was $C_6H_{10}O_5$.

It was found, on further treating any one of these dextrins with fresh malt extract, that its reducing power (K) gradually increased to 66, and its rotatory power fell to $[a]j + 150°$. When this stage had been arrived at all further action ceased. This remarkable stoppage of the action when a K of 66 had been reached was quite inexplicable, and led O'Sullivan to study the characters of the reducing body. The serious discrepancy between the results of experiments a, b, c, with those of d, in which, though the reducing powers of all the four were practically the same, there was a difference of some 6° or 7° in the opticity of d, pointed to the presence of some reducing body with a much higher opticity than glucose. This body proved to be maltose.

In 1875 Bondonneau,[1] presumably in ignorance of O'Sullivan's observations, described three dextrins $a, \beta,$ and γ; but as he was still under the impression that the sugar formed in the reaction was glucose, it is extremely probable, as pointed out by O'Sullivan, that the dextrin γ was an imaginary body, and that its supposed

[1] *Compt. Rend.,* lxxxi. 972 and 1210.

presence could be accounted for by taking into account the correct reducing power of maltose.

Discovery and Rediscovery of Maltose.—De Saussure in 1819[1] isolated a hitherto unknown sugar from the hydration products of starch and described its crystalline nature. In 1847 Dubrunfaut[2] further examined this new sugar and named it "maltose." He found it had an opticity three times as great as that of glucose, and that it was the sugar first formed when starch was acted upon by dilute acids. He described it as a crystalline body much less soluble in alcohol than glucose. Strange to say, these facts seem to have lapsed into comparative oblivion until O'Sullivan rediscovered and fully satisfied himself as to the existence of this sugar in 1872.[3] He prepared it in the following manner, retaining the name "maltose" given it by Dubrunfaut:—

A conversion was made with 100 grammes of malt-starch at 40° C. (104° F.), the action being allowed to proceed for three hours. The solution was then boiled and filtered, the filtrate evaporated at a temperature of 80° C. (176° F.) to 300 c.c. The syrup thus obtained was boiled with two litres of alcohol of a specific gravity of 0.83, and allowed to cool. The clear liquid was then poured off from a syrup which was precipitated, and put on one side in a corked flask. At the end of six days the sides of the flask were covered with a crystalline crust, a portion of which, after being dried at 100° C. (212° F.) and dissolved in water, was found to possess an opticity of $[\alpha]j = +150$ and a reducing power (K) of 65. He further found that the sugar was diffusible, and all attempts to further resolve it into other bodies by dialysis failed.

Further investigation by Schultze and Uhrich[4] confirmed the correctness of O'Sullivan's observations with reference to maltose.

O'Sullivan's Further Investigations.—In 1876[5] O'Sullivan published an account of a further series of investigations on the action of malt-extract on starch. He concluded from a number of experiments, of which the following is an example, that maltose and dextrin are the only products of the action of malt-extract on starch:—

One hundred grammes of potato-starch were intimately mixed with 200 c.c. of water at 55° to 60° C. (131° to 140° F.), and then 400 to 500 c.c. of boiling-water added with constant stirring. In this way a thorough gelatinisation of the starch was secured. The mixture was then cooled to 60° C. (140° F.) and the cold water extract[6] from 10 grammes of pale malt added. In five or six

---

[1] *Ann. Chim. Phys.*, xi. 379. [2] *Ibid.* (3), xxi. 178.
[3] *Journal of the Chemical Society*, xxv. 579.
[4] *Berichte*, vii. 1047.
[5] *Journal of the Chemical Society*, 1876, ii. 125.
[6] Made by mixing finely ground pale malt with water, allowing the mixture to stand for some hours at the ordinary temperature, and then filtering off the clear solution.

minutes the paste dissolved, and the resulting solution was rapidly cooled and filtered. It was now examined, and found to have an opticity of $[\alpha]j_{3.85} + 170.8°$ and a reducing power (K) of 44, which figures are nearly equivalent to those of a mixture of 67.7 per cent. maltose and 32.3 per cent. dextrin. After being slowly evaporated the solid matter in solution had an opticity of $[\alpha]j_{3.85} + 169.1°$ and a K of 45.81, which are nearly equivalent to those of a mixture of 70.5 per cent. maltose and 29.5 per cent. dextrin.

A comparison of the figures actually obtained with those found by calculation for such a mixture show a very close agreement:—

Observed.	Calculated.
170.8°	170.6°
169.1°	168.9°

This tends to the assumption that maltose and dextrin alone are produced.

The solution was then evaporated to a thick syrup, placed, while still hot, over sulphuric acid in the bell jar of an air-pump, and the air slowly exhausted. In a few hours a white, brittle, highly hygroscopic porous mass was obtained. This was powdered and boiled with twenty times its bulk of alcohol of specific gravity 0.83. On cooling the liquid separated into two portions, A a thick syrup which fell to the bottom, and B a clear supernatant solution. This latter (B) was poured off and submitted to distillation. The resulting concentrated solution, after standing for a few days, crystallised to a solid mass, which was further purified by repeated re-crystallisation from alcohol. After being dried in a vacuum and then at 100° C. (212° F.), the crystalline substance was found to have an opticity of $[\alpha]j_{3.85} + 148°$, and a $K_{3.85}$ of 66 to 67. When submitted to fermentation by yeast it was found to be entirely fermentable, yielding 51.5 per cent. of alcohol. The substance was evidently, therefore, maltose contaminated with malt-extract bodies.

The residual syrup had an opticity of $[\alpha]j + 185.5°$ and a K of 26.66, figures which approximate to those for a solution of 41 per cent. maltose and 59 per cent. dextrin. The opticity calculated for such a mixture is $[\alpha]j_{3.85} = +187.76°$, or 2.26° higher than that found; this discrepancy was attributed to the insolubility of a portion of the malt-extract in alcohol.

The syrup A was evaporated in a vacuum to the solid state, and treated with alcohol in a similar manner to B; it separated into a syrup and a solution. The clear alcoholic solution yielded maltose. The solid matter of the precipitated syrup was found to have an opticity of $[\alpha]j_{3.85} + 197.5$ and a K of 15.99, figures equivalent to 24.6 per cent. maltose and 75.4 per cent. dextrin. The opticity calculated for such a mixture is $[\alpha]j_{3.85} + 198.2$.

The same operation was repeated several times with the syrup; the portions taken up by the alcohol were found in every case to be maltose, whilst the resultant syrups appeared to be mixtures of maltose and dextrin, the latter constituent gradually increasing on each successive treatment. When the stage had been reached which showed a K equal to 10 or 12 per cent. maltose, it was found almost impossible to carry the separation further; but the opticities and cupric reducing powers in all cases invariably pointed to the mixture consisting of maltose and dextrin only. This may be seen from the accompanying results, selected from a numerous series:—

Opticity Observed.	$K_{3.85}$	Maltose.	Calculated Opticity.
$[a]_{j_{3.85}} + 204.6°$	9	13.8	205.17°
$[a]_{j_{3.85}} + 205.5°$	8	12.3	206.13°
$[a]_{j_{3.85}} + 203.7°$	10	15.3	204.21°

O'Sullivan concluded from the results of these experiments, and from those of numerous others performed under a great many varying conditions, that as all that portion which was soluble in alcohol was maltose, and since the comparison of the numbers actually found with those obtained by calculation invariably showed that, besides maltose, dextrin is the only other body present in the portion insoluble in alcohol, these two bodies must be, therefore, the sole products of the action of diastase on starch.

Cold Malt-Extract is without Action on Ungelatinised Starch.[1]—When a known weight of potato-starch was mixed with cold malt-extract, and at the end of twenty-four hours the liquid filtered off, it was found unaltered in specific gravity. Had there been any action, maltose and dextrin would have passed into solution, and these would have raised the specific gravity of the fluid. The starch, after being filtered off, washed, and dried in a vacuum for thirty-six to forty hours, was also found to have not lost weight. When, in a similar experiment, water-free starch was employed, a considerable rise of temperature was observed on mixing the starch and malt extract, and the latter was found to have increased slightly in specific gravity. The starch, however, after being dried had not lost weight. When dehydrated starch was mixed with cold water, a similar rise in temperature took place, and this must be attributed to the chemical combination of water with the starch; hence the slight increase in gravity of the malt-extract was due to the loss of the small quantity of water which had entered into chemical combination with the starch.

Starch is Dissolved by Malt-Extract at the Temperature at which it Gelatinises, or a few Degrees Lower.—It was found that on adding malt-extract to starch and gradually raising the temperature, a portion dissolved at one point, a further portion at a higher stage, and so on. Complete solution was effected with potato-starch at 62° to 64° C. (144° to 147° F.). The small-grained starches, such as rice, maize, &c., were found to require a much

---

[1] This has been since found only to be true so far as potato-starch is concerned; other starches are attacked in the cold.

higher temperature for their perfect solution; and in some cases the transforming power of the malt-extract was lost before the whole of the starch granules had been altered; but in these latter cases there was seldom more than 4 per cent. of the starch left undissolved.

Malt-Extract Dissolves Gelatinised Starch almost Completely in the Cold (10° to 20° C., 50° to 68° F.).—15 c.c. of malt-extract were added to 5 grammes of gelatinised starch-paste cooled down to 16° C. (59° F.). After standing a few hours the solution was filtered from an insoluble portion (which did not amount to more than 4 per cent.). The filtrate gave no colour with iodine. After correction for the malt-extract its opticity was found to be $[a]j_{3.85} + 171°$ and its $K_{3.85}$ 43.4. These figures correspond with those of a mixture of 66.7 per cent. maltose and 33.3 per cent. dextrin. The calculated opticity for such a mixture is $[a]j_{3.85} + 171.3°$; thus the agreement is very close. These are the constants which O'Sullivan usually found when the solution was filtered and analysed a few hours after the malt-extract had been added. When the action was allowed to continue for a greater length of time, the reducing power increased and the opticity diminished, especially when the malt-extract was present in large quantity. In another case, where 4 grammes of starch were treated with 15 c.c. malt-extract, and the action allowed to proceed for twenty-four hours, the constants found were equivalent to those of a mixture of 82 per cent. maltose and 18 per cent. of dextrin. In some cases constants equivalent to as much as 90 per cent. of maltose were obtained, in others to quantities varying between this and 67 per cent. The stronger the malt-extract employed and the more prolonged the digestion, the higher was the reducing power and the lower the opticity of the transformation.

Influence of Temperature on the Opticity and Cupric Reducing Power of Starch Conversions effected by Malt-Extract. —O'Sullivan was the first to study in a definite manner the influence of temperature, time, and concentration on these constants in the process of starch conversion by malt-extract.

In making conversions at varying temperatures, care was taken that the malt-extract was not used in excess, that is, in a quantity not more than was sufficient to effect the transformation in from five to ten minutes. He found that the action, which was at first extremely rapid, came to a sudden halt, and when this point had been reached, further action was exceedingly slow. It was found that the temperature at which the action was allowed to proceed decided to a great extent the point at which this halt took place; and by studying conversions conducted at different temperatures, O'Sullivan came to the conclusion that the action proceeded in three distinct phases, to each of which he assigned an equation. These he named the equations A, B, and C.

(A.) When a starch transformation, effected by the agency of malt-extract, is conducted at any temperature below 63° C. (145° F.) and the solution filtered off in five or ten minutes, its opticity and reducing power always approximate closely to $[\alpha]j_{3\cdot 85} + 170.6°$ and $K_{3\cdot 85}$ 44.1 respectively. These numbers correspond nearly with those of a mixture of 67.85 per cent. maltose and 32.15 per cent. dextrin. In the series of experiments which led to this conclusion, and of which a large number were undertaken, the malt-extract was heated to the temperature at which the conversion was to be made, then added to the starch-paste, and the whole maintained at that particular temperature until the termination of the transformation. The filtered solution gave no colour with iodine. The following, taken from a large number of similar experiments, are given as examples. In (a.) the transformation was effected at 30° C. (86° F.); (b.) at 40° C. (104° F.); (c.) at 50° C. (122° F.); and (d.) at 60° C. (140° F.).

	Apparent percentage of Maltose.	Opticity Observed.	Opticity Calculated.
(a)	68.1	$[\alpha]j_{3\cdot 85} = + 169.7°$	$[\alpha]j_{3\cdot 85} = + 170.4°$
(b.)	67.4	$[\alpha]j_{3\cdot 85} = + 170.7°$	$[\alpha]j_{3\cdot 85} = + 170.8°$
(c.)	66.7	$[\alpha]j_{3\cdot 85} = + 169.9°$	$[\alpha]j_{3\cdot 85} = + 171.4°$
(d.)	68.3	$[\alpha]j_{3\cdot 85} = + 170.8°$	$[\alpha]j_{3\cdot 85} = + 170.2°$

O'Sullivan concluded from this that the reaction under those conditions takes place according to a definite equation, which he designates A., thus—

$$\text{A.} \quad 6C_{12}H_{20}O_{10} + 4H_2O = 4C_{12}H_{22}O_{11} + 2C_{12}H_{20}O_{10}$$
$$\text{Starch.} \qquad \text{Water.} \qquad \text{Maltose.} \qquad \text{Dextrin.}$$

and that the product consisted of 67.85 per cent. maltose and 32.15 per cent. dextrin.

When the malt-extract is not in excess, the solution can be kept for four or five hours at the temperature at which the transformation is effected without showing any great change in its opticity and cupric reducing power. In such an experiment, where the solid matter of the malt-extract was equal to 4 per cent. of the weight of the starch taken, the transformation being effected at 55° C. (131° F.), and the solution kept at that temperature for five hours, the opticity was $[\alpha]j_{3\cdot 85} = + 169.5°$ and the reducing power K 44.9. These numbers are nearly the same as those which would be given by a mixture of 68 per cent. maltose and 32 per cent. dextrin. Similar results were obtained in an experiment where the digestion had been carried on for eight hours.

When, however, malt-extract was used in excess, or when it contained more than the usual amount of acid, the maltose on prolonged digestion appeared to increase very slowly at the expense of the dextrin, whilst a small portion of the maltose itself was found to be hydrolysed to glucose.

(B.) When starch is dissolved by malt-extract at any temperature between 64° and 68° to 70° C. (147° and 155° to 158° F.), if the solution be immediately cooled and filtered, the product invariably gave constants equivalent to mixtures of 34.54 per cent. maltose and 65.46 per cent. dextrin; the opticity of such transformations being about $[a]j_{3.85} = +192.9°$ and the K 22.4.

The following experiment is given as an example of numerous others, conducted with varying quantities of starch, and with starch from different sources (barley-malt, rice, maize, &c.), in which the apparent percentages of maltose produced never varied more than 2 per cent.

Malt-extract was heated to 66° C. (151° F.) and added to starch-paste of the same temperature. In four or five minutes the solution was cooled and filtered from a small amount of insoluble matter (0.5 per cent.). The solution gave no colour with iodine; its opticity was found to be $[a]j_{3.85} = +192.9°$ and its reducing power K 22.4, numbers corresponding to those of a mixture of 34.5 maltose and 65.5 dextrin.

O'Sullivan considered that the transformation is expressed by the following equation, which he designates B. :—

B. $6C_{12}H_{20}O_{10}$ + $2H_2O$ = $2C_{12}H_{22}O_{11}$ + $4C_{12}H_{20}O_{10}$
    Starch.   Water.   Maltose.   Dextrin.

When the malt-extract was used in excess, or when it was more acid than usual, and the time of digestion was prolonged to several hours, an increase in reducing power and a decrease in opticity was observed similar to that in the case just described.

(C.) When starch is dissolved by malt-extract at temperatures from 68° to 70° C. (155° to 156° F.) to the point at which the activity of the diastase is destroyed, and the solution is cooled and filtered in five or ten minutes, the products have an opticity of $[a]j_{3.85} = +202.8°$ and a cupric reducing power of K 11.3, agreeing closely with those for 17.4 per cent. of maltose and 82.6 per cent. of dextrin.

The following equation (C.) exhibits the change which is supposed to take place :—

C. $6C_{12}H_{20}O_{10}$ + H_2O = $C_{12}H_{22}O_{12}$ + $5C_{12}H_{20}O_{10}$
    Starch.   Water.   Maltose.   Dextrin.

In numerous transformations conducted at temperatures between 69° to 76° C. (157° to 167° F.), the maltose apparently produced did not differ by more than 2 per cent.

The dextrin produced in this transformation can also be further degraded by employing excess of malt-extract, and by prolonging the time of digestion, as in the former cases A. and B.

O'Sullivan considers that the average quantity of malt-extract necessary to effect a starch conversion should contain solid matter equal to 5 per cent. of the starch taken, although, as he points out, malts differ very much in their starch-converting power.

It is suggested that the cause of this difference in the action of malt-extract on starch at varying temperatures is due most probably to the action of heat on the malt-extract. This may contain three transforming bodies—one which acts upon starch according to the A. equation, and has its activity destroyed at a temperature of 64° C. (148° F.); another which affects starch according to the B. equation, the action of which is arrested when a temperature of 68° to 70° C. (155° to 158° F.) is reached; whilst that which acts in accordance with equation C. is able to survive these temperatures.

In support of this it was found that if a portion of malt-extract was heated to 67° C. (153° F.) for a few minutes, cooled down to 60° (140° F.), and added to gelatinised starch at the latter temperature, the transformation was in accordance with equation B. and not with A. Similarly, by heating a portion of malt-extract to 75° C. (168° F.), and effecting a conversion at 68° C. (155° F.), the products of the action were in accordance with the C. equation.

Views of Musculus and Gruber.—In 1878 these observers published a paper[1] in which they sought to show that the achroodextrin of Brücke, the β of O'Sullivan, and the dextrin II. of Griessmayer were not individual substances, but consisted of three dextrins, which they designated in the following manner:—

Achroodextrin I.	$[a]j = +210°$	$\kappa\ 12$
Achroodextrin II.	$[a]j = +199°$	$\kappa\ 12$
Achroodextrin III.	$[a]j = +190°$	$\kappa\ 12$

When these dextrins were isolated and treated with fresh diastase they showed the following remarkable modifications:—

Achroodextrin I.	$[a]j = +159°$	$\kappa\ 36$
Achroodextrin II.	$[a]j = +168°$	$\kappa\ 20$
Achroodextrin III.	Unchanged.	

They eventually considered the No. II. dextrin to be a mixture of Nos. I. and III., and in the latter portion of their paper only spoke of Nos. I. and III., which they called α and β. They also described another dextrin, which they styled γ. This was produced by adding fresh diastase to a starch conversion made at 50° or 60° C. (122° or 140° F.), and allowing the action to proceed for some time in the cold. It possessed an opticity of $[a]j + 150°$, and a κ of 28. Though these bodies were obtained in an impure condition, still the remarkable difference in their behaviour when further treated with diastase pointed to striking differences in their respective characters. In these experiments diastase prepared from malt was employed.

It was in this paper that Musculus and Gruber first enunciated the theory that when starch breaks down under the agency of diastase, the action proceeds in successive stages; the starch

[1] *Bull. Soc. Chim.*, xxx. 54.

molecule splitting up first into maltose and a dextrin having a molecular weight nearly as high as that of starch. This newly formed dextrin then further splits up into maltose and a dextrin of less molecular weight, and so on, until the last dextrin is produced, which is simply hydrolysed to maltose.

They were led by these considerations to regard starch as a polysaccharide containing the group $C_{12}H_{20}O_{10}$ five or six times.

Investigations of Brown and Heron.—The view of Musculus and Gruber received further confirmation in the results of a series of experiments undertaken by Brown and Heron which were communicated to the Chemical Society in 1879.[1] In these potato-starch was employed, which, after being purified by successive treatment with dilute solution of caustic soda and 1 per cent. hydrochloric acid, was finally washed until free from all traces of acid. Great care was taken to insure thorough gelatinisation of the starch used in the experiments. It was evenly mixed with cold water, and then poured into the necessary quantity of boiling water with constant stirring. After the solution of gelatinised starch had been cooled down to the temperature at which the transformation was to be effected, the necessary quantity of malt-extract was dropped in either from a burette or a pipette, and the whole was then placed in a water-bath maintained at the required temperature by means of a thermostat. Since the authors were not satisfied with the method then in use for preparing diastase, malt-extract was invariably employed. This was prepared by intimately mixing together 100 grammes of finely ground pale malt with 250 c.c. of distilled water, the mixture being allowed to stand for from six to twelve hours. The bright filtrate from this, which was designated normal malt-extract, and which had a specific gravity of from 1036 to 1040 (water = 1000), was used in the majority of the experiments. The specific gravity, opticity, and reducing power were determined either before or after heating, according to the nature of the experiment. At the end of each conversion the quantity of liquid was either measured or weighed, and its volume calculated from its specific gravity. All portions removed for the purpose of testing with iodine were carefully measured and allowed for. In those cases where more than 3 c.c. of malt-extract were used, or in which the transformation lasted more than twenty minutes, a portion of malt-extract was digested alongside the conversion in a separate vessel, and in this way submitted to the same conditions of time and temperature as the starch conversion itself. This precaution was found necessary, since malt-extract was found to change rapidly even at ordinary temperatures; its specific gravity and reducing power increasing, and its opticity diminishing. Malt-extract rapidly enters into a state of active fermentation, caused by the presence of various bacterial organisms. If kept at a temperature

[1] *Journal of the Chemical Society*, 1879, p. 596.

of 40° to 45° C. (104° to 113° F.), bacilli alone appear, and a fermentation accompanied with the copious evolution of carbon dioxide and hydrogen gases makes its appearance, the chief product of the fermentation being butyric acid. But the changes in the opticity and chemical conditions above mentioned do not depend upon the presence of organised ferments, for the same changes take place when the solution has been heated to a temperature high enough to paralyse such organisms. Malt-extract, after having been heated to a temperature of 75° C. (165° F.), may be kept at a temperature of 50° C. (122° F.) without undergoing much change, and this shows that the alteration is not due to hydrolysis caused by the acids present in the malt-extract. The change was partly attributed to the action of an enzyme always present in malt, which possesses the power of hydrolysing the cane-sugar invariably present in the malt-extract; partly also to the action of the diastase on some of the starch granules which are unavoidably ruptured during the process of grinding the malt. The contents of these suffer hydrolysis and are partially converted into maltose.

Action of Diastase on Unruptured Starch.—The previous observations of O'Sullivan (p. 128) were confirmed, it being found that diastase was entirely without action on unruptured potato-starch.

Action of Diastase on Bruised Starch.—When starch was ground up in a mortar with quartz sand or powdered glass, and then exposed to the action of malt-extract, it was readily affected by diastase in the cold. The following are the results of an experiment conducted in this way, the bruised starch being acted upon by a solution containing 10 c.c. normal malt extract in every 100. c.c.:—

Matter found in Solution = 3.155 *grammes*

	$[a]_{3·86}$	$[\kappa]_{3·86}$	Factors corresponding to Mixtures of		
			Maltose.	Dextrin.	Cellulose.
			Per Cent.	Per Cent.	Per Cent.
After 4 hours . . .	+155.4	51.0	83.6	13.9	2.5
After 20 hours . .	+152.0	54.4	89.1	8.5	2.4

All the transformations had very low opticities, and when these were compared with the reducing power, the two did not coincide with the assumption that maltose and dextrin only were present. This discrepancy was attributed to the presence of a small quantity of starch cellulose which had passed into solution.[1]

---

[1] It has been since found (*Journal of the Chemical Society*, 1895, p. 35) that this discrepancy is not due to the cause here stated, but to the fact that the opticity of maltose, when first produced by hydrolysis, is somewhat lower than when its solution has been boiled or has stood for a time (multirotation).

Action of Malt - Extract upon Gelatinised Starch in the Cold.—This was found to be extremely rapid, for on treating a 3 to 4 per cent. solution of gelatinised starch with 5 to 10 c.c. of normal malt-extract, limpidity ensued in from one to three minutes; immediately afterwards iodine ceased to give a blue colour; the brown colour, however, persisted for another five or six minutes. When this point was reached, the solution could be filtered off quite bright and clear, and its opticity was found to be at its lowest point. This and the reducing power do not point to the apparent exclusive presence of maltose and dextrin; evidently a little cellulose also goes into solution. (See previous note.) After a time a deposition of starch cellulose takes place, rendering the liquid turbid. Concurrently with this the opticity and reducing power slowly begin to rise, arriving at their maxima in about three hours, when they now coincide with the apparent presence of maltose and dextrin alone in the solution. The following is a typical experiment:—

3.7 *Grammes of Starch*, 10 c.c. *Normal Malt-Extract in* 100 c.c.

Time.	$[a]_{j_{3.86}}$	$\kappa_{3.86}$	Iodine Reaction.
5 minutes	154.6	...	Slightly brown
15 minutes	145.6	...	None
30 minutes	154.6	...	None
60 minutes	157.5	44.1	None
3 hours	161.6	49.7	None

Action of Malt-Extract, which had been Previously Heated, upon Gelatinised Starch in the Cold.—In this case, though the power of converting starch into soluble starch, as shown by the solution rapidly liquefying, did not suffer much alteration, the further portion of its action, as shown by the iodine test and by the opticity of the liquid, was rendered much slower, and this in proportion as the temperature to which the malt-extract had been previously subjected was increased. Transformations made with malt-extract which had been previously heated to 66° C. (151° F.) were much more transparent than those in which unheated malt-extract was employed.

Action of Malt-Extract upon Gelatinised Starch at Elevated Temperatures.—In order to be able to study the changes taking place at any given point in a starch transformation, it was requisite to discover some method by means of which diastatic action could be arrested at any given point. Boiling the solution, though it completely arrested the further action of the diastase, was unsatisfactory; changes invariably took place during the time required to bring the fluid to the boiling-point. The authors of the paper found that salicylic acid implicitly fulfilled this condition, for when it was added to transformations in quantities of

0.05 gramme to each 100 c.c. of the solution, all further diastatic action was completely arrested. It became now a comparatively easy matter to show by means of a curve the progress of any starch transformation. The authors agree with the previous observations of O'Sullivan (p. 132) that the changes brought about in the malt-extract by the agency of heat are the cause of the differences in its behaviour as a transforming agent, but they were unable to obtain the various halting stages described by O'Sullivan when the processes were carried out at the respective temperatures indicated by him.

Action of Malt-Extract in Transformations conducted at 40° C. (104° F.).—When a conversion was made at this temperature, the process differed in several particulars from one conducted in the cold. Unless the starch-paste was very thick, or the malt-extract very small in quantity, no separation of starch cellulose took place,[1] for, instead of being precipitated, it was converted into maltose and dextrin. The solution consequently remained clear, and its opticity and reducing power coincided with the apparent presence of maltose and dextrin only. The transformation proceeded with great rapidity, and the halting stage, which seemed to be exactly at the same point as when the process took place in the cold, was reached in about thirty minutes. When this stage was attained, further action was exceedingly slow, even when further quantities of the same (i.e. the heated) malt-extract were added. The following are the details of an experiment in which a solution of 5 grs. of starch per 100 c.c. of water were acted upon by normal malt-extract which had been heated for twenty minutes to 40° C. (104° F.), 10 c.c. malt-extract being added for each 100 c.c. of solution:—

Temperature of the Conversion 40° C. (104° F.).

Time.	$[a]j_{3.86}$	$\kappa_{3.86}$	Iodine Reaction.
2½ minutes . . .	+164.1°	...	Full brown
15 minutes . . .	+163.3°	...	Brown
30 minutes . . .	+163.3°	48.8	None

These last figures correspond closely with those of a mixture of 80 per cent. maltose and 20 per cent. dextrin.

After a further addition of 10 c.c. of the same malt-extract per 100 c.c. of solution, the process was allowed to proceed for thirty minutes longer; the opticity was then $[a]j_{3.86} + 161.6°$ and the $\kappa_{3.86} = 49.7$, numbers corresponding with those of a mixture of 81.3 per cent. maltose and 18.7 dextrin. Fermentation had now set in, and this precluded further observation.

Action of Malt-Extract at 50° C. (122° F.).—In a transformation conducted at 50° with malt-extract, which had been

[1] See note, p. 134.

previously heated to that temperature for twenty minutes, and added at the rate of 10 c.c. to each 100 c.c. of 5 per cent. starch solution, the following results were obtained:—

Time.	$[a]j_{3.86}$	$\kappa_{3.86}$	Iodine Reaction.
2½ minutes	+168.1°	...	Full brown
5 minutes	+166.5°	...	Brown
15 minutes	+166.5°	...	Brown
30 minutes	+162.7°	...	No coloration
60 minutes	+162.3°	49.6	No coloration

The last figures correspond with those of a mixture of maltose 81.3 per cent. and dextrin 18.7 per cent.

On the addition of a further quantity of malt-extract there was at first little or no action, but after standing for sixteen hours the opticity had fallen to $[a]j_{3.86}$ +147.7° and the reducing power had risen to $\kappa_{3.86}$ 60.8, numbers which correspond closely with those for maltose only.

Action of Malt-Extract Heated to 60° C. (140° F.).—The following are the results of an experiment similar to the preceding, in which the transformation was conducted at 60° C. with malt-extract which had been previously heated for twenty minutes to that temperature, the further action of the diastase being arrested by salicylic acid at the various points indicated:—

Time.	$[a]j_{3.86}$	$\kappa_{3.86}$	Iodine Reaction.
1 minute	+191.7	...	Pure blue
2½ minutes	+175.6	...	Brown
5 minutes	+166.8	...	Very light brown
15 minutes	+165.7	...	None
30 minutes	+163.7	...	None
60 minutes	+162.9	49.6	None
More malt-extract added at the rate of 4.5 c.c. per 100 c.c.			
90 minutes	+161.3	...	...
More malt-extract added at the rate of 4.2 c.c. per 100 c.c.			
120 minutes	+160.1	52.0	...

An opticity of $[a]j = +162.6$ and a reducing power of κ 49.3 was deduced as the mean of a large number of starch transformations conducted at a temperature of 60° C. (140° F.). In these from 19 to 20 c.c. of malt-extract per 100 c.c. of starch solution were used, and the conversions allowed to proceed for lengths of time varying from five to sixty minutes. These numbers correspond very closely with those of a mixture of 81.3 per cent. maltose and 18.7 dextrin.

This was the temperature at which O'Sullivan states that he obtained his A. equation, the halting stage being reached when the conversion had an opticity of $[a]j_{3.85} + 170$ and a $K_{3.85}$ of 44.1. Brown and Heron were, however, unable to detect any pause in the reaction at this point.

Action of Malt-Extract Heated to 66° C. (151° F.).—If malt-extract be heated rapidly to 66° C. and then immediately added to a solution of gelatinised starch maintained at that temperature, the reaction takes place in its earlier stages very much in accordance with that of a transformation conducted with malt-extract heated to 60° C. (140° F.); but if the malt-extract be kept for ten or fifteen minutes or longer at 66° C., its action is considerably modified. Its general character is much slower, although the formation of soluble starch, as indicated by complete limpidity in the liquid, is accelerated. The following is an illustration of such a conversion in which malt-extract, previously heated to 66° for twenty minutes, was added in the proportion of 8.92 c.c. to every 100 c.c. of a 6 per cent. starch solution. The temperature of the conversion was 65° C., and the action was arrested by salicylic acid at the points indicated :—

Time.	$[a]j_{3.86}$.	Iodine Reaction.
2½ minutes	+204.0°	Full brown
5 minutes	+201.2°	Brown
10 minutes	+195.5°	Light brown
20 minutes	+191.8°	Very light brown
30 minutes	+188.2°	No reaction
60 minutes	+185.0°	No reaction
Same malt-extract added at the rate of 4.3 c.c. per 100 c.c.		
90 minutes	+178.0°	...
Same malt-extract added at the rate of 4.3 c.c. per 100 c.c.		
120 minutes	+178.0°	...

The point at which the iodine reaction for erythrodextrin disappeared was markedly different in transformations conducted at 66° C. and in those at 60° C. In the former the disappearance of the brown colour took place when the opticity had fallen to $[a]j_{3.86} = 188°$ or $189°$, whilst in the latter it was still to be observed when the angle had fallen as low as $[a]j_{3.86} = 165°$ or $166°$. The reason assigned for this was that during the process of heating the various transforming agents in the malt-extract suffer alteration in different degrees; thus in transformations conducted at 60° C. (140° F.) those portions of the malt-extract whose function it is to split up and degrade the higher achroodextrins effect this more rapidly than the erythrodextrins can be produced or transformed into the higher achroodextrins; consequently much maltose and much of the lower achroodextrins may be present in the solution before the erythrodextrins and soluble starch are converted into achroodextrins.

Action of Malt-Extract at Temperatures Higher than 66° C. (151° F.).—When malt-extract is heated to a temperature between 66° C. and 75° to 76° C. (167° to 169° F.) its action becomes further modified. A lower opticity than $[a]j_{3.86} 195°$ is seldom reached, even on the addition of further quantities of the same malt-extract and prolonged digestion.

The following are the results of an experiment conducted at 76° C. with 30 c.c. malt-extract heated to 76° C. for each 100 c.c. of starch solution :—

Time.	$[a]j_{3\cdot 86}$.	$\kappa_{3\cdot 86}$.	Iodine Reaction.
2½ minutes	+205.0°	...	Much erythrodextrin
5 minutes	+200.3°	...	Much erythrodextrin
10 minutes	+196.5°	...	Slight coloration
20 minutes	+195.5°	...	No reaction
30 minutes	+195.5°	...	No reaction
45 minutes	+194.6°	...	No reaction
60 minutes	+194.6°	23.0	No reaction
15 c.c. same malt-extract added per 100 c.c.			
70 minutes	+194.4°	...	...
90 minutes	+194.0°	...	...

The starch solution became limpid very quickly, and the blue or violet reaction with iodine indicative of starch disappeared in less than two minutes, but the brown colour due to erythrodextrin persisted for a considerable time. This latter attained its highest point when the opticity had reached $[a]j_{3.86} + 202°$ to $203°$, and disappeared at 194° or a little lower.

This corresponds very closely with O'Sullivan's C. equation, which has an opticity of $[a]j_{3.85} + 202.8°$, but he describes the dextrin obtained from it to be an achroodextrin. Brown and Heron invariably found it was at this point that the solution gave the strongest brownish-red colour with iodine, characteristic of erythrodextrin. They concluded that in these cases the dextrin formed was invariably an erythrodextrin.

Influence of Neutralisation upon the Action of Malt-Extract.—When malt-extract, after being heated to 66° C. (151° F.), was neutralised with barium hydrate, its action was but little modified, and the curve given by a transformation with malt-extract treated in this way did not differ materially from one made with malt-extract which had been simply heated to 66° C.

If, however, the malt-extract be heated to 66° C., rendered very slightly alkaline with sodium carbonate, and again momentarily heated to 66°, the opticity of the solution was reduced rapidly to $[\alpha]j_{3.86} + 194°$ to $196°$, and it was impossible to push the action beyond this, even by the addition of further quantities of similarly treated malt-extract. The curve obtained in this way did not differ materially from that obtained with malt-extract simply heated to $75°$ to $76°$ C. ($167°$ to $169°$ F.).

When a conversion was made with malt-extract heated to $66°$ C. ($151°$ F.), and then made slightly alkaline with sodium hydrate, the opticity never fell below $[\alpha]j_{3.86} + 202°$. This point corresponds with that at which iodine gives the maximum reaction for erythrodextrin in conversions made with malt-extract heated to from $66°$ to $76°$ C. ($141°$ to $169°$ F.). Further additions of the same alkaline malt-extract failed to bring down the angle farther.

When the malt-extract was rendered still more alkaline with sodium hydrate, its diastatic power was completely destroyed.

The Molecular Transformations of Starch.—The authors consider that the results of these experiments show that there are at least four well-defined molecular transformations of starch, the one most easily obtained being that with malt-extract heated to a temperature not higher than $60°$ C. ($140°$ F.). This has an opticity of $[\alpha]j_{3.86} + 162.6°$, and a reducing power of $\kappa_{3.86}$ 49.3.

The next is that at which the iodine reaction for erythrodextrin disappears in transformations conducted with malt-extract heated to $66°$ C. ($151°$ F.). Its opticity was $[\alpha]j_{3.86} = +188.5$, and its reducing power $\kappa_{3.86}$ 25.0 or thereabouts.

The next is obtained by using malt-extract heated to $66°$ C., $151°$ F., and made slightly alkaline with sodium carbonate. Its opticity is $[\alpha]j_{3.86} + 195°$ to $196°$, and its reducing power $\kappa_{3.86}$ 18.9.

The last is obtained by using malt-extract heated to $66°$ C., and made slightly alkaline with sodium hydrate; also in conversions conducted at temperatures beyond $66°$ C. ($151°$ F.), when the iodine reaction for erythrodextrin is at its maximum. It has an opticity of $[\alpha]j_{3.86} + 202°$ to $203°$, and a reducing power of κ 12.7. This agrees closely with O'Sullivan's C. equation; the dextrin formed is, however, an erythrodextrin.

Transformations from Higher to Lower Equations.—If a little unheated malt-extract be added to any starch transformation having an opticity higher than $[\alpha]j_{3.86} + 162.5°$, and the whole be kept at a temperature of $50°$ to $60°$ C. ($122°$ to $140°$ F.), the opticity of the solution falls very quickly, in favourable cases almost instantaneously, to $[\alpha]j_{3.86} = 162.5°$, the reducing power rising to $\kappa_{3.86}$ 48.3, at which point it remains stationary for some time. For instance, 5 c.c. unheated malt-extract were added to 100 c.c. of a starch conversion which had been boiled, and which had an opticity of $[\alpha]j_{3.86}$ 187.8° and a reducing power of $\kappa_{3.86}$ 28.9; at the end of two minutes the opticity had fallen to $162.6°$ and the reducing power risen to $\kappa_{3.86}$ 49.3.

The remarkable fact that hydrolysis proceeds rapidly to a certain point and then becomes suddenly arrested, points to a marked difference in character between the higher and the lowest dextrin. The most natural explanation is that the dextrins are not metameric but polymeric bodies, those which possess the highest opticities having the largest molecules.

In consonance with the view of Musculus and Gruber (p. 132), Brown and Heron regard soluble starch as $10C_{12}H_{20}O_{10}$. The first step in the breaking down of this large complex molecule is the removal of one of the $C_{12}H_{20}O_{10}$ groups, which, uniting with a molecule of water, becomes maltose; and the remaining nine $C_{12}H_{20}O_{10}$ groups constitute the first dextrin of the series—erythrodextrin α, thus—

1. $10(C_{12}H_{20}O_{10}) + H_2O = C_{12}H_{22}O_{11} + 9(C_{12}H_{20}O_{10})$
   Starch.   Water.   Maltose.   Erythrodextrin α.

This dextrin next suffers decomposition; one of its $C_{12}H_{20}O_{10}$ groups is similarly split off, combines with a molecule of water to form maltose, and the remaining eight $C_{12}H_{20}O_{10}$ groups constitute the second dextrin of the series—erythrodextrin β—

2. $9(C_{12}H_{20}O_{10}) + H_2O = C_{12}H_{22}O_{11} + 8(C_{12}H_{20}O_{10})$
   Erythrodextrin α.   Water.   Maltose.   Erythrodextrin β.

The next stage is—

3. $8(C_{12}H_{20}O_{10}) + H_2O = C_{12}H_{22}O_{11} + 7(C_{12}H_{20}O_{10})$
   Erythrodextrin β.   Water.   Maltose.   Achroodextrin α.

The process goes on in a similar manner until the last $C_{12}H_{20}O_{10}$ group is transformed into maltose.

It is obvious that the number of dextrins possible will depend upon the size of the molecule of the last formed dextrin; if this contains the same number of carbon atoms as maltose, there will be probably nine distinct stages; if it contain twice as many, then the number can be only eight. The following table gives the opticity, reducing power in each of these theoretical stages, and the mixtures of maltose and dextrin with which these factors correspond:—

No. of Equation.	Opticity.	Reducing Power.	Apparent		
			Maltose.	Dextrin.	
	$[α]_{j_{3·86°}}$	$κ_{3·86°}$	Per Cent.	Per Cent.	
Soluble starch	+216.0°	0.0	...	...	...
1	+209.0°	6.4	10.5	89.5	Erythrodextrin α
2	+202.2°	12.7	20.9	79.1	Erythrodextrin β
3	+195.4°	18.9	31.0	69.0	Achroodextrin α
4	+188.7°	25.2	41.3	58.7	Achroodextrin β
5	+182.1°	31.3	51.3	48.7	Achroodextrin γ
6	+175.6°	37.3	61.1	38.9	Achroodextrin δ
7	+169.0°	43.3	71.0	29.0	Achroodextrin ε
8	+162.6°	49.3	80.9	19.1	Achroodextrin ζ
9	+156.3°	55.1	90.2	9.8	Achroodextrin η
Maltose	+150.0°	61.0	100.0	0.0	...

Brown and Heron considered that by their experiments they had unmistakably established the existence of the No. 2, 3, 4, and 8 equations, whilst indications of the others had been observed, though not with the same degree of certainty. Of these equations No. 8 was by far the most stable; and as no pause could be observed in the process of hydration between it and maltose, it favoured the view that there are only eight dextrins. The result of all these experiments tended to show that soluble starch, when hydrolysed by malt-extract, was invariably split up into maltose with an opticity of $[a]j_{3.86} + 150°$ and a reducing power of $\kappa_{3.86}$ 61, and a series of polymeric dextrins, all of which had an opticity of $[a]j_{3.86} + 216°$ and possessed no reducing power.

Diastase has no Action on Maltose.—A number of experiments were made to ascertain if diastase possessed the power of hydrolysing maltose. These were attended with entirely negative results; in no case was glucose found even after the maltose had been subjected to the action of malt-extract for twenty-eight hours.

Investigations of Brown and Morris.—In 1885 Brown and Morris published a paper[1] which may be regarded as a continuation of that of Brown and Heron of 1879. In this they give the results of a number of experiments undertaken with a view of still further elucidating the molecular changes which take place when starch is hydrolysed.

It had been previously pointed out by Brown and Heron that when starch-paste is acted on by malt-extract at any temperature above 40° C. (104° F.), the opticity and reducing power of the products both appear to indicate the existence of maltose $[a]j_{3.86} + 150°$, $\kappa_{3.86}$ 61, and a non-reducing dextrin with an opticity of $[a]j_{3.86} + 216°$. They now showed that the same held good with the various fractions obtained by precipitation with alcohol; the opticity and reducing powers of such fractions invariably pointed to their being apparently composed of maltose and a non-reducing dextrin. A large number of experiments were made in proof of this, of which the following is an example:—A starch conversion was made which gave an opticity of $[a]j_{3.86} + 196.4°$ and a $\kappa_{3.86}$ of 18.6, figures which correspond closely to a composition of maltose 30.4 per cent. and dextrin 69.6 per cent. The conversion was evaporated to a syrup, heated to 100° C. (212° F.), and 90 per cent. of alcohol added until a permanent precipitate was produced. The solution, which was found to contain 48.3 per cent. of alcohol, was then allowed to become cold, and the clear supernatant liquid poured off from the residue. As this latter chiefly consisted of malt-extract bodies, it was neglected. The clear liquid was heated on the water-bath, more alcohol added until a permanent precipitate again formed, cooled, the precipitate allowed to settle, and the clear liquid, which was found to contain 48.34 per cent. of alcohol, poured off. This same process was repeated five times; consequently, five fractions

---

[1] *Journal of the Chemical Society*, 1885, p. 527.

and a mother-liquor were obtained. Each separate fraction was dissolved in water and re-precipitated with alcohol, again taken up with water, boiled to remove alcohol, evaporated to small bulk, and finally dried in a vacuum. The opticity and reducing power of each fraction are given in the following table :—

Fractions.	Strength of Alcohol.	Opticities.	Reducing Powers.	Apparently indicating	
				Maltose.	Dextrin.
	Per Cent.	$[\alpha]_{j\,3\cdot 86°}$	$\kappa_{3\cdot 86°}$	Per Cent.	Per Cent.
1	48.3	+209.8°	5.49	9.0	91.0
2	59.4	+209.6°	8.85	14.5	85.5
3	67.0	+203.1°	9.00	14.7	85.3
4	74.0	+202.4°	13.18	21.6	78.4
5	81.5	+199.1°	14.53	13.8	73.2
Mother-liquor . .	...	+185.6°	29.40	48.1	51.9

The results of this and numerous experiments of a similar nature led Brown and Morris to the conclusion that "just as the total products of all starch transformations brought about by malt-extract always yield numbers indicating the existence only of maltose and a non-reducing dextrin, so it is equally beyond question that the same holds good with regard to any portion of those products which can be separated by fractionation with alcohol."

They consider that they have thus "established a criterion of purity for the products of the action of diastase on starch." And "in those cases in which the observations of other writers do not conform to the above rules," they "have no hesitation in ascribing the discrepancy either to the presence of impurities or to errors in analysis." In more recent times they speak of this as the "law of definite relation"; consequently, if the opticity of a starch transformation is known, the reducing power can be deduced from it, and *vice versa*.[1]

As had been previously shown (p. 140), starch transformations of a higher opticity than 162.6° and a lower reducing power than κ 49.3 are speedily brought down to this stage by being treated with a little fresh (unheated) malt-extract at 50° C. (122° F.). Since maltose is unaffected by this treatment, the change must be owing solely to a degradation of the dextrins.

After making sure that the dextrins suffered no change by the various boilings, evaporations, and precipitations with alcohol necessary to secure their isolation, experiments were made to ascertain if, when isolated, they could be degraded by treatment with fresh malt-extract. It was conclusively shown that the dextrins, when thus isolated, could be degraded as readily as the original substance from which they were prepared. Each dextrin, then, with the exception of that belonging to the No. 8 equation, when treated with fresh malt-extract at 50° C. (122° F.), ought to yield a certain

[1] See Appendix C.

definite amount of maltose. The different quantities which should be yielded in this way are given in the accompanying table :—

No. of Equation.	Constants of Mixed Products.		Apparent Yield of Maltose by 100 Parts of each Dextrin when Degraded to No. 8 Equation.
	$[\alpha]_{j_{3.86}}$	$\kappa_{3.86}$	
Soluble starch	+216.0°	0.0	84.44
1	+209.0°	6.4	82.09
2	+202.2°	12.7	79.20
3	+195.4°	18.9	75.39
4	+188.7°	25.2	70.37
5	+182.1°	31.3	63.33
6	+175.6°	37.3	52.77
7	+169.0°	43.3	35.18
8	+162.6°	49.3	0.00

On ascertaining the amount of maltose yielded by any one of the dextrins, it ought to be possible, by the aid of this table, to fix its position in the series, or with a mixture of dextrins to find their mean position. It was shown by a number of experiments, of which the following is an example, that all transformations which had higher opticities and lower reducing powers than those required for the No. 8 equation, always contained several, and never one individual dextrin. A starch transformation was made with a little freshly prepared diastase, the action being arrested when the maximum of coloration was given by iodine. This possessed an opticity of $[\alpha]_{j_{3.86}} +202.2°$, and a reducing power of $\kappa_{3.86}$ 12.4, which are almost exactly those of the Equation 2, and which closely corresponded to those of a mixture of maltose 20.3 per cent. and dextrin 79.7 per cent. On further degradation with malt-extract at 60° C. (140° F.), 75 per cent. of maltose was yielded, against 79.2 per cent. as required by theory.

A measured quantity of this transformation was then evaporated to a syrup and fractionated with alcohol of gradually increasing strength. The weight of each fraction was taken and calculated as a percentage on the whole, and the amount of maltose apparently yielded by each fraction on subsequent degradation with fresh malt-extract also. These were found to be as follows :—

Fractions.	Strength of Alcohol.	Percentage of Solid Matter to Total Solid Matter.	Maltose apparently Yielded by each 100 Parts of Dextrin.
	Per Cent.		Parts.
1	46.0	43.88	60.8
2	59.3	23.09	70.8
3	75.0	10.64	87.7
4	...	22.39	93.9

The true mean of these figures gives 73.3 per cent. maltose against the 75 per cent. yielded by the original solution, numbers which agree as closely as could be expected in an experiment of this difficult nature. These experiments show that though it is possible to give theoretically well-marked equations representing the hydrolysis of starch, yet in practice, with the exception of the No. 8 equation, no such well-defined equation is met with. The action is evidently exceedingly irregular, some of the $C_{12}H_{20}O_{10}$ groups being hydrolysed more quickly than others; consequently, the opticity and reducing power of any transformation having a higher rotatory power than that of the No. 8 equation does not represent one definite equation, but the mean of several.

It was then suggested that the starch molecule might be even more complex than the authors had hitherto assumed it to be in their former paper, where it was considered to be equal to ten $(C_{12}H_{20}O_{10})$ groups, and that it might be some simple multiple of these ten groups. In such a case the number of dextrins would be correspondingly increased.

Brown and Morris now attempted to obtain the dextrins in a state of purity. The opinions of various observers have differed considerably as to whether the dextrins do or do not possess a reducing action on Fehling's solution. In the experiments of O'Sullivan previously described (p. 124), the reducing power they were found to possess was attributed to traces of maltose which it seemed impossible to completely remove. An unsuccessful attempt to destroy these last traces of maltose was made by Brown and Morris by boiling with Fehling's solution. Bondonneau[1] described a method by means of which he thought he had succeeded in accomplishing this object. It consisted in boiling the dextrin solution for half-an-hour with cupric chloride, to which just sufficient sodium hydrate had been added as would redissolve the precipitated cupric hydrate. The filtered liquid, on being acidulated with hydrochloric acid, was said to yield, on the addition of alcohol, a dextrin entirely free from maltose. Brown and Morris made several attempts to prepare a non-reducing dextrin by this method, but never succeeded in obtaining one with a lower reducing power than κ 0.53. Wiley in 1882[2] designed a method for the analysis of mixtures of maltose, glucose, and dextrin, in which Knapp's mercuric cyanide solution was employed in a novel manner. It was used to destroy the maltose and glucose, under the supposition that the dextrin would be left intact, and that its amount could be afterwards estimated by the polarimeter. By working with this solution, Brown and Morris were able to deprive several of the dextrins of every trace of reducing power, and this they attributed to the complete removal of maltose (*cf.*, however, Scheibler and Mittelmeier's views on p. 155). The following is an experiment of this kind:—

---

[1] *Bull. Soc. Chim.*, xxi. 50 and 159.
[2] *Chemical News*, xlvi. 175.

A dextrin was prepared from a starch transformation and purified by precipitation and re-precipitation several times with 60 per cent. alcohol, the spirit employed being always of the same strength, in order to avoid any alteration in its composition. A slight excess of a solution of equal weights of mercuric cyanide and sodium hydrate [1] was added to a solution of this dextrin, and the whole kept warm until reduction was deemed complete. The solution was cooled, acidulated with hydrochloric acid, and hydrogen sulphide passed through until all the mercury was removed. Ammonia was now added in slight excess to the filtered solution, the whole evaporated to a syrup, this latter was dissolved in hot water, and the dextrin precipitated therefrom with 60 per cent. alcohol. In this way they thought that they had obtained a dextrin entirely devoid of reducing power. This, when degraded with malt-extract at 60° C. (140° F.), gave (after corrections had been made for the salts present) 51.5 per cent. maltose against 49.5 per cent. yielded by the original dextrin; thus it appeared to have suffered no alteration during the process of purification. Its opticity was 215.9°, a value which agrees almost exactly with that which the authors had previously deduced from their experiments on the transformation of starch, i.e. $[\alpha]_{j_{3.86}} + 216.0°$.

Maltodextrin.—The discrepancy in the results of the experiments where the dextrins were degraded with fresh malt-extract (an illustration of which is given on p. 144, where it is seen that while fractions 1 and 2 yield considerably less maltose, 3 and 4 yield considerably more than the quantity required by theory) is explained by the presence of a body which the authors had isolated and named "maltodextrin." It was obtained in the following manner.

Preparation of Maltodextrin.—A starch transformation was made at a temperature of 60° to 65° C. (140° to 149° F.) with a little freshly prepared diastase, the action being arrested by boiling the solution when its opticity had reached $[\alpha]j + 198°$ or thereabouts. The liquid was then evaporated to a specific gravity of 1060, and set to ferment at 28° to 30° C. (82° to 86° F.) with a small quantity of ordinary yeast. When fermentation had ceased, the liquid was filtered, evaporated to a syrup, and digested in boiling 90 per cent. alcohol for a couple of days, with frequent agitation. The object of this treatment was to remove the non-volatile products of fermentation, which amounted to about 5 per cent. of the maltose fermented. The strength of the alcohol was then reduced to 85 per cent., the mixture submitted to further digestion, and the liquid portion, while still hot, decanted from the residue. On removing the alcohol by distillation, maltodextrin was left behind, and this was purified by further treatment with alcohol. Thus obtained, it was found to possess an opticity of $[\alpha]_{j_{3.86}} + 193.6°$

[1] Morris (*Trans. Lab. Club*, i. 70) states that a preferable method is to treat the dextrin solution on the water-bath with mercuric oxide and barium hydrate solution.

and a reducing power of $\kappa_{3.86}$ 20.7, numbers which agree with those of a mixture of 33.9 per cent. maltose and 66.1 per cent. dextrin. Maltodextrin is absolutely unfermentable by the *Saccharomyces cerevisiæ*, and cannot be separated into its constituents by further treatment with alcohol; when acted upon by a little fresh malt-extract it is *completely hydrolysed to maltose*. Here, then, is a body apparently showing by its reducing power that it contains 34 per cent. of maltose which is absolutely unfermentable by the yeast of the primary fermentation.

Hertzfeld's Maltodextrin. — Hertzfeld described in 1879[1] a body which he had found in starch conversions made with diastase at a temperature of 70° C., and to which he gave the name "maltodextrin." He regarded it as a body intermediate between achroodextrin and maltose, and assigned to it the following composition $\left\{ \begin{array}{c} 2C_6H_{10}O_5, \\ C_6H_{12}O_6, \end{array} \right.$ that is, two molecules of dextrin and one molecule of glucose. He found its opticity to be $[\alpha]j + 171.6°$ and its reducing power κ 23.5; but as he was under the impression that one part of glucose gave on reduction 2.362 parts of copper oxide instead of 2.205, a correction has to be applied to his κ, after which it becomes 26.7.

Since this would point to a composition of 43.70 per cent. maltose and 49.12 dextrin, the opticity ought to have been $[\alpha]j + 187.1°$, but as this does not conform to Brown and Morris's law of definite relation, they attribute the difference to impurities introduced into the solution by the large quantity of malt-extract which Hertzfeld used in the experiment; at the same time, they express their conviction that Hertzfeld had obtained in an impure state the body which they had similarly named maltodextrin.

Hertzfeld states that his maltodextrin was completely fermentable by *Saccharomyces cerevisiæ*, but this, it was considered, might be owing to his having made use of a yeast contaminated with other forms of yeast and bacteria.

Unfermentability of Maltodextrin and the Dextrins by *Saccharomyces cerevisiæ*.—The following is an example of a numerous series of experiments which were made by Brown and Morris to show that neither maltodextrin nor the dextrins could be fermented by *Saccharomyces cerevisiæ*[2]:—

The products of a starch conversion which had been separated by alcohol, and which had an opticity of $[\alpha]j + 191.2°$, were dissolved in water; the solution had a specific gravity of 1099.5. A little yeast was then added to this solution, which contained maltose,

---

[1] Inaugural Dissertation, Halle.

[2] As pure yeasts, under the Hansenian acceptation of the term, were then unknown, the *Saccharomyces cerevisiæ* here mentioned is the ordinary yeast of the primary fermentation. It is stated (Moritz and Morris, "Science of Brewing," p. 120, footnote) that on a repetition of the experiments with pure cultures, precisely the same results were obtained.

maltodextrin, and several of the achroodextrins, and the whole left to ferment at 30° C. (86° F.). Brisk fermentation ensued, and this came to an end in three days. The solution had now an opticity of $[\alpha]_{j_{8\cdot 86}} + 194.2°$, and a reducing power of $\kappa_{3\cdot 86}$ 17.7, corresponding to 29 per cent. maltose and 69.7 per cent. dextrin, and 1.3 per cent. inactive matter. When degraded with malt-extract at 60° C. (140° F.), 100 parts of the dextrin yielded 66.8 parts of maltose. The solution was now evaporated to a syrup, and sufficient water added to make a solution having a specific gravity of 1093. It then contained per 100 c.c. 6.986 grammes of maltose, 16.792 grammes dextrin, and 0.315 gramme of inactive matter. A little yeast was added, and the whole kept at a temperature of 30° C. for seven days. No trace of fermentation was to be observed, and the yeast cells were found on examination to be shrivelled up and, to all appearance, dead. At the end of this time a slow fermentation commenced, and, coincident with this, a few cells of *Saccharomyces ellipsoideus* and *Saccharomyces Pastorianus* were detected. These latter increased in quantity, and in proportion as this increase took place, so much livelier became the fermentation. The solution was still in a state of fermentation at the end of forty days, when the experiment was discontinued. An estimation of the amount of maltose fermented away was then made (1) by observing the decrease in the specific gravity of the solution after the alcohol had been driven off by evaporation; (2) by estimating the amount of alcohol which had been formed, correction being made for the non-volatile products of fermentation in both cases.

(1.) Gave a disappearance of 14.398 grammes maltose per 100 c.c.
(2.) Gave a disappearance of 14.410 grammes maltose per 100 c.c.

Since the original solution contained only 6.986 grammes of maltose, the remaining 7.494 grammes must have been derived from the dextrin; for, on subsequent degradation with a little fresh malt-extract, it now yielded only 25.7 parts of maltose per 100 parts of dextrin against the 66.8 parts which it had yielded prior to fermentation.

Brown and Morris's Hypothesis of the Breaking Down of the Starch Molecule by Diastase.—The discovery of maltodextrin led Brown and Morris to a further extension of the previous views of Brown and Heron on the constitution of the starch molecule. They now regarded the starch molecule as consisting of not less than five $(C_{12}H_{20}O_{10})_3$ groups, and considered that when it was acted on by malt-extract, one of the $(C_{12}H_{20}O_{10})_3$ groups was partially hydrated and detached in the form of maltodextrin, the remaining four groups, $4(C_{12}H_{20}O_{10})_3$, forming a dextrin residue. One of the four $(C_{12}H_{20}O_{10})_3$ groups of this residue was similarly partially hydrated and detached, and the remaining three $(C_{12}H_{20}O_{10})_3$ constituted a dextrin residue of less complexity than the preceding one. This action proceeded in a similar manner until the last $(C_{12}H_{20}O_{10})_3$ group was reached, and this formed the stable dextrin of the No. 8 equation (page 141). Maltodextrin, when treated with malt-extract which had not been heated beyond 65° C. (149° F.), was found to be rapidly and completely hydrolysed to maltose. It was, however, possible that the hydrolysis might have taken place in two stages, but so far no indication of this had been observed.

Molecular Weight of Starch.—In a paper read before the Chemical Society in 1889,[1] Brown and Morris describe a number of experiments performed by Raoult's cryoscopic method, in which an attempt was made to ascertain the molecular weight of starch. Ordinary starch, owing to the viscid nature of its solution, was entirely unfitted for an experiment of this nature; but by operating on its near relative, soluble starch, strong indications were obtained that the molecular weight of this body was from 20,000 to 30,000. Not satisfied with this, an endeavour was made to ascertain the molecular weight in another way. Since starch, when hydrolysed, is invariably found to contain a fifth of its weight of a dextrin which is only further hydrolysed with extreme difficulty, it is obvious that if the molecular weight of this dextrin could be established with certainty, then that of starch would be also definitely fixed, since it must necessarily be five times as great as the molecular weight of this dextrin. Experiments made with a number of preparations of the dextrin pointed strongly to its molecular weight being 6480, corresponding to a formula of $(C_{12}H_{20}O_{10})_{20}$. The formula for soluble starch would be consequently five times as much as this, or $5(C_{12}H_{20}O_{10})_{20}$, and its molecular weight 32,400.

All the Dextrins have the same Molecular Weight.— Raoult's method was now tried upon a number of different dextrins to ascertain if they all possessed the same or different molecular weights. The following is an example of one of these experiments:—

A dextrin was prepared which had an opticity of $[a]j_{3.86}$ + 210.7° and a reducing power of $\kappa_{3.86}$ 1.4, figures closely corresponding to those of a mixture of 2.3 per cent. maltose and 97.7 per cent. dextrin. One hundred parts of this dextrin, when degraded with fresh malt-extract, yielded 71.5 parts of maltose; consequently, its mean position was between Nos. 3 and 4 of the series on p. 144. If it were a polymer, or a mixture of polymers of the lowest dextrins, its molecular weight ought to be 22,680, that is, 6480 × 3½. The molecular weight as deduced by Raoult's method pointed strongly to the number 6480. A number of similar experiments made with dextrins in various positions in the series shown in the table on p. 144 gave similar results; hence it was concluded that the view previously held, that the dextrins were polymeric bodies, would have to be abandoned in the face of these fresh facts. Since it

[1] *Journal of the Chemical Society*, 1889, p. 449 and 462.

was now shown that they all possessed the same molecular weight,[1] it became necessary to frame a new hypothesis in accordance with these newly-discovered facts.

Brown and Morris's Second Hypothesis of the Breaking Down of the Starch Molecule by Diastase.—This was described in a paper read before the Laboratory Club in 1890.[2] They now regarded the starch molecule as consisting of four complex dextrin groups, arranged round a fifth similar group. This last, on the breaking down of the molecule, becomes the stable dextrin of the No. 8 equation.

The first act of hydrolysis by diastase is to break up this complex molecule, and liberate all the five dextrin groups, thus:—

$$\begin{cases}(C_{12}H_{20}O_{10})_{20}\\(C_{12}H_{20}O_{10})_{20}\\(C_{12}H_{20}O_{10})_{20}\\(C_{12}H_{20}O_{10})_{20}\\(C_{12}H_{20}O_{10})_{20}\end{cases} = \underset{\text{Stable Dextrin.}}{(C_{12}H_{20}O_{10})_{20}} + \underset{\substack{\text{Readily Hydrolysable}\\\text{Amylin Groups.}}}{4(C_{12}H_{20}O_{10})_{20}}$$
Starch molecule.

One of these five groups is, from some unknown cause, only further hydrolysed by diastase with extreme difficulty, and this forms the stable dextrin of the No. 8 equation (p. 141). The remaining four dextrin groups yield much more readily to the influence of the diastase, and are immediately broken down into a complicated series of amyloïns[3] or maltodextrins of varying composition. The extreme stages and an intermediate one are represented in the following equations:—

$(C_{12}H_{20}O_{10})_{20} + H_2O = \begin{cases}(C_{12}H_{22}O_{11})\\(C_{12}H_{20}O_{10})_{19}\end{cases}$ Amyloïn with lowest $\frac{M}{D}$ ratio $\frac{1}{19}$.

$(C_{12}H_{20}O_{10})_{20} + 10H_2O = \begin{cases}(C_{12}H_{22}O_{11})_{10}\\(C_{12}H_{20}O_{10})_{10}\end{cases}$ Intermediary amyloïn. $\frac{M}{D}$ ratio $\frac{10}{10}$.

$(C_{12}H_{20}O_{10})_{20} + 19H_2O = \begin{cases}(C_{12}H_{22}O_{11})_{19}\\C_{12}H_{20}O_{10}\end{cases}$ Amyloïn with highest $\frac{M}{D}$ ratio $\frac{19}{1}$.

The authors then state, "As the hydrolysis proceeds, the complex amyloïn groups break up into smaller molecular aggregations, which, however, retain all the characteristics of the amyloïns, and this goes on until the maltose stage is reached."

They adduce as examples which have been isolated and examined in a pure state, and the molecular weights of which have been determined, of these latter bodies, maltodextrin $\begin{cases}C_{12}H_{22}O_{11}\\(C_{12}H_{20}O_{10})_2\end{cases}$ (p. 146) and amylodextrin $\begin{cases}C_{12}H_{22}O_{11}\\(C_{12}H_{20}O_{10})_6\end{cases}$ (p. 74).

---

[1] O'Sullivan (*Journal of the Chemical Society*, 1879, p. 783) stated his belief that the first products of the breaking up of the starch molecule were a series of isomeric dextrins.

[2] *Trans. Lab. Club*, iii. 83.

[3] The name "amyloïn" was suggested for these bodies by Professor Armstrong. It had been proposed to terminate the name of all sugars belonging to the $C_{12}H_{22}O_{11}$ group with the syllable *on;* thus maltose would be malton; and as the dextrins already possess the termination *in*, a word ending in *oïn* would signify a compound of amylon with amylin, which latter name has been proposed for dextrin.

Hence, according to this hypothesis, there are present in all starch conversions, when carried out under the ordinary conditions of mashing employed in the brewery, free maltose, free dextrin, which is not further hydrolysed, together with a certain proportion of amyloïns of varying types.

They consider that the free maltose is entirely fermented away during the primary fermentation, and that the amyloïns are gradually broken down and slowly fermented away during the secondary fermentation which takes place after the beer has been stored in casks; the free dextrin persists to the last, and is little, if at all, affected even on prolonged storage.

Discovery of Isomaltose.—The methods employed by Emil Fischer in his remarkable investigations (p. 46) on the constitution of the sugar group, which date from the year 1885, and in which phenylhydrazine played such an important part, could hardly fail sooner or later to throw some light on the nature of the bodies produced during the hydrolysis of starch. They led, in 1891, to the discovery by Lintner of an entirely new sugar amongst these bodies.

Scheibler and Mittelmeier[1] had described, in 1889, an osazone which was obtained from a substance found in the unfermentable residue of commercial glucose, and which, from its composition, was presumably the osazone of a sugar having the formula $C_{12}H_{22}O_{11}$. Shortly afterwards Emil Fischer[2] obtained a substance by dissolving 100 grammes of pure glucose in 400 c.c. of hydrochloric acid of a specific gravity of 1.19, and allowing the mixture to stand at a temperature of from 16° to 20° C. (61° to 70° F.) for fifteen hours. The osazone obtained from this substance melted at 150° to 153° C., and as it resembled in several other respects maltosazone, it was named "isomaltosazone." In 1891 Scheibler and Mittelmeier[3] obtained, by fermenting a 10 per cent. solution of commercial glucose, filtering the solution, evaporating to a syrup, and adding strong alcohol, an amorphous unfermentable substance which had been previously described by Schmitt and Cobenzyl under the name of "gallisin." This body, when purified by repeated precipitation from its aqueous solution by alcohol, was colourless, amorphous, and of a very hygroscopic nature. It reduced Fehling's solution, and, when heated with phenylhydrazine acetate, yielded an osazone melting at 152° to 153° C., which was readily soluble in hot water. This osazone was, they considered, identical with the isomaltosazone previously described by Fischer. In 1891 C. J. Lintner[4] discovered, by the aid of the phenylhydrazine reaction, a sugar in wort and beer which yielded an osazone similar to the one just described; hence he called the sugar from which it was derived "isomaltose." In a subsequent paper published by Lintner and Düll,[5] the results of a further series

---

[1] *Berichte*, xxiii. 3060. [2] *l.c.*, xxiii. 3687. [3] *l.c.*, xxiv. 301.
[4] *Zeit. f. das gesamt. Brau.*, 1891, p. 284.
[5] *Zeit. f. angewand. Chemie*, 1892, p. 268.

of investigations with reference to this sugar were described. They found that after removing the maltose from the products of a starch conversion by a short fermentation conducted at 27° C. (80° F.), they were able to resolve, by repeated precipitation with alcohol, the unfermented matters into dextrin and isomaltose, but they were unable to detect even a trace of any such substance as an intermediary maltodextrin.

Preparation of Isomaltose.—Lintner and Düll prepared isomaltose in the following manner:—

Two hundred and fifty grammes of potato-starch were mixed with 500 c.c. of water containing in solution 0.5 gramme of diastase; the mixture was then added to 2 litres of water having a temperature of 75° C. (167° F.). After liquefaction of the starch had taken place, another 0.5 gramme of diastase was added, and the whole kept at a temperature of 67° to 69° C. (153° to 156° F.) for three hours. The liquid, after being boiled, gave a dark brown reaction with iodine, and had an opticity of $[\alpha]_D = +170°$. The solution was then evaporated to a syrup on the waterbath.

I. The first separation of the starch-conversion products was effected with 80 per cent. alcohol. The syrup was first saturated with alcohol, then poured into hot alcohol, the quantities being so adjusted that to every 10 parts of dry substance there were present in solution not less than 100 c.c. of 80 per cent. alcohol. The whole was allowed to stand until it became cold.

II. The clear alcoholic solution from No. I. was then decanted from the precipitated syrup which had formed, the alcohol removed by distillation, the residue made up to a 20 per cent. solution with water, and set to ferment by the addition of 2 grammes of pressed yeast for every 100 c.c. of solution. After fermenting for twenty hours all traces of maltose and glucose had disappeared. The solution was then boiled, filtered, treated with animal charcoal to remove colour, once more filtered, and evaporated to a syrup.

III. This syrup was again treated with 85 per cent. alcohol in the same way as No. I., the quantities being so adjusted that for every 5 grammes of dry substance at least 100 c.c. of 85 per cent. alcohol were present.

IV. The clear alcoholic solution from No. III., when cold, was decanted from the residue, evaporated to a syrup, and treated in the same way as No. I., using this time 90 per cent. alcohol, the quantities being so arranged that for every 5, or better 3 grammes of dry substance, there were not less than 100 c.c. of 90 per cent. alcohol present. The clear liquid from this, after being decanted from the precipitated syrup, was a solution of isomaltose sufficiently pure for most purposes.

Its weight amounted to 20 per cent. of that of the original starch. The contaminating impurities, which consisted of mineral matters

CHARACTERS OF ISOMALTOSE.

(ash) and acids generated during the fermentation, were only removed with great difficulty. They could only be removed by frequent solution in water and re-precipitation by alcohol.

Characters of Isomaltose.—As thus prepared, isomaltose is a body possessing an intensely sweet taste. Its opticity in a 10 per cent. solution is $[\alpha]_D + 139°$ to $140°$. It has a lower reducing action on Fehling's solution than maltose, only reducing 83 per cent. of the quantity which an equal weight of maltose would, consequently its reducing power is R. 83 or κ 50.63. It has not yet been obtained in a crystalline condition, but it has been prepared, by leaving it for a considerable time under absolute alcohol, in a solid form which could be rubbed to powder. It is exceedingly difficult to free isomaltose from the last traces of alcohol, since before these are completely driven off by heat the isomaltose itself begins to decompose. The weight of dry isomaltose in solution can be determined directly by placing 0.5 to 1.0 gramme of the solution in a weighing-tube with ground stopper, and keeping the whole at a temperature of 100° C. (212° F.) for some time, weighings being made at intervals of an hour until the weight remains constant. The isomaltose assumes a yellow colour, but the decomposition is not sufficiently advanced to affect the accuracy of the result. Isomaltose is of an exceedingly hygroscopic nature. It is slowly fermented by yeast, and, when acted upon by diastase, is completely converted into maltose.

The following is the view adopted by Lintner and Düll with regard to the transformation of starch under the influence of diastase, commencing with amylodextrin (soluble starch):—

I. $(C_{12}H_{20}O_{10})_{54} + 3H_2O = 3[(C_{12}H_{20}O_{10})_{17}.C_{12}H_{22}O_{11}]$.
II. $3[(C_{12}H_{20}O_{10}).C_{12}H_{22}O_{11}] + 6H_2O = 9[(C_{12}H_{20}O_{10})_5.C_{12}H_{22}O_{11}]$.
III. $9[(C_{12}H_{20}O_{10})_5.C_{12}H_{22}O_{11}] + 45H_2O = 54C_{12}H_{22}O_{11} =$ Isomaltose.
IV. $54C_{12}H_{22}O_{11} =$ Maltose.

They did not consider that these stages succeeded one another in regular order, but that the action proceeded irregularly, some of the amylodextrin molecules being much more rapidly broken down than others; consequently, by the time some of these had reached the stage of isomaltose and maltose, others might only be in the earlier stages of degradation. In this way it was sought to explain the simultaneous presence of maltose and the higher dextrins which is found in all high-angled transformations.

Schifferer's Experiments on the Conversion Products of Starch by Diastase.—At the instigation of Lintner, Anton Schifferer carried out a series of experiments on this subject. These were embodied in an "Inaugural Dissertation"[1] delivered at the University of Kiel in 1892.

He first attempted to reproduce the maltodextrin of Brown and Morris (p. 146), and followed the directions given for its

[1] Published by Schmidt and Klaunig, Kiel.

preparation [1] as far as possible; but complains that in some points these are somewhat vague. A starch transformation was conducted at a temperature of 60° to 65° C. (140° to 149° F.), using 500 grammes of potato-starch and 0.416 gramme of diastase. The action was allowed to proceed for fifty minutes, when the conversion showed an opticity of $[a]_D = +180°$. The solution was then boiled and filtered. The osazone obtained from the filtrate sintered at 145° C. and melted at 152° C. This, according to the author, showed that the whole of the transformation products consisted of isomaltose and dextrin, hence the subsequent fermentation for the removal of maltose might have been omitted. To be assured that any traces of maltose, the presence of which might not have been rendered evident by the osazone, were removed, 5 grammes of freshly washed pressed yeast were added to the solution and fermentation allowed to proceed for twenty-four hours at the ordinary temperature. The solution was then filtered, evaporated to a syrup, and poured hot into so much hot absolute alcohol that the quantity of alcohol amounted to 90 per cent. of the whole mixture.

After this had been kept boiling for two days under a reflux condenser, the hot alcoholic solution was decanted from the precipitated syrup, filtered, and the alcohol distilled off. Water was then added to the residue, and the whole evaporated to a syrup. This syrup contained 24 grammes of dry substance, which had an intensely sweet taste, and an opticity of $[a]_D = +142°$; it yielded an osazone which sintered at 142° and melted at 152° C. This product consisted, therefore, of nearly pure isomaltose. A second experiment gave similar results. No trace of a substance similar to Brown and Morris's maltodextrin was observed.

In another experiment, where 1000 grammes of potato-starch were used, a conversion was obtained which, after being boiled and filtered, had an opticity of $[a]_D = +152.01°$ and a reducing power of κ 43.4 (R. 71.12). It was fermented with yeast, and filtered, treated with animal charcoal, and evaporated to a syrup. This syrup was now fractionated with alcohol, much after the manner described before in Lintner's experiments, with the result that the products of the starch conversion were entirely separated into dextrin and isomaltose, no trace of any intermediary maltodextrin being observed. The experiment was so arranged that nothing except the first fraction, which consisted principally of albuminous and mineral matters, was lost.

Another experiment was made with 1000 grammes of potato-starch, in which Hertzfeld's directions for obtaining *his* maltodextrin were exactly observed. The whole of this was fractionated with alcohol, and was found to be separable into dextrin and isomaltose, no trace of maltodextrin being observed.

Schifferer considers that Brown and Morris's maltodextrin was a mixture of about 67 per cent. dextrin and 33 per cent. isomaltose,

---

[1] *Journal of the Chemical Society*, 1885, p. 562.

and Hertzfeld's a mixture of about 74 per cent. isomaltose and 26 per cent. dextrin. Schifferer thinks there is little doubt that Hertzfeld at one time actually had isomaltose in his hands, for in his attempts to obtain crystalline maltose by treating the sweet syrup, from which the erythrodextrin had been precipitated with ethylic and methylic alcohol, he failed to obtain maltose, but succeeded in isolating "a small quantity of a substance, which showed nearly the same reducing power as maltose, but could not be obtained in a crystalline form." This quite coincides with Lintner's observations, that in mashes conducted at 70° C. (156° F.) maltose, if formed at all, is only produced in small quantities. Had Fischer's osazone test been then known at that time, probably the new sugar would have been discovered by Hertzfeld.

Compound Nature of Diastase.—Cuisinier held the view that malt diastase consisted of two enzymes, which he termed respectively "maltase" and "dextrinase." He considered that the latter alone was able to attack starch, which it first liquefied and then converted into dextrin. The maltase then acted on the dextrin and transformed it into maltose. Weijsmann[1] further elaborated this theory of two enzymes, but took a somewhat different view of their respective actions. In his opinion, the maltase yields maltose, and a high erythrodextrin, which he terms "erythrogranulose;" whilst the dextrinase produces maltodextrin, but no maltose. As the action proceeds, the maltase attacks the maltodextrin and transforms it into maltose, the dextrinase meanwhile converting the erythro-granulose into achroodextrin, according to the following scheme:—

Starch.

Maltase yields	Dextrinase yields
Maltose and erythro-granulose.	Maltodextrin.
Converted by dextrinase into achroodextrin.	Converted by maltase into maltose.

This action was demonstrated by a number of very ingenious experiments. Beyerink[2] holds a somewhat similar view: according to this, diastase consists of maltase and granulase; the former is held to convert a portion of the starch first into erythrodextrin, then into maltose, the granulase simultaneously converting another portion into maltodextrin, and afterwards into maltose.

Scheibler and Mittelmeier on the Dextrins.—Since the time that the properties of the gummy substance found in all starch transformations was first critically examined by Biot and Persoz in 1833, who, finding that it rotated the polarised ray powerfully to the right, named it "dextrin," this body, or rather mixture of bodies, has formed the theme of numerous investiga-

---

[1] *Zeit. f. das ges. Brau.*, 1890, p. 186.
[2] *Centralblatt f. Bakt. u. Parasitenkunde*, 1895, p. 221.

tions and of much discussion. Even at the present time it cannot be said that any absolutely definite conclusion has been arrived at as to the real character of these bodies. The question as to whether they possess reducing powers or not has been at one time accorded, at another time denied them. According to the latest view expressed by Brown and Morris, there is but one true dextrin, and this possesses no reducing power; it is the substance that they believed they had obtained in a state of purity by the action of alkaline mercuric cyanide; but the later investigations of Scheibler and Mittelmeier render it extremely doubtful if the body thus obtained was in reality a dextrin at all. It has always been held by the Continental chemists that there were several dextrins, all of which possess a greater or less reducing power, and, according to the light of recent research, this seems to be the more correct view.

In 1890 Scheibler and Mittelmeier[1] pointed out the close analogy between the behaviour of the dextrins and that of the sugars which contain the COH group. For this demonstration they employed commercial dextrin purified by treatment with alcohol. The purified preparation was coloured yellow or brown by heating with a caustic alkali, and it also reduced Fehling's solution, both of which actions are characteristic of the sugars which contain the COH group. They also succeeded in obtaining the phenylhydrazones and phenylosazones of the dextrins, both of which were soluble in cold water. Since the osazones of the monosaccharides are insoluble in water, either cold or hot, and those of the disaccharides are soluble in hot water, it was naturally to be expected that the osazones of the more complex aldoses would be soluble in cold water. By acting on a dextrin with nascent hydrogen, they were able to obtain a body, dextritol, having the characteristics of an alcohol. By the action of bromine on the dextrins bodies were formed which were acid to litmus paper, and which decomposed calcium carbonate with liberation of carbon dioxide; evidently the COH groups of these bodies had been converted into the COOH group, that is to say, the dextrins had been changed into their corresponding carboxylic acids. These acids could be hydrolysed by diastase or by an acid, just as the corresponding maltobionic acid could. They consider these bodies were the same as those which Bondonneau produced by means of cupric chloride and caustic soda, and which Wiley obtained when he employed alkaline mercuric cyanide, and likewise by Brown and Morris when they used this latter reagent for removing what they believed to be the last traces of a reducing sugar from dextrin. In all these cases the assumption that the dextrin remained unattacked was erroneous; it suffered the alteration alluded to above, and, as a consequence, lost its property of reducing Fehling's solution. From these facts they concluded that the non-reducing

[1] *Berichte*, xxiii. 3060.

dextrin of Brown and Morris has no existence. The only objection which can be urged against these experiments of Scheibler and Mittelmeier is that the dextrins operated upon were not obtained from starch by the action of diastase, but Brown and Morris have also adduced an amylodextrin, produced by the action of acid, in support of the amyloïn theory.

Lintner and Düll on the Dextrins.—These investigators, after a long and searching examination into the products of starch transformation by diastase,[1] in which the various bodies were separated by long and tedious fractionation with alcohol of different strengths, and in which the properties of the isolated bodies were controlled by their rotatory and reducing powers, were able to isolate the following substances:—

Amylodextrin.—This is obtained, on precipitating its aqueous solution with alcohol, as a loose white powder, slightly soluble in cold water, readily soluble in hot, and separates from a 20 to 30 per cent. solution in sphæro-crystals. It is insoluble in 40 per cent. alcohol, has a rotatory power of $[\alpha]_D = +196°$, and does not reduce Fehling's solution. It gives a deep blue colour with iodine, and is the principal constituent of soluble starch. They regard its composition as $(C_{12}H_{20}O_{10})_{54}$.

Erythrodextrin.—This body separates from hot solutions of dilute alcohol in sphæro-crystals, is readily soluble in water, very slightly so in 50 per cent. alcohol. Its rotatory power is $[\alpha]_D = 196°$, its reducing power is R. = 1, i.e., 1 per cent. of that of maltose, and it gives a reddish-brown colour with iodine. Its molecular weight was found to be 5786. The authors consider that its composition is $(C_{12}H_{20}O_{10})_{17} \cdot C_{12}H_{22}O_{11}$, and that it arises from the hydrolysis of three molecules of amylodextrin, thus:—

$$(C_{12}H_{10}O_{10})_{54} + 3H_2O = 3[(C_{12}H_{20}O_{10})_{17} \cdot C_{12}H_{22}O_{11}]$$

Achroodextrin I.—This dextrin forms deliquescent sphæro-crystals, is very soluble in water, hardly soluble in 70 per cent. alcohol, and gives no coloration with iodine. It has a rotatory power of $[\alpha]_D = +192°$, and a reducing power of R. = 10, i.e., equal to 10 per cent. of maltose. Its molecular weight was found to be 1963. They regard its composition as $(C_{12}H_{20}O_{10})_5 \cdot C_{12}H_{22}O_{11}$, and consider that it arises from the hydrolysis of three molecules of erythrodextrin, thus:—

$$3[(C_{12}H_{20}O_{10})_{17} \cdot C_{12}H_{22}O_{11}] + 6H_2O = 9[(C_{12}H_{20}O_{10})_5 \cdot C_{12}H_{22}O_{11}]$$

Achroodextrin II. was first found in the transformation products of starch hydrolysed with oxalic acid, and afterwards, though only in small quantities, in those obtained by the action of diastase. It has a rotatory power of $[\alpha]_D = +180°$, and a reducing power equal to 24 per cent. of maltose (R. = 24). Its molecular weight is given as 979.

Ost on the Dextrins.—This observer[2] isolated from those

[1] *Berichte*, xxvi. 2533.
[2] *Chem. Zeitung*, 1895, p. 1501.

portions of a starch transformation soluble in 80 per cent. alcohol, from which the maltose had been removed by long-continued fractionation with 95 per cent., 90 per cent., and finally 85 per cent. alcohol, a dextrin having a rotatory power of about $[\alpha]_D$ 180°, together with maltose, and a considerable amount of other products which were evidently mixtures. He also found that the osazone test afforded a safe criterion of the presence of maltose in mixtures of this nature when present in quantities above 5 per cent. The above dextrin, when precipitated by alcohol and dried, formed a snow-white powder, not in the slightest degree deliquescent. It had a rotatory power of $[\alpha]_D = +180°$ to 183°, and a reducing power equal to 28 to 33 of maltose (R. = 28 to 33). On analysis it yielded figures closely agreeing with the formula $C_{36}H_{62}O_{31}$. Ost believed that it was a definite compound, that it was essentially the same body as Lintner and Düll's achroodextrin II. $[\alpha]_D = +180°$ (R. = 24), and that it was contained in the maltodextrin of Brown and Morris $[\alpha]_{D3.86} = 171°$ (R. = 34). Ost considers that in the preparation of amorphous bodies like the dextrins absolute purity cannot be expected, as it is extremely probable that they may undergo some little change during the frequent evaporations and treatment with alcohol to which they are subjected in the course of isolation.

Investigations of Ling and Baker on Isomaltose.— Much doubt has been thrown on the existence of isomaltose by several observers. The first to bring this subject under notice were Ling and Baker, in a paper read before the Chemical Society.[1] Working with diastase prepared from air-dried malt, and following Lintner and Düll's directions, they obtained a substance having an opticity of $[\alpha]_{D3.86} = +144.5°$ and a reducing power of $R_{3.86} = 79.64$. Its osazone, after being recrystallised from water, had a melting-point between 160° and 170° C. The substance itself, after being kept in an exsiccator for some weeks, became brittle and could be crushed to powder; it was very hygroscopic. Viewed under the microscope, it was seen to contain crystals resembling those of maltose. It had a sweet, but by no means an intensely sweet taste. Both from the results of its ultimate analysis, and of the determination of its molecular weight by Raoult's method, figures indicative of a disaccharide were obtained. When submitted to the action of diastase, the opticity was reduced to $[\alpha]_{D3.86} = +131.2°$ and the reducing power increased to $R_{3.86} = 94.5°$.[2] As had been previously observed by Hiepe, they found that these changes were not accompanied by an alteration in the specific gravity of the solution. Mixed with ordinary brewer's yeast and kept at a temperature of 68° to 86° F., the substance fermented, though not vigorously. They came to the conclusion that the body which

[1] *Journal of the Chemical Society*, 1895, p. 702.
[2] This result, corrected for the error in Wein's table, is R. = 98.9. (See Appendix C).

has been termed isomaltose by Lintner is a mixture of maltose and of a simple dextrin having the formula $C_{12}H_{20}O_{10} + H_2O$, a composition which would account for the absence of a rise in the specific gravity of its solution under hydrolysis. It must, however, be observed that the substance prepared by Ling and Baker had neither exactly the same opticity nor reducing power of the body described by Lintner, besides which it yielded an osazone which had a higher melting-point.

In some further experiments where diastase obtained from high-dried malt was employed, they succeeded in obtaining a body which yielded an osazone having a melting-point identical with that of Lintner's isomaltosazone. An ultimate analysis of this substance yielded figures closely agreeing with those required for the osazone of a trisaccharide. From this they concluded that the body which yields the osazone melting at 151° C. was a trisaccharide. They also noticed that in starch transformations conducted with diastase from high-dried malt a certain amount of glucose was invariably produced, but never when diastase from low-dried malt had been employed.[1]

In a second paper[2] they stated that all the experiments they had made up to that time appeared to favour the view previously put forward, that isomaltose contained a simple dextrin, $C_{12}H_{20}O_{10} + H_2O$ which might be an intermediate product between the higher dextrins and maltose. They found that, by recrystallising the osazone derived from their preparation of isomaltose from water at 35° C., portions crystallised out having considerably lower melting-points than those portions which afterwards formed at the ordinary temperature; and, by successive recrystallisations in this way, they were able to obtain an osazone melting at 182° to 183° C. which closely resembled maltosazone in appearance. A fraction obtained in another experiment, and which melted at 160° C., yielded on analysis results closely agreeing with the composition of a disaccharide, consequently they regarded the fractions which melted between 150° and 160° C. as impure maltosazone. In fact, they were able to obtain a similar osazone by heating mixtures of maltose and dextrin with phenylhydrazine acetate, but this impure osazone could be readily purified by recrystallisation from hot water. They found that when a mixture of maltose and glucose was treated with phenylhydrazine acetate, the osazones of both these sugars could be separated in a state of purity. In working with diastase from kiln-dried malt it was noticed that a small quantity of glucose was invariably formed, and this led to their trying the action of diastase obtained from this kind of malt upon maltose, when it was found that glucose was invariably produced.[1]

---

[1] The authors have since found that the ability to produce glucose which some malts display is in no way connected with the kilning temperature to which they have been subjected (*Trans. Chem. Soc.*, 1897, p. 512).

[2] *Journal of the Chemical Society*, 1895, p. 739.

In transformations of starch conducted with the diastase from kiln-dried malt, they were always able to detect the presence of glucose by the formation of its osazone; but the greater part of the osazone from the products of the transformation, soluble in 90 per cent. alcohol, melted at 151° C., and had the appearance characteristic of Lintner's isomaltosazone. They gave, however, analytical values intermediate between those required for the osazone of a disaccharide and a trisaccharide. From this, and from the fact that apparently no maltosazone was obtained from the product of these conversions, they were at first inclined to think that no maltose was present, but afterwards they were able to show the presence of maltose in similar products. From this it appeared as if a trisaccharide was one of the products of the action of kiln-dried malt on starch; attempts, however, to isolate any other crystalline sugar than maltose entirely failed. Since this osazone of a trisaccharide was only obtained in those cases where glucose was present, they surmised that the trisaccharide might be formed by the condensation of a molecule of the simple dextrin with one of glucose under the influence of phenylhydrazine.

The mixture termed "isomaltose," prepared by means of the diastase from low-dried malt and glucose, yielded a mixture of glucosazone and a soluble osazone; the latter, after recrystallisation, melted at 150° to 160° C., and on analysis appeared to be a mixture of the osazones of a di- and a tri-saccharide, a result which might be expected, considering that maltose is a constituent of "isomaltose." Microscopic examination of the osazone showed that it consisted almost entirely of spherical aggregates of needles, but, under a higher magnification, flat ribbon-like crystals could also be detected.

Brown and Morris on Isomaltose.—Messrs. Ling and Baker's paper was shortly followed by one from Messrs. Brown and Morris,[1] also throwing doubt on the existence of isomaltose. They first point out that this body does not conform to their "law of definite relation," which they consider to be the most valuable criterion of the purity of starch-transformation products. Lintner's isomaltose has an opticity of $[\alpha]_D = +140°$ and a reducing power of R. = 80, equivalent to $[\alpha]j = +156°$ and $\kappa = 48.8$. This opticity, according to their law, requires a $\kappa = 55.3$, and a reducing power of $\kappa = 48.8$ requires an opticity of $[\alpha]j = +163.2$. After an elaborate investigation of a starch transformation conducted on the method described by Lintner, they came to the conclusion that isomaltose was not a chemical entity, but that by fractionation with alcohol, by decomposition, and by fermentation it could be split up in such a manner as to indicate that it was a mixture of maltose and bodies of the amyloïn type. They considered that isomaltosazone was nothing but maltosazone, modified in its crystalline habit and melting-point by the presence of some impurity; and they were able to

[1] *Journal of the Chemical Society*, 1895, p. 702.

compel maltosazone to crystallise in the isomaltose form by the addition of a trace of an amorphous substance, which was produced during the treatment of those fractions of a starch conversion supposed to contain "isomaltose" with phenylhydrazine acetate, and which they regard as a product of the action of this reagent with substances of the maltodextrin or amyloïn class. On fermenting a solution of isomaltose, they found 50 to 60 per cent. of this body fermentable, and from a calculation of the opticities and reducing powers of the solution before and after fermentation, they show that the portion fermented is simply maltose, and that the unfermented portion approaches very closely in composition that of an amyloïn containing two amylon groups and one amylin group. They consider that the only *crystallisable* body formed in starch transformations by diastase is maltose, and point out that the osazone alone cannot be relied on as a sufficient means for identifying a sugar.

Lintner's Rejoinder.—In this the discoverer of isomaltose considers[1] that Brown and Morris set about their course of investigation in an incorrect way, and could consequently only arrive at erroneous conclusions. They evidently worked with impure materials, and could in this way readily obtain osazones resembling that of isomaltose. That they had a mixture of maltose and dextrin in their hands was shown by the isolation of the maltose which they accomplished on further fractionation, and also by the residue left behind after fermentation, which had a higher opticity and a lower reducing power than isomaltose, and which, therefore, he believes to have been dextrin. He further thinks that it was a very unscientific proceeding to endeavour to form isomaltose by making mixtures of maltodextrin and maltose, because naturally mixtures of such a nature would yield osazones having a lower melting-point than maltosazone. He himself had observed the same phenomena with mixtures of maltose and isomaltose, but had never found it possible to transform isomaltosazone into the well-characterised maltosazone, nor to raise its melting-point.

Ost's Investigations on Isomaltose.—In these an attempt[2] was made to prepare isomaltose according to the directions given by Lintner and Düll. The fractions obtained with 85 to 90 per cent. alcohol, and which had a rotatory power of about 140°, were collected, dissolved in methylic alcohol, and separated into three or four portions. The treatment with methylic alcohol was found unsatisfactory, since fractions obtained in this way, and which had a rotatory power of 140°, did not prove to be homogeneous, and could be further separated by ethylic alcohol. A successful result could be obtained by the employment of ethylic alcohol, but the operation was a most tedious one, on account of the ready solubility of maltose in dextrin syrups, and of the dextrins in maltose solutions. Owing to this, the further prosecution of

---

[1] *Zeit. f. Brau.*, 1895, p. 233.
[2] *Chem. Zeitung*, 1895, p. 1501.

the process was handed over to Herr Ulrich, who was unable to detect the presence of isomaltose amongst the starch-transformation products.

Ost then tried another method for removing the maltose. It consisted in evaporating the alcoholic extracts with sufficient water to bring the maltose into the hydrated condition, in which state it could be readily induced to crystallise; since isomaltose is said to be uncrystallisable, it was thought that this method would afford a ready means of separating the two bodies. According to a statement made by Soxhlet, crystallised maltose is but slightly soluble in methylic alcohol, but maltose and dextrin syrups are soluble in all proportions; hence, by bringing maltose into the crystalline condition, it should be possible to remove it completely in this way. After the maltose had been removed from the products of a starch transformation in this manner, there only remained small quantities of syrup, which had a rotatory power of less than 145°. These were tested for isomaltose by the phenylhydrazine method. Osazones, very similar to those said by Lintner and Düll to be characteristic of isomaltose, were obtained from them; but, on further study, they proved to be maltosazone. This body was found to be of very changeable nature, it could be readily made to alter its crystalline form, melting-point, &c., by re-crystallisation, by heating, or by a difference in the conditions under which it was formed. Especially interesting is the fact (first observed by Ling and Baker) that mixtures of maltose and dextrin, which latter body itself yields no crystalline osazone, when treated together yield a crystalline osazone simulating isomaltosazone.

Experiments were also made in which part of the starch-transformation products had been removed by fermentation with yeast. In those where the whole of the maltose had been removed, no osazone could be obtained; those, in which a portion of the maltose had been left unfermented, yielded osazones similar to isomaltosazone.

Ost came to the conclusion that isomaltose has no real existence, and that the isomaltosazone of Lintner and Düll is simply modified maltosazone. He considers that the altered rotatory and reducing power, the uncrystallisability, and the browning of the original substance which ensued on heating, simply depended on impurities, and the reason why Lintner and Düll did not obtain pure maltose is to be attributed to their using methylic alcohol in the fractionation.

Jalowetz[1] also came to substantially the same conclusion as Ost on this subject.

Ost's Conclusions with Regard to the Action of Diastase upon Starch.—After an investigation into the products of starch transformation by diastase,[2] to which reference has been already made (p. 157), Ost concludes that starch itself

---

[1] *Chem. Zeitung*, 1895, p. 2003.   [2] *Ibid.*, p. 1501.

has the composition $(C_{12}H_{20}O_{10})_n$, the n indicating a large but as yet unknown number; that the first products of the action which possess a reducing power, though their rotatory power is either unaffected or affected in a degree so small as to be unrecognisable, contain chemically combined water, and that dextrins of the composition $(C_{12}H_{20}O_{10})_n$ have no existence. He considers that the non-reducing dextrin of Brown and Morris is non-existent, the substance obtained by treatment with alkaline mercuric cyanide consisting, as had been previously shown by Scheibler and Mittelmeier, of the carboxylic acids of the dextrins. He points out that O'Sullivan was never able to obtain a dextrin entirely free from reducing power, and that the law of definite proportion between the rotatory and reducing powers enunciated by the English observers is not always fulfilled even in their own experiments, and that it is certainly not in the achroodextrin II. which he himself had prepared. The method of determining the reducing power employed by the English chemists is, Ost says, a fallacious one, and so is the method of estimating the quantities of the analysed material from the specific gravity of the solution instead of by actually determining the weight of the matter in solution. Though this may suffice for approximate determinations, it is insufficient for the deduction of important laws. The theory of Brown and Heron arose at a time (1879) when the reducing power of the dextrins was solely attributed to the maltose with which they were contaminated. Ost considers that, after maltodextrin and amylodextrin had been recognised by Brown and Morris as individual substances with independent reducing powers, the old hypothesis, instead of being altered to the amyloïn theory, should have been entirely discarded, since it is neither in harmony with the real facts nor is it without internal contradictions. He is unable to decide whether the different erythro- and achroo-dextrins of Lintner and Düll are individual substances, but considers they were obtained in as pure a condition as our present means will allow. Ost thinks the molecular weight determinations, especially those of the higher dextrins, are of a very doubtful value, an opinion also shared by Kuster. He is disposed to regard the erythrodextrins not as individual substances, but as mixtures of starch and achroodextrin, in accordance with the view which had been put forward by Musculus and Meyer, who found that achroodextrin, to which half a per cent. of soluble starch had been added, gave a red colour with iodine. Should this prove to be the case, then all the dextrins would be achroodextrins.

Mittelmeier's Views on Starch Degradation.[1]—According to this observer, the first action of diastase on the starch molecule is to split it up into two amylodextrins, which exhibit considerable differences in their further behaviour with diastase, one of them being much more resistant to its action than the

[1] *Zeit. angewand. Chem.*, 1895, p. 552.

other. By the time taken for the first to reach the stage in which it gives a red colour with iodine, the second has passed through all the intermediary phases of starch degradation down to maltose. In this way an explanation is readily afforded of the simultaneous presence of the highest dextrins and maltose in the same starch conversion. That the two amylodextrins differ chemically is shown by the fact that the dextrins which they respectively yield differ. Mittelmeier thinks it probable that there is a greater number of chemically different dextrins than has hitherto been assumed.

The erythrodextrin which forms the more quickly he terms the primary; it is distinguished from the secondary by its comparative insolubility in cold water, and by being precipitated from its hot solution on cooling as a pulverulent precipitate, which increases on standing. It is insoluble in cold water, but soluble in water having a temperature of 60° to 65° C. Both the primary and secondary erythrodextrins reduce Fehling's solution and form compounds with phenylhydrazine, thus showing their aldehydic nature. Their opticity is about $[\alpha]_D = +170°$. Under the influence of diastase, a primary achroodextrin arises from the primary erythrodextrin, which reduces Fehling's solution still more powerfully than the dextrin from which it is derived. It is soluble in water, but can be precipitated from its aqueous solution by the addition of alcohol, when it assumes the form of star-shaped crystalline aggregates. On further hydrolysis it yields maltose. A secondary achroodextrin is also derived from the secondary erythrodextrin. This is characterised by its different behaviour with diastase, and also by the fact that its hydrolytic products yield jelly-like osazones. (See *Metamaltose*, next paragraph.)

Metamaltose.—The osazone of this body was found by Mittelmeier[1] amongst the jelly-like substances which are invariably formed when the products of a starch transformation are treated with phenylhydrazine acetate. The secondary achroo- and erythrodextrins when treated with malt-extract yield an abundance of this substance, whilst the primary ones under similar treatment do not. These jelly-like osazones are only purified with the greatest difficulty, and on analysis gave figures corresponding with those of the osazone of a disaccharide. The osazone when heated began to sinter at 120° C., melted at 145° to 148°, and commenced to decompose at 160° to 165°. It evidently differed considerably from the osazones hitherto obtained from starch-transformation products, and as the body which yields it is produced in considerable quantity by diastatic action, and cannot be found in beer, Mittelmeier thinks that it probably plays an important part in fermentation.

Prior's Achroodextrin No. III.—This body was isolated in 1896[2] from the products of a starch transformation which gave the reaction for erythrodextrin, and from which the more

---

[1] *Zeit. angewand. Chem.*, 1895, p. 552.
[2] *Bayer. Brauer.*, 1896, p. 157.

easily fermentable bodies had been removed by fractionation with alcohol. It had an opticity of $[a]_D = 172.1°$ and a reducing power of R. 42.5.[1] It was considered that this substance probably had the composition $(C_{12}H_{20}O_{10})_2 + H_2O$. It was fermentable with difficulty, and then only partially by the Frohberg or Saatz yeast, the fermentation being carried to a somewhat higher degree by the former. Mixed with 20 per cent. of maltose, it yielded an osazone which exactly resembled the so-called isomaltosazone.

Ling and Baker on the Dextrins.—These investigators, in a third paper read before the Chemical Society,[2] stated that they had been able to isolate, from the products of a starch conversion which were soluble in alcohol, a substance agreeing in its properties with the maltodextrin of Brown and Morris, and also another corresponding to the achroodextrin III. of Prior; they propose to term these bodies respectively "maltodextrin a and β." They also obtained, from those unfermentable portions of the products of a starch transformation soluble in alcohol which were said by Lintner to consist principally of "isomaltose," a body isomeric with maltose, and which had an opticity of $[a]_D = 156°$ and an R. of 62.5. This they thought might be the simple dextrin of the formula $C_{12}H_{20}O_{10} + H_2O$, the existence of which they had surmised in their previous investigations (p. 158). It yielded a small amount of an osazone melting at 120° to 130° C., and this probably pointed to slight contamination with maltose. It was soluble in alcohol.

ACTION OF THE PROTEOLYTIC ENZYMES ON THE PROTEIDS.

Since it has not yet been found possible to isolate the proteolytic enzymes from barley or malt, we have to content ourselves with studying the action of the two proteolytic enzymes, pepsin and trypsin, which can be obtained in a state of comparative purity from the stomach and pancreas of animals. Krauch unsuccessfully attempted to isolate a proteolytic enzyme from a number of seeds, bulbs, tubers, &c. Gorup-Besanez considered he had succeeded in extracting enzymes of this nature from tares, hempseed, linseed, and germinated barley; but this is somewhat doubtful, since the method he employed for detecting the products of their action is open to criticism. A powerful enzyme of this nature has been isolated in a fairly pure state from the latex of the tropical plant *Carica Papaya*. It digests proteids with great facility, converting them into proteoses and peptones, a small quantity of amides being simultaneously formed. Certain plants, which possess organs peculiarly arranged for entrapping insects,

[1] *Bayer. Brauer.*, 1896, p. 385.
[2] *Journal of the Chemical Society*, 1897, p. 509.

are also provided with a powerful proteolytic enzyme, by means of which the insects so caught are digested and utilised in the nutrition of the plant. The fluid contained in the pitchers of a genus of insect-eating plants, the *Nepenthes*, when slightly acidified, instantaneously dissolves coagulated egg albumin.

Action of Pepsin on Proteids.—The action of this enzyme can be readily studied on the proteid, egg albumin, which is a substance easily obtained in a state of comparative purity. It is only necessary to boil an egg until its contents are completely hard, remove the white from the yolk, and cut the former into thin slices, or finely divide it by rubbing through a sieve. The finely divided albumin is then placed in a very dilute solution of pepsin which contains 0.2 per cent. of hydrochloric acid, and kept at a temperature of 38° C. (100° F.) for two or three hours, when the bulk of the albumin will have by this time dissolved. If a little of the solution is now taken out, filtered, and exactly neutralised with an alkali, a precipitate falls. This consists of syntonin or acid albumin, and represents the first stage of the action of pepsin on a proteid. If the original mixture is allowed to stand at the temperature before indicated for several hours longer, a portion of the solution removed, and, as before, exactly neutralised with an alkali, not nearly such a large precipitate will form. The bulk of the syntonin has by this time been converted into a group of bodies collectively called "proteoses," together with a little peptone. The presence of the proteoses may be detected by removing the syntonin from the neutralised liquid by filtration, and saturating the filtrate with ammonium sulphate. The heavy precipitate which is thrown down consists of the proteoses. Nitric acid also precipitates these bodies from a fairly strong solution as white flocks, which dissolve when the mixture is heated, and reappear as it cools. The group of the proteoses may be separated into its three members, the proto-proteoses, the hetero-proteoses, and the deutero-proteoses, in the following manner. The fluid which contains all these three members is saturated with sodium chloride; this precipitates the proto-proteose and hetero-proteose. The latter of these two bodies is insoluble in pure water; the precipitate obtained above, which contains these two bodies, is therefore dissolved in a dilute solution of sodium chloride and dialysed. When the salt has been completely removed by dialysis, the hetero-proteoses are precipitated, and the proto-proteoses remain in solution. In order to obtain hetero-proteose in considerable quantity, the digestion of the original mixture is allowed to continue for some time longer, when the transformation of a large portion of the proto-proteoses and hetero-proteoses is carried a stage further, and they are converted into deutero-proteoses and peptones. In order to obtain the deutero-proteoses, the solution is first saturated with sodium chloride, then acetic acid added drop by drop until a precipitate ceases to be formed. The proto-

ACTION OF PEPSIN ON PROTEIDS.

and hetero-proteoses are not completely removed from a solution by saturation with salt, it is only after the addition of acetic acid that the last portions are precipitated; these carry down with them a little deutero-proteose. The deutero-proteose can then be obtained by dialysing the acid salt solution until it is quite free from sodium chloride, evaporating to small bulk, saturating with ammonium sulphate, and filtering off the precipitate. This precipitate consists of the deutero-proteoses; the peptones remain in solution. The precipitate thus obtained is then dissolved in a little water, the solution dialysed, evaporated to small bulk, and poured into strong alcohol; this re-precipitates the deutero-proteoses, which are then washed with alcohol and dried over sulphuric acid. If the digestion of the original mixture is carried on for a much longer period, say several weeks, a large quantity of the original albumin will have been converted into peptones. In order to obtain these bodies, the solution, after concentration, is saturated with ammonium sulphate, which throws down all proteids out of the solution with the exception of the peptones. In order to obtain the peptones from the saturated ammonium sulphate solution, the precipitated proteoses are filtered off, and the filtrate concentrated so as to remove the bulk of the salt by crystallisation. The solution is then boiled with barium carbonate to remove the remainder of the ammonium sulphate; the ammonia volatilising, and the sulphuric acid being precipitated as barium sulphate. The boiling is continued until the liquid ceases to evolve ammonia; it is then filtered, concentrated to a syrupy consistence, and poured into strong alcohol. Peptone thus prepared is a gummy substance, which on dehydration with alcohol is obtained as a yellowish powder of a very hygroscopic nature and having a bitter taste. When perfectly dry it dissolves in water with a hissing sound, and its solution is accompanied with evolution of heat. From a large number of experiments, which we cannot enter into here, it has been shown that the proteid molecule consists of two distinct groups or radicals, from which the various products of proteolytic action are derived. The following schematic arrangement will, without going into too minute distinctions, give a general idea of the series of changes:—

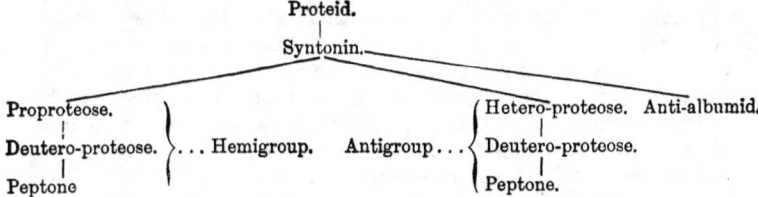

The great distinction between these two groups, which are called the hemi- and anti-groups, and which appear to exist in about equal quantities, is that the peptone of the former can be further broken down by the action of trypsin into a number of amide bodies, whilst the peptone of the latter cannot be so broken down. In every digestion a variable quantity of a substance is left undissolved, which is called anti-albumid; it is extremely resistant to the further action of the proteolytic enzymes, but, by treatment with strong solution of pepsin and acid, may be partially converted into anti-deutero-albumose, and finally into anti-peptone.

Action of Trypsin on Proteids.—Trypsin, which is obtained from the animal pancreas, is a much more powerful proteolytic agent than pepsin, and capable of carrying on the breaking-down process to a much greater extent. As a consequence of this, the products of trypsin proteolysis are much more complicated than those of pepsin proteolysis. Trypsin can act in an alkaline, neutral, or slightly acid medium, but its action is most energetic in an alkaline one. In its natural state, it occurs as a constituent of the pancreatic juice, which is a slightly alkaline fluid. When caused to act upon a proteid in a solution containing from 0.5 to 1.0 per cent. of sodium carbonate, the stage of the proto- and hetero-proteoses is passed through so rapidly that it can seldom be observed, and the proteid matter is quickly converted into deutero-proteoses. These are rapidly transformed into hemi-peptone and anti-peptone in nearly equal quantities. The hemi half is now quickly broken down into a number of amide bodies, such as leucin, tyrosin, &c., whilst the remaining half suffers no further cleavage. The great peculiarity with regard to trypsin is its power to break down the hemi portion of the proteid molecule into a number of crystalline bodies. These are leucin or amido-caproic acid, $CH_3CH_2CH_2CH_2CH(NH_2)COOH$; tyrosin or hydroxyphenylamido-propionic acid (page 100), aspartic and glutamic acids. These bodies have long been known to be products of trypsin proteolysis, but more recently two other bases have been discovered, lysin or diamido-caproic acid, $C_6H_{14}N_2O_2$, and lysatin, $C_6H_{13}N_3O_2$, the constitution of which is not yet known. Small quantities of ammonia have also been detected amongst the products of the action of trypsin on the proteids.

Proteolysis during the Germination of Barley.—It is quite certain that proteolysis goes on to a marked extent during the germination of barley, since a portion of the insoluble proteids are converted into soluble proteoses, and amide bodies are formed as well. The vegetable proteolytic enzyme or enzymes resemble trypsin in their action when we regard the profound changes which are brought about in the proteid molecule under their influence, and yet they resemble pepsin in acting best in a slightly acid medium. Proteolytic action seems to be once more aroused, and to proceed to a limited extent during the process of mashing.

Proteolysis of the Proteids a Necessity.—In the germination of the barley, since the proteid matter of the grain has to travel from the endosperm to the germ, it is necessary that those proteids which are in an insoluble condition should be rendered both soluble and diffusible, because, in passing from cell to cell, they have to traverse the cell-walls. The wort which the brewer produces must also contain proteid bodies of a nature readily assimilable by yeast, or that organism would speedily become so weakened as to be incapable of fulfilling its functions properly. The diffusible and assimilable bodies requisite for this purpose are yielded by the proteolysis of the proteids originally contained in the barley, effected for the most part during its germination, afterwards to a slight extent during the mashing process. The albumin and globulins which occur in the barleycorn seem completely indiffusible, the proteoses are very slightly diffusible, whilst the peptones have very considerable powers of diffusion, and the amides are highly diffusible bodies.

The Nitrogenous Constituents of Malt.—These are partly of a proteid and partly of an amide nature, and though they have often formed the theme of investigation, as yet no thoroughly reliable process has been discovered for their separation and quantitative estimation. The usual plan has been to extract the malt with water, and boil the aqueous solution after the addition of a little cupric hydroxide or ferric acetate. The precipitate thus formed was filtered off and considered to be albumin. It contained, however, part of the proteoses as well. The filtrate was then acidified with hydrochloric acid and solution of phosphotungstic acid added, or tannin without previous addition of acid, and the precipitate thus obtained was called "peptone." It consisted, however, mainly of proteoses, very little, if any, true peptone being found in malt. The remainder of the nitrogenous matter left in solution was assumed to represent the amides. The writer[1] has proposed the following process for the estimation of the proteids in malt: The malt is extracted with cold water, filtered, and the albumin in the filtrate coagulated at a temperature of 60° C. (140° F.). The coagulum is filtered off and weighed. The filtrate is then boiled for a short time, when a second precipitate forms, which is filtered off and weighed. This was considered to represent the globulin, but from the recent experiments of Osborne it is extremely doubtful if all the globulin can be separated by boiling. The filtrate was then evaporated to small bulk and saturated with ammonium sulphate, when a precipitate consisting of the proteoses (and probably some globulin) comes down. This is filtered off, washed with saturated solution of ammonium sulphate, dried and weighed, the weight of the ammonium sulphate adhering to the precipitate and filter-paper being determined and deducted. The filtrate from this precipitate is then diluted with an equal quantity of water and solution of tannic acid added. This would precipitate peptone, if it were present; but no precipitate of any moment was ever observed, and this

[1] *Trans. Inst. of Brewing*, vol. iv. p. 173.

led to the conclusion that real peptone does not occur in malt. Szymanski[1] considered that he had separated peptone from malt, but from the process he adopted for its isolation, the substance which he obtained and mistook for real peptone was evidently deutero-proteose.

The author of this work, in experimenting on the proteids of malt, was never able to detect a hetero-proteose on dialysing the precipitate thrown down by ammonium sulphate. A precipitate was always yielded on saturating malt-extract with sodium chloride and adding a little acetic acid, and the body is left in solution which Szymanski mistook for peptone. In all probability, therefore, the two proteoses are proto- and deutero-proteose. (See also proteids of malt, page 96.)

Bungener and Fries have given the following analysis of the soluble nitrogenous constituents of a sample of barley and of the malt made from it:—

	Barley. Per Cent.	Malt. Per Cent.
Total nitrogen	1.690	1.580
Nitrogen as albumin	0.161	0.230
Nitrogen as peptone	0.040	0.060
Nitrogen as amides (precipitable by mercuric acetate)	0.052	0.182
Nitrogen as amides (not precipitable by mercuric acetate)	0.154	0.352
Total soluble nitrogen	0.355	0.642

Hilger and Van der Becke give the following figures of the soluble nitrogenous constituents of a barley and of the malt made from it, all of which are calculated on the dry substance:—

	Nitrogen of Albumin.	Nitrogen of Peptone.	Nitrogen of Ammonium Salts.	Nitrogen of Amido-Acids.	Nitrogen of Amides.
	Per Cent.	Per Cent.	Per Cent.	Per Cent.	Per Cent.
Barley	0.0600	0.0046	0.0169	0.0417	...
Barley steeped	0.0354	0.0009	...	0.0294	...
Green malt	0.1671	0.0058	0.0290	0.1417	0.0505
Dried malt	0.1194	0.0233	0.0057	0.2257	0.0029

HOPS.

Hops give to beer that pleasant, bitter, aromatic flavour which characterises it from all other alcoholic beverages. They have, however, several other important functions to fulfil. During the boiling process they precipitate albuminoids, and so assist in the clarification of the wort, and, by virtue of the resins which they contain, contribute essentially to the production of a stable beer.

[1] *Landwirthsch. Vers.-Station*, vol. xxxii. p. 389.

Hops are not only indispensable on account of these preservative properties, but also for the peculiarity of the bitter which they contain. This is of a transient nature, that is, so soon as a beer flavoured with hops is swallowed, the sensation of bitterness immediately passes away, while with almost all other vegetable bitters the taste is persistent, and clings to the mouth for a considerable length of time. The aromatic quality of the hop is essentially distinctive; although many other plants possess aromatic flavours, none of them exactly resembles that of the hop. Hops possess distinctly narcotic properties, and these appear to be greatest in the coarser varieties.

The Hop Plant.—This, the *Humulus lupulus* of the botanist, is a perennial climbing plant, often met with in the wild state. It belongs to the *Urticaceæ* or Nettle family, and is diœcious, that is, the male flowers appear on one plant and the female on another. In English and American hop-gardens it is usual to grow a sprinkling of male plants, but on the Continent these are rigorously excluded. In the former case the female flowers become impregnated and form seeds; in the latter they do not. It is thought by some that seed formation ought to be prevented, since the seeds are useless to the brewer, and only add weight to the hops. Hop plants are not raised from seeds, but by cutting off and transplanting portions of their underground stems. They are proverbially a risky crop, are subject to the attack of numerous insects and of various mould fungi, and are exceedingly susceptible to climatic conditions. Consequently, throughout the whole period of their growth they need great care and attention, such as constant washing to keep off insects, and sulphuring to keep off mould; and often, after the expenditure of an immense amount of trouble and care, a promising crop may be completely ruined by a few weeks of unpropitious weather just before the hop harvest.

The ripe female flowers are the parts of the hop plant used in brewing. Each of these consists of a number of bracts or small leaves attached to the extremity of the stem in such a way as to form a cone. The bracts, which are shaped something like roofing tiles, are from a half to three-quarters of an inch in length, and each hop cone contains from twenty to forty of them. The lower portion of each bract is twisted on itself, and in the interior of this fold the female flower develops, to be followed later on by the seed.

The fruit of the hop is a little hard corn, about eight to twelve hundredths of an inch in diameter. Fig. 8 exhibits one of these fruits in the ripe condition; Fig. 9 the same with the exterior skin removed; and Fig. 10 the seed taken out of the seed-vessel. The opinion that there is a relation between the size of the fruit and the quantity of flour which the hop contains is erroneous.

The peculiar yellow dust known as the hop-flour, or the "lupulin," which is contained in less or greater quantity in all hops, is of great importance to the brewer, for it chiefly contains those resins, volatile oils, and bitter substances which are of such an exceedingly valuable nature. The yellow powder, or "lupulin," as it is termed, is secreted by certain glands which are found on the inside of the bracts of the hop cone, upon the stem which bears the fruit, and upon the outside of the fruit. The quantity of lupulin varies considerably in different varieties of hop, and also in the same kind of hop in different years, even when grown in the same district. The respective proportions of the different constituent parts of the hop cone, mechanically separated, are given by Haberland as follows:—

	Per Cent.
Lupulin (hop-flour)	7.92 to 15.70
Leaves	69.79 to 78.36
Stems	8.50 to 17.54
Ripe seeds	0.02 to 7.80

In a sample from Wisconsin the last item amounted to 20 per cent.

In practice, the quantity of the hop-flour present—the "condition," as it is called—is roughly estimated by rubbing the hops

FIG. 8.   FIG. 9.   FIG. 10.

Seed of the Hop.

between the hands, the amount of stickiness left forming an index as to the quantity of lupulin present in the hops.

Varieties.—Hops are of many varieties. To the finer kinds, distinguished by their delicate aroma and pale colour, belong the Goldings, Worcesters, and some few others; to the less-prized varieties, the Jones, Fuggles, &c., while the Colegates represent what may be termed a coarse kind. Besides our native produce, excellent hops are now imported from Germany, America, and Belgium.

Valuation.—Hops should be of a bright greenish-yellow colour, rather more inclined to yellow than to green. When rubbed between the hands, they should yield an agreeable and powerful aromatic odour, and leave yellow sticky traces, which show that they contain an abundance of flour or "lupulin." They should also rub to a fine chaff, and leave no residue of fibrous matter behind. A mass of hops, when properly dried, is extremely elastic; when released after being compressed, it immediately resumes its former bulk. Hops which are used for dry hopping should not disintegrate too readily, or, as it is termed, be "mashy," since the detached fragments float in the beer, and are drawn off along with it. Mashiness is either caused by over-ripeness or some defect in the drying process.

It is, however, only long experience in the practical handling of samples which will enable a person to form a correct judgment as to the value of any particular sample of hops.

Drying and Sulphuring.—The hops, after being plucked, are dried in a kiln resembling in some respects the drying kiln of the maltster. Much care is needed during the process, or they may be easily damaged. When the moisture has been completely removed, sulphur is placed on the fire, and this has the effect of brightening the colour of the hops; the sulphurous acid evolved also acts as an antiseptic, destroying to some extent the germs of mould fungi and other organisms attached to them. The question as to whether the sulphuring of hops has a prejudicial effect on the wort during its subsequent fermentation has often been a theme for discussion. The bulk of the evidence, however, goes to show that it is practically harmless. Probably this is not the case with regard to the particles of sulphur which are left on the hops after the sulphuring for mildew. When powdered sulphur is mixed with yeast, it exercises no prejudicial effect upon its fermentative action, but causes during the fermentation the evolution of a small amount of the offensive gas, hydrogen sulphide.

The Chemical Constituents of Hops.—Hops contain, in addition to the ordinary constituents of plants, such as cellulose, woody fibre, &c., several other substances which are of the greatest importance to the brewer. These are the essential oil, the resins, the tannin, and the diastase.

Essential Oil.—On this constituent depends the characteristic agreeable aromatic flavour of the hop. Hop-oil is readily soluble in alcohol and ether, but only slightly so in water, one part of hop-oil dissolving in from 600 to 700 parts of water. The boiling-point of the essential oil is about 125° C. (257° F.), but when hops are distilled with water the oil passes over with the steam. According to Personne, hop-oil is a mixture of a hydrocarbon, $C_{10}H_{16}$, and of another body having the composition $C_{10}H_{16}O$. Chapman[1] obtained, by distilling 80 kilogrammes of hops with water, 140 c.c. of essential oil. After this had stood for ten or eleven months, he obtained from it, by fractional distillation over sodium, 40 c.c. of a sesquiterpene of the formula $C_{15}H_{24}$, which had a boiling-point of from 261° to 265° C., and a specific gravity of 0.8987. Its opticity was $[a]_D = +1.2°$. Another oil, similarly obtained by distilling hops with water, and which was investigated while fresh, was found to have a slightly lower boiling-point. It consisted of lower terpenes, of a sesquiterpene, and another constituent which contained oxygen. Upon further investigation the greater part of this oil was found to distil over at from 168° to

---

[1] *Trans. of the Chemical Society*, 1894, p. 54.

173° C. in a partial vacuum (60 m.m.), and to consist of a sesquiterpene having the formula $C_{15}H_{24}$. Its corrected boiling-point was between 263° and 266° C., its specific gravity at 15° C. was 0.9001, and its molecular refraction 66.2. The oil was optically inactive, and united with four atoms of bromine to form an oily bromide. Since the sesquiterpene was not identical with any of those already known, it was named "humulene." A number of its derivatives have been described.[1] A hydrocarbon was found in the remaining portions of the distillate which rapidly polymerised, and probably consisted of two bodies having the formulæ $C_{10}H_{16}$ and $C_{10}H_{18}$. The former of these probably belongs to that group of bodies called "olefineterpenes" by Semmler, and the latter is probably tetrahydro-cymene. An oil was obtained from another fraction, which had the agreeable odour of geranium oil, and which closely resembled geraniol. On account of its volatile nature, the greater portion of hop-oil is lost in the boiling of the wort. It has hitherto been supposed that in hops which are stored for a considerable length of time, the hop-oil becomes converted into valeric acid, and that it is this acid which gives the peculiarly cheesy odour to old hops. Bungener has found that this is not the case; hop-oil can be kept for a long time exposed to the air without its odour being materially altered.

Bitter Principles.—Issleib extracted from hops about 0.004 per cent. of a bitter substance, soluble in water, alcohol, and ether. It had an agreeable, aromatic, intensely bitter taste, possessed a slightly acid reaction, and had the composition $C_{29}H_{46}O_{12}$. Lermer and Etti also isolated crystalline bitter principles from hops. Bungener prepared a bitter principle of an acid nature which melted at 92° or 93° C., was soluble in alcohol and ether, but quite insoluble in cold water. It had the composition $C_{24}H_{35}O$, and appears to have had the properties of an aldehyde and a weak acid. It was readily acted upon by oxidising agents with the formation of a large quantity of valeric acid. A small quantity of this principle dissolved on prolonged boiling with water, to which it communicated a bitter taste. On evaporation of the solution a resinous residue was left, which was probably an isomeric modification of the bitter substance. The bitter principle, when left exposed to the air, rapidly became resinous, and had then the odour of old hops; hence it was assumed that this peculiar odour was caused by the oxidation of this principle, and was not due, as had been formerly supposed, to an alteration in the hop-oil. Though hops owe a certain portion of their bitterness to these principles, by far the larger amount is due to the hop-resins.

The Hop-Resins.—These are very important constituents, for to them much of the bitterness of hops is to be attributed, and probably the whole of their preservative power. The hop-resins have been investigated by Hayduck, who divides them into three groups. In order to obtain these bodies, the powdered hop cones

[1] *Trans. Chem. Society*, 1895, p. 780.

are first extracted with ether, the solvent being afterwards removed by distillation. The residue from the distillation is then treated with alcohol, which dissolves the resins and leaves substances of a fatty nature behind. To the alcoholic solution of the resins an alcoholic solution of lead acetate is added; this precipitates one of the resins (No. 1). The lead is removed from this precipitate by hydrogen sulphide, and the resin recovered by solution in ether. After filtering off the precipitated No. 1 resin, the filtrate is evaporated to dryness, and the residue treated with petroleum spirit; this dissolves the second resin (No. 2), and leaves the third resin (No. 3) undissolved. Resin No. 3 is contained in the hops in much larger quantity than the other two. Nos. 1 and 2 are soft resins, and although they are comparatively insoluble in water, yet on prolonged boiling a sufficient quantity dissolves to communicate to the water an intensely bitter taste. It is on these two soft resins that the preservative powers of the hops depend; they are distinctly prejudicial to the growth of the butyric acid and many other bacteria, but do not appear to have much action upon the acetic acid bacteria or upon sarcinæ. No. 3 is a solid resin, only slightly soluble in water, to which it communicates a pleasant bitter taste. It is considered that the two soft resins separate for the most part from the wort during fermentation, where they form a layer on the surface of the yeast, but that more of the third resin remains in solution. The third resin, which is not precipitable by lead acetate, and which is soluble in petroleum spirit, is closely related to the acid bitter principle of Lermer, which, from the researches of Bungener, appears to be the mother substance from which the second soft resin is formed.

Hop-Tannin Bodies.—Hop-tannin, which was first isolated in a fairly pure state by Etti, is contained in quantities of from 2.5 to 6.0 per cent. in hops. Etti extracted hops first with ether and then with absolute alcohol, by which solvents the hop-oil, the resins, and the fatty substances were removed. The hops were treated with dilute alcohol to dissolve the tannin, which was precipitated therefrom by an alcoholic solution of lead acetate. The precipitate thus obtained was decomposed by hydrogen sulphide, when the tannin was left in solution, from which, by subsequent filtration and evaporation, it was obtained in the solid state. Prepared in this way, hop-tannin is a fawn-coloured powder, of the composition $C_{25}H_{24}O_{13}$, soluble in water and dilute alcohol. Its aqueous solution precipitates albumin, but not gelatin. Etti also found that the hop-tannin was accompanied by another substance, phlobaphen, having the composition $C_{50}H_{46}O_{25}$. Hayduck [1] has recently examined both these bodies with great care. Working by Etti's method, he extracted hops with ether and absolute alcohol, and then with dilute alcohol. This latter solution, which contained the tannin and phlobaphen, was fractionally precipitated with alcoholic solu-

[1] *Woch. f. Brau.*, 1894, p. 409.

tion of lead acetate. The first fractional precipitate consisted chiefly of phlobaphen, the last of tannin, the others of mixtures of the two. The separation of the two bodies was effected by means of acetic ether, which dissolves the tannin and leaves the phlobaphen behind. The purified hop-tannin was obtained as a pale brown amorphous powder, soluble in water, insoluble in ether, and nearly so in absolute alcohol. Dissolved in water, it had a slightly bitter, afterwards somewhat astringent taste. It showed an acid reaction with litmus-paper, and gave a dark green coloration with ferric chloride, but no precipitate. It is absorbed by powdered hide, and this has an important bearing on its quantitative estimation. Especially remarkable is the readiness with which it becomes transformed into phlobaphen; even in the evaporation of a solution of hop-tannin part of the tannin suffers this change. If hop-tannin is heated in the dry state to $140°$ C., or if its solution is evaporated with a little sodium carbonate, the whole of the tannin is quickly and completely converted into phlobaphen.

Phlobaphen was obtained in the form of a reddish-brown powder, only partially soluble in boiling water or in absolute alcohol. Its aqueous solution is of a brownish-yellow colour, and has a disagreeably harsh, astringent taste. It gives with ferric chloride a dirty dark-green precipitate. Albumin is precipitated by phlobaphen, and the latter is removed from its aqueous solutions by powdered hide. The most noticeable point with regard to this substance is that it forms with the albuminous matters of the wort a compound which is completely insoluble, while the tannin forms a compound which is soluble in boiling water or wort, and also soluble to some extent in these fluids when cold.

Alkaloids of Hops.—Griessmayer detected the presence of an exceedingly small quantity of an alkaloid in hops, which he terms "lupuline," and which is said to be a narcotic. It seems to be present only in the coarser varieties of hops, some of the finest specimens being quite free from it. American hops are said by Williamson to contain 0.15 per cent. of an alkaloid which he has named "hopeine." This, according to Williamson and Leuken, can be separated into two bodies by means of amylic alcohol, the one identical in composition with morphine, and called by them "isomorphine," the other being pure hopeine. Another basic substance is the cholin found in hops to the extent of 0.02 per cent. by Griess and Harrow. Griessmayer is of opinion that cholin is not contained in the hops as such, but as a constituent of lecithin, a complicated fatty body found in the brains of animals.

Beer, when concentrated in a vacuum and partaken of, is said by Langer to exhibit a pronounced narcotic action similar to that caused by morphine.

Other Constituents.—Besides the constituents mentioned above, hops contain a gum, which is soluble in water; a soluble red colouring matter, which has an influence on the colour of the beer; proteid bodies, and some other substances which are extracted on prolonged boiling with water, and which have a rank, coarse, disagreeable, bitter flavour. Hops, like all other plants, contain a certain amount of mineral matter, which chiefly consists of the phosphates of potassium and magnesium.

Chemical Analysis of Hops.—Although it is possible to estimate with a fair degree of precision the various constituents of hops, it has not been so far found possible to establish any very definite relation between the value of the hops and the amounts of essential oil, resins, and tannin which they contain. Consequently, up to the present time, chemistry has not afforded much assistance in this direction. Recently Briant and Meacham[1] have proposed a method for the valuation of the preservative power of hops, which consists in estimating the amount of the soft resins they contain. So far as can be seen at present, the method appears to be capable of yielding satisfactory results. Heron[2] proposes to estimate the value of hops by the amount of tannin they contain; he considers that the larger this amount is, the better the sample. Barth[3] confirms this view to a certain extent; he finds that all good hops contain high percentages of tannin, although there are cases in which hops of very different quality contain nearly the same quantities of tannin. Contrary to the statement of Etti (p. 175), both these observers find that hop-tannin is precipitated by gelatin, and that this substance is a much more efficient reagent in the estimation of the tannin in hops than hide-powder.

Diastase of Hops.—Hops, like all other plants, contain diastase, but this point appears to have been overlooked until attention was called to the fact by Brown and Morris.[4] The enzyme could only be extracted by water after removal of the tannin contained in the hops. A similar phenomenon was observed by Baranetzky[5] with reference to acorns. The presence of diastase in hops is important as regards its influence on the secondary fermentation in cask, since it there helps to degrade the higher dextrins and bring them into a state amenable to the action of the yeast.

[1] *Journal of the Federated Institutes of Brewing*, 1897, p. 233.
[2] *Ibid.*, 1896, p. 164.
[3] *Zeit. f. d. gesammt. Brauwesen*, 1897, p. 168.
[4] *Transactions of the Institute of Brewing*, 1893, p. 94.
[5] Sachsse, *Agriculturchemie*, p. 364.

PART II.

CHAPTER III.

THE MICROSCOPE.

THE microscope in its simplest form is merely a magnifying-glass, and, as such, has undoubtedly been known for a very long period. We meet with frequent references in ancient literature to burning-glasses, which are convex lenses; and as all such lenses possess the power of magnifying objects, this property could not have failed to attract attention and receive practical application. The compound microscope, which seems to have been invented in the sixteenth century, was at its birth, and for a long time afterwards, a very imperfect instrument. For a considerable period it seems to have been used for little else but purposes of amusement and curiosity, such as the observation of small insects, until it was first practically applied by Leeuwenhoek to purposes of scientific import in 1675. This pioneer in microscopical science found that while fresh water contained no living organisms, an infusion of ground pepper, which had been allowed to stand for a few days, teemed with them. He also expressed the opinion that these organisms did not arise spontaneously, but were developed from germs which previously existed in the atmosphere, and which had fallen into the liquid; he considered that the reason they grew abundantly in an infusion of pepper or other vegetable substance, and did not appear in pure water, was because there was an abundance of food in the infusion. Since then the question as to the origin of the living organisms which appear in this way in infusions has been the subject of many discussions and contentions, and these have ended in absolutely confirming the truth of Leeuwenhoek's remarkable surmise. This observer also recognised the peculiar forms of starch and pollen grains; he also observed that yeast consisted of a number of round or oval bodies. His observations were soon followed by those of Hook, Malphigi, and Grew, by whom the cellular nature of the tissues of living organisms was recognised. The microscopes used at that time were necessarily very imperfect instruments; the art of grinding lenses correctly to shape was not then understood, nor was the method of constructing a lens with two different kinds of glass, each having a different dispersive power, and in this way getting rid of chromatic aberration. The images of the objects under observation were not only indistinct, but all lines or points in them were bordered by a series of coloured rings. It was only at the commencement of the present century that these matters came to be better understood, the result being that in 1830 microscopes capable of magnifying up to 400 times, and of a much more perfect construction, were obtainable. From this time forward a list of the names of the microscopical workers who have enriched our knowledge, not only of the minute structure of animals and plants, but also of the processes going on in them, would in itself form a long catalogue. As time went on, the study of the lower organisms, such as the mould and yeast fungi, became themes for frequent investigation. Bail studied the organisms met with in fermenting worts, and surmised that there existed a family relationship between them and the mould fungi. The yeast fungi were also minutely investigated by Rees, and the mould fungi by Brefeld and De Bary. Pasteur extensively employed the microscope in his researches on fermentation. With his important discovery that many of the disasters of the brewery were caused by organisms foreign to the true ferment of yeast, and that their presence could be readily detected by the aid of the microscope, this instrument at once took its place amongst the other instruments of precision employed in the brewery, such as the thermometer and the hydrometer. The more recent investigations of Hansen have rendered the microscope of still greater importance to the brewer, for without its aid the cultivation and practical employment of single-cell yeast would be an impossibility. In more recent years the study of the bacteria, which are the smallest organisms met with in nature, has occupied much attention, and the results already obtained in this field are of immense importance. Continued investigation in it is likely to lead to discoveries the significance of which can be scarcely conjectured. Apace with this, and partly impelled by the necessity for increased perfection in the instruments specially designed for the observation of these minute organisms, such enormous improvements have been made both in the optical and mechanical parts of the microscope, that it is now possible to obtain colourless and absolutely distinct images of objects under magnifications of two or three thousand diameters.

Description of Microscopes.—Microscopes are of two kinds, the simple and the compound. The former in its most rudimentary form consists of a single lens; it is, in fact, simply a magnifying-glass. As such, it is occasionally used in the brewery for the examination of malt, hops, &c.

The compound microscope, which is, however, much more extensively employed by the brewer, consists of two portions, the optical and the mechanical. Of the two, the former is by far the most important, for though it is possible to do fairly good work with inferior mechanical arrangements providing the lenses are good, it is impossible to work with any degree of satisfaction if the optical portion of the microscope is bad. For comfortable, con-

DESCRIPTION OF THE MICROSCOPE. 181

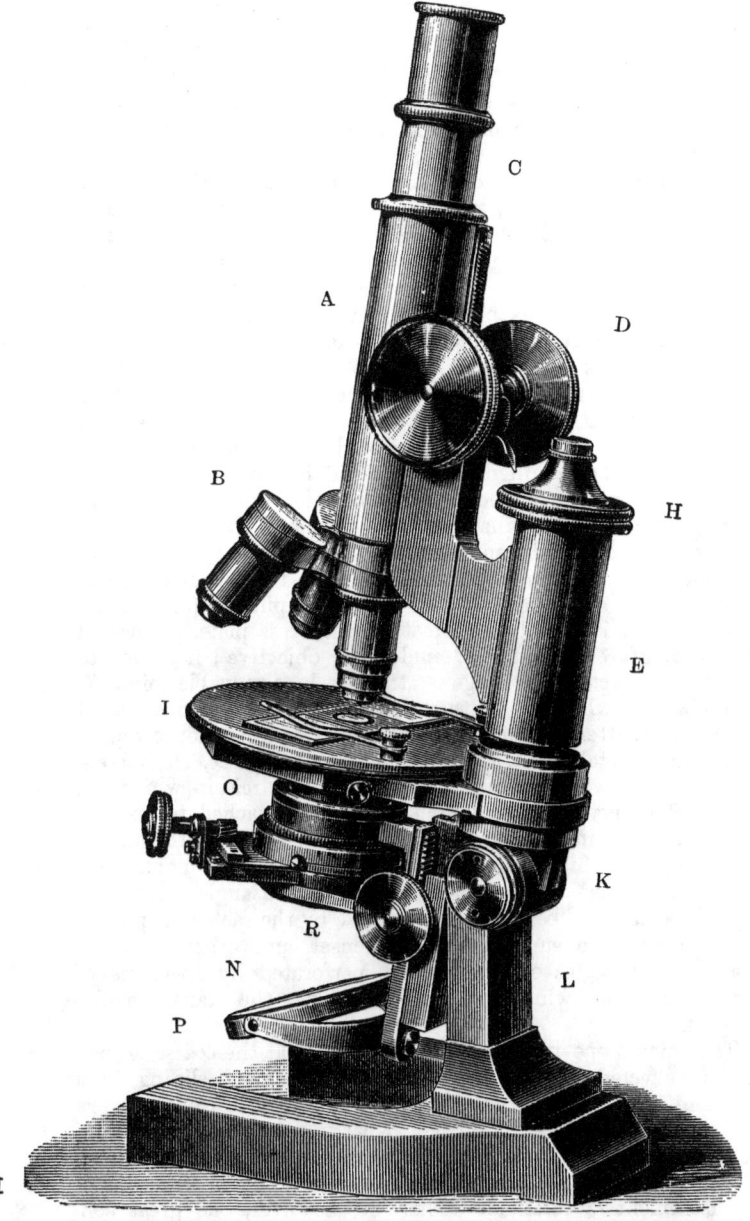

FIG. 11.—Microscope (Leitz's Stand I*a*, with triple nose-piece and Abbé's Condenser).

venient, and accurate working, it is necessary that both portions of the instrument should not only be of good workmanship, but also specially adapted to the class of work for which it is required. For this reason it is extremely desirable, when purchasing an instrument, only to go to first-class makers who guarantee their work, and also, if possible, to obtain the assistance of some one who is an adept at the instrument.

The mechanical portion of a microscope consists of a tube, A (Fig. 11), called the body-tube, upon the lower end of which, technically known as the "nose," an internal thread is chased, into which either the various objectives are screwed in turn, or the nose-piece which carries two or more objectives; this latter is shown attached at B. Into this body-tube there slides another tube, C, of slightly less diameter, termed the "draw-tube." This is completely open at its upper end, and into this opening the various eye-pieces are inserted. By means of the draw-tube the length of the instrument can be varied; this alters the distance between the eye-piece and the objective. The body-tube is moved up and down by rotating the large milled heads, D, and by this movement the "coarse adjustment" of the distance between the objective and object is effected. The pillar, E, has a triangular bore, and slides on a triangular pillar, the movement being effected by an exceedingly fine screw actuated by the button, H; this arrangement is called the "fine adjustment." The triangular pillar is fixed to the stage, I, which is usually a perfectly level brass plate, perforated with an aperture immediately under the objective; it serves to support the object. The stage is attached by a movable joint, K, to the pillar, L, which, in its turn, is fixed to the horseshoe-shaped foot, M; this latter is generally loaded with lead in order to render the instrument stable. Under the stage the tail-piece, N, is fixed, which on its lower extremity carries the half-circle in which the mirrors, P, swing. The semicircular piece is attached to the tail-piece by a screw in such a way that it can revolve on the axis of the screw, and this motion, combined with the swinging one, enables the mirrors to be placed at any required angle. The sub-stage, O, which can be raised or lowered by the rack and pinion, R, is intended to support the condenser and other accessory apparatus. A series of diaphragms, perforated by apertures of different size, and which can be fixed under the stage, are also provided.

The microscope is often constructed so that the coarse adjustment is effected by the body of the instrument sliding in a tube, and this arrangement is sufficient for the ordinary work which the brewer has, as a rule, to carry out; yet, when really fine work has to be done, the rack-work adjustment is almost a necessity, since by its employment a more accurate centering of the instrument is retained. There is, however, one great convenience about the ordinary sleeve adjustment, which is, that the tube can be taken bodily out of the stand when an objective has to be changed.[1]

The length of the body-tube varies considerably in the microscopes of different makers; the English ones provide a wide tube, 10 inches in length, the Continental makers a narrower tube, 6 inches in length. This difference in length becomes a matter of considerable importance if the observer uses both English and Continental objectives, since the former are optically corrected for the long tube, the latter for the short one. In such a case it is better to have a microscope with a 6-inch tube, which can be extended to 10 inches by means of the draw-tube. The shorter and smaller instruments are undoubtedly more convenient to handle, and can be worked with a greater degree of comfort than the larger instruments, which are somewhat unwieldy. In the instruments with the shorter tube, the hands, when moving the object about on the stage, are nearer the eye; and delicate movements, such as those required for moving the object about on the stage, are more easily performed under these circumstances than when the hands are a considerable distance away from the eye. Since this fact has become universally recognised, most of the English makers manufacture instruments with the body-tube of the Continental length.

The nose of the microscope should be provided with a screw-thread of the pitch known as the "Microscopical Society's standard," as all English and most Continental objectives are provided with this standard thread. In case an objective is not fitted for this gauge, an adapter which will enable the objective to be screwed on to the nose of the instrument can be procured from the optician at a trifling cost.

Nose-Piece.—This arrangement (Fig. 11, *B*) screws on to the nose of the microscope, and is made to carry two, three, or more objectives. These can be successively brought into line with the instrument by revolving the lower plate of the nose-piece, which is then secured in position by a catch. This appliance is a great convenience, and saves much time which would otherwise be lost in screwing and unscrewing the objectives on and off. The objectives, when attached to the nose-piece, are covered and protected from dust.

The Stage.—This, in its simplest form, consists of a perfectly level brass or glass plate mounted so as to be at right angles to the axis of the body of the microscope. In its centre is a circular opening for admitting light to the object. A pair of spring clips serve to keep the glass slip closely applied to the stage, while, at the same time, they allow it to be freely moved about; for most purposes these are all that is necessary. Some stages are provided with a sliding bar, which can be moved to and from the observer, the bar always maintaining a position at right angles to the sides of the stage. Some microscopes are provided

---

[1] When doing this, care must be taken that the eye-piece does not fall out of the tube and become damaged.

with a mechanical stage, by means of which the object can be moved in a straight line from side to side by turning a milled head, while another causes the object to travel from before backwards. This arrangement adds considerably to the cost of the instrument, and for the purpose of the brewer is quite unnecessary. With a plain stage the hands soon become accustomed to move the object about with the greatest degree of nicety. All objects when seen under the microscope appear upside down, and consequently all movements of the object appear reversed. This reversal of the movements of the object causes a little difficulty at first, but this soon vanishes.

The Diaphragms.—These consist of thin plates of metal perforated with apertures of various sizes, and are attached to the stage in various ways; they serve to regulate the amount of light which obtains access to the object. The diaphragms are often attached to short cylinders which accurately fit into the circular opening in the stage; or the cylinders may be attached in turn to a plate which slides into a dovetailed groove underneath the stage. These forms are often replaced by a circular plate of metal, perforated near its margin by circular apertures of varying sizes, which is fixed under the stage by a screw passing through its centre, around which the plate rotates. The apertures, as the plate is revolved, successively come into position, each being then exactly in the axis of the microscope; a little catch retains the plate in this position. This is not such a good arrangement as the former, for to obtain the best effects the diaphragm should be immediately under the object. In some instruments the diaphragm-plate is attached to the sub-stage, and is then a considerable distance below the stage; this is the worst form of all.

Iris Diaphragm.—An exceedingly convenient form of diaphragm called the "iris" diaphragm is obtainable. It is provided with a number of plates of a peculiar form, so slotted and pivoted that when these are actuated by a lever the aperture in the diaphragm is widened or contracted.

Sub-Stage.—This is an arrangement underneath the stage, which generally consists of a brass ring of about $1\frac{1}{2}$ inch in diameter, fitted to the tail-piece of the microscope in such a way that it can be moved to and from the stage. This movement is often obtained by a slide working in a dovetailed groove, but it is much better effected by a rack-and-pinion movement. The principal use of the sub-stage, so far as the work of the brewer is concerned, is to support the condenser. The best instruments are provided with an arrangement for bringing the axis of the condenser accurately into line with that of the body of the instrument. In the instruments of Zeiss the condenser is hung on a side-pin attached to a piece which slides up and down the tail-piece; this permits of the condenser being temporarily turned out of the way when not wanted, and it can be completely removed by simply lifting it off the pin.

Cheaper instruments have, instead of a sub-stage, a short plain tube fixed under the stage, into which the condenser slides; the tube should be fitted so as to be truly central, or it is comparatively useless.

Objectives.—These are by far the most important portion of the microscope, for on their perfection depends, in the greatest measure, the clearness and distinctness of the image viewed. Great care should, therefore, be exercised in their choice. Objectives generally consist of three small achromatic lenses mounted in brasswork, the upper portion of which carries an external screw-thread by which the objective is attached to the microscope. Each achromatic lens consists of two lenses of different kinds of glass cemented together; by this arrangement an image free from colour is obtained. Many lenses have what is called a "correction mount," by means of which the front lens can be removed nearer or farther from the other two lenses by turning a collar. This arrangement is for making a correction for the thickness of the cover-glass employed. All objectives without a correction mount, homogeneous oil immersions excepted, yield their best results with a cover-glass of a definite thickness, and this is generally indicated on the objective. For instance, if an objective is marked 0.18, this means that a cover-glass of the thickness of 0.18 millimetre is to be used with that particular objective in order to obtain the best results with it. When, however, the objective is provided with a correction mount, the observer is to a great extent independent of the thickness of the cover-glass; he has simply to focus the object, and turn the correction collar backwards and forwards until he finds the position which yields the best and clearest image. The correction mount is generally provided with an index, every division of which corresponds to a 0.1 of a millimetre thickness of cover-glass; consequently, if the thickness of the cover-glass is known, the index is simply set to the figure corresponding to that thickness.

A certain amount of correction for thickness of cover-glass can also be made with the draw-tube of the microscope: when the cover-glass is too thin, the draw-tube is to be pulled out; when too thick, it is to be pushed in.

The magnifying power of objectives is not, as a rule, directly stated, but their focal length is given instead. The magnifying power of the combination is equivalent to the enlarging power of a simple lens of the same focus. A simple lens of 1-inch focus has, for ordinary eyesight, a magnifying power of ten times, a half-inch of twenty diameters, and so on, the magnifying power being in inverse proportion to the focal length of the objective. An objective having a focal length of a quarter of an inch will, therefore, have a magnifying power of 40.

Aperture.—The amount of detail which an objective is capable of showing depends upon what is called the "angle of aperture" of the objective. Of two objectives of the same focal length but different apertures, that with the wider angle will permit the entrance of a larger cone of rays than that with the narrower angle; the former will consequently have a higher resolving power. Such objectives are considerably more difficult to manufacture than those of a narrow angle, and are consequently much more expensive; they are not, however, required for the purposes to which the brewer ordinarily applies his microscope. Moreover, they possess but little "depth of focus," that is, the thickness of the plane which is in focus at one time is exceedingly thin, and the wider the aperture with objectives of the same focus, the thinner this plane becomes. The same happens with objectives of high power; the shorter the focus, the thinner the plane in focus. This is a source of some difficulty to the beginner, and is remedied by perpetually altering the focus of the instrument. In this manner a number of different planes are brought successively into view, and, by combining these mentally, a general conception of the structure of the whole thickness of the object is obtained. This cannot, however, be pushed too far, because, if the object under examination is thick, its deeper layers appear misty and indistinct.

Apochromatics.—A new series of objectives has been introduced of late years which are called "apochromatic." By the employment in their manufacture of special kinds of glass, and the introduction of a lens of the mineral fluorite, and also by means of the most careful mathematical calculations, these objectives have been brought to a higher degree of perfection than any hitherto produced. The apochromatics are exceedingly expensive, and will never be required by the brewer unless he intends to become an expert in bacteriology; even in this case he will manage for a long time with the ordinary objectives, since at first he would probably be unable to appreciate the difference between the two. The apochromatics possess one marked advantage; they can be used with eye-pieces of very high magnifying power. Thus a quarter-inch objective may be used with an eye-piece magnifying 27 diameters, which gives an enlargement of 1080 diameters, as well as with an eye-piece magnifying only 2 diameters, or with any other eye-piece intermediate between these two. This, which gives magnifications of from 80 to 1080 by simply removing the eye-pieces, saves much trouble in screwing and unscrewing the various objectives.

Immersion Objectives.—In addition to the ordinary objectives, which are termed for the sake of distinction "dry," another series is manufactured which are termed "immersion." When using these, a drop of water is placed between the cover-glass and the front lens of a " water immersion " objective, or a drop of cedar-wood oil in the case of an "oil immersion." The object of using an immersion fluid is to bring a larger cone of rays into the objective. Light in its passage from the cover-glass to a dry objective has to traverse a layer of air, and rays of light in passing from glass to air suffer refraction, that is, they are bent away from the perpendicular. If, however, a layer of a medium such as water, which has a higher refractive power than air, is inserted between the cover-glass and the front lens, then the rays are not bent so far away from the perpendicular, and consequently a larger bundle enters the objective. When a layer of cedar-wood oil, which has the same refractive power as glass, is inserted in this way, the rays suffer no refraction at all; they continue straight along in their course, consequently a still larger bundle is able to enter the objective. This is not only equivalent to increasing the angle of the objective, and thus giving better resolving power, but, as there is always a certain loss of light when the rays pass through layers of media of differing refractive powers, a gain in the direction of illumination is also obtained. It follows, as a consequence of the oil employed being of the same refractive power as the cover-glass, that an oil immersion is independent, within wide limits, of the thickness of the cover-glass, and for this reason such objectives are called "homogeneous."

Objectives Necessary for the Brewer.—For most purposes two will suffice, a half-inch and a tenth of an inch. The former will be found useful where low magnification is required, or for finding the position of small objects, and the second is sufficiently powerful for the examination of yeast, whilst bacteria are readily seen with it.

Should it be intended to enter more deeply into the study of bacterial organisms, a higher power will be required, and the most useful all-round objective for such work is a twelfth of an inch oil immersion.

Care of Objectives.—Microscopical objectives are very delicate and fragile articles; consequently great care must be taken not to subject them to any rough usage in handling; a fall on to the floor would be almost certain either to fracture the lenses or dislocate them from their settings. If the surfaces of any of the lenses, with the exception of the front surface of the lowermost one, get contaminated with dirt, the objective should be sent to the optician to be cleaned. After working with an immersion objective, it should be carefully freed from adherent fluid and polished with a soft cloth. A little chloroform or benzine will assist in the removal of the last traces of the cedar-wood oil used with oil immersions. When using an immersion objective, care must be taken that the immersion fluid does not get mixed with the fluid oozing from between the glass-slip and cover-glass; also, when observing objects mounted in Canada balsam, that the oil does not get contaminated with the balsam.

Eye-Pieces.—The most common and generally useful form of microscopic eye-piece is that invented by Huyghens, and called, after him, the "Huyghenian." It consists of two plano-convex lenses, having their convex sides directed towards the objective, with a diaphragm between the lenses. It is the simplest and least expensive form of eye-piece, and is suitable for all ordinary purposes; in fact, it is the best form to use with ordinary objectives, for, as these are over-corrected and the eye-pieces under-corrected, the two balance one another and give a correct image. When, however, apochromatic objectives are made use of, what is called "compensating" eye-pieces must be employed. These are over-corrected, and, as the apochromatics are under-corrected, the two balance one another. As compensating eye-pieces are constructed with achromatic lenses, they are considerably more expensive than the ordinary Huyghenian.

The various magnifying powers of the eye-pieces are denoted in England by the letters A, B, C, &c., beginning with the lowest power; on the Continent by numbers, 1, 2, 3, &c. The magnifying powers corresponding to these will generally be found in the maker's catalogue.

With a microscope having a 10-inch tube, the total magnifying power of any combination is found by multiplying the magnifying power of the objective by that of the eye-piece. Thus, taking a tenth of an inch objective, which magnifies 100 diameters, and an eye-piece magnifying 5 diameters, we obtain a total magnification in diameters of 100 × 5 = 500. With a 6-inch tube the magnifying power is found by the formula—

$$\frac{\text{objective power} \times \text{ocular power} \times \text{tube length}}{10};$$

with the before-mentioned combination this would be 300 diameters $\left(\frac{100 \times 5 \times 6}{10}\right)$.

Illuminating Apparatus.—In work where objectives of no higher power than a quarter of an inch are employed, the light reflected by one of the mirrors under the stage will be sufficient to fully illuminate the object. One of these mirrors is plane (that is, perfectly flat), the other concave; the latter brings a conical bundle of light rays to a focus on the object, and consequently gives a greater illumination. The former mirror is employed when viewing objects with low magnifying powers, the latter when higher-power objectives are being used. When, however, it is required to work with powers higher than the quarter of an inch, or with an immersion objective, it is absolutely necessary to employ a condenser which brings a very large bundle of light rays into the objective. The most common form of condenser is that invented by Abbé (Fig. 11, O), which consists of two or three convex lenses of differing curves. Since it is made with simple lenses, it is not achromatic; but as it is comparatively inexpensive and fulfils most ordinary requirements, it is frequently employed. Achromatic condensers are also made on a similar principle, and, though considerably more expensive, undoubtedly possess advantages over the cruder form. The condenser, whether of the achromatic or non-achromatic form, is either provided with a series of diaphragms, with a revolving plate, or with an iris diaphragm. An arrangement is provided on the sub-stage by which the condenser can be moved nearer or farther away from the object, the best form for this being a rack-and-pinion movement. An arrangement for centring the condenser truly is also necessary where fine work has to be performed.

One of the lower-power objectives, half- or quarter-inch, makes a very efficient condenser, if an arrangement is provided for attaching it to the sub-stage.

Choice of a Microscope.—For the ordinary work of the brewery, which chiefly consists in examining yeast, one of the smaller and less expensive forms of microscope stand is sufficient. These are to be obtained at prices ranging from £1, 10s. to £5. Such a stand, fitted with a half-inch and a tenth of an inch objectives, and a couple of eye-pieces magnifying five and ten times, will be ample for this purpose.

The microscope stand should be thoroughly rigid, stable, and free from all vibration; the form of foot such that when the body-tube of the instrument is placed in a horizontal position there is no tendency to topple over. All the movable parts should work smoothly and without any tendency to jerkiness, and should fit so accurately that the body of the instrument, wherever placed, is never thrown out of line. This should be especially the case with the fine adjustment. When an object is being focussed, either with the coarse or the fine adjustment, it should not change its position in the field. Instruments which do not comply with this condition are particularly irritating to work with.

The Objectives.—These should give a bright, well-illuminated field; the image of the object should be clear and distinct, without the slightest tendency to haziness; its contours should be cleanly cut, and not surrounded by fringes of various colours. In testing objectives it is usual to employ what is called a "test-object." This, for objectives up to 1-4th of an inch, generally consists of the scales of the wings of moths, many of which exhibit delicate markings. The *Pleurosigma angulatum*, a diatom which is readily obtainable, forms an excellent object for testing the highest objective. Its markings should be distinctly brought out by the 1-10 inch objective, their honeycomb formation being rendered distinctly visible.

The field of view should be perfectly flat. When a perfectly flat object, such as the impression of the tip of the finger on a glass slip, is viewed, the whole impression should be in focus at the same time. This should, at all events, be the case with the lower objectives, with which objects of some size are generally examined. Unfortunately, with objectives of a higher power than the quarter-inch, this does not seem to be so readibly attainable; but as with these either very minute objects are examined in the middle of the field, or only very small portions of larger objects, it is not a very serious source of trouble, since a very slight motion of the fine adjustment serves to bring the different portions of the field successively into focus. The objective should be free from spherical aberration, a defect depending upon the lenses not being ground to the proper curves, or a distinct image is unobtainable. If the images of objects viewed under a particular objective are fringed with a border of several colours, such an objective is not properly corrected for colour—it exhibits chromatic aberration, which is a very bad fault. All objectives, apochromatics excepted, show some little colour, which in the best glasses is of a reddish-purple tint (port-wine colour), and this arises from what is known as "irrational dispersion." It is impossible to correct the whole of the colours of the spectrum with lenses made of different kinds of glass, but by introducing into the combination a lens made of the mineral substance *fluorite*, as is done in the apochromatics, colour can be totally suppressed.

Illumination.—The microscope may be used either with natural or artificial light. In the former case we utilise the light reflected from the sky, or, better, that from a white cloud. Direct sunlight must never be used. In choosing a room for microscopical work, one with a window looking towards the north should be secured if possible, for here we are never troubled with direct sunlight. In case we are unable to obtain a room with a northern aspect, then the window should be provided with a roller-blind, to shut out the direct rays of the sun. The best material for such a blind is tracing-linen. This, when the sun shines on it, gives a bright white light well adapted for microscopic observation.

A small flat-flamed paraffin lamp is one of the most convenient and readily obtainable sources of artificial light for microscopic work after dark. It should be provided with a shade, so as to protect the eyes of the observer from the glare of the direct light. A chimney of a light-blue tint, or a piece of glass of the same colour placed between the light and the mirror, takes off the yellow glare of the lamp-light. The light should be sufficiently modified by introducing discs of optically worked tinted glass between the mirror and condenser, so as not to be glaring, for nothing fatigues the eye so much as a dazzling field.

Exceedingly convenient lamps for microscopical work are supplied by the opticians. In these the chimney is made of metal, and as it has only an aperture in front to permit the exit of light, the remainder of the room is kept in comparative darkness. In front of this aperture a slide is provided, into which fits loosely an ordinary glass slip. This is an exceedingly convenient arrangement, since, if the glass breaks owing to the heat, it can be immediately replaced by another. The light from such a lamp may be modified at will by inserting a slip of blue glass of the size of a cover-glass. Such slips, of varying depths of tint, should be kept at hand for this purpose.

The Diaphragms.—When using the higher powers, a diaphragm perforated with a small aperture must be employed. This should be slightly less in diameter than the front lens of the objective, but the field of view should be illuminated up to its border. If too small a diaphragm is used, only the central part of the field will be illuminated.

Drawing Apparatus.—Nothing conduces so much to produce a lasting impression on the mind as drawing objects seen under the microscope. Although any person with a competent knowledge of drawing may do this without assistance, yet, if absolutely accurate pictures are to be produced, such as will allow of comparison with one another, then one of those arrangements must be used which, while permitting the observer to see the image of the object under the microscope, also allows him to see the point of his pencil and the surface of the paper. Several such appliances may be had.

Glass Slips.—A number of these will be required; they serve to support the object during observation under the microscope. They are thin level slips of glass, with ground edges; the size most commonly employed being three inches by one inch. A quantity of hollow-ground slips should also be provided; these are somewhat thicker, and have a circular or oval hollow trough ground on one of their sides. They are very frequently employed in brewing work.

Cover-Glasses.—These are very thin slips of glass, varying from 0.1 to 0.3 millimetre ($\frac{1}{250}$ to $\frac{1}{80}$ of an inch) in thickness. They are either cut into squares or circles of from three-eighths to three-quarters of an inch in diameter, but larger sizes are occasionally required. Cover-glasses differ much in thickness, and as "dry" or "water immersion" objectives are constructed to yield their best results with a cover-glass of a definite thickness, care should be taken to obtain glasses of the various thicknesses suitable to the different objectives the observer employs. Unfortunately, cover-glasses of definite thicknesses are not sold by English opticians; they stock only two kinds, No. 1 thin, and No. 2 thick, the individual members of which vary considerably in thickness. Cover-glasses of a measured thickness are obtainable at most of the Continental houses.[1]

Micro-Chemical Reagents.—A numerous class of sub-

[1] Zeiss has now a branch establishment at 29 Margaret Street, Cavendish Square, London.

stances of this kind are used in microscopical manipulation; they cause certain reactions, coloured or otherwise, with the various constituents of animal or vegetable tissues, by means of which these constituents can be identified. For our present purpose only two of these will be necessary: iodine solution, made by adding 0.5 gramme of iodine and 2 grammes of potassium iodide to 150 c.c. (cubic centimetres) of water, the mixture being frequently shaken until perfect solution is effected; caustic potash solution, made by dissolving 0.5 gramme of that substance in 100 c.c. of water.

Staining Fluids.—A number of these, which are generally solutions of the aniline colours, are also employed in microscopic work. The various tissues stain in different ways, or take one stain and reject others; consequently, much assistance is obtained in this way in their differentiation. One staining fluid will suffice for present purposes, and for this methylene-blue is chosen, because its solution keeps indefinitely, which is not generally the case with the rest. The stain is made by dissolving one part of the dye in 100 parts of water.

Microscopical Manipulation.—In commencing, a low-power objective is screwed on to the nose of the microscope. If the body-tube slides in a sheath, then it may be entirely removed for this purpose, care being taken that the eye-piece does not fall out during the manipulation. This is best avoided by holding the eye-piece with the index finger of the left hand, while the thumb and remaining fingers grasp the tube. In case the instrument is provided with a rack-and-pinion coarse adjustment, the tube must be racked out as far as it will go, and the objective then screwed on. In the former case the tube is replaced and pushed down to within the distance of a quarter of an inch from the stage; in the latter, racked down to the same distance. The instrument is then placed before a window, the body-tube slightly inclined at an angle, and the mirror (the plane one this time, as a low power is being used) moved about until the field of view is equally illuminated. Care must be taken that the tail-piece which supports the mirror is perfectly central, and does not hang to one side or the other, as we now wish to observe by direct, and not by oblique light. If, on looking through the instrument, black spots of irregular outline are often seen in the field, these are caused by dust or dirt on one or both of the lenses of the eye-piece, and, if this be slowly rotated, the specks will be seen to move round with it. In this case the upper lens must be unscrewed, the dirt on it removed with a piece of soft linen rag or chamois leather, replaced, and looked through to see if the objects have disappeared. If not, it is the lower lens on which the dirt is lodged; this must be removed and similarly cleaned.

Interpretation of the Appearance of Objects as seen under the Microscope.—Objects observed by the unaided

OBJECTS AS SEEN UNDER THE MICROSCOPE.

vision are almost invariably seen by reflected light, rarely by transmitted. In microscopic observation these conditions are reversed; here it is comparatively rare that we use direct light in the examination of objects; they are in most cases examined in a thin layer of some fluid, generally water; consequently, their appearance as seen under the microscope is entirely different to that of objects as we usually see them. For this reason it is necessary that the microscopic observer should familiarise himself with the appearance of objects as seen under the microscope, so that he may be able to interpret their forms correctly. In order to do this, it is best to begin by examining a few objects of known shape, first by direct, then by transmitted light. For this purpose take a piece of thin glass rod, heat about half an inch of it in the flame of a Bunsen-burner or blow-pipe until softened, then draw it out quickly until a strand of glass about the thickness of a hair is obtained. Repeat the process with a piece of thin glass tube; in this manner a fine tube the thickness of a hair will be obtained. Now place fragments of these, about an inch long, on a glass slip; block up the aperture in the stage by the proper appliance for that purpose, and examine these two objects by direct light, using the lowest power objective, and keeping the microscope in a vertical position. In order to find the correct focus, the body of the instrument is pushed or racked down until the objective nearly touches the object, and, while looking down the microscope, the body is slowly raised until the object comes into view. Should the objective have been raised some considerable distance, say an inch, without the object being seen, then the correct point has been overstepped, and the process must be commenced anew. This is the best way of finding an object for a beginner; but with practice, and a correct appreciation of the distance (technically called the "working distance") which the various objectives require, the worker will be able to push the tube down until the object is found. Great care must be taken in these manipulations, or the objective may be brought in forcible contact with the object and damaged; this is especially liable to occur with a rack-and-pinion adjustment. Having caught sight of the object, the final focussing is proceeded with; this is effected by rotating the milled head of the fine adjustment first in one direction, then in the other, until the object is seen sharply defined. We now see these objects under the usual conditions of ordinary vision, that is, by reflected light, and shall have no difficulty in detecting which is the solid rod and which is the tube. Now place a drop of water on these two objects and cover them with a cover-glass, avoiding the introduction of air-bubbles. This is best effected by holding the cover-glass at an angle, bringing one of its sides into contact with the glass slip, and then slowly lowering it into position. Move the minute tube and rod so that they occupy a position at right angles to the long axis of the glass slip; leave the aperture in the stage completely open, that is, without diaphragm; place the mirror in position, and focus. A difficulty will now be experienced in detecting which is the rod and which is the tube. This may be instantly dispelled by swinging the mirror to one side, at the same time adjusting it so that there is always some light in the field. The white strip of light in the rod moves as the mirror moves, and follows the direction of its motion; thus if the mirror be moved to the left, the strip of light in the rod will move to the left. The strip of light in the middle of the tube moves also, but in a direction opposite to that in which the mirror moves. Now place the mirror in its former central position, and note the difference in the appearance of the two objects; the rod has a single bright line running down its centre, with two dark borders; the tube has a number of alternate bright and dark lines. Touch the end of the tube with a drop of water—water will enter it by capillary attraction; stop when the tube is half filled; examine and note the difference in appearance between the portion containing water and the empty one. That portion which is filled has lost its complicated series of lines, and now looks like an empty tube. Swing the mirror, and note the appearances under oblique illumination. Now take off the objects and dry them externally. Having also cleaned the slip and cover-glass, examine the two objects in a drop of glycerin. The rod has lost nearly the whole of its black border, and appears like a flat strip of glass. The walls of the tube also look like strips of flat glass, and the bands of white and black in its empty portion are much more strongly pronounced than they were when it was viewed in water; they are slightly more so in the part filled with water. Examine a little wheat-meal in a drop of water, and note the circular starch cells, which are seen with perfect distinctness. Now examine another portion of the meal in a drop of turpentine; the starch cells are almost invisible.

Now place a drop of wort, mucilage, or white of egg on a glass slip, and beat it up with a penknife so as to introduce bubbles of air; place on a cover-glass and examine under the microscope. Get a perfectly circular bubble in the centre of the field, and examine with central light. The bubble, when in focus, appears as a black disc with a small bright spot exactly in its centre when the light is central.[1] Slowly elevate the body of the microscope; the centre gradually becomes dim, and the bubble loses its sharp outline. After focussing again, swing the mirror to one side; the bright spot moves in a direction contrary to that of the mirror; under very oblique illumination an air-bubble appears as a black thick ring with a grey centre, having a very bright spot at its side. Now beat a drop of oil of any kind into one of the liquids mentioned above, cover, and examine. The bubble

---

[1] Thus the appearance of the bright spot in an air-bubble forms a convenient means of ascertaining when the light is central.

resembles to some extent an air-bubble, but its dark border is not so wide. Focus up, and notice that, as the tube rises, the centre of the oil-globule is the last to disappear. Swing the mirror to one side; the bright spot follows the motion of the mirror. The cause of these differences in the appearances in the same objects, when seen under the varying conditions of observation, is owing to the difference in their refractive power and in that of the various media in which they are being examined. The oil-globule and the air-bubble were very distinct, because each differs greatly in refractive power from the medium in which it was surrounded; oil has a much higher refractive power than the liquid medium, air a much less. Glass has a refractive power higher than either of them, and glycerin has a power nearly as high as that of glass. Turpentine has almost the same refractive power as starch; hence starch is almost invisible when viewed in that fluid. Had we examined the glass rod in a drop of cedar-wood oil, which has exactly the refractive power of glass, it would have been perfectly invisible.

The results of these few simple observations, if carefully borne in mind, will afford considerable help in assisting the observer to the proper interpretation of the forms and nature of the objects he has to examine under the microscope.

Microscopic Examination of the Starches.—We now proceed to the examination of some natural objects, and for the first of these we choose potato-starch, as it is always readily obtainable. A drop of clean water is placed on a clean glass slip by means of a glass rod, and after cutting off a slice from a well-washed potato with a sharp knife, a trace of the substance which is scraped from the cut surface is mixed with the drop of water on the slide. A clean cover-glass is then placed over the drop, which, if of the right size (and with a little practice the right quantity can generally be hit), there will be no exudation from the sides of the preparation. If there is, the superfluous fluid must be removed by absorbing it with a small piece of blotting-paper. The preparation is now placed under the microscope, and the diaphragm suitable to the objective under employment having been put into position, we proceed to focus. A number of colourless bodies, such as those represented in our illustrations of the starches (Fig. 7e, p. 67), will be seen, varying much in size, the larger ones being more or less oval, the smaller ones nearly round; these are the starch granules. In order to examine them more minutely, the glass slip is moved about until a portion of the field is found on which the grains do not lie too closely together. The weaker objective is now removed and replaced by the stronger, the diaphragm changed for a smaller, and the focussing performed as before. This time, however, it will be necessary to move the microscope much more slowly upwards in order to catch sight of the object. When in focus, the larger starch grains are seen to be covered with a series of concentric rings, which show that each starch grain is built up of a series of layers. The rings are arranged excentrically around a point situated at one end of the grain, called the "hilum," which often exhibits a number of minute rifts. The smaller grains are rounder; they do not show the stratification nearly so distinctly, and are much more centric in construction, the hilum being in the centre of the grain and the rings more circular. Besides these, compound grains will be occasionally met with; these consist of two or more starch grains, between which a dark dividing line is generally discernible.

It sometimes happens, in examining a preparation straight away with a higher power, that no trace of the object can be found, because in that portion of the field immediately under the objective no object or portion of one happens to be present. In this case the body of the instrument must be lowered, and, while it is being drawn up to focus, the glass slip must be moved about on the stage. As soon as shadowy images are detected, the glass slip is allowed to remain at rest, and the focussing with the fine adjustment proceeded with. If there are only one or two very small objects in the preparation, it is better to find them with the lower power first, and then exchange the objective for the stronger. It is in such cases as these that the nose-piece is especially convenient.

Dirt on Objective.—Sometimes after focussing, or even moving the object about, especially when using one of the higher powers, the field of view appears hazy, and the object blurred and indistinct. This arises from the objective having come into contact with the fluid on the glass slip. In such a case the tube must be elevated, and the adherent fluid removed from the surface of the front lens with a piece of soft linen rag or wash-leather.

Effect of Micro-Chemical Reagents on Starch.—We now proceed to observe the effect of our two reagents on the starch granules. We place a drop of the dilute caustic potash solution on the side of the cover-glass, with its border touching that of the film of fluid under the cover-glass. The potash solution will gradually penetrate under the cover-glass and come into contact with the starch grains. Looking down the tube, we soon detect the progress of the drop, for as it touches the individual granules, these, after showing their lamination much more clearly for a short time, commence to swell out, the hilum of each enlarges and becomes a mass of rifts, and, finally, each granule attains a size ten or twelve times larger than it originally had, the contents of the granules become perfectly transparent and their borders scarcely visible.

The same appearance can be brought about by the agency of heat, and when they are produced in this way the starch is said to be gelatinised. In order to observe this phenomenon, a fresh preparation is made as before, and carefully heated over a small

EFFECT OF MICRO-CHEMICAL REAGENTS.

gas-flame, moving the glass slip to and fro so as to apply the heat equally, or the glass may crack. Even with the naked eye a change is seen to take place as the heating proceeds, and when the object is examined under the microscope, the starch granules will be found to present the same swollen and modified appearance that they did when treated with potash solution. The preparation is now to be put on one side for a time, as it will be wanted again. Gelatinisation of potato-starch takes place at a temperature of about 70° C. (160° F.), but the gelatinisation temperature of the different starches varies considerably. An appliance has been contrived by means of which a preparation of starch can be gradually heated up to a known temperature while under observation on the stage of the microscope, and by its aid the temperature at which the various starches gelatinise has been ascertained with great exactitude.

In order to note the behaviour of starch with iodine, a fresh preparation is made, focussed under the microscope, and a drop of the iodine-potassium-iodide solution placed on the border of the cover-glass. As this penetrates and comes into contact with the starch granules, we observe that each of these turns first a pale blue; this colour then gradually deepens until the grains appear blue-black. The progress of the iodine solution through the preparation can be followed by moving the glass slip about. The preparation in which the starch has been gelatinised by heat can now be similarly treated; as the iodine reaches the almost invisible swollen starch granules, they turn a deep blue colour, and although hardly discernible before, they now stand out prominently. The starch treated with the potash solution might have been similarly treated, and a similar result obtained; but in this case a much larger quantity of iodine solution would have been required to colour them, because caustic potash combines with iodine, and it is only after the potash solution has been saturated with iodine, and some of the free element is present in the solution, that the coloration can take place.

The general appearance and behaviour with micro-chemical reagents may be studied with the other starches commonly met with. A little wheat, rye-flour, or oatmeal may be placed in a drop of water and each of these starches examined in turn. In order to obtain barley, malt, maize, or rice starches, a grain of these substances is cut in two with a sharp penknife, a little of the white interior scraped off and examined in water. Figures of these starches will be found on pages 66 and 67.

Particularly interesting is the starch of green malt. When this is examined, it is found that many of the individual starch grains are corroded; a number of canals will be found running from the outside to the inside of the grains; these are caused by the action of the diastase. The same appearance may be brought about by mixing a small quantity of starch on a glass slip with a few drops of saliva, and keeping the whole at a temperature of 38° C. (100° F.) for a little time. Saliva contains the enzyme ptyalin, which acts upon starch in a manner similar to that in which diastase does. This experiment may be rendered particularly instructive if the action be allowed to proceed for a considerable time; all the granulose of the starch grains is dissolved out, and skeletons of starch cellulose are left, which accurately preserve the size and shape of the original starch grains.

Examination of Yeast.—Next place a minute trace of yeast on a glass slip, mix it intimately with a drop of water, and examine under a high power. The yeast cells appear as round or slightly oval bodies, much smaller than starch grains. Treated with iodine, the contents of the cells are stained yellow, or brown when the cells are rich in glycogen; no blue colour is observed, hence they do not contain ordinary starch. If a fresh preparation be treated with caustic potash solution, the cells do not swell up; they gradually lose their granular appearance, become clear and transparent, because the potash solution dissolves the substance within the cells. Next make a fresh preparation, and while observing it under the higher power, press on the cover-glass with the point of a dissecting needle,[1] and notice that the cells are elastic. Press still harder until some of the cells are ruptured; notice that the contents which exude are of a tough, elastic, jelly-like nature; this semi-solid substance, which fills each cell, is the protoplasm. Those cells from which the protoplasm has exuded now look like little sacks, which, on emptying, contract to a slight extent, thus showing that before being ruptured they were to a certain extent distended by internal pressure. Now run in a little of the methylene blue solution.[2] The contents of all those cells which are disorganised stain a deep blue; the uninjured cells remain, as before, colourless. This experiment exhibits one of the peculiar properties of protoplasm: when alive it refuses to be stained by any of the dyes; when dead it evinces a powerful attraction for these substances.

Place a little yeast and water on a glass-slip and heat it nearly to boiling temperature; afterwards add a little of the stain and examine it; all the yeast cells now take the dye—they are dead. This behaviour of living protoplasm with reference to tinctorial substances affords an excellent method of detecting the dead yeast cells in a sample of yeast; it is only necessary to place a little of the yeast on a glass-slip, mix it with a drop of water, and then run in a small quantity of methylene blue solution, or almost any other dye, when the dead cells are readily detected by their coloured appearance. Now make a fresh preparation of yeast, and, by pressing on the cover-glass with the finger-nail, rupture some of

[1] These are made by pushing a medium-sized needle held with a pair of pliers point foremost into a soft wooden penholder, withdrawing the needle, and then pushing it in head foremost for half an inch.

[2] Too much of the stain must not be used or it will kill the yeast cells.

the cells; then grind their contents between the cover-glass and glass slip. On examination, a number of exceedingly minute bodies, the microsomes, will be seen. These are in a state of lively movement, which conveys the impression that we are looking at living organisms. This, however, is not the case; all very minute bodies when examined in a state of suspension in a liquid have this peculiar motion, first called attention to by Brown, and hence called after him the "Brownian movement." When examining very minute organisms, such as bacteria, it is often difficult to distinguish the living organisms from the minute particles of dead matter; but, if careful attention is paid to the nature of the movement, the two objects may generally be differentiated. In the Brownian movement the motion is more localised, the objects rotating in sinuous lines about an imaginary centre; with bacteria the movement is a more independent one, they rush about all over the field, and often out of it.

The Hanging-Drop Method.—We will now proceed to examine yeast by another method, using one of the glass slips with a circular hollow ground in it. After being cleaned, paint, with a small camel-hair pencil, a ring of vaseline about the eighth of an inch wide round the border of the hollow. Place about 10 c.c. of wort in a test-tube, or, if wort is not at hand, of a 10 per cent. solution of cane-sugar; mix intimately with this a piece of stiff yeast about the size of a pin's head. Next place a minute drop of the mixture in the centre of a clean cover-glass,[1] turn the glass upside down and breathe into the hollow of the glass slip so as to slightly moisten its surface, and lay on it the cover-glass so that the drop occupies the centre of the hollow; press down gently. The vaseline ring makes an air-tight seal, the drop is now in a closed chamber, and could be kept for weeks or months without suffering loss from evaporation. Place under the microscope, and if the result is successful, not more than two or three yeast cells will be seen in the drop. If more than three cells are present in the drop, it is best to begin again, and dilute the mixture in the test-tube with more wort or sugar-water until a drop is obtained containing the right number. Make a rough drawing of the appearance of the yeast cells, and keep the preparation under the microscope, observing it from time to time. Note how the yeast cells first throw out buds, how these increase in size, and finally become detached from the parent cell. The preparation, after being examined throughout the day, can be placed on one side and examined occasionally during several subsequent days. In a few days the two or three cells will have multiplied to hundreds. It is obvious that by using for the preparation a fluid entirely free from other organisms, and taking care that there is only one yeast cell present in the drop, we are able, in this way, to secure a pure

[1] This is best done with one of the steel goose-quill pens which are sold for drawing maps.

culture of yeast from a single cell. This method also affords an exceedingly convenient way of studying the growth of many minute organisms. Growths of the various mould-fungi made in this manner are described in the chapter devoted to those organisms.

Objects often met with in Polluted Water.—When water is polluted with refuse from human habitations, portions of articles of food, clothing, &c., are generally to be found in it. It is well, therefore, for the microscopist to make himself acquainted with the appearance of such objects as seen under the microscope. For this purpose, shreds of cotton, woollen, and silken clothing should be teased out with a pair of needles in a drop of water, examined under the microscope, and their general appearance noted. These will almost invariably be coloured, since the majority of materials made from these substances are dyed. Portions of lean meat should be similarly treated—here we note the appearance of muscular fibre; also portions of vegetable tissues, such as leaves and stalks of cabbage, the husks of corn, &c. Human hair and the hair of animals is also frequently found in water, portions of the feathers of birds, epithelium scales; the last may always be seen in a drop of saliva.

Microscopic Examination of Bacteria.—As these are the smallest organisms known, we require for their examination the best and most powerful objectives and the most favourable conditions of illumination. Though their presence may be readily detected with a tenth of an inch dry objective, a homogeneous immersion objective becomes a necessity if their complete investigation is to be undertaken. Water immersions may be employed, but, as stated before, the twelfth of an inch oil immersion is the best lens for this class of work.

Methods of Obtaining Bacteria for Examination.—As bacteria, or their germs, are omnipresent, specimens of these organisms are very easily obtained for examination. It is only necessary to boil a potato, cut it into slices, expose the freshly-cut surfaces to the air for a few hours, and then place the slices in a moist chamber, which latter may be extemporised with an ordinary basin and a dinner-plate, two or three sheets of blotting-paper (cut into a circular shape and moistened with water) being placed on the plate. The slices of potato are laid on the moist filter-paper, covered up, and allowed to stand at the ordinary room temperature for several days, being examined at intervals. During the time they were exposed to the air, germs and bacteria in greater or less numbers will have fallen upon the slices; and since the conditions mentioned above are favourable for their growth and development, the organisms and germs will in a few days have germinated, multiplied, and formed patches which are technically called "colonies," and which appear as small, whitish, and occasionally coloured gelatinous masses. A small portion of any one of these colonies may be removed by means of a platinum wire, mixed intimately with a drop of water on a glass slip, covered with a thin cover-glass, and examined under the microscope in the usual manner. Bacteria may also be obtained in abundance by making an infusion of lean meat and allowing it to stand in a warm place for from twenty-four to forty-eight hours, when it will be found to be literally swarming with bacteria. Vegetable infusions made from hay, turnips, peas, &c., when similarly treated, yield bacterial organisms in immense numbers. If a small quantity of malt or barley be covered with distilled water and allowed to stand for a week, the liquid will be found to swarm with bacteria and yeasts. It is only necessary to take a small drop of such an infusion, place it on a glass slip, put on a cover-glass, and examine it under the microscope, when large numbers of these organisms will be seen of various sizes and shapes, some quiescent, some in a lively state of motion.

It is interesting to examine a culture which contains only one single species of bacteria. The *Bacillus subtilis,* or hay-bacillus, owing to the resistance of its spores to boiling water, readily permits this to be effected. In order to obtain a pure culture of this organism, dry hay is soaked in the smallest possible quantity of water and allowed to stand at a temperature of about 36° C. (97° F.) for a few hours, the infusion so obtained is poured off without filtering, and diluted with water until it has a specific gravity of 1004. The infusion is placed in a flask, and after a plug of cotton-wool has been inserted in the neck of the flask, gently boiled for an hour, cooled, and kept at 36° C. (97° F.) until a film of bacteria appears on its surface. This will generally be the case in twenty-four or thirty-six hours.

We take a minute piece of this pellicle off by means of a wire, and place it on a glass slip, together with a minute drop of the fluid; place on a cover-glass, and examine with the 1-10 inch dry objective. The pellicle is seen to be formed of long wavy threads lying parallel to one another, each of which is divided into a number of segments, each segment being a separate individual. The segmentation can be rendered more distinct by running a little iodine solution under the cover-glass, which stains the organisms a brownish-yellow colour. The threads are held together in an immovable position by a mass of invisible jelly, produced by the swelling of the slimy coats of the organisms. Bacteria, in this state, are said to be in the "zoogleal" condition.

Being certain that we have the object in the field, we proceed to its examination with the immersion objective. If the condenser is not already in position, it must be so placed, using one of the smallest diaphragms and the plane mirror. The objective having been screwed in its place, a small drop of the cedar-wood oil,[1]

---

[1] Care must be taken that the fluid which oozes from between the slip and cover-glass does not get mixed up with the cedar-wood oil, or, when examining an object mounted in Canada balsam, that the oil does not get contaminated with the balsam.

which was supplied with the lens, is placed in the middle of the cover-glass, the microscope pushed or racked down until the nose of the objective dips into the oil-drop and nearly touches the cover-glass. Focussing is proceeded with as usual, taking care to move the microscope body very slowly upwards. It is here that the benefit of a rack-and-pinion movement is experienced. The condenser must now be focussed. If we are using daylight, we move the mirror about until a shadowy glimpse is caught of one of the window bars, then the condenser is moved slowly up and down until the bar is seen distinctly in focus. The mirror is then moved until the field is equally illuminated. When working by lamplight, the lamp is adjusted so that the edge of the flame is turned towards the microscope, and the condenser focussed until the image of the flame appears clearly defined. The best conditions of illumination are when a perfectly flat superficies of achromatic light is in the focus of the objective at the same time that the object itself is in focus, and the diameter of the section of the cone of light which enters the objective is about three-fourths the diameter of the front lens of the objective. To obtain the very best possible conditions of observation, the condenser requires as much care in its construction and use as the objective itself.

We now proceed to examine the hay bacilli again, first using the lower-power eye-piece and then proceeding to the higher-power one. It must be remembered that an oil immersion will bear a much higher power, or, as it is often called, a much deeper eye-piece, than a dry objective will. The bodies of the individual bacilli are seen to be composed of a colourless, homogeneous, somewhat highly refractive substance, in which, even with the highest powers at our command, no trace of structure or of a nucleus can be detected.

We then make a hanging-drop preparation, taking for this a minute portion of the pellicle and a very small drop of the fluid. By carefully noting the appearance of such a preparation (which is best effected by drawing it with the camera), and by keeping the preparation for some time under observation, we are able to observe the process by which the bacilli grow. We see that each segment gradually increases to a certain length, while preserving the same thickness; then a partition-wall gradually forms which divides the organism into two. It is owing to this growing in length that the bacilli come to be arranged in lines. After the lapse of six or eight hours the nutriment in the drop becomes exhausted; the bacilli then cease to grow, and begin to form spores. When this commences, bright spots make their appearance in many of the bacilli, which gradually increase in size, until at last each presents the appearance of an ellipsoidal, bright, strongly refractive body with sharp outlines; these are the "spores." As the process is proceeding the remaining contents of the organism gradually disappear, the cell-wall finally disintegrates, and the spores are set free.

THE STAINED COVER-GLASS PREPARATION. 203

These, when introduced into a fresh nutritive medium, soon commence to germinate, especially if they have been boiled for a few minutes; they then commence to germinate in two or three hours. The envelope of the spore ruptures on one side; from this protrudes a bud, which gradually enlarges in length to form a bacillus, which divides by segmentation for the first time in about twelve hours. The organisms do not immediately enter into the zoogleal state, but pass through what is termed the "swarming stage." In this, pairs or chains of individuals are seen moving about in a very lively manner. They are propelled by delicate hair-like appendages at each end, technically called "flagella," which can be readily seen when the organism has been stained in a particular manner. The swarming bacteria, after living in the body of the fluid for some time, ascend to its surface and enter into the zoogleal condition. This bacillus has recently been studied by Adrian Brown in relation to wort and beer, who finds that it exerts no prejudicial influence on either. Its chief interest resides in the extreme resistance of its spores to heat.

Bacteria in the swarming stage may be very conveniently examined by the hanging-drop method; in this form of preparation the organisms are under conditions closely resembling those in which they naturally exist. At the sides of the drop they will be in a more or less quiescent state, and permit their size and shape to be readily observed; whilst in the middle of the drop they have more room for their evolutions, and are uninfluenced by the capillarity conditions which exist at its border. By examining the organisms in the upper layer of the middle of the drop, which are sufficiently near the surface to be within the range of focus of the object-glass, we are able to form some opinion of their behaviour under conditions closely approximating to those in which they naturally exist.

The Stained Cover-Glass Preparation.—Bacteria are much more easily seen when stained, and in this condition, being dead, we are no longer inconvenienced by their lively movements. In making such a preparation, a small drop of the fluid containing the bacteria is placed on a thin cover-glass by means of a sterilised platinum wire, and spread out as much as possible with the wire into a thin layer.[1] It is now left until the fluid has evaporated and the surface appears dry. The bacteria are now to be "fixed," or they would be washed away during the subsequent staining and rinsing. This is effected by exposing the cover-glass and its adherent bacteria to a temperature of 120° to 130° C. (248° to 266° F.) for five minutes in a hot-air oven. In

---

[1] In order to make a successful preparation in this way, it is necessary that the cover-glass be absolutely clean, or the drop will not spread evenly. The cover-glasses, after having been washed with water, should be heated in a mixture of strong sulphuric acid and potassium bichromate, afterwards well rinsed in distilled water, and then kept in alcohol until required for use.

this way the mucilaginous coats of the bacteria become adherent to the cover-glass. The process of fixing may be readily executed in the following manner :—The cover-glass is taken up with a pair of tweezers, the surface to which the bacteria are attached being kept upwards; the cover-glass is then passed three times through the flame of a Bunsen-burner or that of a spirit-lamp. The heating must be carefully performed; if insufficient, the bacteria will be afterwards washed away; if overdone, the bacteria refuse to take the stain well. The happy medium is obtained by moving the hand vertically in a circle a foot in diameter; the time taken to complete the circle should be one second. When traversing the lower portion of the circle the preparation is momentarily introduced into the flame, but care must be taken that no halt takes place here, or the preparation will be overheated. The movement in describing the three circles must be a constant and uninterrupted one, so that the preparation is exposed momentarily thrice to the heating influence of the flame. When the cover-glass has become cold, its upper surface is covered with a large drop of staining solution (see below), which is allowed to remain there for several minutes. In case the stain do not take the stain freely, the preparation may be held over a small flame until the drop of staining fluid steams. The cover-glass is now gently rinsed with clean water to remove the staining fluid (which is best effected by a gentle stream of water blown from a pipette), then placed with the side on which the bacteria are fixed downwards on a glass slip, its upper surface dried with blotting-paper and examined under the microscope. When viewing stained preparations, the diaphragm of the condenser is opened to its full extent, so as to admit as much light as possible.

Conversion into a Permanent Preparation.—The preparation obtained in the foregoing manner can be easily converted into a permanent one. For this purpose the cover-glass is removed from the slip and allowed to dry. A drop of a mixture of equal parts of Canada balsam and xylol is placed in the middle of a clean glass-slip, the cover-glass taken up with a pair of tweezers (care being taken that the side to which the bacteria are attached is downwards) and slowly and gently dropped down on the drop of balsam, in such a manner that no bubbles of air are included. A little gentle pressure applied to the cover-glass will cause the balsam to spread out to the margin of the cover-glass, and, if the proper quantity of balsam has been taken, it will be just sufficient to do this and no more. If, however, as often happens, rather too much balsam has been used, the excess must be carefully removed by means of a piece of linen rag moistened with benzol or xylol, or, when the cover-glass is afterwards covered with cedar-oil in order that it may be viewed with the immersion lens, the Canada balsam will dissolve in the oil and alter its refractive index, and it will also soil the lens. In a few days the balsam dries and secures the cover-

BACTERIOLOGICAL METHODS.

glass; after a few weeks the balsam becomes thoroughly hardened, and in this way a permanent preparation is secured. All preparations coloured with aniline colours should be carefully kept excluded from the light, otherwise they are apt to fade.

In order to investigate the different species of bacteria, it is necessary to have recourse to certain methods which have for their object the isolation of the different species; this forms the subject of the following section.

BACTERIOLOGICAL METHODS.

In order to identify and investigate the nature and properties of the various bacterial organisms, it is necessary to first isolate them. This having been done, it then becomes possible to study the behaviour of each species under certain specified methods of treatment, by means of which their identification is effected. The effects produced on various media by different species can also be studied; for instance, we may endeavour to ascertain if a particular organism turns wort or beer sour, if it renders either fluid turbid or ropy, or if it produces a pronounced flavour in one or both of these liquids.

Of the various methods for effecting the isolation of the individual bacteria which have been proposed from time to time, that invented by Dr. Koch is the one almost universally employed. The principle of his process is to widely separate the individual bacteria from one another in a solid nutrient medium. This is spread out in a thin layer on a plate of glass, and then placed under conditions that are favourable to the growth and development of bacterial organisms. In this way each individual bacterium increases and multiplies, in other words, it forms a colony, the members of which are retained in position, and not permitted to wander about, by the solid nature of the culture medium. In the course of a few days the colonies become sufficiently large to be detected by the naked eye; they can then be examined in various ways.

Sterilisation.—In carrying on bacteriological work, it is not only necessary that all the utensils and apparatus employed should be in a state of absolute cleanliness, but that they should be also absolutely sterile, that is, completely freed from living organisms in any shape. The media made use of in such work must also be rendered absolutely sterile. No pains should be spared to secure these conditions, otherwise the results of any investigation in which they are not rigidly adhered to will be valueless.

Several methods are in use for securing absolute sterility of the apparatus. Many chemical substances are known, such as corrosive sublimate, salicylic acid, sulphurous acid, &c., which act as powerful bactericides, and these are occasionally used for the object in view; but the agency chiefly relied upon is heat, applied in one form or another. Heat applied in the form of steam saturated with aqueous vapour is a much more powerful steriliser than what is technically known as "dry heat," such as that obtained by employing dry air or superheated steam. No organisms or their germs are able to withstand a prolonged exposure to dry air heated to 150° C. (300° F.), and but one germ, the spore of the *Bacillus subtilis*, is known which can survive prolonged boiling in water or lengthened exposure to the heat of steam saturated with moisture having a similar temperature. Liquid culture media, such as wort, grape-juice, &c., are generally boiled in order to effect their sterilisation.

Sterilising by the Direct Heat of a Flame.—This is a very convenient way of sterilising small articles, such as platinum wires, microscopic slips and cover-glasses, tweezers, &c. The articles are simply passed through a Bunsen flame or that of a spirit-lamp; in this way any organisms or germs present on such articles are absolutely consumed. In case of emergency, quite large articles, such as cultivation-plates, Petri-dishes, &c., may be similarly treated, but great care must be taken with glass articles, more especially with the larger ones, as they easily fly to pieces.

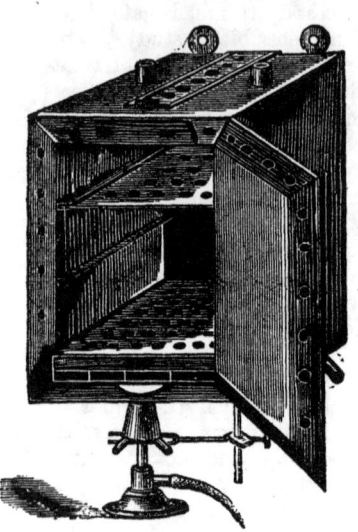

FIG. 12.—Hot-Air Oven.

The Hot-Air Oven.—Much use is made of this form of apparatus for the sterilisation of such articles as glass plates, test-tubes, Petri-dishes, cotton-wool, &c., which are able to withstand the comparatively high temperature required without injury.

An apparatus of this kind generally consists of a rectangular box of sheet metal with double walls, and having a door in front such as is seen in Fig. 12. In its roof are two apertures—the one intended for the insertion of a thermometer; the other for that of an arrangement called a "thermostat," for keeping the oven at a certain definite temperature. The oven is heated by a Bunsen or other form of gas-burner, the gas supply to which passes through and is regulated by the thermostat. Shelves are provided for supporting the various articles, and trays of wire-gauze for holding such small articles as test-tubes, &c. When a number of glass plates, intended for plate-cultivations, are to be sterilised, they are usually placed in a sheet-metal box (Fig. 17, p. 212) provided

THE THERMOSTAT.

with a lid, which, after being filled with plates, is placed in the sterilising oven, the lid of the box serving to protect the plates from subsequent contamination. After the oven has been filled with the articles to be sterilised, the gas is turned on, then the thermostat and the sliding door in the roof of the oven manipulated until a constant temperature of 150° C. (300° F.) is indicated by the thermometer. It is maintained at this degree for two or three hours, then the gas is turned off, the oven allowed to cool down to the room temperature, and the articles it contains removed. If the oven door is opened whilst the temperature inside is still high, there is great danger of the glass articles cracking.

The Thermostat.—There are several forms of this apparatus; one of these is shown in the accompanying illustration, Fig. 13. It consists of a glass tube divided into two chambers, a and b, by a partition, which is perforated by a central aperture, to which is fused a thin glass tube reaching nearly to the bottom of chamber b. In the upper chamber a side tube, c, is inserted; a perforated cork closes the upper part of the apparatus, and through this passes the bent tube, d, which can be raised or lowered within the chamber a as circumstances demand. A small quantity of mercury is introduced into the chamber b. The apparatus is inserted into the oven through the aperture specially provided for that purpose; one extremity of an indiarubber tube is attached to the tube c, the other to the burner underneath the apparatus. Another indiarubber tube coming from the gas supply is attached to the tube d. The action of the apparatus is as follows:—When the air in the oven becomes heated, a portion of its heat is communicated to the air in the chamber b, causing it to expand. In doing so the mercury is driven up into the tube a, and as the temperature rises, so does the column of mercury in a. When this has risen to a certain height it covers the end of the tube d, and gradually cuts off the principal supply of gas, so that at last no gas can pass to the flame except through a minute opening in the tube d. As a consequence, the flame is nearly extinguished. The temperature in the oven then begins to lessen, the air in the bulb to contract, the mercurial column lowers, and the orifice at the bottom of d is once more opened, and gas admitted to the flame. After a time a sort of equilibrium becomes established between the amount of gas necessary to keep up the heat required, and then just sufficient gas passes through the regulator to maintain the oven at the required temperature. It is obvious that by adjusting the quantity of mercury in the bulb, and regulating the position of the tube d, any required temperature within certain limits can be steadily maintained.

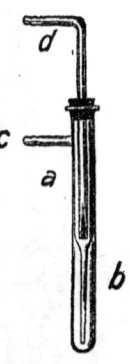

FIG. 13.—Thermostat.

208   THE MICROSCOPE.

Steam-Steriliser.—Cultivation media, especially those containing gelatin, are generally sterilised by the agency of moist steam. The apparatus used for this purpose consists of a tall cylindrical vessel constructed of sheet metal clothed with felt, which can be closed with a lid pierced by several apertures, one of which is intended for the insertion of a thermometer, the others for the escape of steam, Fig. 14. A removable vessel (Fig. 15) serves to hold the different articles which are to be sterilised. A water-gauge indicates the level of the water in the apparatus. When about to be used, water is placed in the bottom of the vessel to a depth of

FIG. 14.—Koch's Steam-Steriliser.   FIG. 15.—Basket for Steam-Steriliser.

three or four inches, the basket with the various articles inserted, the cover placed on, and a Bunsen-burner or other source of heat placed underneath. The water is kept boiling for two or three hours at such a rate that there is a free issue of steam through the orifices in the lid; the apparatus is then allowed to cool, and the basket removed. The water-gauge must be examined from time to time, to see that all the water does not boil away; should it get low, the vessel is replenished with fresh boiling water.

The Autoclave.—In this apparatus the sterilisation is effected through the agency of steam under pressure, in which condition it possesses a higher temperature than steam existing at the ordinary

THE AUTOCLAVE.

atmospheric pressure. Steam when used in this way exercises a much more powerful bactericidal action than when its temperature is only that of boiling water; for instance, Globig found that certain bacteria, which survived a three hours' exposure to steam at the temperature of 100° C. (212° F.), readily succumbed to steam at a temperature of from 110° to 120° C. (230° to 248° F.) in an autoclave. This apparatus, which is illustrated in Fig. 16, resembles the ordinary steam-steriliser in shape, but since its walls have to bear considerable internal pressure, they are made correspondingly strong. The lid is made to screw down steam-tight, and is provided with a safety-valve and a pressure gauge. A removable shelf-holder serves to carry the articles to be sterilised. When used, water is placed in the bottom of the vessel to a depth of a few inches, the shelf-holder with its articles introduced, the lid screwed down, and heat applied from below by means of a large Bunsen-burner. Since water always boils at a certain temperature for a particular pressure, the heat of the interior is ascertained from the indications of the pressure gauge. Thus when a pressure of 14.6 lbs. on the square inch is indicated, the temperature will be 122° C. (252° F.). By a little regulation of the gas supply a point is soon found where the required pressure and corresponding temperature are constantly maintained. The apparatus is then kept at this temperature for one or two hours, when the source of heat is removed, the steriliser allowed to cool, and the shelf-holder removed.

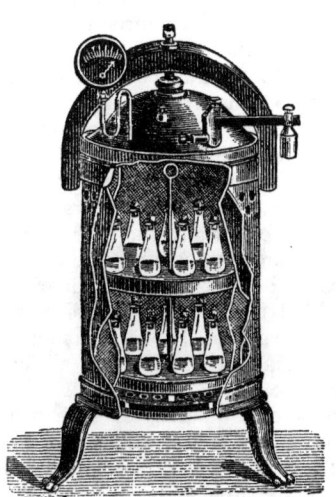

FIG. 16.—Autoclave.

The safety-valve is generally loaded to a slightly higher pressure than that required in the interior of the chamber, consequently there is neither escape of steam nor loss of water. This form of apparatus is useful, in so far as the bacteriological work connected with brewing is concerned, when samples of beer have to be sterilised. If such samples are treated in the ordinary steam-steriliser a loss of alcohol takes place. By conducting the operation in an autoclave, and using beer of the same alcoholic strength as the samples, freed from carbon dioxide in order to prevent frothing, instead of water for charging the apparatus, such samples may be sterilised without loss of alcohol.

Preparation of the Solid-Culture Medium.—This is prepared as follows:—One pound of fresh lean beef, finely divided by chopping or running through a mincing-machine, is added to a litre of cold distilled water, and well stirred in. The mixture is allowed to stand for twenty-four hours in a cool place, an occasional stir being given, after which it is poured into a clean linen cloth placed over a large glass funnel, and the meat-juice strained off, the pulpy mass in the cloth being squeezed until a litre of meat-juice is obtained. The following substances are then added to it—100 grammes of gelatin, 10 grammes of peptone in powder, and 5 grammes of sodium chloride. The mixture, after being allowed to stand until the gelatin has swollen, is then gently warmed until this substance has dissolved. Meat-juice is always slightly acid, because it contains a small quantity of lactic acid; and since bacteria flourish best in a neutral or slightly alkaline medium, the meat-juice is carefully neutralised with sodium carbonate solution until it is just neutral or faintly alkaline to litmus-paper. This neutralisation of the medium has a most important influence on the number of bacteria which afterwards develop in it. Reinsch, when experimenting with river-water, found that by employing un-neutralised gelatin meat-broth only 475 colonies per 0.5 c.c. of the water were formed; when 0.01008 gramme sodium carbonate had been added to 10 c.c. of the same medium, 2976 were formed; whilst the addition of 0.3024 gramme entirely suppressed the development of the colonies. He recommends, when experiments of a delicate nature are to be carried out, that just sufficient solution of caustic soda should be added to the medium (to which a little phenolphthalein solution has been added) to turn it pink. The necessary quantity of alkali is found by titrating an aliquot part of the medium with dilute caustic soda solution, throwing this away and then adding the calculated quantity of soda solution requisite to bring the bulk of the medium to the same state of neutrality as the aliquot part. The whole mixture is then placed in a large flask, the whites of two or three eggs are well beaten up, added, thoroughly mixed in by shaking, the flask and its contents placed in the sterilising apparatus, and exposed to a temperature of 100° C. (212° F.) for an hour. The object of the addition of the white of egg is to promote clarification.

The hot gelatin meat-broth is then filtered through fine filtering-paper, when it should run off as a perfectly bright fluid. As it rapidly solidifies on cooling, the filtration must be either conducted in a hot chamber, or the funnel which contains the filter-paper surrounded by a hot-water jacket, or the operation may be performed inside the steam-steriliser. As it passes through the filter the gelatin meat-broth is caught in quantities of about 10 c.c., either in small flasks or in test-tubes, care being taken that it does not soil the neck of the flask or the upper portion of the test-tube. Plugs of cotton-wool are inserted into the mouths of the flasks or the test-tubes, to prevent the access of germs from the atmosphere. The tubes or flasks are then placed in the steriliser, the water in it brought to the boil, and maintained at that temperature for a quarter of an hour, this treatment being repeated every day for three days in succession, the object of this procedure being to secure effectual sterilisation. It has been shown that the spores of bacteria are much more resistant to the sterilising effect of heat than the bacteria themselves, therefore a period is allowed between the consecutive heatings to afford any spores which may be present an opportunity of developing into the less resistant stage of bacteria.

This method, which is called "intermittent" sterilisation, is of great use in the technical application of the process to such purposes as the preservation of meat in hermetically closed tins, for in this way the over-cooking of the meat, which took place under the older methods of sterilisation, is to a great extent avoided.

The gelatin meat-broth tubes are then kept for a few days to see if the sterilisation has been complete. If such has been the case, the jelly will remain perfectly bright and clear; should signs of turbidity appear in any of the tubes, these must be rejected, for they are evidently still contaminated with living organisms.

Preparation of the Cultivation-Plate.—The actual cultivation of the bacteria is performed on glass plates, hence the name "plate-culture." For this purpose plates of ordinary glass are used, of such a size as can be conveniently examined under the microscope, their breadth being determined by the distance between the centre of the objective and the pillar which supports the body of the instrument, and they may be about twice as long as they are wide. Three inches by five is a size frequently used; the ordinary quarter-plates used in photography are also of a handy size for this purpose.

The plates are first thoroughly cleaned, and then sterilised by exposing them for an hour to a heat of 150° C. in the hot-air oven (Fig. 12, p. 206). During this process the glass plates are enclosed in the covered box of sheet metal (Fig. 17).

In spreading the gelatin meat-broth on the glass plates a levelling apparatus, such as that shown in Fig. 18, is generally used. This consists of a wooden tripod stand, the feet of which are formed of three screws by means of which the apparatus can be accurately levelled. The shallow glass tray, shown on the top of the glass plate in the figure, is filled with ice and water and placed on the tripod stand; this is covered with the glass plate on which the culture-plate is to rest; and on this the bell-glass, for covering the cultivation-plate and protecting it from stray germs.

When the apparatus is to be used, the tray is filled with water and ice, covered with the large plate of glass, and carefully levelled by adjusting the screws until the small spirit-level indicates that the apparatus is level. These arrangements having been made, a sterilised glass plate is removed from the box and carefully placed on the glass plate, and, in handling it, more contact with the fingers than is absolutely necessary should be avoided. It is then covered with the bell-glass.

Several of the gelatin meat-broth tubes or flasks are heated in the water-bath until their contents have become fluid, care being taken that the heat does not rise beyond 35° C. (95° F.), or some of the bacteria might be injured; the plug is carefully removed, and held between the fingers of the right hand in such a manner that that portion of it which enters the test-tube is not soiled by the fingers. If, for instance, we wish to investigate the organisms which have developed in meat infusion, we take out a minute drop of this by means of a platinum loop,[1] and stir it into the gelatin meat-broth contained in one of the tubes (the medium should measure

FIG. 17.—Plate-Box.

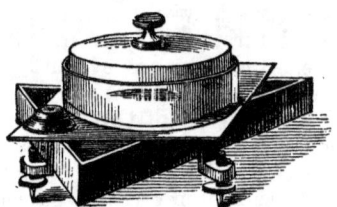

FIG. 18.—Levelling Apparatus.

approximately 10 c.c.), insert the plug, and, by a rolling and twisting motion applied to the tube, mix the infusion intimately with the medium. The plug is once more removed, the upper portion of the neck of the tube passed several times through the flame of a Bunsen-burner or that of a spirit-lamp, so that any adherent organisms may be destroyed. When the neck of the tube is sufficiently cool its contents are poured out, at one sweep, on to the middle of a cultivation-plate, which is placed in position on the levelling apparatus, and spread out into a thin layer with the assistance of the neck of the test-tube. The film must not be allowed to approach the edges of the plate within a distance of half-an-inch, as this portion may have been contaminated by the fingers during the previous manipulation. The glass cover is placed over the cultivation-plate, and the whole allowed to remain until the gelatin film has become solid; this, under the cooling

---

[1] Made out of a piece of platinum wire about one-fortieth of an inch in thickness, one end of which is fused into a glass tube, the other end twisted into a loop about one-sixteenth of an inch in diameter.

influence of the ice-cold water, takes place in one or two minutes. When solidification has taken place, the cultivation-plate is removed to the moist chamber.

The Moist Chamber.—This, which usually consists of two glass dishes, is represented in the following illustration (Fig. 19). The lower dish has, as a rule, a diameter of about 8 inches and a depth of 4 inches. The other dish is slightly wider, and is used for the cover. The dishes are first thoroughly cleaned by washing, then smeared inside with glycerin, which will retain any bacteria which may fall upon them. A layer of filter-paper, moistened with thirty or forty drops of a saturated solution of mercuric chloride (corrosive sublimate, a powerful bactericide, and also a most poisonous substance), is placed in the bottom of the narrower dish. The water contained in this solution serves to keep the air in the apparatus saturated with moisture. A little stand, which will hold several cultivation-plates, is often employed in the apparatus, so that several cultivations may be simultaneously carried on in it (Fig. 20). It is constructed of glass plates, to the extremities of the longer sides of which strips of plate-glass about

FIG. 19.—Moist Chamber.

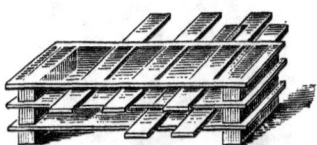

FIG. 20.—Support for Cultures.

three-eighths of an inch in thickness and half-an-inch in breadth, and of the same length as the width of the cultivation-plate, are affixed either with sealing-wax or mineral glue. When a number of these are superimposed they form, as it were, the shelves of a small open cupboard, on which the cultivation-plates can be placed. A slip of paper, bearing a reference number or letter, is placed under each cultivation-plate, so that the plate can be afterwards identified. The moist chamber containing the cultivation-plates is then placed in the incubator.

The Incubator.—This consists of a vessel of sheet metal (Fig. 21), generally made of copper, having double walls, and provided with a door in front; the space between the walls is filled with water. The apparatus is kept at the required temperature by means of one or more gas-flames placed underneath it, the supply of gas to which is governed by a thermostat. In plate-cultivation experiments this is so adjusted that there is steadily maintained within the incubator a temperature of 20° C. (68° F.), as indicated by the thermometer, the bulb of which is inside the apparatus.

The moist chamber and its contents are introduced into the incubator, the door closed, and the whole left for three or four days, the culture-plates being occasionally examined. At the end of two or three days the individual colonies will have become visible to the naked eye, and can be further examined by the aid of a hand magnifying-glass. They will differ considerably in appearance, since the meat infusion would be sure to contain numerous forms of bacteria. Some of the colonies will appear as white slimy-looking spots resting on the surface of the gelatin, others will have spread more considerably amongst the surrounding gelatin, some may have an iridescent appearance, others may be brilliantly coloured. Many of the colonies will be found to have left the gelatin film intact, others to have liquefied it, and various degrees of liquefaction may be observed amongst the different colonies.

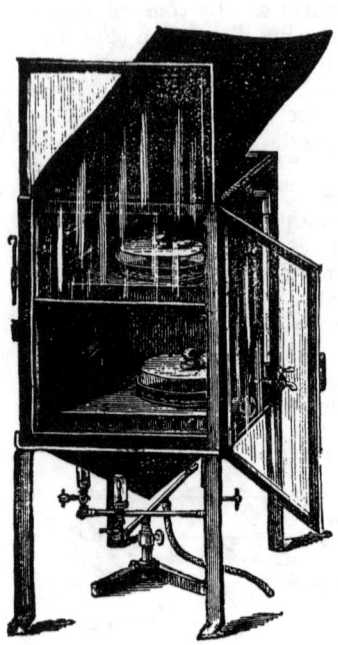

FIG. 21.—Incubating Chamber.

Observation of the Colonies under the Microscope.—For this purpose the cultivation-plate is placed on the stage of the microscope, and viewed under a power of about 100 diameters. Some of the colonies will have developed inside the gelatin film, and these will appear as dark round bodies. Others will have vegetated on the surface of the film, or have reached its surface in growing. Remarkable differences in the form, and especially in the borders, of the colonies will be observed. Some may be surrounded by a well-defined line, some by a notched and indented one, some are bordered by projections having the appearance of locks of hair, some throw out hair-like shoots, while in others the margins are, as it were, beset with fine prickles. These differences in appearance show that the bacteria belong to different species.

Observation of the Individual Bacteria forming the Colonies.—The bacteria may be further examined by removing a portion of the colony with a straight platinum wire attached to a glass handle, which, just before being used, is passed through the flame of a Bunsen-burner, in order to destroy any adherent bacteria. The "fishing," as this process is called, is performed with the culture-plate under the microscope, the particular colony being kept in view under a low magnifying power. The bacteria adhering to the platinum wire are now to be mixed with a small drop of water on a cover-glass; this is then placed on a glass slip and the bacteria examined either in the ordinary manner, or by the hanging-drop method (p. 199); afterwards it can be made into a permanent preparation (p. 203).

"**Impression Preparations.**"—These are easily made, and serve for keeping a permanent record of the forms of colonies. In order to obtain an "impression preparation," a perfectly clean cover-glass is laid over the colony of which it is wished to obtain a permanent record, allowed to remain in contact with it for a moment, and then removed from the surface of the gelatin with a pair of tweezers. By this means an impression is obtained of the colony. The preparation is now allowed to become perfectly dry, and is then stained and mounted in the same manner as a cover-glass preparation. Not only can such a preparation be examined with low powers for the shape of the colony, but the highest magnifying powers may be applied so as to examine the individual bacteria. Young colonies lend themselves more readily to treatment in this way than older ones. It is not well adapted for colonies where a large amount of liquefaction of the gelatin has taken place; in order to make successful impression preparations of the liquefying bacteria, they must be made while the colonies are quite young.

Substitutes for the Cultivation-Plate of Koch.—Petri replaces the plate by small dishes (Fig. 22), one of which is slightly larger and fits over the other. The dishes are about four inches in diameter and half-an-inch high. After they have been sterilised, the warm infected gelatin meat-broth is run into the lower dish, and spread over the bottom of the dish in a thin layer;

FIG. 22.—Petri-Dishes.

the upper dish, which forms the cover, is put in its place, and the apparatus simply left to itself until the gelatin has solidified, no artificial cooling being required. When the colonies are formed, they can be examined under the microscope, and preparations made from them in exactly the same manner as when a cultivation-plate is employed. Esmarch has proposed another modification, in which a small quantity of the infected gelatin meat-broth is run into a sterilised test-tube, a sterilised plug of cotton-wool inserted, and, by a gentle rolling motion, the gelatin is spread in an even layer on the inside of the tube, the tube meanwhile being held under a stream of cold water, care being taken that the gelatin does not touch the plug. The colonies develop on the film in the inside of the tube, and can be counted, examined under the microscope, "fished" with a bent platinum wire, and preparations made from them. This modification obviously does not permit impression preparations being taken from the film. G. H. Morris slightly extends this method by running the infected material into a globular sterilised flask, inserting a plug of sterilised cotton-wool, and distributing the gelatin in a thin layer by rotating the flask.

Cultivation of Anaërobic Organisms.— Organisms of this nature, such as the butyric acid ferment, which cannot live in the presence of oxygen, must be either cultivated in an atmosphere deprived of this gas, or in one of an indifferent gas such as hydrogen. One of the simplest methods for performing a culture of this kind is to infect the contents of a gelatin meat-broth test-tube, plugged with cotton-wool, in the ordinary way, and insert this into a much larger test-tube, which contains a small quantity of an alkaline solution of pyrogallol, the larger tube being closed with an indiarubber cork. As alkaline pyrogallol solution rapidly absorbs oxygen, the enclosed air is entirely deprived of this gas in a very short time.

Bacteriological Examination of Water.—A sample of water may be examined bacteriologically, either by Koch's cultivation-plate method, or by that in which Petri-dishes are used, or by the Esmarch tube-cultivation method. Of the three, the second method, in which Petri-dishes are employed, is the easiest and simplest to manipulate, since no levelling apparatus nor ice is required; but whichever method is employed, we commence in the same way. A tube containing approximately 10 c.c. of sterilised gelatin meat-broth is warmed just sufficiently to render its contents liquid. The sample of water to be examined is shaken up well, so as to evenly distribute the organisms it contains, and, if the sample be a fairly pure one, 1 c.c. of it is measured out by means of a sterilised pipette, and added to the gelatin medium. The plug is then inserted in the tube, and the water intimately mixed with the bouillon by turning and twisting the tube about. A sterilised glass plate is then coated with one-half of the contents of the tube in the manner described on p. 212; when this has solidified, a second plate is similarly coated with the remaining half of the bouillon. If, instead of using cultivation-plates, we employ Petri-dishes, two sets of these are sterilised in the hot-air oven and allowed to grow cold. Into the lowermost dish of one of these sets half the contents of the infected tube are poured, and the remaining half into the lower dish of the second set. As the top dish is only removed for the few seconds it takes to pour out the contents of the tube, the chance of organisms falling in from the air is well-nigh excluded. The gelatin is then distributed in a thin even layer on the bottom of the dish by gentle rotation. Esmarch's tubes may be coated in the manner previously described on p. 215; they are not so suitable for this purpose as plates or dishes, for, if many liquefying bacteria are present, the colonies are very apt to run together.

In case the sample of water is somewhat impure, a smaller quantity than 1 c.c. must be taken for infecting the gelatin bouillon, say 0.5 c.c. When very impure samples have to be examined, it becomes necessary to employ a considerably less quantity than half a cubic centimetre. As smaller quantities than this cannot be conveniently measured, in these cases it is better to begin by diluting the water. For this purpose we take a small stoppered bottle which has been sterilised, place in it 4 or 9 c.c. of sterilised water, and then add 1 c.c. of the sample of water. The mixture is thoroughly incorporated by shaking, and 1 c.c. of it taken for infecting the gelatin tube. Obviously in the first instance we use 0.2 c.c., and in the second 0.1 c.c. of the sample of water for the infection, and these quantities must be taken into account in the subsequent calculation to 1 c.c.

Standard for Comparison.—The number of colonies which 1 c.c. of water yields is taken as a convenient standard for the purpose of comparison. Thus, if, on examining a certain sample of

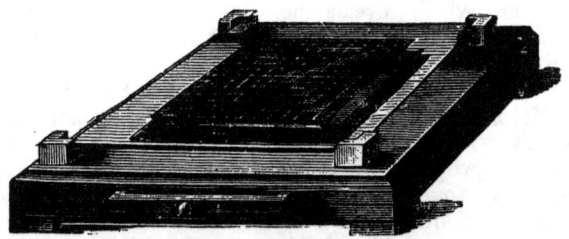

FIG. 23.—Wolfhügel's Counting Apparatus.

water, of which 1 c.c. had been taken to infect the gelatin tube, and if, of the two plates which had been run with it, the one yielded 218 colonies, the other 224, then such water would be said to contain 442 organisms per cubic centimetre. Obviously, if the sample had been previously diluted to such an extent that only 0.1 c.c. had been taken, then the numbers found on the two plates would require to be multiplied by 10, in which case the number of organisms contained in such a water would be 4420 per cubic centimetre.

Counting the Colonies.—For this purpose the apparatus invented by Wolfhügel (Fig. 23) is generally employed. It consists of a wooden base, surmounted by a blackened ground-glass plate, on which the cultivation-plate is placed, and over this a glass plate marked into squares with a diamond, so supported at the corners that it does not come in contact with the gelatin film on the cultivation-plate. The sides of the squares engraved on the counting-plate are generally a centimetre (about three-eighths of an inch) long, and several of these squares are still further subdivided. The counting is performed with the assistance of a magnifying-glass; the number of colonies in a certain number of squares are counted, and the average contained in each square estimated from these. The number so found, multiplied by the number of squares which are equal to the size of the gelatin film, gives the number of bacteria contained in the quantity of the water used for infecting the gelatin. The colonies formed in a Petri-dish may be similarly counted by placing the dish on a sheet of black paper ruled in squares with Chinese white.

Biological Examination of Filtered Waters.—Where the water employed in a brewery is of such an impure nature as to necessitate its filtration prior to use, the biological examination of such a water, after filtration, affords valuable information, in the first instance, as to whether the filter is properly constructed, and capable of efficiently dealing with the quantity of water required; afterwards it is of great value in ascertaining if the filter is performing its functions properly. This is the one case where that cultivation method in which meat-broth gelatin is employed, and which is the one which reveals the total number of bacteria contained in a water, may be used with advantage in the practical working of a brewery.

Hansen's Method for the Bacteriological Examination of Brewing Waters.—It has been pointed out by this distinguished physiologist, to whom we owe so much for the elucidation of many points connected with the science of brewing, that the vast majority of the organisms found in water by Koch's method (meat-broth gelatin) have no significance whatever for the brewer. Very few of those which are capable of living and multiplying either in peptone meat-broth or on a gelatin meat-broth plate are capable of vegetating in such fluids as wort or beer. The conditions in these two last fluids are extremely unfavourable to the life of many bacterial organisms, since they are always slightly acid, and also contain some of the constituents of the hop, which are directly bactericidal in their action. Endeavours have been made to estimate the quantity of those bacteria contained in a water which are injurious to wort or beer, by the plate method of Koch, using, instead of meat-broth gelatin, ordinary boiled wort in which 10 per cent. of gelatin has been dissolved. It was found, however, that many organisms which were capable of developing on a gelatin-wort plate would not grow in wort itself; hence this method, also, did not afford reliable information as to the number of bacteria which were capable of growing in wort. As Dr. Hansen suggests, the real question which the brewer has to decide is, "How does a water behave with reference to the wort and the beer; how many micro-organisms does it contain which are able to develop in those two liquids; and are there amongst them any species capable of causing disturbances in the brewing operations?"

EXAMINATION OF WATER.

The brewing chemist obviously holds a much more advantageous position in the study of those bacteria which concern him than the hygienist does. The latter would be able to obtain much more precise and satisfactory information concerning the action of the individual organisms which he isolates on the human economy, if he were able to try their effect on the human body itself; but since he is obviously debarred from such a proceeding, he has to content himself with inventing methods for their identification, and trust to chance for the occurrence of cases where he can study their action on the human body. The brewer, in investigating the action of micro-organisms on wort and beer, is able to experiment directly with these very fluids themselves, and carry out his experiments under conditions which resemble as closely as possible those which actually exist in the usual routine of the working of a brewery.

In the actual examination of a water by Dr. Hansen's method, fifty flasks, each having a capacity of about 20 c.c., are taken and divided into two groups, A. and B. Into each of the 25 flasks of group A. 10 c.c. of wort are introduced; and into the remaining 25, which constitute the B. group, the same quantity of beer. All the flasks are plugged with cotton-wool and sterilised. When the flasks and their contents have become cold, one measured drop (0.04 c.c.) of the water under examination is added to each of fifteen of the flasks belonging to group A., and also to fifteen of the flasks of group B. To each of the remaining ten flasks of each group 0.25 c.c. of the water is added. The whole of the flasks are then shaken, so as to distribute the organisms, and kept at a temperature of 24° to 25° C. (75° to 77° F.) for fourteen days. At the expiration of this period the flasks are examined for signs of turbidity, and those which exhibit this sign of infection counted. Taking into account the total quantity of water added to the ten flasks, the number of flasks infected, and the total number of flasks taken, we find the number of organisms presumably present in this quantity of water, and from this is calculated the number present in each cubic centimetre of the water. Thus, if five out of the ten wort flasks (group A.) to which 0.25 c.c. of the water had been added have become turbid, then, as the total amount of water added to these ten flasks was $0.25 \times 10 = 2.5$ c.c., this quantity presumably contains five organisms[1] capable of developing in wort; consequently, the water contains two such organisms per cubic centimetre. If, as sometimes happens, the whole of these ten flasks have become turbid, no conclusion can be formed from them. We then turn to the remaining fifteen

---

[1] This is under the supposition that only one germ or organism is introduced into each flask with the 0.25 c.c. of water, which, in the vast majority of cases, is undoubtedly true, when a large number of the flasks remain unaffected. There is always, however, the *possibility* that two or more germs or organisms may have been introduced into one flask.

flasks of each group, to each of which only 0.04 c.c. of water had been added, and apply the same method of calculation. Thus, supposing that twelve out of the fifteen show signs of infection, then $0.04 \times 15 = 0.6$ c.c. of the water presumably contains twelve germs; consequently, one cubic centimetre of such a water will contain twenty organisms capable of developing in wort. The same process of calculation is applied to the flasks of group B., and the number of organisms per cubic centimetre of the water capable of developing in beer similarly found.

A comparative experiment performed by Dr. Hansen with the water supplied to the Old Carlsberg Brewery will serve to illustrate these differences. A plate-cultivation was made after the method of Koch (gelatin meat-broth), using $\frac{1}{2}$ c.c. of the water. At the end of fourteen days there were found on this 111 colonies, that is, 222 organisms per 1 c.c. water, all of which consisted of bacteria. Very few of the colonies had liquefied the gelatin. Another plate-cultivation was made simultaneously, using wort gelatin. It was found at the end of fourteen days that fifteen colonies had developed on it, and as $\frac{1}{2}$ c.c. of water was used, this would be at the rate of thirty colonies per c.c. At the same time one drop (0.04 c.c.) of a mixture of equal parts of the water and sterilised wort was added (*i.e.*, 0.02 c.c. water) to each of a series of fifteen wort flasks and to an equal number of beer flasks. These were kept for fourteen days and then examined, when it was found that the contents of none of the flasks showed signs of infection. In a second similar experiment, where the water yielded by Koch's method 1000 colonies per c.c., and the wort-gelatin plate showed thirty-four colonies, but two of the wort flasks became infected. One of these had developed a growth of bacteria, the other a growth of *Penicillium glaucum* (one of the commonest mould fungi), that is, 6.6 organisms per c.c. of water. The whole of the beer flasks remained unaltered.

Holm gives the following results of three comparative experiments of a similar nature, all of which are reckoned upon 1 c.c. of water :—

	I.	II.	III.
Koch's method	8000.0, mostly bacteria.	350.0, bacteria.	370.0, bacteria.
Wort-gelatin plate	14.0, mould fungi.	8.0, mould fungi.	4.0, bacteria.
Wort	5.4, bacteria and mould fungi.	5.8, bacteria and mould fungi.	1.1, mould fungi and *Torulæ*.
Beer	0.8, mould fungi.	0.8, mould fungi.	0.4, mould fungi.

The last example is worthy of note, since it shows that those species which developed in the wort and beer entirely differed from those that developed either in gelatin meat-broth or wort-gelatin plate-cultivations.

The following results are taken from a large number of analyses of brewing water which were made by Professor Schwackhöfer. They likewise serve to show the remarkable differences in the behaviour of the organisms contained in the same water under the various methods of culture. The results of the experiments with wort and beer are given in percentages of the number of flasks which became infected. As 4 c.c. of water was added, collectively, to each hundred flasks, the numbers contained in columns three and four, divided by four, will give the number of growths per c.c. water :—

No. of Colonies per c.c. Water.		Per Cent. of Flasks Showing Growths.		Opinion Formed from the Biological Examination.
Koch's Method.	Wort-Gelatin.	Wort.	Beer.	
56	0	0	0	Good.
165	0	0	0	Good.
609	0	0	0	Good.
3,260	0	0	0	Good.
4,134	0	8	0	Good.
6,120	0	8	0	Good.
6,000	0	80	8	Unfavourable.
8,456	0	20	19	Unfavourable.
46,700	0	70	0	Permissible.
121,687	0	100	60	Bad.
769,780	0	100	4	Bad.

Wichmann's Method.—By this method, which is a modification of Hansen's, it is sought to give a numerical expression for the degree of vital energy possessed by the organisms existing in a water. In examining a number of samples of water by Hansen's method, in some cases it is found that the whole of the flasks become turbid on the first day, in other cases only after the expiration of three days, and in others only after a longer period than this. In those cases where the turbidity appears most rapidly it is assumed that the organisms possess the most vital energy, and that they possess lesser amounts as the length of time increases before which they afford visible indications of their presence. Hansen's method affords no information as to the amount of vital energy the organisms possess, and this deficiency Wichmann endeavours to remedy.

He proceeds in the following manner:—Twenty-five small flasks, each of which contains 10 c.c. of sterilised wort, are taken (group A.), and a like number, each of which contains the same quantity of sterilised beer (group B.). To twenty flasks of each of these groups a measured drop (0.025 c.c.) of the water under examination is added, and to the remaining four flasks of each group respectively 1.0, 0.75, 0.5, and 0.25 c.c. of the water; the remaining flask of each group being kept as a check on the sterility of the fluids employed.

These latter four flasks are numbered respectively 1, 2, 3, 4, and the time is noted at which each becomes turbid. If, with a certain sample of water, all these four flasks become turbid within twenty-four hours, then such a water is considered to possess a degree of impurity expressed by 100. This number is arrived at by multiplying the numbers of each flask 1, 2, 3, 4, which stand in a sort of inverse ratio to the varying quantities of water added, by 10, and this is the factor with which the number of any flask is to be multiplied by in which turbidity shows itself in one day. Thus, where all are affected on the first day, the calculation is $(1 \times 10) + (2 \times 10) + (3 \times 10) + (4 \times 10) = 100$. In those cases where turbidity appears in any flask in two days, the factor to be employed is 8; if in three days, 6; in four days, 4; and in five days, 2. The general method of calculation is shown in the appended example:—

No. of Flask.	Appearance of Turbidity.	Factors.
1. Infected with 1.00 c.c. water	2 days	$1 \times 8 = 8$
2. Infected with 0.75 c.c. water	3 days	$2 \times 6 = 12$
3. Infected with 0.50 c.c. water	3 days	$3 \times 6 = 18$
4. Infected with 0.25 c.c. water	4 days	$4 \times 4 = 16$

The infective energy of the water towards wort is 54.

The remaining twenty flasks, to each of which 0.025 c.c. of the water was added, are calculated by Hansen's method.

As beer is supposed for this purpose to have a resistance to the vitality of micro-organisms $1\frac{2}{3}$ greater than that of wort, in calculating the results with beer the number of the flask is multiplied by $1\frac{2}{3}$ (1.67) in addition to the factor for the length of time elapsed before turbidity makes its appearance. For example:—

No. of Flask.	Appearance of Turbidity.	Factors.
1. Infected with 1.00 c.c. water	3 days	$1 \times 6 \times 1.67 = 10.0$
2. Infected with 0.75 c.c. water	4 days	$2 \times 4 \times 1.67 = 13.4$
3. Infected with 0.50 c.c. water	5 days	$3 \times 2 \times 1.67 = 9.9$
4. Infected with 0.25 c.c. water	5 days	$4 \times 2 \times 1.67 = 13.2$

The infective energy of the water towards beer is 47.2.

Investigation of the Air for Living Organisms.—Many forms of complicated apparatus have been devised for this purpose from time to time, some of which are described in the chapter dealing with the number and variety of the organisms found in the atmosphere (p. 310). Now much simpler methods are employed, and these are, as a rule, sufficient for the purposes of the brewer. When the air of any particular locality, such as that of the cooler-room, the fermenting-rooms, or the store-cellars, is to be examined, a number of glass cylinders (or wide-mouthed bottles) are thoroughly cleaned, labelled, and the superficial area of the mouth of each ascertained and noted on its label. They are then closed with cotton-wool plugs, and sterilised either by steam heat or in the hot-air oven. Caps of sterilised paper are then tied over the plugs, and the cylinders or bottles taken to the locality where the observation is to be made. The caps and plugs are here removed, the plugs being placed temporarily inside the caps and the cylinders left standing open for an hour. The plugs and caps are then replaced, the particulars concerning each cylinder noted on its label, and the cylinders taken back to the laboratory, where the germs which have fallen into them may be brought to development in several ways. Each cylinder may be filled one-fourth full of sterilised meat-broth gelatin, or of gelatin wort, either of which has been just warmed sufficiently to render it fluid. The cylinder is then held under a stream of cold water and turned and twisted about until the gelatin is spread over its interior in a thin even layer and become solidified; or the gelatin, after admixture with the germs, may be made into a plate-culture or poured into a Petri-dish. The Petri-dishes, after being sterilised, may be used to catch the germs in the first instance, sterilised gelatin being added afterwards. In whichever way the culture is made, it is placed in the incubating chamber and allowed to remain for two or three days, when the various organisms will have formed colonies. These are counted, and, by taking into consideration the area of the mouth of the particular cylinder or dish and the number of colonies formed, a simple calculation gives the amount of germs falling per square foot from the air in a particular locality. The individuals composing the colonies can be investigated by fishing any particular colony with a platinum wire and examining the organisms by the usual bacteriological methods, or they may be cultivated by the hanging-drop method, or their action on such liquids as wort or beer studied.

The colonies of mould fungi are readily recognised by their branching mycelium, which spreads out from the borders of the colony in all directions, giving it a star-shaped appearance. Many of them throw up aërial spore-bearing hyphæ, which are distinctive for the various species. The yeasts form small compact colonies about the size of a pin's head, generally having well-defined borders; they are nearly always colourless, though occasionally colonies of the so-called red yeasts are met with. The bacterial colonies are smaller than those of the yeast fungi; they are not so compact, and present a glutinous appearance; some of them may be coloured (chromogenous bacteria), while others may have an iridescent appearance. Many species of bacteria secrete a proteolytic enzyme and as a consequence liquefy the gelatin medium.

CHAPTER IV.

VEGETABLE BIOLOGY.

THE brewer is, in the course of his operations, brought so constantly in contact with processes of a purely biological nature that it is highly necessary he should possess some knowledge of this branch of science. A short account of the elementary principles of biology is therefore given, together with a brief biological description of those organisms with which he is chiefly concerned.

The Living Cell.—This constitutes the foundation, or, as it may be termed, the unit of all living organisms. Among the lowest and simplest forms of vegetable organisms many are known which, though they consist of but one single cell, yet these are able to carry on all the functions essential to plant life, such as respiration, nutrition, reproduction, &c. As we ascend in the scale of vegetable life, the larger becomes the number of cells which enter into the structure of the plant; and very soon, instead of all the cells being alike and performing similar functions, they become divided into groups, to each of which a particular set of functions is assigned. Thus a highly organised plant, such as the hop plant, may be regarded as a vast congregation of groups of different cells, each group constituting a different tissue in the plant, and each having its own special set of duties to perform in the collective economy.

In commencing the study of the problems of plant life, we first devote our attention to those organisms which consist of a single cell, for here the conditions of life are presented in their simplest form.

Protococcus pluvialis.[1]—This simple unicellular plant (Fig. 24) is described first, because, though it consists of merely a single cell, it performs all those biological functions which are essentially characteristic of the chlorophyll-bearing plants.

This organism is commonly met with in considerable quantities in the mud which collects in roof gutters, water butts, &c. When submitted to microscopic examination, the protococci are seen to consist of small round or oval bodies, each of which is surrounded by a tough resistant membrane called the "cell-wall," and this, by treatment with appropriate reagents, can be shown

---

[1] Another closely allied species, *Protococcus viridis*, often found as a green film on the bark of trees, will serve equally well for examination.

to consist of cellulose. Each of these little bags of cellulose is filled with a semi-fluid, jelly-like substance called "protoplasm," and this constitutes the essential portion of the cell; for living cells are known which consist of a mass of protoplasm without a protecting envelope. This remarkable substance is found in all growing cells, where it forms the seat of those vital actions which distinguish living from lifeless organic matter. Through its agency chemical decompositions and re-compositions of the most profound and complicated nature are constantly being carried on, with the most perfect smoothness and apparently with the greatest ease; for protoplasm possesses the power of absorbing many compounds, the elements or elementary groups of which it unlocks and rearranges into fresh states of chemical combination. These masses of protoplasm, though in themselves extremely minute, are contained in such countless myriads in the members of the vegetable kingdom distributed throughout the world, that,

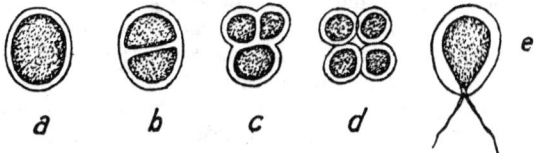

FIG. 24.—Protococcus pluvialis.
(a) Adult condition; (b) Reproduction by division, first stage; (c, d) Second and third stages; (e) Zoospore.

taken collectively, they constitute one of nature's greatest laboratories, in which transformations, the magnitude of which is immense, are continuously being carried on.

Protoplasm, when moderately magnified, is seen to be of a more or less granular nature,[1] according to the age of the cell, being less granular in young cells, more so in old. It is not homogeneous throughout the cell; the portion nearest the cell-wall is somewhat denser and more translucent than the portion occupying the interior of the cell. The composition of protoplasm is not known with exactitude; it is undoubtedly extremely complex, and is of a proteid nature. In the cells of the *Protococcus*, and in almost all other vegetable cells, a small round body, the nucleus, may be observed. This seems to be concerned, in some obscure way, in guiding and controlling the functions of the cell; for instance, the division of the cell is always preceded by the division of its nucleus. Chemically it differs in composition from the protoplasm by containing nuclein, a substance into the composition of which phosphorus largely enters. In the interior

[1] Protoplasm is much more easily examined in the organism next described, the yeast plant.

of most active cells, especially when fully grown, there are to be seen one or more round or oval spaces, the vacuoles. These are filled with the cell-sap, which fluid also permeates the protoplasm and the cell-wall. A number of green or red granules are diffused throughout the protoplasm of the *Protococcus;* these are the chlorophyll cells, which play such an important part in the nutrition of this and other green plants.

Food.—When the substance of a chlorophyll-bearing plant is analysed, it is found to consist of the elements carbon, hydrogen, oxygen, nitrogen, phosphorus, sulphur, together with minute quantities of potassium, magnesium, calcium, and iron. This leads at once to the supposition that all these elements must be present in the food of such a plant, if it is to carry on its vital functions efficiently. It has been proved experimentally that this is so, for if a plant like the *Protococcus* is grown in a solution which lacks any one of these elements, it gradually sickens and dies. In addition to this, we shall find, as we proceed further, that it is also requisite that these elements should be presented to the plant in such forms of combination as the plant can assimilate, and plants differ considerably with regard to the compounds they are able to appropriate for the purposes of nutrition; like animals, they exhibit considerable choice in the nature of their diet.

Assimilation.—Plants which possess chlorophyll cells are able to obtain the carbon necessary for their nutrition from the carbon dioxide of the air. The process by which they accomplish this is effected through the agency of the chlorophyll cells, and is called "assimilation." That this process goes on in the *Protococcus* may be demonstrated by the following experiment:—A tolerably large strong test-tube is three parts filled with mercury, the remaining fourth being filled with water containing sufficient *Protococci* to render the liquid distinctly green; the whole is then inverted and placed in a dish of mercury. Carbon dioxide gas is now passed into the tube in quantity sufficient to displace about a fourth of its contents. The whole arrangement is placed *in bright sunlight* for several hours, at the expiration of which it will be found that the greater portion of the carbon dioxide originally present in the tube has been replaced by an equal volume of oxygen. This is shown by passing a piece of caustic potash into the tube, which dissolves in the layer of water, and absorbs the small quantity of carbon dioxide which has not been decomposed by the plant. If a solution of pyrogallol be then passed into the tube, the remaining gas, which is nearly as large in volume as the gas introduced into the tube, will be absorbed; this proves that it is oxygen. The little plant evidently takes in carbon dioxide, deprives it of its carbon, which it holds back for its own use, and rejects the oxygen. It is known that, in order to decompose carbon dioxide into its constituent elements, a large amount of energy is required. This energy the plant derives from the luminous rays of the sun;

hence the transformation only proceeds in the light. We know that the peculiar agents which bring about this change are the chlorophyll corpuscles, because all plants unprovided with them are incapable of effecting this transformation. The process is supposed to take place in the following manner by double decomposition:—

$$\underset{\text{Carbon dioxide.}}{CO_2} + \underset{\text{Water.}}{H_2O} = \underset{\text{Formaldehyde.}}{CHOH} + \underset{\text{Oxygen.}}{O_2}$$

A number of formaldehyde molecules may then suffer condensation, one of the hydrogen atoms of one of them being simultaneously transposed, and in this way a carbohydrate formed. For instance, we may imagine the formation of glucose to take place in the following way:—

$$\underset{\text{Formaldehyde.}}{6(CHOH)} = \underset{\text{Glucose.}}{CH_2OH-CHOH-CHOH-CHOH-CHOH-COH}$$

That some reaction of this kind actually does take place is rendered extremely probable by the researches of Emil Fischer, who has actually produced a sugar by the condensation of formaldehyde; and Brown and Morris[1] have pretty conclusively shown that the first products of assimilation in plant life consist of one or more sugars. The protoplasm is able, in some unknown way, to transform these first-formed carbohydrates into other bodies having a much larger molecule, such as starch, cellulose, lignin, fat, &c.

Every pound of carbohydrate formed in this way represents so much stored-up energy, a portion of which is afterwards liberated and employed in performing the internal work of the plant, such as the building up of its tissues, effecting chemical transformations in the protoplasmic contents of its cells, &c.; and the amount of carbohydrate employed in this way has been estimated to be about a thirtieth part of the total amount formed by assimilation.

In passing it may be mentioned that an immense portion of the work performed in the world is obtained from the combustion of the carbon which has been stored up by plants through the agency of their chlorophyll cells. Many animals derive their energy from vegetable food alone; the carnivorous animals also obtain theirs in an indirect manner from the vegetable world, since they feed on their herbivorous fellow-creatures. Those immense stores of energy, our coalfields, represent the energy of the sun's rays, which has been stored up through the agency of the chlorophyll cells of plants which lived countless ages ago.

Formation of Proteids.—In addition to the formation of carbohydrates, the protoplasm of the *Protococcus* performs another important function, viz., the building up of the proteids, or the very

[1] *Journal of the Chemical Society*, 1893, p. 604.

bodies of which it is itself composed. From what has gone before it will be remembered that these bodies contain, in addition to carbon, hydrogen, and oxygen, also nitrogen and sulphur; and the protoplasm of many plants is able to build up its proteids from the carbohydrates and such compounds of nitrogen as nitrates or ammonia, which are of an entirely inorganic nature. The *Protococcus* flourishes in rain water, which always contains minute quantities of ammonia and nitrates. Hardly anything is definitely known as to how this is accomplished, but it is supposed that the ammonia or nitrate is first reduced in a manner somewhat analogous to that in which carbon dioxide is, and that in this reduced form it unites with a portion of the carbohydrates to form amides such as asparagine, glutamine, &c.; the amide then combining with a further portion of a carbohydrate to form a proteid.

This power of being able to construct proteid matter from purely inorganic substances, though peculiar to vegetable cells, is not possessed by all cells. No animal cell is capable of effecting this transformation, hence all animals have either directly or indirectly to depend upon vegetable organisms for their supply of proteids as well as for that of their carbohydrates.

Respiration.—All living beings breathe, that is to say, they take in oxygen and give out carbon dioxide. They derive in this way the energy requisite for performing the various functions of their life from the combustion of carbon. In the construction of a building a certain amount of force has to be expended; this is supplied by the labourers engaged in the work, who derive it from the combustion of the carbonaceous matter taken into their bodies as food. Many other kinds of work are performed by the energy liberated in the combustion of coal in the boiler of the steam-engine. Similarly, a certain amount of work has to be performed in the building up of a plant, and this is obtained by the combustion of the carbon of some of its own carbohydrates. That the *Protococcus* respires may be shown by a modification of the experiment just described. If the test-tube be filled three parts full with mercury, the remaining fourth part with water containing *Protococci*, the whole inverted in a dish of mercury, and kept for several hours *in the dark*, a certain amount of gas will be evolved. If a piece of caustic potash be passed up the tube, the whole of the gas will be absorbed in a short time, showing that it is carbon dioxide. Since the gas is obtained from the combustion of the carbon of a portion of the carbohydrates contained in the plants by the oxygen dissolved in the water, its quantity may be taken as a measure of the energy expended in carrying on the collective life of the plants during the period of the experiment.

Excretion.—In addition to these constructive processes, a set of destructive processes are continually going on in the living cell. In these the protoplasm is broken down into simpler bodies, which, being of no further use to the plant, are thrown off or excreted. It

MINERAL FOOD.

is supposed that the oxygen taken in by respiration combines with the protoplasm, renders it unstable, and finally leads to its resolution into simpler substances.

Mineral Food.—Every plant when burnt leaves behind a certain quantity of mineral matter, technically known as "the ash." This consists chiefly of the elements potassium, magnesium, calcium, iron, sulphur, and phosphorus, and it has been determined experimentally that all these substances are indispensable for the proper nutrition of all green plants. These elements, in the case of the *Protococcus*, are in the most suitable state for its nutrition when combined as the nitrates, sulphates, and phosphates of potassium, magnesium, calcium, and iron. The last element is only necessary to those plants which contain chlorophyll, iron being a constituent of that substance.

Osmosis.—Since all such plants as the *Protococcus*, which consist of a single cell, are closed in on every side by a cell-wall unprovided with any visible opening, it seems at first sight difficult to understand how such organisms obtain their supply of food from without. This they are enabled to do by the peculiar property which their cell-walls, and which membranes generally possess, of permitting various substances in solution to pass or diffuse through them by a process called "osmosis." If a dry bladder is filled with water, the previously hard and rigid membrane becomes soft and pliable owing to the penetration of water into its substance; but the water does not pass through the bladder, it simply wets it. This peculiar property, exhibited by animal and vegetable membranes, of allowing liquids to permeate, without actually passing through them, enables such membranes to set up osmosis.

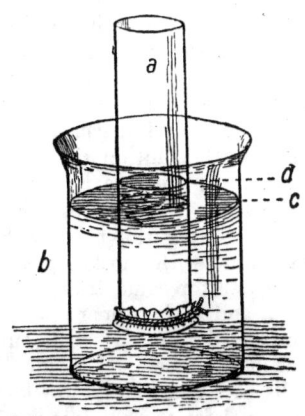

FIG. 25.—Osmotic Apparatus.

This process may be demonstrated in the following manner:— A wide glass tube (Fig. 25, a) is taken, and its lower end closed with a piece of bladder or parchment-paper. It is then partially filled with a strong solution of common salt or sugar, and supported in the vessel of water (b) at such a height that the level of the solution within the tube and that of the surface of the water outside (c) coincide. The whole arrangement is left at rest for a few hours, at the expiration of which the level (d) of the liquid in the tube will have risen considerably. The reason for this is that the water outside the tube permeates the membrane, and as the molecules of the salt present in the solution within the tube have a great attraction for the water molecules in the membrane, they draw them from the membrane into the tube. This passage of a fluid inwards is called "endosmosis." The membrane is also permeable to the salt solution, but to a much less extent than to water; consequently, the membrane is also wetted to a certain degree with the salt solution. The salt molecules present in this are attracted and drawn from the membrane by the water outside. This, which is the reverse of endosmosis, is called "exosmosis." What the relative proportions of these two actions shall be entirely depends upon the extent to which the membrane is permeable to the solutions with which it is in contact. Thus, in the preceding case, endosmosis is far greater than osmosis, because the membrane is much more permeable to water than to salt solution.[1]

If we conduct a similar experiment, closing the tube with a thin membrane of indiarubber, and using alcohol instead of salt solution, we shall find that the alcohol will pass through the membrane into the water, but that no water will pass through the membrane to the alcohol; indiarubber is permeable to alcohol, but absolutely impermeable to water. In the former experiment, as the salt solution rises in the tube, it is evident that a certain amount of pressure is developed, and in some cases this may be very great. This pressure can be more clearly shown by completely filling the tube with salt solution and closing its upper extremity with another piece of bladder. After the tube has stood in water for some time, both membranes will bulge out considerably, and if the upper one is pricked with a pin, a small fountain of salt solution will gush out, showing that considerable pressure has been developed inside the tube. It is in this way that the cells of living plants are kept in that state of turgidity which is essential for the efficient performance of their vital functions.

Crystalloids and Colloids.—Those bodies which are able to pass through a membrane by osmosis are termed "crystalloids;" amongst these the crystalline bodies diffuse very rapidly, others, such as peptone, with extreme slowness. Many substances when in solution are unable to diffuse through a membrane; for instance, if a solution of albumin (white of egg) is placed in the tube, and the tube placed in water as before, none of the albumin will pass through the membrane into the outside water. All those substances, such as gelatin, gum, &c., which behave in a similar manner are called "colloids" (glue-like bodies). The difference in the nature of bodies in this respect is one of great importance in plant life, for crystalloid bodies taken into the living cell are often transformed in its interior into colloids, in which state they

[1] The solution of substances in fluids is also due to the mutual attraction between their respective molecules. When a lump of sugar is placed in a vessel of water, and the whole allowed to remain at rest for some time, a perfectly homogeneous solution results; the sugar molecules rise up and become evenly distributed through the water molecules by virtue of their mutual attraction, in spite of the fact that the sugar molecules are held down by the force of gravity.

are incapable of passing out of the cell until they are again changed into crystalloids. It is by osmosis that the *Protococcus* is able to obtain its nutriment from the outside; since its cell-sap is richer in substances in solution than the water in which the plant is immersed, the molecules of the substances dissolved in the cell attract and draw in the molecules of water which permeates the cell-wall, along with the carbon dioxide and the salts dissolved in this water. On the other hand, the substances that are to be excreted are dissolved in the cell-sap which permeates the cell-wall, from which they are attracted and withdrawn by the water outside; but as the balance between endosmosis and exosmosis is in favour of the former, the cell is kept in a state of turgidity. The transference of carbon dioxide from without to within, and the converse with regard to oxygen, is explicable on the same principle.

It is, however, extremely questionable if the processes by means of which solid, liquid, and gaseous bodies are transferred from the outside of the living cell, and *vice versâ*, can be completely explained on purely physical principles; most probably some of those unknown manifestations of force which at present we designate vital are also at work. In experiments we can only employ dead membranes; the cell-walls of growing plants are living tissues.

Reproduction.—In the resting stage, which may be looked upon as the adult condition of the *Protococcus*, reproduction takes place by cell-division (see Fig. 24, p. 225). First the nucleus divides into two, then a new cell-wall commences to form, which eventually divides the organism into two halves (*b*). The nuclei of each half again divide, another cell partition being formed at right angles to the first (*c*), and the original cell is thus divided into four (*d*) cells, which then separate and become rounded off until they resemble the parent cell (*a*). It is in this way that cell-division usually takes place in the vegetable kingdom.

The *Protococcus* can also reproduce itself in an entirely different manner. When this is about to take place, the protoplasm contracts and shrinks from the cell-wall, and then commences to divide first into two, then into four, and sometimes into eight portions, each of which finally assumes a rounded form. These "zoospores,"[1] as they are termed, are then liberated by the rupture of the wall of the parent cell. Each zoospore becomes pointed (*e*), and throws out from the point fine hair-like projections called "cilia," which, by their rapid lashing movements, propel the organism along. Sooner or later each zoospore draws in its cilia, secretes a cellulose covering, and comes to a state of rest, when reproduction takes place in the ordinary manner by cell-division.

Dormant Vitality.—If the *Protococcus* is carefully and slowly dried, it preserves its vitality for a considerable length of time; and

[1] These afford instances of naked cells, they consist of a mass of protoplasm without any external covering whatever.

VEGETABLE BIOLOGY.

in this condition it is very readily disseminated by currents of air, hence its almost universal presence.

The Yeast Plant.—When a trace of yeast is mixed with a drop of water and examined under the microscope, it is seen to consist of a number of round or oval bodies, either isolated or joined together in groups, and varying in length from 3 to 10 micromillimetres (Fig. 26). Each of these separate bodies is an individual plant, which consists of a single cell, surrounded by a thin transparent cell-wall of cellulose or some closely allied substance.

The interior of the cell is almost completely filled with protoplasm, which is almost transparent in very young cells, but becomes more or less granular as the cells grow older. The protoplasm is soaked through with the cell-juice, which also permeates the cell-wall. It collects in one or more cavities in

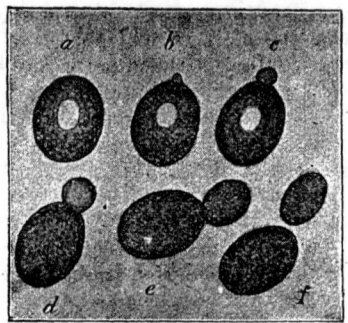

FIG. 26.—Budding Yeast.

the protoplasm, forming spaces having the appearance of small bubbles, which are termed "vacuoles." Unlike the *Protococcus*, the yeast plant contains no chlorophyll; consequently its protoplasm is much more easily examined. By carefully crushing yeast diluted with water between a cover-glass and slip with the blade of a knife, some of the cells will be ruptured and their protoplasmic contents squeezed out. By applying a little pressure to the cover-glass with the point of a needle when the preparation is under the microscope, it will be seen that the escaped protoplasm is of an elastic jelly-like nature. It may also be noted that the cell-wall, after the rupture of a cell, contracts, and this shows that, when the cells are entire, a certain amount of internal pressure is exerted on the cell-wall by the contents of the cell. If a little staining reagent, such as indigo or methyl blue solution, is run under the cover-glass, the dead protoplasm of the ruptured cells will take the stain freely, while that of the majority of the unruptured cells remains uncoloured. This rejection of staining matter is peculiar to living protoplasm, for this substance when dead exhibits a powerful attraction for many dyes. The number of dead cells in a sample of yeast may be readily detected by this curious difference in the behaviour of living and dead protoplasm. It is only necessary to place a little of the yeast mixed with a drop of a dilute staining fluid, such as methyl blue, under the microscope, when the dead cells are readily detected by the deep blue colour they assume. The yeast cell also contains a nucleus, but this is by no means such a prominent object as it is in most vegetable cells. It can, however, be rendered visible by staining with osmic acid or hæmatoxylin and picric acid; and Hansen has distinctly seen it without any special treatment, such as staining, in yeast cells which have been growing for some time on the surface of wort. According to the observations of Jansens, the nucleus of the yeast cell divides previous to the division of the protoplasm of the cell taking place, either in the formation of a new bud or of spores.

Nutrition.—If the yeast plant is placed in a solution similar to that in which the *Protoccocus* flourishes, but to which a little ammonium tartrate has been added instead of nitrate, though it will live for a time, it eventually perishes from starvation. Yeast possesses no chorophyll cells, and cannot therefore obtain the carbon necessary for its nutrition from the carbon dioxide dissolved in the liquid. Here, then, is a most important distinction between plants which bear chlorophyll cells and the Fungi, to which yeast belongs, which do not. Carbon must be supplied to them in the form of a substance, such as a carbohydrate, which they are able to assimilate. If to the former solution a little sugar be added, the yeast will then grow and multiply freely, showing that it is now provided with all the elements necessary for its nutrition. The Fungi, being devoid of chlorophyll cells, are dependent upon those plants which bear chlorophyll cells for their supply of carbonaceous nutriment, and in this respect they resemble animals. In ordinary brewing operations the yeast obtains its carbonaceous food supply from the sugar in the worts, and this is derived from the starch which has been manufactured by the chlorophyll cells of barley or some other green plant.

As the protoplasmic contents of the yeast cell contain nitrogen, this element must also be supplied to the plant in a form which it can assimilate, and yeast manifests certain peculiarities in this respect. The *Protococcus*, and the chlorophyll-bearing plants generally, are able to derive their supply of nitrogen from a nitrogenous compound of a purely inorganic nature such as potassium nitrate, but the yeast plant is quite unable to do this. If, however, ammonium nitrate be substituted for this salt, then the yeast increases and multiplies, showing that its wants in this direction are satisfied. If, instead of the nitrate, a compound of ammonium with an organic acid such as tartaric acid is employed, then the yeast grows much more luxuriantly; and a still further improvement in this direction is exhibited when an amide body, such as asparagine, is substituted for the ammonium tartrate. It has also been found that peptones, one of the series of products of the hydrolysis of the proteids, are readily assimilated and utilised by yeast; and here again the plant resembles an animal organism. Yeast, when growing in brewer's wort, obtains its nitrogenous food supply chiefly from the amides present in the wort; probably also to a small extent from some of the products of the hydrolysis of the proteids of the malt. It also contains phosphorus in considerable amount; this it derives from phosphates, an abundant supply of which are always present in wort. In addition to the foregoing substances, small quantities of the elements magnesium, potassium, and calcium are indispensable, and wort generally contains these bodies in quantity over and above the requirements of the yeast.

Respiration.—The yeast plant, in carrying on its vital functions, expends a certain amount of energy, which it obtains by the oxidation of carbon through the instrumentality of the protoplasm. For this purpose a supply of oxygen, which the plant obtains by respiration, is necessary; it takes in oxygen and gives out carbon dioxide; but with reference to this process it exhibits a certain marked peculiarity. Since yeast has to spend a considerable portion of its existence in fermenting wort where it is entirely precluded from obtaining a supply of free oxygen, it is endowed with the power of taking in a store of this gas in order to bridge over those intervals. During these it derives the energy necessary for fulfilling its vital functions from the carbon burnt up by this stored-up oxygen. When a fresh opportunity arrives, it again absorbs a fresh supply of the gas. In this respect, only in a greater degree, it resembles some of the aquatic air-breathing animals, such as the whale, which are able to take in a supply of air sufficient to enable them to remain under water for a considerable time. When yeast is grown under the conditions of complete exclusion of air, its growth gradually becomes more and more languid, and the yeast finally dies.

Excretion.—The products of the breaking down of the protoplasm of the yeast, being of no further use to the plant, are excreted, and remain dissolved in the fluid in which the yeast has been growing. These, so far as has been ascertained, consist of leucine, tyrosine, xanthine, guanine, and sarcine, bodies which have all been derived artificially from the proteids by hydrolysis. The several enzymes which yeast secretes, and which have the properties, individually, of hydrolysing maltose to glucose, cane-sugar to invert sugar, &c., are also probably decomposition products of the protoplasm.

Fermentation.—Yeast possesses the remarkable property of exciting the alcoholic fermentation, that is, of converting sugar into alcohol, carbon dioxide, and several other products. This faculty, which renders yeast indispensable to the brewer, is only incidentally referred to here, but it is one on which we shall have much to say afterwards.

Reproduction.—Yeast has two distinct methods of reproduction; the first of these, that is, by budding, is by far the more usual one; hence the name "*Budding Fungi*," which is often applied to the members of this class. The second, and less usual method, is by the production of spores, which form in the interior of the cell.

When yeast grows in a fluid containing an abundance of food, it invariably increases by budding. As more food is absorbed by the yeast cell than is again destroyed in yielding the energy required for carrying on its vital functions, there is a gradual accumulation of substance in the interior of the cell. Its contents, therefore, gradually increase in bulk, and as the cell cannot grow beyond a certain size, some provision must necessarily be made for this surplus. This is effected by the protrusion of a bud, which

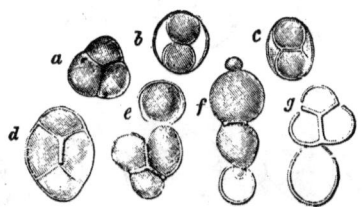

FIG. 26*a*.—Yeast Spores (*Hansen*).

gradually enlarges, and finally separates from the parent cell. When budding is about to take place in a cell, a small portion of the cell-wall becomes softened at some particular point, probably by the action of an enzyme secreted by the protoplasm. The protoplasm then pushes the softened portion of the cell-wall before it, and forms a small prominence (Fig. 26, p. 232) which gradually increases in size to form a bud, *c*. This bud, when it has nearly attained the size of the original cell, becomes detached, *f*, and forms a separate individual, which, in its turn, reproduces itself by budding. In some species of yeast the newly developed plant does not separate from the parent cell immediately it reaches its maximum size, hence a number of cells remain attached to one another for longer or shorter periods, forming cluster-like groups.

Sporulation.—When healthy and vigorous yeast is suddenly deprived of nutriment and kept in a moist condition, and is abundantly supplied with air, it forms spores. This method of reproduction may be illustrated by placing the yeast in a thin layer on the surface of a slice of potato, or on the surface of a block of plaster of Paris, kept damp by having its lower portion immersed in water. Sooner or later, according to the nature of the yeast, the temperature, &c., a series of changes begins in the protoplasmic contents of the cell, which ends in the formation of from one to six or eight glistening cells which are called "spores" (Fig. 26a). These, when placed in a suitable nutritive medium, begin to germinate by throwing out one or more buds, which eventually become yeast cells. The newly-formed yeast cells then reproduce themselves in the ordinary manner by budding (Fig. 26b).

Species and Varieties of Yeast.—For a long period the yeasts were arranged in various species according to the differences in

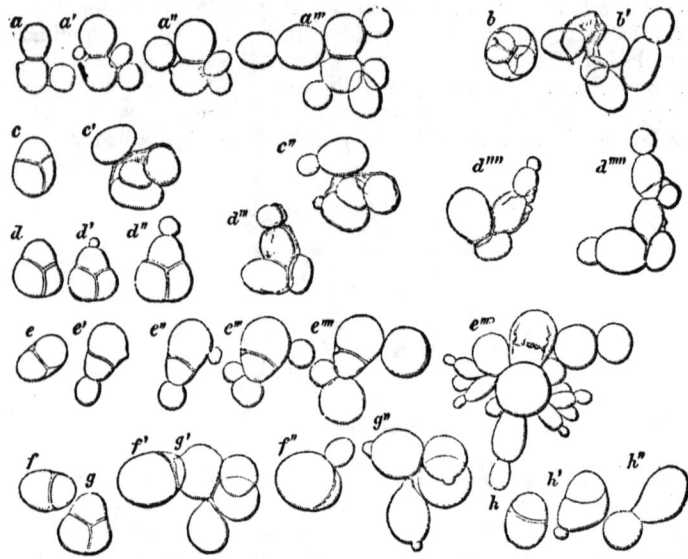

FIG. 26b.—Germination of Yeast Spores (*Hansen*).
Each letter in the figure refers to the same cluster of spores in the different stages of development.

form which they exhibited when viewed under the microscope. Those of a more or less spherical shape were termed *Saccharomyces cerevisiæ;* those of an elliptical form, *Saccharomyces ellipsoideus;* those having a sausage-shaped form, *Saccharomyces Pastorianus.* After Hansen had introduced his beautiful method of growing yeast from a single cell, by means of which it first became possible to secure cultivations which consisted absolutely of one variety, it was found that any method of distinguishing the species of a yeast from its shape alone was misleading, for each species of yeast was able, under different conditions of cultivation, to assume the forms of the others. Hence other and more perfect methods of discrimination had to be devised, such as that which depends on the differences in the time taken by the various species of yeast to form spores at the same temperature, described farther on. It was also found that many organisms which did not form spores had, from their close resemblance in appearance to the yeasts, been included under the denomination *Saccharomyces*. It was now, however, determined to restrict this designation to those organisms only which were able to produce spores.

Saccharomyces cerevisiæ.—Under this term are included those yeasts which have been employed from time immemorial in brewing operations, and which are, therefore, spoken of as the "cultivated yeasts." They are divided into two great classes—the top fermentation and the bottom fermentation yeasts, so called because the former rise and collect on the surface of the fermenting fluid, while the latter fall to the bottom. They are evidently distinct species, for, after much experimenting, it has not been found possible to convert the one into the other. Of each of these two species there are many varieties, which differ remarkably in their respective properties, such as the degree of attenuation they are able to carry the wort down to, the flavour of the beer they yield, &c. The top fermentation yeasts readily form spores, the bottom ones only with difficulty.

The Wild Yeasts.—In contradistinction to the cultivated species, there are numerous other yeasts met with in Nature, some of which occasionally intrude themselves into the brewery, and there cause trouble. Among these are Hansen's *Saccharomyces ellipsoideus* I. and II., also *Saccharomyces Pastorianus* I., II., and III., and several others. Some wild yeasts exhibit a great tendency to grow on the surface of the liquid in the form of a film, hence it would appear that they require a free supply of air for their welfare. Many of them are able to ferment maltose, glucose, and cane-sugar; others cannot. One variety of the film-forming yeasts, the *Saccharomyces anomalus*, readily causes fermentation in wort, to which it communicates a fruity odour. This yeast produces peculiarly shaped spores, which have the form of a hat.

Saccharomyces apiculatus.—This yeast (Fig. 27) is readily distinguishable by its lemon-shaped cells, in fact, it is the only yeast which can be recognised with absolute certainty from its form alone. The buds invariably form at the pointed ends of the cells, and the newly detached cells preserve for some time the oval shape of an ordinary yeast cell. According to Grönlund, these yeasts, which, when living in their natural state, pass a portion of their time embedded in the soil, during that period produce spores; they do not, however, yield spores when cultivated under artificial conditions. *Saccharomyces apiculatus* contains no inverting enzyme, consequently it cannot ferment any of the disaccharide sugars, but it readily ferments glucose and lævulose. This yeast occasionally

238   VEGETABLE BIOLOGY.

invades the worts, especially in autumn, but it rarely produces trouble there unless it happens to be present in very large quantity.

The Schizo-saccharomycetes.—These yeasts do not throw off buds after the manner of the ordinary yeasts, but rather by a process of fission, such as that which obtains with the bacteria. The elongated cells, which somewhat resemble those of an *Oidium*,

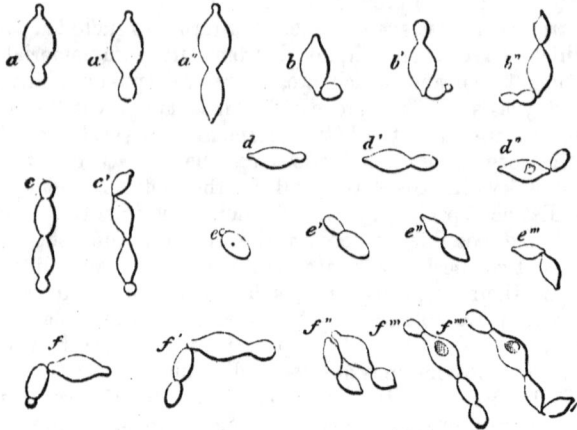

FIG. 27.—Saccharomyces apiculatus (*Hansen*).

FIG. 28.—Schizo-saccharomyces pombe (*after Lindner*).

when about to divide, form a partition-wall at right angles to the long axis of the cell; the cell then breaks into two portions at this division-wall, and the square ends thus left become afterwards rounded (Fig. 28). To this genus belong the *Schizo-saccharomyces pombe*, isolated from millet-beer which is produced in Central Africa; and the *Schizo-saccharomyces octosporus*, so called from its tendency to form eight spores in each cell. These yeasts appear to be able to ferment some of the dextrins, which the cultivated yeasts are unable to do.

The Mycoderms.—These organisms are very similar in appearance to the film-forming yeasts, but since they do not form spores, they are not true yeasts. They grow with enormous rapidity on the surface of wort or beer, provided they are supplied with the necessary amount of air, of which they require a considerable quantity, and there form white or greyish films, which often have a wrinkled appearance. When growing on the surface of an alcoholic fluid, they decompose the alcohol into carbon dioxide and water. When immersed in a saccharine fluid, some of them, such as the *Mycoderma cerevisiæ*, are able to cause alcoholic fermentation, small amounts of alcohol (under 1 per cent.) being produced. As a rule, they appear to do no harm when growing on the surface of beer, though occasionally they cause turbidity.

The Torulæ.—These organisms were first described by Pasteur; they increase by budding after the manner of the *Saccharomycetes*, which they much resemble in shape; their cells are, as a rule, nearly spherical. The *Torulæ* are generally very much smaller than yeast cells, though occasionally they attain the same size. None of those described by Pasteur were able to induce the alcoholic fermentation; but of the seven varieties studied by Hansen, some secrete invertase, and some few of these, though not all of them, are able to ferment cane-sugar.

THE BACTERIA.

The bacteria, which are often termed "micro-organisms" or "microbes," are a class of exceedingly minute organisms met with in the greatest abundance in nature. When an organic infusion of almost any kind, vegetable or animal, is left exposed to the air for a short time, it becomes invaded by hosts of these organisms. They are objects of considerable import to the brewer, for, as we shall see farther on, they are capable of causing him an immense amount of damage in many ways.

Bacteria vary much in shape, and may be divided, according to their form, into three principal classes. To the first of these belong those micro-organisms which in a perfect state of development have the form of little spherules; they are called *Cocci* (Fig. 29, *a*). Those of the second class have the shape of cylindrical rods, and are distinguished as *Bacilli* (*e*). The third class have the form of a spiral, and are called *Spirilla* (*f*). The three classes may be likened in form, as De Bary suggests, to a billiard-ball, a lead-pencil, and a corkscrew. The bacteria are, so far as is known at present, the smallest organisms in nature, and require for their examination the highest powers of the microscope. Being much smaller than yeast cells, the two are readily distinguishable from one another.

Structure.—Bacterial organisms consist of a mass of proto-

240   VEGETABLE BIOLOGY.

plasm surrounded by a cell-wall of cellulose; the outside of this envelope is coated with a mucilaginous layer more or less swollen by the absorption of water. In consequence of this, the outline of these organisms, as seen under the microscope, never presents that sharply defined border that a yeast cell does. The protoplasmic contents of bacteria are, as a rule, colourless; but to this there are exceptions. Some bacteria contain starch (granulose), and these are coloured blue on the application of a solution of

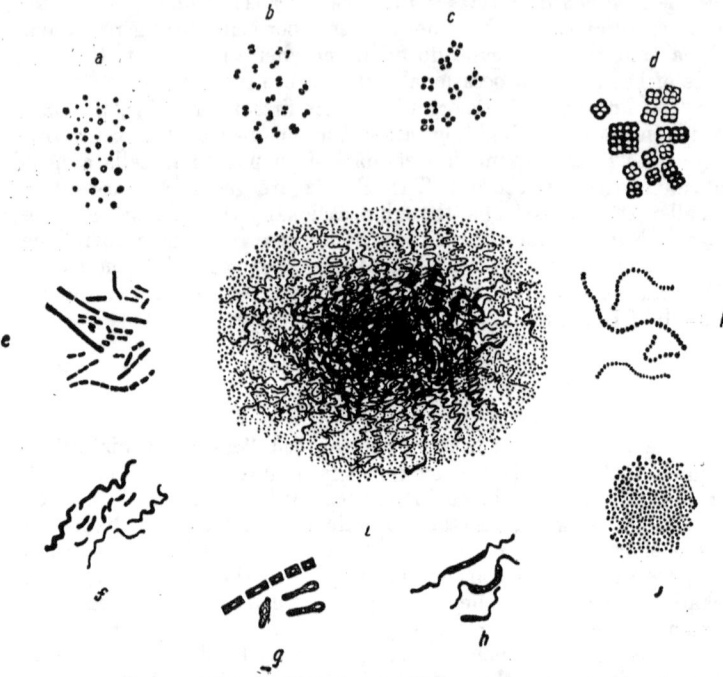

FIG. 29.—Bacteria (*after Baumgarten*).
(*a*) Cocci; (*b*) Diplococci; (*c*) Tetracocci; (*d*) Sarcinæ; (*e*) Bacilli; (*f*) Spirilla; (*g*) Bacteria with spores; (*h*) Bacteria with flagella; (*i*) Zoogleæ; (*j*) Staphylococci; (*k*) Streptococci.

iodine. The protoplasm of some bacteria is studded with grains of sulphur, and ferric oxide is deposited in the cell-walls of others.

Reproduction by Fission.—Bacterial organisms, like yeast, have two separate modes of reproduction. The first, and by far the more common mode, is by division, and this takes place in the following manner:—When a bacterium has grown to a certain length, a constriction, situated at right angles to the long diameter of the organism, appears in its middle; this gradually deepens, and at

REPRODUCTION BY FISSION.

last divides the microbe into two separate individuals. Even in the micrococci a slight lengthening takes place before the organism divides. Owing to this peculiar method of reproduction, by splitting, as it were, the bacteria are collectively known as the *Schizomycetes* or *Fission Fungi*. In the *Bacilli* and *Spirilla* division always takes place in a direction at right angles to the long diameter of the organism. In some of the micrococci the splitting-up proceeds simultaneously in two directions, the plane of the second division being at right angles to that of the first; in this way one micrococcus is converted into a group of four, arranged in the form of a square and called a *Tetracoccus* or *Pediococcus* (Fig. 29, c). In a few other species cell-division takes place simultaneously in three planes, all of which are at right angles to one another, and, as a consequence of this, small packets, each of which consists of eight micrococci, are produced (d); these are known as the *Sarcinæ* or *packet* organisms. In many bacteria which divide in the most usual way, that is, in a direction at right angles to their long axis, the two cells after division do not separate at once, but remain attached to one another for longer or shorter periods. In this way chains of bacteria are formed, each of which consists of at first two, then four, then eight individuals, and so on. When large numbers hang together in this way, the chain often has the appearance of a long thread or filament. When micrococci remain attached to one another after division, the chain has the appearance of a necklace; such forms are called *Streptococci* (k). Sometimes the tendency to adhere only extends to the pair first formed; in this way we get the forms called *Diplococci* (b). When the micrococci separate immediately after they have divided, and lie alongside one another haphazard, they present the appearance of a bunch of grapes; such varieties are entitled *Staphylococci* (j). In describing the bacilli we often speak of the organisms as long or short, thick or thin bodied. Bacteria, especially when grown on the surface of fluids, often become united to a tough gelatinous mass. In assuming this state the exterior mucilaginous coatings of the organisms become considerably swollen and finally fuse together, the bacteria being retained in a perfectly regular manner throughout the mass. This is known as the zoogleal state (i).

Reproduction by Spores.—In addition to the method of reproduction by cell-division, bacteria, under certain conditions, form spores (Fig. 29, g). One of the chief of these is the exhaustion of the nutritive materials of the medium in which the organisms have been growing; but other special conditions are also necessary, and these vary considerably with the different species. The power of forming spores is almost entirely confined to the bacilli; it is only occasionally met with in the spirillæ, and some of the sarcinæ are also said to possess it. When a bacillus begins to sporulate, a portion of its protoplasmic contents assumes a granular appearance, and in this a cell-wall gradually forms, which encloses a portion of the protoplasm; this gradually becomes more and more homogeneous, and eventually forms the spore. The contents of the spores are of a highly refractive nature, and consequently present a brilliant, glistening appearance when seen under the microscope. The motile bacteria invariably become motionless when about to sporulate. In some bacilli the spore forms in the middle of the protoplasm, the remainder of which is afterwards absorbed; in consequence of this the organism becomes spindle-shaped. In other bacilli the spore forms at one extremity, and when this is the case the organism acquires the form of a drumstick. When the spore has completely formed, the remainder of the organism disintegrates and disappears, leaving the spore isolated. It remains in this condition until chance places it in a suitable nutritive medium, where, after a shorter or longer period of rest, it germinates, reproduction being afterwards carried on in the ordinary manner by cell-division. The membranes which invest the spores of bacterial organisms are of an exceedingly tough and resistant nature, and this enables the spores to withstand the action of agents which would rapidly prove fatal to the organisms when in their fully-developed state. Tyndall found that the spores of the hay bacillus (*Bacillus subtilis*) were capable of germinating after they had been boiled for several hours in water. As the spores of the bacterial organisms are extremely minute, they are disseminated and carried to considerable distances by air-currents, hence their universal presence in the atmosphere.

Nutrition.—As the vast majority of bacteria possess no chlorophyll, they are unable to derive their supply of carbon from the carbon dioxide of the atmosphere. According to Nägeli, they can appropriate the carbon of almost all compounds which contain this element, providing these are soluble in water and are not actually poisonous. Some bacterial organisms met with in the soil can even obtain their supply of carbon from carbonates. The bacteria are able to derive their supply of nitrogen from an enormously large variety of substances. One of the most suitable of these is a solution of a proteid, or, in those cases where the organism does not secrete a proteolytic enzyme, a solution of the hydrolytic products of the proteids. Substances of this nature may be replaced by amides such as asparagine, leucine, &c., or by ammonia and its salts, and in most cases an inorganic salt of nitric acid may take the place of ammonia. Some varieties of bacteria found in the soil are not only able to derive their own supply of nitrogen from the air, but also to hand over a certain quantity of it to certain crops, such as peas, beans, clover, vetches, &c. These plants form nodules on their roots in which these nitrifying organisms congregate. Consequently a soil is richer in nitrogen after a crop of any of these leguminous plants has been grown on it than it was before.

Respiration.—Many of the bacteria cannot carry on their vital functions without a supply of free oxygen, but with others a small amount of this gas immediately impairs their functions, and a larger quantity proves fatal to the organisms. The former are termed "aërobian," the latter "anaërobian." A third class, called "facultative aërobians," seem to occupy a middle position; they are able to functionate either with or without a supply of oxygen.

Excretion.—It is certain that the substances formed by the breaking down of the protoplasm of the bacterial organisms are excreted. Little is known as to the nature of these products.

Secretion.—Many of the bacteria are able to secrete enzymes of various kinds, proteolytic, amylolytic, &c. A number also secrete pigments of various colours, such as yellow, green, blue, orange, &c. The *Micrococcus prodigiosus*, for instance, when grown on a starchy medium (such as a slice of moist bread), secretes a blood-red pigment.

Fermentation and Putrefaction.—The most remarkable property possessed by bacterial organisms is the power of inducing a number of profound changes in the media in which they grow. By virtue of this they are able to decompose an amount of matter enormously in excess of that which they require for their own nutrition. This extraneous chemical work is called "fermentation" when it is unaccompanied by an offensive odour, and especially when the products yielded are of a useful nature, as in the fermentation caused by the *Bacterium aceti*, where alcohol is transformed into acetic acid. When, on the other hand, the change is accompanied by an offensive odour, the process is called "putrefaction," and, as a rule, this takes place when the substratum is of an albuminous nature. In putrefaction a number of very malodorous compounds are usually formed, and in the case of putrefying animal substances, as is occasionally seen in tinned meats and fish, in cheese, &c., bodies called "ptomaines" are produced, which are of an extremely poisonous nature. But the nature of the chemical changes which bacteria are able to effect in the medium in which they flourish are nearly as numerous as the organisms themselves, and this is truly remarkable when we consider that the essentials of plant life, that is, the protoplasm and cell-wall, are apparently identical in every organism. Thus we have bacteria which transform sugar into lactic acid, others which transform lactic acid into butyric acid, others convert alcohol into acetic acid; some effect the transformation of the ammonia present in the soil into nitrites; while another set carry this transformation a stage further, and convert the nitrites into nitrates. When organic substances have been broken down to a certain stage by one class of micro-organisms, the products of these are seized upon and further broken down by micro-organisms of another class, and this proceeds until at last the original matter is entirely converted into ammonia, water, and carbon dioxide, substances which are again available for the food supply of a new generation of plants. The micro-organisms are, in fact, nature's scavengers, for, as Sachs observes, "If there were no fungi, the entire surface of the earth would be covered with dense layers of the bodies of plants and animals which had accumulated for thousands of years." If such layers had been permitted to form, they would long since have locked up in themselves the elements necessary for the building up of new plants and animals, and life, ages since, would have become an impossibility. Were micro-organisms suddenly annihilated, the return to the atmosphere and to the mineral kingdom of all that has ceased to live would be totally suspended. It is by means of this peculiar power of effecting the decomposition of masses of matter enormously in excess of their own nutritive requirements, which micro-organisms are endowed with, that they are able to perform this absolutely necessary and indispensable work in the economy of nature. If they only appropriated the quantity of matter necessary for their own nutrition, the result would simply be an immense accumulation of the dead bodies of these organisms.

Bacteria are omnipresent, and have the unfortunate habit of intruding into places in the brewery where their presence is highly undesirable. They are invariably present on the surface of barley and malt, the former of which, when in a moist condition, forms an excellent breeding-ground for bacterial and other organisms. They also invade the wort and the finished beer. As the changes which bacteria bring about are often of an extremely prejudicial nature, one of the brewer's principal cares is to keep these invaders away from his materials as far as possible, and to hold those organisms in subjection which unavoidably do gain access to them. All accumulations of moist organic materials, such as damp malt, spilled yeast, moist grains, &c., form excellent breeding-grounds for bacteria. Such masses of matter afterwards become dry and get more or less pulverised. In this state they are liable to be carried away by currents of air and disseminated in all parts of the brewery, carrying with them the dried organisms, which often become deposited in places where they may cause considerable mischief. Cleanliness is one of the brewer's most powerful aids in keeping down these objectionable intruders. Like yeast, bacteria preserve their vitality for considerable periods when slowly dried, and, on account of their exceedingly minute size, they are especially liable to become suspended in the air and wafted about by air-currents.

Locomotion.—Many bacterial organisms possess the power of locomotion, but this is almost exclusively confined to the bacilli and spirilla, only one motile micrococcus, the *Micrococcus agilis*, being known. The motion is brought about by one or more whip-like filaments called "flagella," which are generally attached to the ends, more rarely to the whole surface of the organism (Fig. 29, *h*). Under the ordinary conditions of microscopic observation the flagella cannot be seen. They can, however, be rendered visible by particular methods of staining. They propel the bacteria along by

BACTERIA MET WITH IN BREWING.

their rapid undulating movements, and in some of these organisms the rate of motion is so great as to cause considerable inconvenience to the observer when examining them in the living state; one is no sooner got fairly into the field of the microscope than it rushes out of sight.

THE PRINCIPAL BACTERIA MET WITH IN BREWING OPERATIONS.

Fortunately for the brewer, of the large number of bacteria known, but few are able to live in hopped wort, and many of those which can are unable to survive the fermentation with yeast which the wort is subjected to in the production of beer; a few of the latter are able to exist in beer if they are added to it after it has passed through the primary fermentation. The reason for this is that yeast, when in an active state of fermentation, possesses the peculiar power of protecting itself against the attacks of bacterial organisms, and so much is this the case, that at the end of the primary fermentation very few living organisms beyond the yeast itself are to be found in the beer. The case is, however, exactly the reverse when the yeast is in a resting condition, such as it is in when stored for subsequent use. Those organisms, the attacks of which the yeast was so well able to resist when in the active state, now obtain the mastery; they exercise their fermentative actions upon it and set it into a state of decomposition. Those bacteria which are able to survive the fermentation appear to have little or no action upon yeast when in the resting stage. Of the bacteria met with in wort and beer, some are aërobic, others anaërobic. The latter are the more dangerous class, since the conditions of rigorous exclusion of air under which beer is kept are without influence on them.

Bacteria which can Exist in Wort, but cannot Survive the Primary Fermentation.—Many of these, to which Hansen gave the generic name of "wort bacteria," are only able to thrive in unhopped wort. The bactericidal action possessed by some of the constituents of the hop, notably the soft resins, completely arrests the growth and development of many micro-organisms.

Bacterium termo.—Under this designation a number of organisms (Fig. 29a) very similar in appearance, which invariably appear in great abundance in any vegetable or animal infusion that is left exposed to the air for a short time, have been confused. They are all motile. A number of these have been examined and described by Lindner[1] under the following names:—

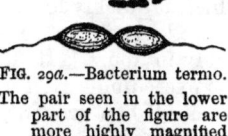

FIG. 29a.—Bacterium termo. The pair seen in the lower part of the figure are more highly magnified and stained, so as to show the flagella.

---

[1] *Mikroskopische Betriebskontrolle Gährungsgewerben*, p. 243.

Termobacterium lutescens.—It had the form of short rods about 1μ in breadth, pointed at the ends and slightly contracted in the middle. Under certain culture conditions it developed involution forms, which consisted of indented chains or of long filaments.

On meat-broth gelatin plates, the colonies had at first sharp, well-defined borders, but afterwards threw off radiating limbs. The gelatin was liquefied in five or six days, the culture being kept at the ordinary temperature. They grew rapidly in wort at a temperature of 15° C. (59° F.), and caused it to become turbid in twenty-four hours; a film then formed on the surface of the fluid at its margins, and a pronounced odour of celery was developed. In four weeks the pellicle had increased to a thick, yellow film, the wort was then clear, but a considerable glutinous deposit of a dirty brownish-yellow colour had formed. In beer which had been infected with a vigorous culture of this bacterium, and which had been preserved in a well-corked bottle for two and a half months at a temperature of 50° F., some of the organisms were found alive, and this shows that, when added to beer after fermentation, they are able to survive. When, however, they were added to wort previous to fermentation, all were found to have perished at its termination. A few of the bacteria added to pressed yeast caused it to putrefy in a few days.

Termobacterium fuscescens.—This bacterium closely resembled *T. lutescens* in appearance, motility, and the power of causing turbidity in wort and in developing an odour of celery. It was specially characterised by being able to form lactic acid in hopped wort and beer. Old cultures on wort and gelatin had a brownish colour, surrounded by a zone of similarly coloured gelatin; hence the name of the bacterium.

Termobacterium iridescens.—This bacterium, when cultivated in wort, caused turbidity and a celery-like odour; hydrogen was also evolved in small bubbles. Unlike its congeners, it was immotile. On meat-broth gelatin it formed colonies which had a bluish, iridescent colour, somewhat in appearance like fish-scales. It did not grow nearly so vigorously on wort gelatin.

Termobacterium erythrinum.—Grew luxuriantly in wort, on which it quickly formed a reddish film. In shape it resembled the *T. iridescens*, the cells being generally grouped in pairs, more rarely in fours. It did not grow readily on meat-broth gelatin, but exceedingly well on wort gelatin, which it liquefied in a short time. When growing in wort or glucose solution, a slight evolution of gas was observed.

The Butyric Acid Ferments.—These rod-shaped organisms (Fig. 30) were first studied by Prazmowski, who, under the supposition that they were of one species, termed them *Clostridium butyricum*. They readily form spores. From the more recent observations of Beyerink,[1] there are several species of

---

[1] Koch's *Jahresbericht*, 1893, p. 258.

butyric acid ferments, of which he describes several under the designations *Granulobacterium butyricum*, *G. saccharo-butyricum*, *G. lacto-butyricum*, *G. polymyxa*. They are all facultative anaërobic organisms, and, when grown under exclusion of air, become either partially or completely filled with granulose; hence their name. When in this condition they are coloured blue by iodine. An exceedingly unpleasant rancid odour is developed in fluids in which they are present. As they are very sensitive to the bactericidal action of the hops, and require a somewhat high temperature for their development (about 122° F.), they are but rarely met with in brewing operations.

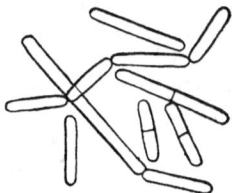

FIG. 30.—Butyric acid Bacteria.

Bacillus viscosus, No. 1 and No. 2.—Under these names Van Laer has described[1] two species of rod-shaped organisms which are able to induce what is termed "ropiness"[2] in wort and beer. They resemble one another closely in appearance, and, when grown on meat-broth gelatin, form brownish-coloured colonies with well-defined borders. They do not liquefy the gelatin. No. 1, when introduced into sterilised beer wort, induces viscidity in twenty-four hours, and in forty-eight hours the wort is converted into a viscid mass as coherent as white of egg, carbon dioxide being meanwhile evolved. The surface of the wort becomes studded with pale-yellow, glutinous, branched patches. In cultivations of No. 2 this last appearance is absent, and there is less evolution of gas. In closed bottles the first variety forms mucilaginous masses, in which the rods sporulate; the second variety does not behave in this manner. The ropy condition is not induced in worts at temperatures below 44° F.; the action proceeds most rapidly at 91°, and is arrested at 107° F.

Bacteria which are Able to Survive the Alcoholic Fermentation.—The Lactic Bacteria.—These, of which there appear to be several species, are anaërobic rod-shaped organisms, much resembling *B. subtilis* in appearance, and are often found in beer which has turned sour. They were first described by Pasteur. One of these has been specially studied by Van Laer, who named it *Saccharobacillus Pastorianus*. It is able to grow both in wort and beer, the latter it first renders turbid, then acid, and at the same time communicates an unpleasant odour and flavour to it. This bacterium ferments some of the maltose and dextrin of the

[1] *Zeit. f. Brau.*, 1890, p. 11.
[2] When beer enters into this condition, it at first becomes turbid, and afterwards so viscid that it resembles oil when poured from one vessel to another. Later on the viscidity increases to such a degree that the vessel containing it may almost be inverted without the beer running out. On introducing and withdrawing a rod, the portion of beer which remains attached to the rod is drawn out into a long string.

wort, yielding lactic, formic, and acetic acids, together with ethyl alcohol and some of its higher homologues. Another probably different species has been isolated by Lindner from German white beer, and another by Hayduck from distillery mashes. The development of the lactic ferments is either partially or completely arrested by the bactericidal action of the soft resins of the hop; consequently they are most likely to appear in lightly hopped beers.

The Acetic Bacteria.—These organisms, which occasionally make their appearance in beer, are not nearly so sensitive to the

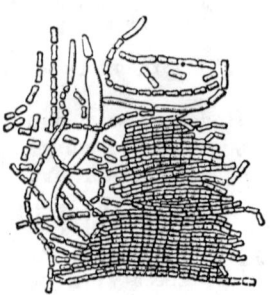

FIG. 31.—Bacterium aceti (*Hansen*).

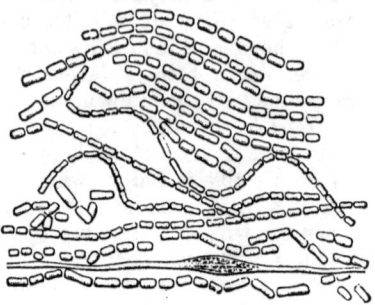

FIG. 32.—Bacterium Pastorianum (*Hansen*).

antiseptic influence of the hop as the lactic ferments. Their presence may be recognised by the vinegar-like odour which they communicate to the beer. Being highly aërobian organisms, they cannot develop to any extent without a liberal supply of air. The acetic ferments have been very thoroughly investigated by Hansen,[1] who was able to isolate three varieties, which he respectively named *Bacterium aceti*, *B. Pastorianum*, and *B. Kützingianum* (Figs. 31, 32, and 33). When cultivated in beer, at a temperature of 34° C. (93° F.), they all form pellicles on its surface, which differ markedly in appearance. Beers infected with any of the three bacteria remain clear

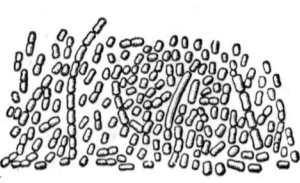

FIG. 33.—Bacterium Kützingianum (*Hansen*).

at the cultivation temperature; also, in the case of the two former, after the beers have been cooled down to the ordinary temperature, those infected with the last bacterium become turbid on cooling. All three bacteria assume very varied involution forms, the shapes being influenced to a great extent by the

---

[1] *Compt. Rend. Lab. Carlsberg*, 1894, p. 182.

temperature at which the bacteria are grown, as may be seen from the accompanying figures. When portions of the several films are placed under the microscope and treated with iodine solution, the jelly-like substance in which the *B. aceti* are imbedded is simply stained yellow; that of the other two is coloured blue. Zeidler[1] has described two acetic bacteria which he distinguishes as *B. aceti albumosum* and *B. aceti friabile.* The former is probably identical with Hansen's *B. aceti*, but the second does not seem to resemble any of Hansen's varieties. Both are coloured yellow by iodine. The former has a strong tendency to produce involution forms, though the latter yields them much less readily. Both forms, when cultivated in hopped wort, form pellicles, that of the former being firm, tough, mucilaginous, coherent, and not easily broken to pieces on shaking the vessel; that of the latter is of a white, loose, flocculent nature, readily breaking to pieces on the slightest movement. When *B. aceti albumosum* is added to wort, either previous to or during fermentation, or to finished beer, it causes at first turbidity; after the beer has been kept for some time it becomes acid, and assumes a viscid condition, much deposit of a tough mucilaginous nature being formed. Similarly, fluids infected with *B. aceti friabile* exhibit turbidity, but not to such a marked extent; less acid is formed, and there is an entire absence of viscidity.

The Sarcinæ.—This name was first applied by Hansen to a number of organisms met with in brewing operations. They have been investigated by Lindner, who divided them into three classes. The first of these grow by division in two planes, and are termed *Pediococci;* four varieties are described, *P. cerevisiæ, P. acidi lactici, P. albus,* and *P. viscosus.* The second class includes the *Sarcina candida, S. rosea,* and *S. aurantiaca,* which usually increase in a similar manner, but when grown under certain conditions increase by division in three planes. The last class, which form the true *Sarcinæ*, invariably grow by the last method of division. Three varieties of these are described, *Sarcina flava, S. lutea,* and *S. maxima.* Film-formation has been observed with *P. cerevisiæ* and *S. aurantiaca;* and all, with the exception of *S. maxima* (which has not been studied), produce lactic acid, *P. acidi lactici* being the most energetic in this respect. *P. viscosus* produces at times ropiness in beer. Another variety, *P. sarcinæformis*, has been described by Richard,[2] which closely resembles *P. cerevisiæ* in appearance. It produces lactic acid in considerable quantity, and also small quantities of alcohol; it may also cause the beer to become turbid and to acquire an unpleasant flavour. This, the *P. cerevisiæ*, and *P. viscosus* are regarded by Lindner as the dangerous species; and it is asserted by Lasché that *S. flava, S. aurantiaca,* and *S. alba* have caused trouble in American breweries. The former three varieties, when cultivated

[1] *Woch. f. Brau.*, 1890, p. 1213. [2] *Zeit. f. Brau.*, 1894, p. 257.

on meat-broth gelatin, form colourless colonies, which grow very slowly, and do not liquefy the gelatin. The other varieties either grow much more luxuriantly or form coloured colonies. The sarcinæ are motile organisms, though it is not known with absolute certainty whether they possess flagella or not; from the lively movements of some of them, it is extremely probable that they do.

THE MOULD FUNGI.

A moist organic substance of almost any kind, such as bread, boiled potato, jam, fruit, damp grains, damp barley, &c., when exposed to the air for a time, becomes covered, sooner or later, by the fluffy silky coating, often of various colours, which is familiarly known as "mould." Such layers consist, in reality, of a number of microscopical plants, which are of an exceedingly simple structure, and which, as they do not contain chlorophyll, are included amongst the fungi.

Mucor mucedo.—This mould makes its appearance on many moist organic substances when they are left exposed to the air. It is, however, most readily obtained by placing a few balls of fresh horse-manure on a sheet of wet blotting-paper under a bell-jar, and keeping the whole in a warm place for a few days. The surface of each ball will then usually be found to be covered with a luxuriant vegetation, consisting of long delicate silky filaments (Fig. 34, *a, a*), each of which ends in a rounded head that gives it the appearance of a pin. The body of the plant consists of a number of branched filaments (*b, b, b, b*), or "hyphæ," as they are called. Some of these ramify into the substratum on which the fungus grows, and the mass so formed is termed a "mycelium." The mycelia of contiguous plants interlace with each other and form a felted mass. The whole of the plant, notwithstanding its complex configuration, consists, until the period of spore formation, of a single cell, formed externally by a thin, much-branched tube of elastic, transparent, and colourless cellulose. The inside of this tube is more or less completely filled with protoplasm permeated by the cell-juice, and this latter, collecting at various points, forms vacuoles.

Growth.—The appearance of the spores and of their growth during germination can be best studied in a hanging-drop culture. For this purpose one of the ripe little round heads (Fig. 34, *c*), called "sporangia," is detached from the plant with a platinum wire and immersed in a little sterilised water contained in a watch-glass, and if the sporangium does not rupture spontaneously, it is crushed with the wire. A small quantity of sterilised wort is then added to the contents of the watch-glass, the whole covered up, and allowed to remain at rest for half-an-hour, after which the fluid is vigorously stirred up, so as to distribute the spores evenly. From this one or more hanging-drop cultures are made,

MUCOR MUCEDO.

taking care that each drop does not contain more than one or two spores. The culture is then placed under the microscope, and its progress watched from hour to hour. By the time the preparation has been completed, each spore will have swollen to about ten times its original size, become circular, and, as the result of active endosmosis, developed a large vacuole. In the course of a very few hours it will push out one or more buds,

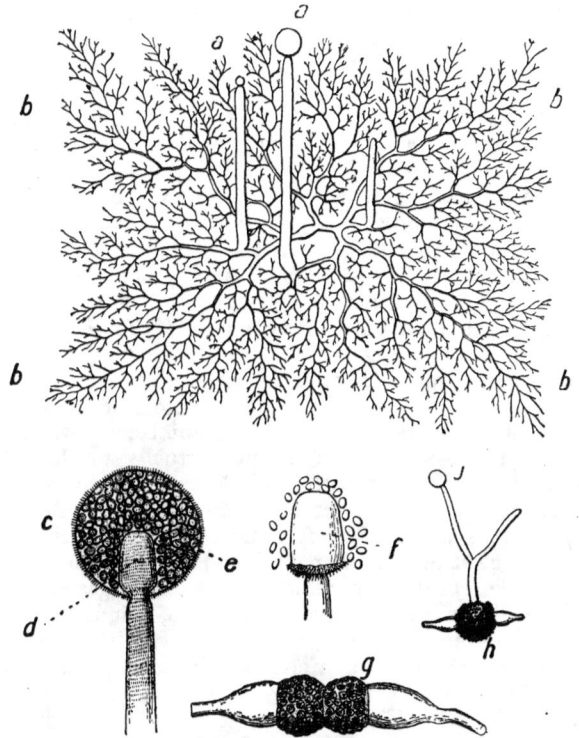

FIG. 34.—Mucor mucedo (*after Brefeld and Kny*).

a, a. aërial hyphæ; *b, b.* branched hyphæ (mycelium); *c.* sporangium; *d.* columella; *e.* spores; *f.* columella; *g.* cell-fusion; *h.* ruptured zygospore; *j.* formation of sporangium.

which, gradually increase in length and become hyphæ. As the plant absorbs more nutritive material than is required to supply the energy necessary for the performance of its vital functions, an accumulation of substance takes place in its interior, and space is provided for this first by the hypha increasing in length, and afterwards by its throwing out side-branches. When the lengthening process has proceeded for some time, the cellulose at some point on the side of the tube becomes softened, and is pushed out by the protoplasm. In this way a bud appears, which gradually enlarges to form a hyphal branch. This growth and subdivision of the hyphæ goes on until the many-branched mycelium is formed. In old plants there is a tendency for the hyphæ to become divided into distinct cells by the formation of partition-walls, as is the normal course with many other mould fungi. As the hyphæ grow the protoplasm secretes new layers of cellulose, so that the cell-wall is kept intact and of about the same thickness.

Sporulation.—When the plant has attained a certain age it puts forth, at various points, upright hyphæ, somewhat thicker than those that form the body of the plant. These are called the "aërial hyphæ" (Fig. 34, a, a), and their function is to bear the organs of fructification. When each of them has attained a certain length it ceases to grow, and its upper end becomes swollen into a small knob, the "sporangium," on which minute drops of water often appear.[1] A partition now commences to form at the junction of this knob with the stem, the end of which assumes the form of a little knob situated within the larger (d). A portion of protoplasm is in this way cut off from the rest of the plant, which, notwithstanding its complicated figure, had up to this time consisted of a single cell. The exterior of the sporangium becomes covered over with minute spicules of calcium oxalate, which give it a rough appearance. The protoplasm in the sporangium now becomes marked out into a number of oval masses, which, though close together, do not actually touch. These eventually become invested with a cellulose membrane, and gradually develop into perfect spores (e). The remainder of the protoplasm which is not used in forming the spores is then transformed into a gelatinous substance which has a powerful attraction for water. The sporangium, which in its early stages is of a pale colour, gradually, as it ripens, turns first brown, then black, its outer wall at the same time becoming very thin and brittle. When the sporangium is ripe, the contained gelatinous substance absorbs moisture, swells up, and the cell-wall suddenly bursts, causing a miniature explosion. This scatters the spores in all directions, and they may then be carried by currents of air to considerable distances. The interior knob, or "columella" (Fig. 34, f), as it is called, remains for some time attached to the end of the aërial hypha, often with a portion of the covering of the sporangium attached, which forms a little collar round its base. The detached spores now travel until chance brings them in contact with a medium favourable for their nutrition, when they germinate and form new plants.

Sexual Reproduction.—The mucors are the lowest members of the vegetable kingdom which exhibit a method of reproduction

---

[1] This is not condensed on the outside of the plant, but actually pushed through the cell-wall by the osmotic pressure inside.

SEXUAL REPRODUCTION.

which is universal in the higher forms of plant life. In this form two cells are always concerned, one representing the male element, the other the female; hence the process is called "sexual." The mucor, so long as it grows on the surface of a fluid and has a plentiful supply of nourishment at its command, only reproduces itself in the asexual manner just described; but when it grows on a solid substratum, as is almost invariably the case in its natural state, it then occasionally reproduces itself in an entirely different manner. When this is about to take place, two short club-shaped branches grow out from contiguous hyphæ; these become filled with protoplasm, and gradually approach one another until their extremities touch. A cell-wall, which cuts off a portion of protoplasm from that of the branch, is now developed in the end of each of these short branches. Those portions of the cell-walls that are in contact then dissolve, and the contents of the two cells fuse together (Fig, 34, g), after which they grow into an enlarged cell called a "zygospore;" this, as it ripens, turns black, and its exterior becomes covered with warty excrescences. The zygospore is then detached from the rest of the plant, and when placed in a suitable medium, after the lapse of a period of six or eight weeks[1] commences to germinate. On germination the cell-wall of the zygospore (Fig. 34, h) ruptures, and an unbranched hypha grows out from it; this sends up an upright stem into the air, at the extremity of which a sporangium is formed (Fig. 34, j), the spores of which when ripe escape, and reproduce the plant in the asexual manner.

Nutrition.—The tissues of the mucors and of the other mould fungi, when analysed, are found to consist of the elements carbon, oxygen, hydrogen, nitrogen, sulphur, and phosphorus, together with minute quantities of potassium, magnesium, and calcium. In order that the plants may carry on their vital functions perfectly, these elements must be supplied them in the form of such chemical compounds as they can assimilate. As the mucors possess no chlorophyll, carbon must be presented to them in the form of sugar or some analogous carbohydrate; hence they are dependent on chlorophyll-bearing plants for their carbonaceous diet. Nitrogen may be supplied to them in many forms, such, for instance, as potassium nitrate, or an ammonium salt such as ammonium

---

[1] This peculiar character of not being able to germinate until after the lapse of a certain time has led to the employment of the term "resting spores," oftener applied to the spores of the bacteria. It seems to be a provision of nature to tide the plant over a period of scarcity of food, for resting spores are, as a rule, only formed when the supply of food runs short. Analogous instances of a resting period are found in the higher plants; for instance, the tubers of the potato, the bulbs of the onion, hyacinth, and many other plants, remain perfectly quiescent during the winter months, and only begin to show signs of life in the spring, though the conditions as to warmth, moisture, &c., under which they have been kept may not have materially differed during the two periods.

tartrate; but they can assimilate amides, peptones, and even crude proteids. The sulphur, phosphorus, and mineral bases are best supplied in the form of the phosphates and sulphates of the metals potassium, magnesium, and calcium. The following forms an excellent medium for their cultivation :—

	Grammes.
Cane-sugar	15.00
Ammonium nitrate	1.00
Potassium phosphate	0.50
Calcium phosphate	0.05
Magnesium sulphate	0.25
Water	100.00

If it is attempted to grow a mould fungus in a solution from which any one of the elements forming compounds is withheld, though it may contain all the rest, the plant sooner or later pines and dies.

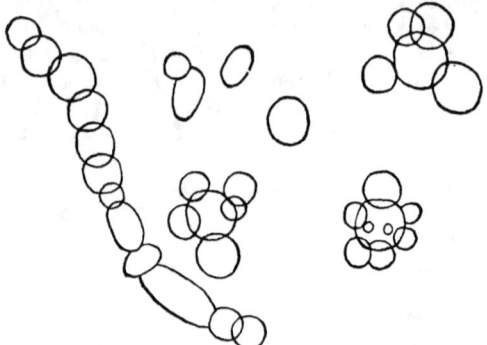

FIG. 35.—Gemmæ of Mucor.

Respiration.—The mould fungi respire freely, taking in oxygen and giving out carbon dioxide; consequently, unless liberally supplied with air, they soon perish.

Mucor racemosus.—This fungus is distinguished from *Mucedo* by its much smaller size, by the upright hyphæ (which are short and bear mouse-grey sporangia) being branched, and by the columella, which is round or pear-shaped. The external covering of the sporangium is tough and resistant, and is not soluble in water. As regards respiration, nutrition, and reproduction it resembles *Mucedo*. It has, however, a third method of reproduction which, so far, has not been observed in all the other members of the family; this is by the formation of gemmæ. When this process is about to take place, the protoplasm in some of the hyphæ of the mycelium collects at certain points in the tube; transverse walls then form, which divide these portions from the rest of the contents of the tube and convert them into distinct cells, in the interior of which fatty substances are deposited.

The cell-walls then thicken considerably, the intermediate portions of the hyphæ decay, and the gemmæ become detached; each of these under favourable circumstances can develop into a new plant. But the most curious fact with regard to these gemmæ is their power of setting up alcoholic fermentation when they are compelled to grow *under the surface* of a saccharine fluid. When cultivated under these conditions they throw out buds of a spherical or oval shape, which often hang together for a time in strings or clusters (Fig. 35). In this form they bear such a strong resemblance to yeast that they have been called "mucor yeast," and this resemblance has often led observers to erroneously conclude that they had been able to transform the mucors into yeast.

Other Mucors.—In addition to these two, several other members of the family are of frequent occurrence. The *Mucor stoloniferus*, so called from the stolons or runners which it throws

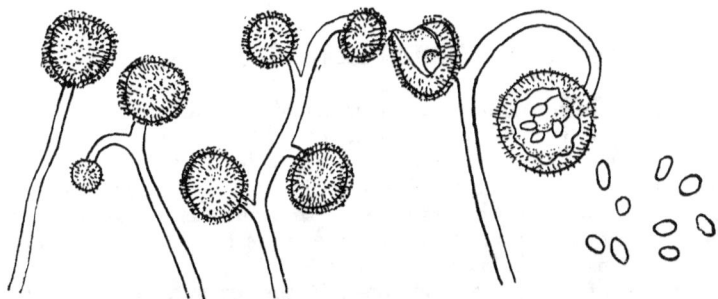

FIG. 36.—Mucor circinelloides (*after Van Tieghem and Gayon*).

out, something after the fashion of the strawberry plant; these bend down towards the substratum on which the plant is growing, and when they touch it take root again, as it were, and form new plants. Several aërial hyphæ arise at one point in the substratum, each bearing a black sporangium. The columella has the shape of half a sphere. In *Mucor circinelloides* (Fig. 36) a remarkable branching of the aërial hyphæ takes place, and the small branchlets which bear the sporangia become considerably curved during development; hence the name of the plant. Like *racemosus*, it forms gemmæ. *Mucor spinosus* has chocolate-brown sporangia, and the upper portion of its columella is studded with pointed excrescences. *Mucor erectus* resembles *Mucor racemosus* very much in form, but has, like *spinosus*, spicules on the top of its columella.

Alcoholic Fermentation.—As has been incidentally mentioned, some of the Mucor family have the property of exciting alcoholic fermentation. The *Mucor mucedo* does not form gemmæ, but its mycelium when submerged in wort is able to excite a feeble alcoholic fermentation, yielding, according to Hansen, in two months and three-quarters 1 per cent. by volume of alcohol, and in six months 3 per cent. The power of causing fermentation, however, seems to be greater in those mucors which form gemmæ; thus *Mucor erectus*, when grown in the submerged condition, yields, in a wort of a specific gravity of from 1056 to 1860, as much as 8 per cent. of alcohol by volume; it can also ferment a solution of dextrin. The fungus secretes an enzyme which is capable of hydrolysing starch. *Mucor spinosus*, under similar circumstances, yields 5.5 per cent. by volume of alcohol; *racemosus*, 7 per cent. This last secretes an enzyme capable of inverting cane-sugar, and is the only member of the family which can ferment this sugar. *Mucor circinelloides*, though unable to ferment cane-sugar, rapidly attacks this sugar after inversion.

SIMPLE MULTICELLULAR ORGANISMS.

Penicillium glaucum.—This mould is of almost universal occurrence; it is met with on jam, cheese, bread, decaying fruit, &c. In the malt-house it frequently makes its appearance on the barley when on the floor, and is especially prone to attack damaged corns. In warm weather it grows with enormous rapidity. In the brewery it is often found on the walls and ceiling of the various rooms, on the sides of vats, the outside of casks, also on the insides of such as are left open to the air, and in any other place where masses of damp organic matter, such as portions of spilled grains, yeast, &c., collect. Its green-blue colour is due to the myriads of spores, which form with the most extraordinary rapidity.

Growth.—If a little of this greenish-blue powder is taken up on the end of a platinum wire, stirred into a drop of wort and a hanging-drop culture made from this, the spores may be examined and the subsequent development of the plant studied under the microscope. The spores (Fig. 37, *a*) are minute spherical bodies, about 2.5 micromillimetres in diameter, which when seen singly are colourless; it is only when viewed in mass that they exhibit their peculiar colour. Each spore consists (Fig. 37, *a*) of a thin cell-wall filled with protoplasm, which, when the spore comes in contact with water, rapidly absorbs moisture, the spore swelling out to about three times its original bulk. It then commences to germinate; little projections make their appearance at one or more points (Fig. 37, *b*) in the cell-wall, and these gradually develop into hyphæ (Fig. 37, *c*), which do not extend to any great length before transverse partitions make their appearance. These divide the hyphæ into a number of distinct cells, and in this way convert

PENICILLIUM GLAUCUM.

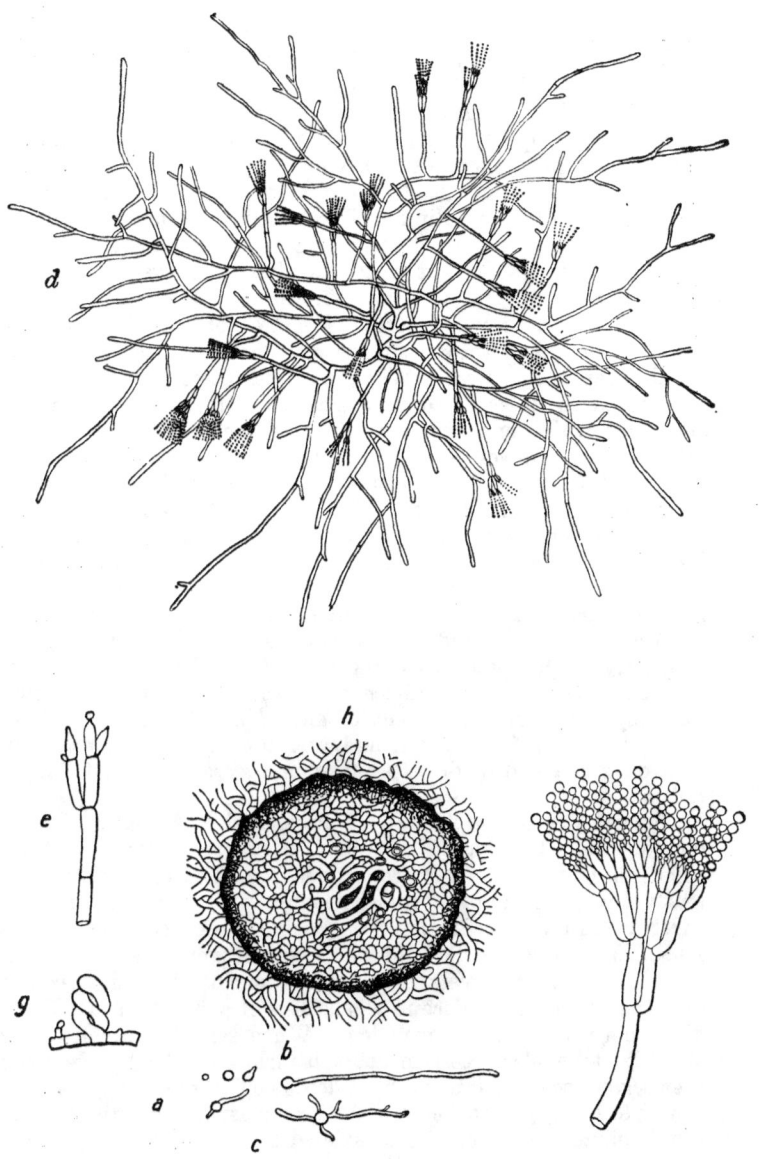

FIG. 37.—Penicillium glaucum (*after Brefeld and Zopf*)

a. spores; *b.* and *c.* incipient hyphæ; *d.* multicellular hyphæ; *e.* hypha forming spores *f.* fully-formed head; *g.* conjugation of hyphæ; *h.* sporocarp (seen in section).

the plant into a multicellular organism (*d*). Many of the hyphæ, after having become divided into segments in this way, throw out a side-bud, which is generally situated immediately behind a partition-wall. These buds increase in size and become hyphæ, which extend in length, and then form branches in a similar manner. The tips of many of the short branches meet and fuse together, and in this way a true network is formed; these frequent branchings and fusions lead at last to the formation of a dense, complicated, felted mycelium. The mycelia of contiguous plants closely interlace, and often form a mass of considerable size on the surface of the fluid on which the fungus grows, of a nature so tough and coherent that its removal from the flask or bottle in which it has grown is often a matter of considerable difficulty. The plant throws up erect filaments, or aërial hyphæ, which bear the spores. When each of these has attained a certain length it ceases to grow, and the cell at its apex throws out a bud which enlarges to a cell. On the top of this another cell (Fig. 37, *e*) develops, which swells at its apex, and just below the swelling a constriction forms. The portion thus marked off becomes rounded and forms the first spore. The cell again swells, another portion is similarly cut off, and the second spore produced, the process being repeated until a string of spores resembling a chaplet of beads is formed. In the meantime side-cells, which form strings of spores in a similar manner, are put out from the cell at its apex, and in this way the brush-like arrangement (Fig. 37, *f*) is produced. The intermediate portions which connect the spores then decay, and leave the spores free to be wafted away by the slightest current of air. The spores are covered externally by a fatty layer, and are, consequently, not readily wetted; when a drop of water falls on a mass of them it simply rolls off.

Sexual Reproduction.—In addition to this asexual method of reproduction, the *Penicillium* can also propagate itself sexually when placed under certain conditions, the chief one being apparently the absence of a sufficient supply of oxygen. Brefeld effected sexual reproduction by placing the mycelium of the *Penicillium* between slices of bread, which were then kept in a dark, damp place, access of air being excluded. Lindner has seen the process take place in a gelatin-wort cultivation in a Petri-dish from which air was excluded. When sexual reproduction is about to take place, certain short branches of the hyphæ twist themselves together into a sort of double spiral (Fig. 37, *g*), and as it is exceedingly probable that the protoplasmic contents of each couple mingle together, it is considered that one hypha represents the male element, the other the female.

A series of remarkable changes now take place, which end in the formation of a small round body, the "sporocarp" (Fig. 37, *h*), which passes several weeks in a resting condition. If, at the end of this period, the sporocarp is placed in a damp situation, it takes on a series of changes, at the termination of which its interior is filled with an enormous number of spores, together with crystals of calcium oxalate, which latter salt also collects in the cell-wall and renders it extremely brittle. When this ruptures the spores are set free. The whole process is exceedingly slow, requiring for its completion six or eight months. Each of the spores produced in this manner is somewhat spindle-shaped in form, with three or four ridges on its external covering, and a circular groove running round its margin. The sexual spores when placed under suitable conditions germinate at once, but if kept in a dry condition in the unruptured sporocarp, they remain dormant, but preserve their vitality for several years.

Respiration.—The *Penicillium* respires, that is, it takes in oxygen and gives out carbon dioxide. That it requires a considerable amount of this gas may be seen by watching a hanging-drop culture from day to day; the plant being in an air-tight cavity, soon begins to languish, and finally perishes from want of air.

Nutrition.—This fungus is by no means choice in the matter of diet. Since it contains no chlorophyll, it is unable to obtain its supply of carbon from the carbon dioxide of the atmosphere, but it is able to appropriate the carbon of a vast number of other substances, such as the various sugars, mannitol, glycerin, acetic, tartaric, citric, or succinic acids, asparagine, leucine, alcohol, and many other bodies. When grown on a solution of racemic acid, which is composed of equal molecules of the dextro- and lævo-tartaric acids, the *Penicillium* feeds upon the dextro variety and leaves the other almost intact. This property was discovered by Pasteur, who utilised it for obtaining lævo-tartaric acid. The *Penicillium* exhibits an equal indifference as to the source of its nitrogenous supply. It obtains it most readily from a peptone, but it can derive its nitrogen from almost any salt of ammonium, such as the carbonate, lactate, tartrate, phosphate; from ethylamine, propylamine, trimethylamine, or from the amides—acetamide, oxamide, asparagine, leucine, &c. It is able to obtain its supply of sulphur not only from sulphates, but also from sulphites, hyposulphites, and from proteids. The organism produces oxalic acid during its growth, and this often combines with calcium to form crystals which are deposited in the solution on which the plant grows. Occasionally the oxalic acid combines with potassium, and in this case the resulting salt remains in solution. Calcium oxalate is very frequently met with in plant life; Sachs considers that it is secreted in order that the plant may be able to decompose calcium sulphate, and in this way obtain the supply of sulphur that it requires for the formation of its proteid constituents.

Penicillium when grown submerged in a saccharine fluid is unable to excite the alcoholic fermentation. When placed under these conditions it ceases to grow normally, its hyphæ assume

VEGETABLE BIOLOGY.

peculiar and irregular forms, and large quantities of fat become deposited in their protoplasmic contents; indeed, in such cases the fat has been found to amount to three times as much as that which the plant contains when grown under normal conditions.

Temperature.—The *Penicillium* is able to grow within very wide limits of temperature. Its spores will germinate at any temperature between 2° and 43° C. (35° and 110° F.), but 22° to 26° C. (72° to 79° F.) seems to be the most favourable temperature. Pasteur found that the dry spores could withstand a temperature of 108° C. (226° F.) without losing their vitality, but that they were soon killed when immersed in boiling water.

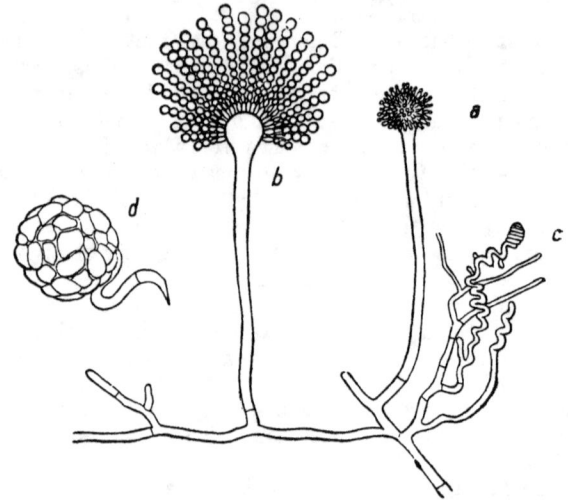

FIG. 38.—Eurotium aspergillus glaucus (*after De Bary*).
a. hypha commencing to form spores; *b.* fully-formed head (seen in section); *c.* conjugating hyphæ; *d.* sporocarp.

Eurotium aspergillus glaucus.—This exceedingly common mould fungus (Fig. 38) is met with under much the same conditions as the *Penicillium glaucum*. It closely resembles this latter in the shape of its spores, their germination, and the subsequent formation of a mycelium therefrom; the principal difference observable being that the partition-walls of its hyphæ are not so numerous, and consequently the cells are somewhat longer.

Growth.—This is best studied in a hanging-drop culture. When a certain amount of mycelium has formed, the plant begins to throw out upright stems which contain very few partition-walls, and these eventually bear the spores.

Aërial Sporulation.—The method of fructification, however, differs considerably from that of the *Penicillium*. Each upright

SEXUAL REPRODUCTION.

hypha, after having attained a certain length, ceases to grow, and its upper extremity expands into a pear-shaped head (Fig. 38, *a*). All over the surface of this, little protuberances, like so many pegs, arise, the protoplasm of each of these being continuous with that of the head. The end of each of these pegs then puts forth a little round bud, under which a constriction forms, and this gradually tightens until a little round knob is formed hanging by an exceedingly delicate neck. The end of the peg then similarly throws out another bud, the process of constriction is repeated, and this goes on again and again until a chaplet of cells resembling a string of beads is produced, of which the oldest are the most distant from, the youngest nearest to the head. These are the spores, and when the growth of a head is completed, these numerous chains of spores have the appearance of a brush-like ball (Fig. 38, *b*). The head, when all its protoplasm has been exhausted, gradually decays, and the spores fall off; but those which have ripened before this takes place, may be removed by the slightest touch or by a slight breeze; they then float in the air, and may be widely disseminated by air-currents. The plant, in the earliest stages of its growth, consists of a white patch of mycelium, which gradually turns grey; it is only after the spores are formed that it assumes a green appearance. Each spore as seen individually under the microscope is colourless, and it is only when a large number are seen together, as when growing on the plant, that they exhibit their peculiar green colour.

Sexual Reproduction.—In addition to this asexual method of reproduction, the *Eurotium* can produce an entirely different set of spores by a sexual process. When this is about to take place, two of the short hyphæ coil and interwine together (Fig. 38, *c*), and their protoplasmic contents become fused together. A complicated series of changes then ensues, which finally terminates in the formation of a spheroidal body called the "sporocarp" (*d*), of a yellow colour, which contains a host of spores, each of which is in shape something like two watch-glasses joined by an enveloping band. When placed in a suitable medium, the protoplasm of the spore swells up, pushes aside the two concave valves, and throws out a bud. This gradually enlarges, forms partition-walls, and becomes a hypha, which grows and eventually forms a mycelium. From this the upright stems for bearing the asexual spores arise, while sporocarps may also form in other portions of the mycelium.

Respiration and Nutrition.—These processes are carried on in a manner analogous to those of the *Penicillium*, though the *Eurotium* is unable to derive its supply of food from such a considerable number of substances as the former.

Temperature.—The *Eurotium* grows best at a temperature of from 10° to 15° C. (50° to 59° F.), and is said to be unable to withstand a higher temperature than 30° C. (86° F.). This question has, however, been very incompletely studied.

Botrytis cinerea.—This is another fungus commonly found growing on decaying organic matter, in the form of small greyish-yellow patches, from which aërial hyphæ of the length of about a millimetre ($\frac{1}{25}$ of an inch) arise, each of which bears a tufted head on which are seated a mass of spores (Fig. 39, a). If a few of these are removed and placed in a hanging-drop culture, their appearance and subsequent development may be observed in the usual way.

Growth.—The spores germinate (Fig. 39, b), and form a dense network of mycelium, similar to that of the *Penicillium* or *Eurotium*. In a few days the cell-walls of the hyphæ of the mycelium become swollen and gelatinous, and, when lifted out of the fluid in which the plant is growing, appear glairy and feel slippery, as if they were coated with mucilage. When the mycelium is some days old, some of the hyphæ, which in the course of their growth have come in contact with a solid body, such as the walls of the vessel the plants are grown in, rapidly throw out a number of small branches, which rapidly subdivide, and at last assume the form of little tassels, which gradually turn black. These develop a sort of mucilaginous matter by means of which they become firmly adherent to any object with which they come in contact.

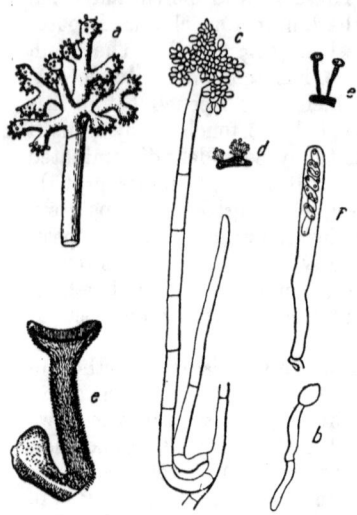

FIG. 39.—Botrytis cinerea *(after De Bary).*

a. tufted head of hypha with spores; b. germinating hypha; c. hypha with bunched spores; d. sclerotia; e, e. cup-shaped sclerotia; f. ascospore.

Aërial Sporulation.—From the mycelium, which gradually assumes a greyish-brown colour, upright hyphæ (which are divided by septa and often branch) arise; these are destined to bear the spores. After these aërial hyphæ have grown to the height of a millimetre, they throw out stout, short branches nearly at right angles to the principal stem, and each branch subdivides again into several shorter twigs. The ends of each of these swell, and on its swollen surfaces an egg-shaped spore gradually develops. The spores, when ripe, are about 8 to 9 micromillimetres long and 6 to 7 broad, pale reddish-brown in colour, and connected to the stem by a delicate neck, only one spore being formed at the end of each twig. When growth is completed the head of spores somewhat resembles a bunch of grapes (Fig. 39, c); hence

SEXUAL REPRODUCTION.

the name of the plant, *Botrytis*, from the Greek word signifying such a bunch.

Sexual Reproduction.—When this takes place in the *Botrytis*, round or irregular shaped, hard, black bodies, called "sclerotia" (Fig. 39, *d*), form in the mycelium. The size of these varies from half a millimetre to several millimetres in diameter; they are made up of an extremely dense and complicated mass of interlacing hyphæ. If, directly after their formation, they are removed from the plant and placed in a nutrient fluid, or on a nutrient substratum, they at once commence to throw out upright hyphæ which bear spores. If, however, they are dried, kept for at least a year, and then buried under the surface of soil which is constantly kept moist, small buds protrude from their surfaces. Each of these grows into a stout, hairy, cylindrical stem (*e, e*), which at length pushes its tip through the soil and then ceases to grow. The apex of each of these stems expands into a little cup-shaped formation, the interior of which eventually becomes lined with a velvety layer, each strand of which is an elongated cell. Some of these cells remain barren; in others the nucleus divides, first into two, then into four, then into eight (*f*); each of these divisions of the nucleus attaches to itself a portion of protoplasm, and after investing itself with a cell-wall, finally becomes an "ascospore."[1] The elongated cells in which the ascospores have formed absorb water, and the effect of this is to gradually push the spores towards the upper extremity of the cell; when the pressure becomes sufficiently great the end of the cell ruptures, and the spores are driven out with an amount of force sufficient to carry them to some distance. The spores, if they happen to alight in a suitable medium, germinate in a few hours, develop into a mycelium which eventually throws out spore-bearing aërial hyphæ and forms sclerotia.

Nutrition.—The plant will grow on almost any kind of organic substance, if sufficiently provided with moisture and oxygen, and provided that the temperature is a few degrees above the freezing-point. It is able to obtain its supply of carbonaceous diet from the sugars, or from organic acids, such as tartaric, citric, &c. Its tissues, upon analysis, are found to contain large quantities of potassium phosphate; the plant, during its growth, produces much oxalic acid, and this combines with potassium. It secretes two enzymes; one of which dissolves cellulose and the other inverts cane-sugar; but it is quite unable to excite the alcoholic fermentation.

The *Botrytis* also grows on living plants, its hyphæ piercing the cuticle and penetrating into their cells; in this way it often causes considerable damage to fruit and other crops. The peculiar flavour and bouquet of certain Rhine wines are caused by modifications which this fungus induces in the contents of the over-ripe

---

[1] So called from being formed inside a cell or sack (*ascus*, a sack).

grapes before their juice is subjected to the vinous fermentation. *Botrytis* is sometimes the cause of a smoky flavour in wines, and it is probable that a similar flavour occasionally met with in beer is also produced through its agency.

Aspergillus oryzæ. — A kind of fermented liquor made from rice, and known by the name of "saki," has long been made in Japan by the agency of this fungus. It is manufactured in the following manner:—A quantity of rice is first steamed, allowed to become cold, and a portion of rice grains, kept over from a previous brewing and on which the mycelium of the fungus is growing, is mixed with it. In a few days the mycelium will have spread over the new rice, this is then mixed with a further portion of steamed rice, the process being repeated until a sufficient quantity of infected rice is obtained. Under this treatment the mass of rice develops an odour similar to that of apples or pine-apples. In the actual brewing 21 parts of this infected rice are mixed with 68 parts of steamed rice and 72 parts of water, and the semi-solid mass thus obtained is kept at a temperature of 20° C. (68° F.) for several days, at the expiration of which it becomes considerably liquefied, the starch being hydrolysed by an enzyme secreted by the fungus. In about three days alcoholic fermentation sets in spontaneously, and lasts for about three weeks. A beverage is obtained in this way which, after filtration, is of a bright, sherry-wine colour, and contains from 13 to 14 per cent. of alcohol. The Koji-ferment has been much studied; it has been found to contain various enzymes, all of which are able to hydrolyse starch, dextrin, maltose, and cane-sugar. Hydrolysis of the proteids also appears to take place, for much insoluble proteid matter is converted into soluble during the process.

Juhler, who recently investigated the action of the Koji-ferments, was led to the conclusion that the alcoholic fermentation is brought about by the ripe spores of the aspergillus. These, when grown at a suitable temperature and with a free supply of air, develop a mycelium in the ordinary manner, but when, after having liquefied the starch, they sank into the fluid, and were excluded from contact with the air, Juhler thought that he had been able to observe a series of extraordinary changes, which ended in their conversion into *Saccharomycetes* or true yeast. Recent investigations have, however, disproved the correctness of these conclusions.

In addition to the *Aspergillus oryzæ* two other fungi have been found in the Koji-ferment, a white mould very similar to that described by Calmette as *Amylomyces Rouxii*, and a variety of *Mucor*, but these are not nearly such energetic agents in effecting the hydrolysis of starch as the *Aspergillus*.

Chalara mycoderma. — This is another mould fungus which is able to grow on wort or beer, on which it forms a greyish film, which consists of a branched mycelium (Fig. 40). The contents of some of its hyphæ have a brilliant refractive appearance, and from these conidia of different shapes and sizes arise. The fungus is met with very abundantly on the surface of grapes.

Oidium lactis.—This fungus (Fig. 41) is often found on the surface of milk, hence its name. It is also often met with in air analyses, where it forms on the thin gelatin layer a bright, silky mass of mycelium, the hyphæ of which radiate outwards from a common centre. Grown on a thick layer of gelatin, it forms large colonies having a radiating form, and a dry, powdery appearance on their surface. A considerable quantity of gas is liberated by each colony, which forces up the gelatin into a bubble; this may become as large as a walnut, but does not burst. The fungus is met with on malt, on leaky places in brewing vessels, and is sometimes found growing on yeast. When vegetating on wort it forms white, floating masses having a woolly appearance, under which bubbles of carbon dioxide collect. The woolly mass consists of transparent, branched, forked and thin-walled hyphæ, the extremities of which are divided into separate cells by numerous transverse septa. These cells form the spores of the plant; they finally become detached, and, when placed in conditions favourable to their growth, germinate. The fungus is not, as was at one time supposed, the cause of the lactic fermentation in milk; it can, however, excite a feeble alcoholic fermentation in wort. It grows freely on yeast, and though the organism is not known to exercise any prejudicial influence on either yeast or beer, yet its presence there is undesirable. When the spores, which consist of short cylinders, are cultivated in a hanging-drop, they swell up, become egg-shaped, and then throw out one or more hyphæ, which grow rapidly and branch freely. Numerous partition-walls then appear, and the separate cells formed in this way break off with a kind of convulsive movement, which can often be induced by tapping the stage of the microscope.

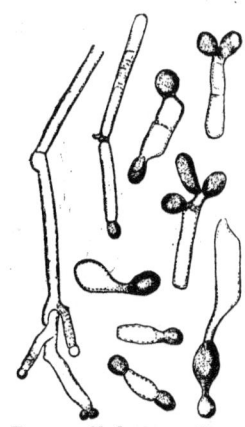

FIG. 40.—Chalara mycoderma (after Hansen).

Oidium lupuli.—Under this name Matthews and Lott have described a fungus somewhat akin to the *Oidium lactis*, which they found growing on spent hops in the form of a reddish-brown or salmon-coloured dust. When cultivated in a hanging-drop it forms a luxuriant aërial mycelium, the hyphæ of which become branched at their apices. Each branch then divides up by the formation of transverse septa into individual cells which constitute the spores of the plant.

Lindner describes another *Oidium*, provisionally named *Pullulans*, which he has often found in air analyses, and in samples of lager beer removed from the storage casks by spiling. It is able to grow on the surface of wort, but does not seem capable of exciting alcoholic fermentation. Its influence on beer is unknown.

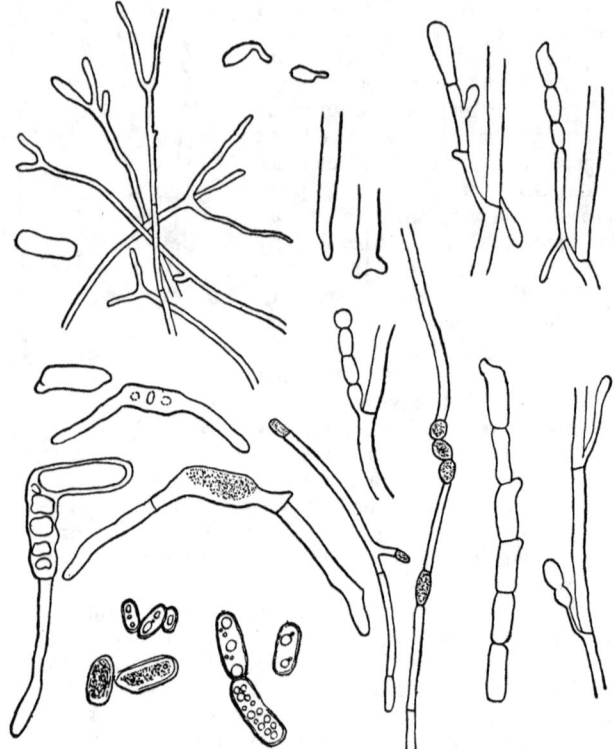

FIG. 41.—Oidium lactis (*after Hansen*).

Dematium pullulans (Fig. 42).—Another fungus also frequently met with on the surface of grapes and other fruits. It is frequently found in air analyses, and its presence can be readily detected by the rapidity with which it liquefies the gelatin layer. According to Lindner, who has studied the habits of this fungus, it grows rapidly in hopped wort, in which it throws off cells strongly resembling those of yeast. In a few days numerous flocculent slimy masses make their appearance under the surface of the wort, which cling tenaciously to the sides of the vessel. The liquid itself, although not rendered turbid, becomes ropy, and can be pulled out into long strings by means of a platinum loop. A microscopic examination of a drop of the wort, when it has arrived at this stage, reveals nothing but the presence of a number of isolated cells resembling long budding yeast-cells. After some days the surface of the wort becomes covered with long threads in which the resting spores of the fungus may be seen; these, which are at first of a dirty green colour, gradually turn black. The ropiness of the fluid is caused by the deposition of a transparent mucilaginous matter on the walls of the cells, which keeps them at some considerable distance apart. During the time the wort is turning ropy it develops a paste-like flavour, and this persists after it has undergone the

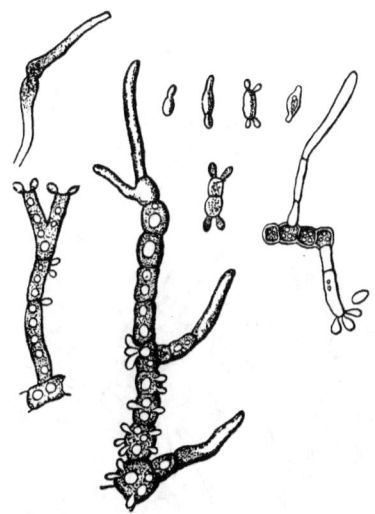

FIG. 42.—Dematium pullulans (*after Loew*).

alcoholic fermentation. During the latter process the ropiness disappears and the *Dematium* cells die. The acidity of the beer does not seem to be influenced by an invasion of this organism. The fungus can induce a similar ropy condition in a solution of cane-sugar.

Apparent Origin of Wine Yeast.—Jörgensen, incited by the supposed discovery of Juhler that the spores of the *Aspergillus oryzæ* could be transformed into yeast-cells, thought that wine yeast must have some similar origin. After much fruitless experimenting, he found, amongst the mould fungi growing on the surface of grapes which had been kept for some time in a moist atmosphere under a bell-glass, a vegetation consisting of numerous branched hyphæ, which subsequently formed conidia. On cultivating these conidia upon the surface of sterilised grapes, he thought he had observed a complicated series of changes in the vegetation thus produced, which ended in the formation of true yeast. Other observers, notably Klöcker and Schiöning, have repeated these experiments with an entirely negative result; so this is almost certain, that the views of Jörgensen were erroneous on this subject.

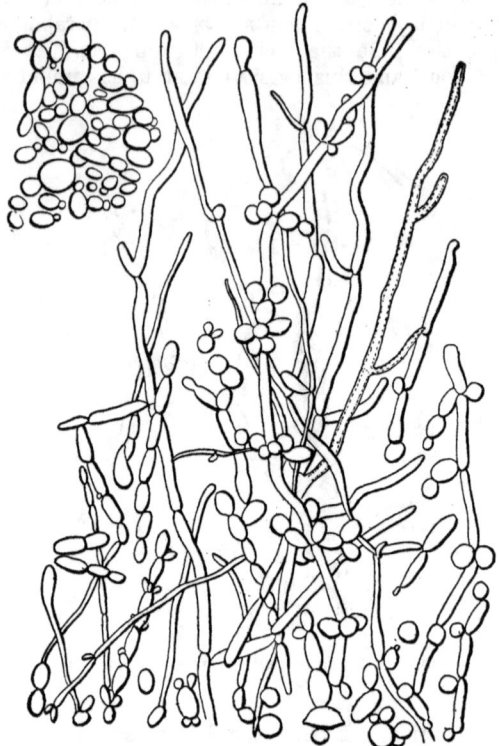

FIG. 43.—Monilia candida (*after Hansen*).

Monilia candida.—This fungus (Fig. 43) occurs as a white layer on cow's dung, and on the surface of sweet succulent fruits. When submerged in wort it excites the alcoholic fermentation, and develops cells much resembling those of yeast. It ferments maltose and cane-sugar, and has recently been shown by Fischer, in contradiction with the views previously held, to secrete invertase.

Mould Fungi Dangerous on Brewing Premises.—The presence of mould fungi, either in the malthouse or the brewery, is highly undesirable, since, as they grow and multiply with extreme rapidity, and are always accompanied by bacteria, they may at times become a source of real danger. Barley, on the growing floor, is especially liable to be attacked by *Penicillium glaucum*, *Mucor mucedo*, and *Aspergillus glaucus*, all of which secrete diastases that convert a portion of the starchy constituents of the barley into sugar, which eventually becomes transformed into acids. In this way a portion of the malt, which should have contributed to the production of extract, is not only lost, but converted into substances of a prejudicial nature. The moulds communicate an objectionable flavour to the growing barley that is not lost on kilning and which persists in the finished beer. On rare occasions the barley becomes infested with a *Fusarium* which gives the resultant malt a reddish tint and renders it quite unsuitable for the production of beer.

Worts, especially during hot weather, may become invaded by some of the mould fungi. The smoky flavour due to the presence of *Botrytis cinerea* has already been mentioned. Beers weak in alcohol may be infested by *Monilia candida*, which, when growing under these conditions, has very much the appearance of yeast. *Oidium lactis* sometimes makes its appearance during hot weather on the surface of beer, over which it rapidly spreads as a thin skin resembling powdered starch; breweries situated in close proximity to stables are especially liable to the invasions of this fungus. Hops which have been imperfectly dried, or which are kept in a damp place, are liable to the attacks of some of the *Mucedo* species. This causes them to become heated and take on a peculiar fermentation, in which trimethylamine and butyric acid are formed, bodies that communicate an exceedingly unpleasant flavour to the hops.

As the presence of moulds in the malthouse or brewery is so highly undesirable, they should be destroyed as soon as patches of them make their appearance. A solution of chloride of lime (bleaching-powder) is a very effective mould-destroyer for those places where it can be used. For the interior of vessels, &c., a strong solution of sulphurous acid or bisulphite of lime, which acts as a powerful fungicide, may be used. A preparation sold under the name of "Antinonnin" is also a very efficient fungicide. It is a pasty mass, consisting of soap, glycerin, and potassium dinitro-cresol ($C_6H_2(NO_2)_2CH_3OK$), and is generally mixed with whitewash and applied to walls and woodwork. Its nitro-cresol constituent is an exceedingly poisonous substance.

THE HIGHER PLANTS.

These may be regarded as vast congregations of cells, all of which, at an early period in their existence, have had the same simple structure as the yeast-cell. In this simple organism we have seen that one portion of its tissues performs one, another portion another function; thus the cell-wall, in addition to being a protective layer for the protoplasm, carries on the important function of osmosis, and the protoplasm itself performs several different functions. A certain amount of differentiation of function is therefore observable even in the structures of the simplest vegetable organisms, but, as we proceed higher in the scale of plant life, this differentiation of function is not merely confined to the contents of the individual cell, but is extended to the cells themselves. We then find special cells set apart for the performance of certain functions, and the higher we proceed in the scale, the more complex the organisation of the plant becomes, and the greater the differentiation of function between its various groups of cells. Though the embryo of the barley-plant as it exists in the barley-corn consists of an aggregation of simple cells; yet even at this early stage of its existence there is no difficulty in recognising, from the way in which these cells are arranged, that they already are marked out for different purposes. The portion of the germ which is intended to form the root is readily distinguishable from that which is intended to form the stem. But as growth proceeds and the cells multiply by division, the cells which are concerned in the formation of the root become further subdivided into distinct groups; one of these eventually forms the epidermis or exterior covering of the root, the cells of another group coalesce to form vessels for the conveyance of water, together with those constituents of the soil which are dissolved in it, from the root to the leaves. Other cells form delicate prolongations called the "root hairs," the function of which is to absorb water from the soil. Similarly, as growth proceeds, changes take place in the cells intended to form the stem; cellulose of a tough resistant nature is deposited in the walls of those cells intended to give rigidity to the plant and enable it to preserve its shape against external influences. Other sets of cells become elongated and unite at their ends to form vessels for the transmission of the various fluids of the plant. Another series unite and form the leaves, in the exterior cells of which the chlorophyll granules are met with in the greatest abundance. When the period of flowering approaches, certain cells in the apex of the stem commence to form the complicated series of tissues which first constitute the various parts of the flower and afterwards those of the fruit. In the case of such a plant as the barley-plant, which only lives for a year, during the period of its fructification nearly the whole of the starch and other carbohydrate matter contained in the plant, with the exception of the cellulose, is conveyed to the seed and there stored up as starch. The whole of the carbon of this starch, with the exception of the minute quantity derived from the germinating seed, has been separated from the carbon dioxide of the atmosphere by the chlorophyll granules of the plant's leaves and stem, and as these granules are most abundant in the leaves, it follows that the bulk of the carbon must have been disassociated by them, afterwards to be conveyed down the conducting tissues of the leaves in the form of sugar, then up the stem, at the apex of which it is finally deposited as starch in the seeds.

Like the *Protococcus*, the higher plants require carbon dioxide and water for their nutrition, also nitrogen, sulphur, and phosphorus for the construction of their proteid constituents, and also the mineral elements potassium, magnesium, calcium, and iron. These latter elements are derived from the nitrates, phosphates, and sulphates, which are invariably present in the soil, and which are dissolved by the water that is absorbed by the root hairs of the plant, which insert themselves for this purpose into the interstices between the particles of the soil. It is then passed by them into the root, from which it gradually ascends through the stem of the plant to its leaves. The amount of water above and beyond that required for the actual nutrition of the plant is returned to the atmosphere as vapour, leaving behind in the leaves the various salts it previously held in solution.

The Substances Produced by the Cells of Different Plants.—Hitherto only those products of plant life which are indispensable to the life of the plant have been mentioned, but very many plants form a variety of other substances from the four elements carbon, hydrogen, oxygen, and nitrogen, which are often exceedingly complex in composition, and differ with each plant. The number of the substances so formed is almost incalculable. For example, the hop plant, in addition to carrying on those functions that are absolutely necessary for ordinary plant life, secretes the various volatile oils upon which its aromatic flavour depends, the resins which are of such importance to the brewer, the bitter principle called lupulin, the tannin, &c. The number of different substances which have been isolated from plants, especially of those known as medicinal, is simply enormous, and even this large number is probably but a small fraction of those which are still unknown.

Germination of Barley.—This, and the complicated series of physiological changes which accompany it, are subjects of the deepest interest to the brewer, for during the process of germination, barley, which, as such, is totally unfitted for brewing purposes, is converted into malt. The grain of barley contains, in addition to the germ or embryo, a large store of material in the shape of starch, proteids, mineral matters, &c., which are destined to serve as nourishment to the young plant before it has thrown out its

272 VEGETABLE BIOLOGY.

first green leaves, after which it is capable of maintaining an independent existence. This store of reserve material, as it exists in the barleycorn, is in a state quite unfitted for the nutrition of the growing embryo; the starch and a large part of the proteid matters are in an insoluble condition, and the greater portion of the proteid matter which is already soluble is in a non-diffusible form, in which condition it cannot be utilised by the young plant until after considerable modification. Everything is, however, ready for altering this condition of things; enzymes, which are destined to induce

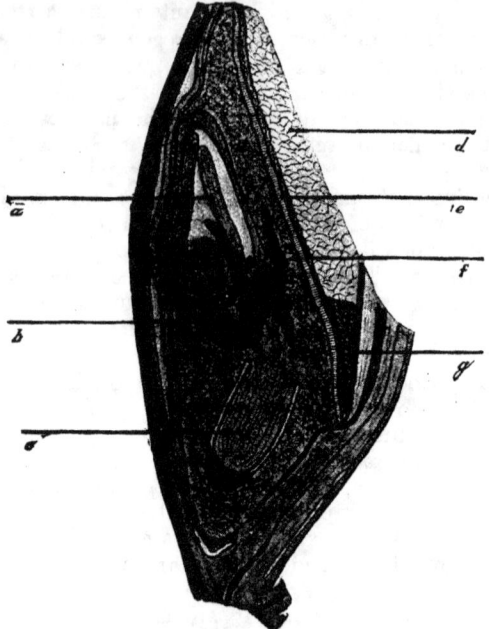

FIG. 44.—Sections from a Grain of Barley.

A. The germ and its appendages.—*a*. rudimentary leaves; *b*. rudimentary stem; *c*. rudimentary root; *d*. starch cells (empty); *e*. epithelial layer; *f*. layer of emptied cells; *g*. starch cells (filled).

the necessary metamorphoses, are already to some extent present in the seed, and the material is at hand for the production of those other enzymes which are necessary to complete the transformation. All that is required to set in action the vital machinery of the barleycorn is the presence of water in sufficient amount, a suitable temperature, and a supply of oxygen. When vital activity has been fully aroused, an abundance of those enzymes are secreted whose function it is to convert the reserve materials into soluble and diffusible forms fitted for the plant's nutrition during

GERMINATION OF BARLEY.

the earliest stages of its existence. But in germinating barley for malting purposes, as soon as a certain amount of change

FIG. 44.—Sections from a Grain of Barley.

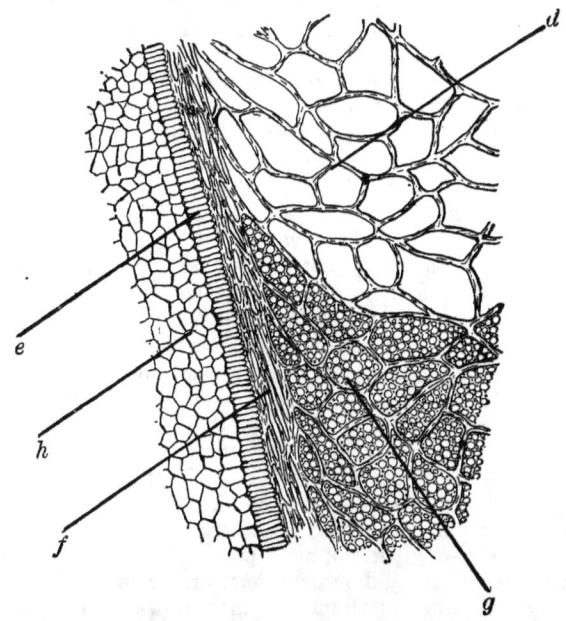

B. Section showing secretory layer of epithelium (more highly magnified).—
d. walls of starch cells; *e.* secretory layer; *f.* layer of emptied cells;
g. cells filled with starch granules; *h.* cells of the scutellum.

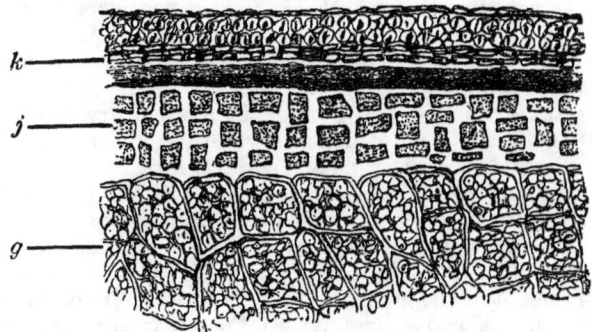

C. Section showing aleurone layer.—*g.* starch cells; *j.* aleurone layer; *k.* layers which collectively constitute the husk.

has been secured, the maltster steps in, and, by withdrawing water from the grain, puts an end to further vital action. His object is

s

to secure certain important modifications in the grain which are brought about by enzymic action, and also to obtain an abundance of those enzymes which convert the starch of the malt in the mash-tun into substances that are for the most part fermentable, and which transform the albuminoids into derivatives assimilable by the yeast.

Structure of the Barleycorn.—As may be seen by examining the accompanying illustration (Fig. 44, *A*), which represents the germ end of a grain of barley cut in two through the furrow which extends down the face of the corn, the seed is divided into two chambers by a sort of partition-wall (*e*), the lower one containing the germ, which would, in the natural course of circumstances, develop into the barley plant. Even at this stage of its existence the rudimentary stem and root (*b* and *c*) are distinguishable. The germ lies embedded in a mass of cells, the uppermost layer of which consists of the series of elongated cells (*e*) termed the "secretory layer;" it has certain very important functions to fulfil during the process of germination. The endosperm is filled with a network of thin-walled cells (*Ag* and *Bg*) closely packed with starch granules and small granules of proteid matter. Immediately above the secretory layer are several rows of empty cells (*Af* and *Bf*). The walls of this chamber are lined with three or four layers (*Cf*) of somewhat square-shaped cells, known as the "aleurone cells;" these, which collectively constitute the "aleurone layer," are filled with the aleurone or proteid grains, together with small quantities of fat and mineral matter. The aleurone layer, when it reaches the scutellum or partition-wall, diminishes in thickness to a single layer of small cells, and these gradually disappear towards the base of the grain. Several layers of various other tissues surround the seed and are closely adherent to it; these collectively constitute the "husk," and are of considerable importance to the brewer, for in the mash-tun they form a mass of filtering material which completely removes suspended particles from the wort and permits it to flow off bright and clear.

Dormant Vitality.—The germ of the grain of dry barley may be said to be in a state of dormant vitality. When the barley grain is planted in the soil, it finds there all the conditions necessary for bringing its dormant state of vitality into one of activity; these are the presence of water, a suitable temperature, and free supply of air. The maltster imitates these conditions artificially; he supplies the requisite water by soaking the barley in the cistern, subsequently he keeps the grain at a suitable temperature on the floor of the malthouse, and secures its aeration by frequently turning. The barley grain, even when in a state of dormant vitality, manifests to a very small extent the actual phenomena of life; it very slowly respires, absorbing oxygen and giving off carbon dioxide, in consequence of which there is a small but appreciable loss of weight in barley during storage. It has been estimated from this cause that the best and driest barleys lose $1\frac{3}{10}$ per cent. of their weight in the first year, $\frac{9}{10}$ per cent. in the second year, and $\frac{1}{2}$ per cent. in the third year; and this loss is considerably greater with inferior and damp samples.

Germination.—When in one or the other of the above-mentioned ways the dormant vitality of the germ is roused into activity, that complicated series of processes collectively comprised under the general term of "germination" commences. Starch begins to appear in the tissues of the germ, which previously were almost or absolutely starch free, and the first rootlet begins to elongate. The cells of the scutellum, which are as yet free from starch, now show signs of activity, the proteid matter deposited in them disappears to a considerable extent, and the protoplasmic contents of the cells assume a very granular condition and secrete much fat. As germination proceeds, the pointed mass of cells which surrounds the rudimentary root, and which constitute the root-sheath, is pushed forward by the developing root, and protrudes through the base of the grain. At this stage the barley is said to "chit." The first rootlet soon breaks through the end of the sheath, and is followed in due course by other rootlets. The cotyledonary sheath, or mass of cells surrounding the stem, commences to elongate on the third or fourth day of germination, and in so doing ruptures the proper covering of the seed, it then grows upwards between this and the husk, and forms the acrospire or "spire" of the maltster. After growing to about fifty times its original length, its apex is broken through, and the first green leaf makes its appearance, but in the process of malting the growth of the grain is arrested long before this takes place.

First Signs of Enzymic Action.—When the first rootlet is breaking through the root-sheath, starch begins to appear in the tissues of the germ, also in the protoplasm of those cells of the scutellum which are nearest the secretory layer, and it gradually invades the deeper-seated cells. At this period the contents of the cells of the secretory layer have assumed those appearances which indicate that they are now in a vigorous state of activity.

First Signs of Dissolution.—These first appear in the cellulose walls of the empty cells of the endosperm, which are situated immediately above the secretory layer; these become softened and partially dissolved, the dissolved matter passing into the cells of the scutellum, there to be transformed into starch. The dissolution process gradually extends to the cellulose walls of the whole of the starch-containing cells of the endosperm, and until these are affected no evidence of a solvent action can be observed in the starch grains themselves. Thus it was supposed that the first formed enzyme was solely one which dissolved cellulose, and it was consequently termed a "cytohydrolysist." As the solution of the cellulose proceeds, the tough resistant contents of the endosperm gradually become softened, and if rubbed between the finger and thumb when this stage is completed, appear and feel like a moist powder. This so-called "mealy" or modified condition the maltster endeavours to bring about to its fullest extent in the process of malting, for the degree of friability of the finished malt depends on the extent to which the cellulose has been affected. The cellulose walls of the starch-containing cells of the endosperm of different samples of barley exhibit considerable variations in their readiness to submit to enzymic influence, variations which materially depend on the influence of the soil and climate where the particular samples were grown. Those samples in which this action takes place most readily are the most suitable for malting purposes.

Grüss, in a series of recent observations, found that the cell-walls of the endosperm, after having been subjected to the action of the enzyme, behaved differently towards certain staining agents, and that by employing these he was able to trace the progress of the action of the enzyme with considerable exactitude. When a section of a grain of germinating barley is stained with Congo red and examined under the microscope, those cells which have suffered modification are found to be much more deeply coloured than the remainder. Grüss found that action on the cell-walls of the starch cells of the endosperm, in progressing from the secretory layer, proceeds more rapidly in the neighbourhood of the aleurone layer, and that after eight or nine days' germination the modification is complete throughout the whole endosperm, with the exception, in many cases, of a small portion at the apex. These unaffected portions form the "hard ends" so common in malt. The modification of the cell-walls appears to be only a partial solution, and this is readily explicable when it is considered that the cell-walls consist, according to the researches of Schulze, of two substances— araban and xylan, and that the enzyme, though able to effect the complete solution of the first of these, is unable to dissolve the second. This limited action is, however, sufficient to render the cell-walls permeable to such enzymes as diastase. Reinitzer[1] concludes from his experiments that there is no special cytohydrolytic enzyme secreted, but that it is the ordinary diastase which dissolves the cell-walls. He considers that the cellulose which constitutes these is of a much less resistant nature than the ordinary hemicelluloses, but that it is probably closely allied to them.

Solution of the Starch of the Endosperm.—When the cellulose walls of the starch-containing cells are dissolved, another enzyme begins to make its appearance. This is the "diastase of secretion," so called by Brown and Morris to distinguish it from another diastase which is already present in the barley grain. It is through the agency of the diastase of secretion that the starch conversion in the mash-tun is effected; consequently this enzyme is a body of considerable importance. As in the former case, the

[1] Hoppe-Seyler's *Zeit. f. Physiolog. Chem.*, xxiii. p. 175.

first traces of the action of this diastase are to be found in those starch grains which are situated in close proximity to the secretory layer, and are first observed when the rootlet has grown to the length of the fiftieth of an inch, and the acrospire to about three-quarters that length. In the natural course of growth of the barley plant this solution of the starch grains proceeds until the whole of the endosperm is emptied, but in malting it is restricted within as narrow limits as is compatible with securing the proper modification of the grain. The way in which the starch granules are dissolved by the diastase of secretion entirely differs from that in which the diastase of translocation affects them. The former causes minute pittings on the surface of the granules, which increase in number and depth, and eventually give to the starch grains a very irregularly corroded appearance. A series of star-shaped fissures then appears, the layers of starch of which the grain is built up become detached, and the whole starch grain dissolves, with the exception of some of the more resistant layers which persist for some time.

A diastase exists in fairly large quantities in ungerminated barley (see p. 111), its function apparently being to effect the transportation of starch, during the development of the barleycorn, while it is still attached to the plant. Hence it was termed by Brown and Morris "translocation diastase." The collection of materials which make up the endosperm are the first to be deposited in the seed, the development of the germ taking place at a later stage; and it is from the materials deposited in the endosperm that the germ obtains the matter necessary for the building up of its tissues. The empty cells that are situated immediately above the secretory layer, and which were originally filled with starch, are supposed to have been emptied for this purpose. This diastase dissolves the starch grains without previous corrosion and pitting. By treating microscopic sections of germinating grains of barley with solutions of guaiacum and hydrogen peroxide, Grüss was able to locate the diastase by observing the results it gives with those reagents (see p. 110).[1] In this way he found that, in the ungerminated barleycorn, the cells of the embryo, and especially those of the scutellum, contained diastase, but that it is very rarely present in the starch-containing cells of the endosperm, and then only in traces. When the grain has been steeped, the diastase is found to have left the interior of the cells and passed into their walls. As germination proceeds, the presence of diastase becomes apparent in those cells of the endosperm nearest the scutellum, and in its further progress exactly follows the course of the modification in the cell-walls. Consequently, when barley is ready for the kiln, the diastase ought to have reached every cell of the endosperm,

[1] Pawlewski (*Berichte*, 1897, p. 1313) expresses considerable doubt as to whether the reaction is so characteristic of enzymes as has hitherto been supposed.

and, when sections of such grains are stained with guaiacum and hydrogen peroxide, every cell should be coloured. Also, when the sections are stained with Congo red, all the cell-walls should take this stain deeply, and so show that they are properly modified. Grüss does not agree with the view that the enzyme which has been termed the diastase of secretion is secreted by the germ, for he found that the separated endosperm was able to secrete diastase, and this accounts for the results obtained by Pfeffer and Hanstein (see below).

Some considerable light was thrown, during the course of Grüss's experiments, on the effect of aeration during germination. When barley was germinated in the absence of air, it was found that in eight days the diastase of translocation, though it had permeated the cell-walls of the endosperm, had been unable to effect any modification in them. But the more favourable the conditions of aëration under which the grain was grown, the more was the diastase of secretion produced, and the more was modification of the cell-walls induced. Apparently this diastase is formed by some process of oxidation.

Parasitic Nature of the Germ.—Many plants are known which, being devoid of chlorophyll, are obliged to depend upon other plants for their supply of carbohydrate food. Organisms of this kind which live on living members of the vegetable or animal kingdom are called "parasites;" the various fungi found growing on living trees are instances of these. Many fungi only vegetate on dead organic matter; these are termed "saprophytes." It has been considered that the germ of barley in the earlier stages of germination behaves like a saprophyte; Brown and Morris concluded from their experiments that the endosperm of the barley-corn was a mass of dead matter. It was found possible to remove the germs of barley and of some other seeds from their respective endosperms, and grow them on a substratum of a starch derived from another plant, or even on a layer of gelatin, moistened with a solution of such sugars as glucose, lævulose, maltose, cane-sugar, &c., this last sugar yielding the best results. The secretion of the diastase seemed to be a starvation phenomenon, for the enzyme was only secreted when the embryo was without food; when the separated germ was abundantly supplied with cane-sugar, but trifling quantities of diastase were secreted; when fed on starch, the enzyme was secreted in abundance. Endosperm, from which the germs had been removed, when kept in a moist condition, never evinced the slightest signs of vitality. According to this view, the germ of barley, until it has thrown out its first leaves and is able to manufacture its supply of carbohydrate and proteid material from the constituents of the air and soil, lives the life of the saprophyte, obtaining its whole supply of nutritive material from the dead substances of the endosperm. During this portion of its existence it simply breathes, that is, absorbs oxygen from the atmosphere and gives off carbon dioxide. During the whole of the malting process its life is carried on under these conditions.

Some doubt has recently been thrown upon the view that the emptying of the endosperm is entirely effected through the agency of the enzymes secreted by the germ. Pfeffer and Hanstein have shown that the endosperms of barley, from which the germs had been removed before the commencement of germination, and had been replaced by little pillars of plaster-of-Paris, the lower ends of which were immersed in water, emptied themselves in thirteen days. Precautions were taken to exclude the action of bacteria. Puriewitsch found that the plaster pillars were unnecessary, and that the same object could be effected by keeping the ends of the endosperms in contact with sufficient water. Too little water or the presence of an anæsthetic arrested the action.

Action of the Proteolytic Enzymes.—At the time that the carbohydrate portions of the endosperm are being broken down there can be no doubt that a similar action takes place in the proteids, evidently through the instrumentality of one or more proteolytic enzymes, but we are still in ignorance as to where or how these originate. The proteid grains of the cells of the scutellum disappear at a very early stage in germination; and, subsequently, solution and absorption of the entire contents of the endosperm, which include a considerable amount of proteid matter, takes place. Hence it is only fair to presume that a solvent action analogous to that which occurs with the carbohydrates is at work. We know that much of the insoluble proteid matter of the barleycorn is converted during the malting process into a soluble condition, and that a large proportion of it is also broken down to the amide form, in which state it is not only soluble but eminently diffusible.

The proteid grains of the aleurone layer are the last to be affected; no trace of solvent action can be detected in them until the acrospire projects some one-fourth or one-fifth of an inch beyond the end of the barleycorn; consequently, they are only appropriated at a late stage of germination. These cells apparently undergo no change during the limited time which the germination is allowed to proceed in malting.

Since the barleycorn, during all those stages of its growth which concern the maltster, derives the energy expended in the building up of its tissues from the combustion of carbon, the seed loses weight, and as the loss of weight falls principally on the starch, it is evident that all unnecessary growth implies a waste of starchy material. The barley germ, during that period of its growth in which it is unable to manufacture its own carbohydrate food from the carbon dioxide of the atmosphere, is indebted to the green leaves of the plant on which it was formed for its supply of this nutriment, since these were the agents concerned in the manufacture of the starch stored in its endosperm.

CHAPTER V.

FERMENTATION.

Antiquity of Fermentation.—The process of fermentation, with its accompanying phenomena, was evidently known at a period long antecedent to that of which we have any historical record, for the earliest known writings distinctly show that, at the time they were composed, not only was the comparatively simple method by which wine is produced from the grape in full operation, but that the more complicated processes by which fermented liquors are made from barley and other cereals were also in common use. The ancients knew, also, how to manufacture vinegar, a substance produced by submitting an alcoholic liquid to a subsequent fermentation of an entirely different nature. The practical application of fermentation to bread-making also dates from a very early period.

Beer, and the employment of barley in its manufacture, are frequently mentioned in the ancient Egyptian *papyri*. For instance, the Westcar Papyrus,[1] now in the Berlin Museum, contains frequent allusions to beer and barley being given as presents to the living or as offerings to the dead. This papyrus was most probably written in the twelfth dynasty, and the tales it contains were presumably folklore at a period long antecedent to this; without doubt, they date back to several thousands of years before the commencement of the present era. It is impossible to say what the nature of the beverage brewed from barley in these days was, but that it was a liquid, and not a semi-solid article, as has been suggested, is shown by a remark which occurs in the Ebers Papyrus, where directions are given to infuse a certain quantity of a particular herb in a certain quantity of beer.

Although the ancients possessed an extensive practical knowledge of fermentation, and though we cannot but feel certain that many an intelligent individual, even in those primeval times, must have been struck with its singular and almost mysterious attendant phenomena, and have been induced to reflect on the subject, yet only in comparatively recent times do we find attempts made to give a theoretical explanation of the process. One of the first known instances of this occurs in Pliny's "Natural History," in which a definite statement is made that the fermentation of dough in bread-making is effected through the instrumentality of an acid

---

[1] For an exceedingly interesting translation and account of this papyrus see "Egyptian Tales" (First Series), by W. M. Flinders Petrie.

body. As history progresses, however, we meet with frequent references to the theoretical aspect of the subject, in which opinions of the strangest and most varied nature were propounded from time to time, and which were materially influenced by the prevailing phase of chemical theory of the period in which they were enunciated.

Views of Fermentation during the Period of the Alchemists (about from 350 to 1525 A.D.).—After the fall of the Roman power the seat of science became transferred to Arabia. During the earlier portion of the period which we have under discussion, the Arabians devoted their scientific energies chiefly to observing and recording natural phenomena, as well as to making experiments, and even now we must regard with astonishment the vast and varied amount of knowledge accumulated by this people at that remote period of history. Already in the eighth century Geber taught the method of preparing vinegar, and in the eleventh century Alzaharavius knew how to separate spirit from a fermented liquor by distillation. Traces of this knowledge are still met with in our common orthography; many of those words which begin with *al*, such as alcohol, algebra, alchemy, &c., are derived from the Arabic, and were expressions evolved during the period of which we speak.

This period of observation and experiment was succeeded by that of alchemy pure and simple, and though it was an age of scientific error, still it was one in which much experimentation took place, and this led to the compilation of an enormous mass of chemical facts, as well as to the discovery of many new bodies. As astrology was the precursor of astronomy, so may alchemy be regarded as the foundation upon which the science of chemistry was subsequently built. The alchemists fondly imagined that it was possible to find a substance, vaguely termed "The Philosopher's Stone," which would be able to turn the baser metals into gold, to heal every disease, and confer omniscience on its possessor. Led by the pursuit of this chimera, thousands of the best minds in Europe expended their skill and energy for centuries, and performed countless numbers of experiments for the furtherance of this vain object.

In the writings of this period we often meet with the expression *fermentatio*, but it seems used in a somewhat vague way, and appears to be frequently employed indiscriminately with the words *digestio* and *putrefactio*. Any chemical reaction which was accompanied by external phenomena of a marked character, such as effervescence, was called a fermentation. The action which ensues when an acid is poured over chalk was called a fermentation, since the action was accompanied by the copious evolution of gas. No distinction, however, was made between the reactions which took place between inorganic bodies and those which occurred between organic substances.

The word "ferment" is also frequently met with, and seems to have been similarly used in a vague and dubious sense. Most frequently it appears to imply a substance capable of exciting some kind of germination, and generally seems intended to denote a body endowed with the power of causing a metamorphosis (*transmutatio*) in other bodies. The "Philosopher's Stone" was even looked upon as a kind of ferment, which was to set up a fermentation in the commoner metals, such as lead or quicksilver, and, as it were, awaken in them that dormant germ which, when aroused, should, by a species of germination, ennoble those metals, and finally convert them into the noblest of metals—gold. The view that the process of fermentation had a purifying and elevating effect on the bodies which had been submitted to its influence seems to have been held not only by the alchemists, but also by the philosophers of a later age. Thus Basil Valentine, in the fifteenth century, considered that when yeast was added to wort an internal inflammation was communicated to the liquid, which determined a perfection and separation of the clear parts from the turbid. The fact that alcohol was present in a fermented wort was well known to him, but he imagined it to have been pre-existent in the decoction of germinated barley, and that by fermentation it was freed, as it were by a refiner's fire, from all the contamination and impurities which accompanied it, and which had hitherto prevented it from exhibiting its true properties. This was supposed to be the reason why alcohol could not be obtained by distilling the wort before it had undergone fermentation. It was held back by its accompanying impurities; when freed from these, it could be obtained by distillation.

Basil Valentine may be regarded as the last of the alchemists, for, at the time he lived, knowledge of the facts of nature had progressed to such an extent that philosophers could not help perceiving the utter futility of their search after the "Philosopher's Stone," and when this came to be fully recognised, they began to turn their efforts in other directions, where there seemed to be a greater probability of their being rewarded with substantial results.

The Period of Iatro-Chemistry.—During this period, which immediately succeeded that of the alchemists, science was regarded as being merely subservient to medicine, and the investigations which were undertaken during this time kept this particular idea in view. Fermentation seems to have largely occupied the attention of scientists, and we find many new hypotheses proposed for its explanation.

Libavius (1595) no longer regarded digestion and fermentation as the same, but as allied processes. He was of opinion that the ferment must have a close natural relation to the fermentable body; he also showed that a body will not ferment unless it is in a state of solution or of fine division; he also noted the effect of temperature on fermentation.

In 1648 Van Helmont published several works on fermentation. His views on the subject were of the most remarkable nature. He considered that not only were all such purely chemical phenomena, as the effervescence which ensues when an acid is poured on an alkaline carbonate, processes of fermentation, but that every physiological action carried on in the animal economy, such as the formation of intestinal gases, the formation of the blood itself, &c., were similar processes, and even that small animals could be formed by its agency.[1] Amidst these extravagant views we find others of a more reasonable nature, as, for instance, that in which he states that out of the ferment something passes into the fermenting fluid which grows in it as a seed vegetates in the earth, and through its agency the fermentation is brought about. This is to a certain extent in agreement with modern views of the subject, and had Helmont discovered the fact that it was the ferment itself which grew, and thereby caused fermentation, he would have anticipated the modern view of Pasteur. Helmont also recognised the fact that the escaping gas was an entirely different substance from the alcohol left behind; and that this gas, which he calls "gas vinorum," was derived from something which, without fermentation, would have appeared as carbon.

Sylvius de la Boë, in 1659, distinctly perceived the marked difference in the nature of the two processes, fermentation and the effervescence which occurs on the addition of an alkaline carbonate to an acid. He says that though an evolution of gas occurs in both instances, in the former case the compound acted upon suffers decomposition; in the latter, a new compound is formed.

The views expressed on the subject of fermentation in the writings of Becker show a marked advance on those of his predecessors. He points out that fermentation can only be induced in sweet liquids—that the alcohol is not pre-existent in the fermentable fluid, but is formed during the fermentation. He regards fermentation and combustion as analogous processes, a view which has only been abandoned in quite recent times. He seems also to have had an indistinct idea that the presence of air was necessary to fermentation. According to his views, there are three distinct kinds of fermentation; the first that which we call now effervescence, the second the alcoholic fermentation, and the third the acetic fermentation. He looks upon fermentation and putrefaction as similar but distinct processes, and distinguishes them by

---

[1] Helmont believed that the effluvia arising from morasses produced frogs, snails, leeches, &c. But still more curious is the receipt which he seriously gives for producing a pot of live mice. To effect this, he says, it is only necessary to close the mouth of a pot containing a little corn with a dirty shirt. The ferment evolved from the shirt, modified by the odour of the corn, effects in about three weeks' time the transformation of the corn into live mice. These, according to Helmont, who states that he has actually witnessed the phenomenon, are born full grown, and of both sexes, and that, in order to continue the progeny, it is only necessary to pair them.

the character of their products; in the first case the substance is improved in its nature, in the second it is deteriorated. He made the important observation that the juice of grapes, the skins of which had not been ruptured, never entered into fermentation, and this he considered to be due to the fact that the juice never came into actual contact with air. He recognised the fact that alcohol and salt of tartar, when present in sufficient quantity, impeded and finally arrested fermentation.

Willis, an English physician, who wrote in 1659, attributed much importance to fermentation; he considered that all the processes carried on in living bodies are caused by its agency, and that diseases are caused by fermentations which have deviated from their ordinary healthy course. He regards a ferment as a body which is able to communicate a portion of its own internal movement to another body, and thus to induce in it fermentation. This view of the subject, which is known as "the mechanical theory of fermentation," afterwards, with certain elaborations, formed the basis of the theories of Stahl and Liebig.

Views on Fermentation during the Age of the Phlogiston Theory.—With the enunciation of the theory of phlogiston by Stahl in 1697 there commenced a new era in philosophy. The chief merit of Stahl and his school was that they no longer cultivated science as a subject dependent on some object, but for its own sake alone. Natural phenomena were no longer investigated in the light of their utility or otherwise to mankind, but simply in order that new truths might be established.

The doctrine of phlogiston, which was propounded by Stahl to explain the phenomena of combustion, dominated the scientific world for a very long period. According to this theory, every combustible substance was a compound of phlogiston and some other body. The phenomena of combustion in a body were caused by the escape of the phlogiston which it contained, and the more of this hypothetical substance entered into its composition the more combustible was that body. Such a highly combustible body as alcohol was supposed to contain a large amount of phlogiston. The adherents of the theory were well aware of the fact that the weight of a dephlogisticated body was often greater than its original weight, and they sought to explain this anomaly by assuming that phlogiston was a body possessing negative gravity, and that, though every other body in nature was attracted to the earth by the force of gravitation, it was repelled. Consequently, phlogiston, when present in a substance, tended to buoy up and render it lighter; hence, after its escape, the substance became heavier.

Stahl paid much attention to the subject of fermentation, and recorded his peculiar views on it in a work, *Zymotechnica Fundamentalis*, which was, however, only published after his death. He regarded fermentation and putrefaction as analogous processes, and considered that a body which is itself in a state of putrefaction can readily cause another body, capable of undergoing a similar change, to assume a like condition; such a body, itself in a state of active internal motion,[1] can readily communicate this motion to another body, which, though its particles may be at present in a state of rest, are easily set in motion. Stahl seeks to explain the reason why bodies are produced during fermentation which differ from those originally present in the liquid by assuming that the ultimate particles [2] of which the original substances were composed (salt, acid, combustible substance, water) are first disassociated by the internal motion excited by the ferment, these then re-combine to form more stable compounds, in which the same elements exist, but are combined in different proportions. If the fermentation took such a direction that the combustible particles predominated in the newly-formed body, then it was an alcoholic one; if the acid particles predominated, then such a fermentation was an acid one. For more than eighty years the views of Stahl were held by the members of the scientific world, amongst whom we find some of the most honoured names; and though modifications were proposed from time to time by his immediate successors, they were of an unessential nature. One of these, enunciated by Boerhave, a celebrated scientist and physician, was, that substances of animal origin alone were capable of undergoing putrefaction, whilst those derived from the vegetable kingdom were alone capable of entering into fermentation. A German chemist, Wiegleb, endeavoured to show that the products of fermentation were already present in the liquid before fermentation, but this was merely a return to the older doctrine. An attempt was made to obtain a more intimate knowledge of the nature of fermentation by a study of the gases which were evolved during the process. The English chemist Cavendish, in 1766, not only established the identity of the gas evolved during fermentation with that evolved when an acid is added to an alkaline carbonate, but also actually measured the gas. He found that when 100 parts of sugar were fermented, 57 parts of carbon dioxide were evolved; and this may be looked upon as the first experiment in which an attempt was made to quantitatively estimate the products of fermentation.

Views on Fermentation during the Period of the Foundation of Modern Chemistry.—The distinguished French chemist Lavoisier, who was born in 1743 and died in 1794, may be regarded as the founder of modern chemistry; he and his associates effected a complete reform in chemical philosophy, and in this way effectually banished the theory of phlogiston.

---

[1] This expression, "internal motion of the particles of a body," and the previous one of Willis, must not be confounded with that modern view of the vibration of the ultimate atoms of matter, as we now understand it, when speaking of the mechanical theory of heat.

[2] It must be remembered that at this period the elements, as we regard them at the present day were unknown.

The doctrines of Stahl and his school, both on this subject and on that of fermentation, then gradually fell into oblivion. By the extensive application to his work of a systematic method of measuring and weighing, as accurate as the means at his command would permit, Lavoisier introduced methods of precision into the domain of chemistry which had been hitherto unknown. He determined the relative proportion of the elements in water and in the atmosphere; discovered the important part played by oxygen (which element had been previously discovered by Priestley in 1774) in the process of combustion, and this had a profound influence on the views then held with regard to the phenomena of fermentation; he also showed that oxygen was absolutely necessary for the acetic fermentation. He paid considerable attention to the subject of fermentation, and estimated quantitatively the elementary composition of sugar and alcohol. By weighing the sugar introduced into a liquid previous to fermentation, and estimating the respective quantities of the alcohol and carbon dioxide formed during that process, he found that the weights of these two latter bodies when added together were equal to that of the original sugar, and hence came to the conclusion that, in fermentation, the sugar was simply converted into alcohol and carbon dioxide. Though the figures he obtained are now known to be far from accurate, this cannot be wondered at when we consider the primitive nature of the apparatus at his command, and it was only by a singular series of compensating errors that he obtained this result. He believed that the yeast suffered no chemical change in the process, for, as the whole reaction was explained quantitatively in the above manner, he did not see any place for the participation of the yeast. In his first formula, published in 1789, Lavoisier does not seem to have regarded the formation of acetic acid, which body he invariably found to be produced in all his fermentations, as an essential part of the process; afterwards, however, he came to a different conclusion, and propounded formulæ in which the formation of acetic acid also was accounted for. Fermentation from his point of view being purely a chemical process, he regarded sugar as an oxide, which during fermentation was split up into two other oxides; the one (carbon dioxide) richer in oxygen, the other (alcohol) poorer in this element.

The next important theory on this subject was advanced by Fabroni in 1787, and this excited great attention at the time it was propounded. Fabroni, who regarded fermentation as a purely chemical process, considered that yeast was very much allied in nature to the gluten of plants, and he demonstrated by experiment that fermentation could be induced in a saccharine liquor by the addition of gluten. He also proved experimentally that fermentation could not be induced in uninjured fruits, and drew from this the correct conclusion, that, in order that fermentation may be induced, the ferment and the fermentable body must

VIEWS ON FERMENTATION.

be in actual contact. He imagined that the carbon dioxide gas evolved during fermentation was evolved from the ferment, and not from the fermenting substance; also that the alcohol was not pre-existent in the fermented fluid, but only arose during the distillation.

Although many chemists continued to turn their attention to the phenomena of fermentation, nothing of a particularly important nature was contributed to the subject until we come to Gay-Lussac, who propounded a fresh theory. In this, he considered that oxygen was the sole cause of all fermentation phenomena, be they those of the alcoholic fermentation, of putrefaction, or of decay. According to his view, fermentation was caused by a body which, though not itself possessed of as much stability as the fermentable matter present in the solution, was acted upon by oxygen, under the influence of which its particles were set in motion. This motion being communicated to the particles of the other and more stable body, they were likewise set in motion, and thus the whole body was brought into a state of fermentation. Gay-Lussac performed many experiments which seemed to fairly prove his conclusions. In one of the most important of these, he enclosed some uninjured grapes (after having washed them in an atmosphere of hydrogen) in a vessel, which was then filled with air-free mercury. The grapes were then pressed in such a way that the grape-juice did not come in contact with air. The juice obtained under these circumstances did not enter into a state of fermentation, whereas it was notorious that when grapes were expressed in the usual manner in the presence of air, the juice speedily fell into a state of active fermentation. When, however, the smallest quantity of pure oxygen was admitted to the grape-juice floating on the surface of the quicksilver in the former case, fermentation speedily commenced. The well-known action of sulphurous acid in arresting fermentation also lent a powerful support to this theory. This acid has a powerful affinity for oxygen, and its marked effect on the process of fermentation was attributed to the complete manner in which it removed the oxygen from the fermenting fluid. A method for preserving provisions had been introduced to public notice by Appert about this time (1810), which apparently added considerable weight to Gay-Lussac's view. By this process provisions of the most varied nature, such as soups, meat, vegetables, &c., were preserved for an indefinite period by enclosing them in a perfectly air-tight vessel which was subsequently exposed for a certain length of time to the temperature of boiling water.[1] The method is very similar to the one of which such frequent use is made at the present day, and by means of which much valuable food is imported from distant countries.

It was sought to explain the rationale of the process by assuming that the small quantity of oxygen present after the vessel had been

[1] An exactly similar process for the preservation of vinegar had been proposed by the celebrated Swedish chemist Scheele in 1782

sealed up combined with the substance contained therein, and being in this way withdrawn from the sphere of action, was not permitted to exercise its power of inducing fermentation. Another curious fact had been observed which seemed to add considerable support to Gay-Lussac's theory. A bottle of grape-juice which had been preserved for a year by Appert's process was uncorked and transferred to another bottle, when it soon fell into a state of fermentation. It was supposed that the fermentation was caused by the short contact of the fluid with the oxygen of the atmosphere which took place during its transference from one vessel to the other.

Thénard, to whom we owe some important contributions to the science of fermentation, framed the following theory on the subject, in which oxygen played an important part. He assumed that the yeast withdrew oxygen from the sugar, and in this way effected its conversion into alcohol and carbon dioxide. He also demonstrated the incorrectness of the theory of Fabroni, by showing that purified gluten was unable to excite alcoholic fermentation in a saccharine fluid.

In the meanwhile, from the time of Lavoisier to that of Gay-Lussac, chemical apparatus had been considerably improved, and chemists were enabled to obtain much more accurate analyses of many of the commoner organic substances met with; amongst these were alcohol and sugar. It was also shown that alcohol was present in the fermented fluid before distillation, a matter on which some uncertainty had previously prevailed. After studying the composition of the substances taking part in the process of fermentation, in the light of the more correct knowledge of their actual composition, and drawing his deductions from a consideration of the decomposition which ought to occur rather than by direct experiment, Gay-Lussac was led to propose the following formula for the change which takes place when sugar is subjected to the alcoholic fermentation :—

$$\underset{\text{Sugar.}}{C_6H_{12}O_6} = \underset{\text{Carbon dioxide.}}{2CO_2} + \underset{\text{Alcohol.}}{2C_2H_6O}$$

This formula is found in the text-books of the present day, and is so near the truth that it is the one we now invariably employ when we require a short and concise expression for the reaction which takes place during the alcoholic fermentation of sugar.

Between the time of Gay-Lussac and Liebig, fermentation continued to engage the attention of many men of science, and various views were propounded by Trommsdorf, Foucroy, and others. Amongst these, Kamtz expressed the peculiar view that fermentation was an electrical phenomenon, that sugar formed the negative and the yeast the positive element in an electric couple, and, under the influence of the electric tension, the sugar was split up into the negatively-electric carbon dioxide and the positively-electric alcohol. Berzelius invented the name "catalysis" to explain a number of phenomena which it was not possible to account for at that time by the ordinary laws of chemistry. The characteristic of such a body is the power which a small quantity possesses, without itself undergoing any change, of inducing a change in a large quantity of another substance. He sought to explain the process of alcoholic fermentation by assuming that yeast was a substance endowed with an immense catalytic power, in virtue of which it was able to set up decomposition in large quantities of a fermentable body. With the progress of time many of the processes which were referred to catalysis by Berzelius have found their true explanation, and the use of the word is now all but abandoned.

The Views of Liebig on Fermentation.—We now come to the theory of Liebig, propounded in 1839, that ruled the scientific world for thirty years, and which has undoubtedly much in common with that proposed by Stahl a hundred years before. In both these theories the ferment is regarded as a body which is itself in a state of decomposition, and which is able to communicate this peculiar state to other bodies of a more stable nature; it is, in fact, a sort of infection. Both theorists assume the hypothesis that the smallest particles of a decomposing body are in a state of movement, and that when such a body exists in solution along with another body, also to some extent of an unstable nature, the former is able to communicate this motion to the latter. Bersch,[1] in illustrating the action which, according to this theory, takes place during fermentation, mentally pictures a single snowflake which commences to roll down a mountain-slope covered with snow. In rolling, it carries along with it another snowflake, and these in turn communicate motion to others, until at last the flakes set in motion are so numerous that an avalanche is the result, which, from the collective energy of the enormous number of particles of which it is made up, is capable of exhibiting the mightiest effects of force. So, in like manner, the decomposition actually taking place in the smallest trace of a ferment may be transferred to, and occasion the decomposition of, almost unlimited quantities of a fermentable substance. This view also yielded a rational explanation of the well-known fact that an almost imperceptible trace of yeast is able to ferment away an infinitely large amount of fermentable matter.

But while Stahl only proposed this action as an explanation of the phenomena of fermentation, Liebig endeavoured to show that it was merely one of the ordinary manifestations of the chemical laws common to the inorganic as well as the organic realms of chemistry, and he brought forward innumerable examples in attestation of this view; such, for instance, as the solution of platinum when alloyed with silver in nitric acid; the action of hydrogen peroxide in solution, which, when brought into contact with many metallic oxides, not only gives off its own oxygen, but

[1] *Die Hefe u. d. Gahrungs-Erscheinungen*, i. 28.

also induces the metallic oxides to yield up a portion of their own oxygen; the action of nitric oxide in the manufacture of sulphuric acid, &c. He also adduced the action of enzymes, which at that time were all classed together under the category of ferments, and considered that when a true explanation of the action of these should be found, it would also elucidate the action of the organised ferments. Bodies of an explosive nature were also adduced; these have their particles in such a loose state of combination, that moderate heating or a slight blow causes their separation. He held that, to a certain extent, the fermentable bodies were similarly constituted, their particles only being held together by the "force of inertia," or, as we should now say, by a weak chemical affinity. The slight impulse given to such compounds by the ferments sufficed to upset their equilibrium, and to induce in them a decomposition of a nature somewhat akin to an explosion.

Certain observations which had been made before Liebig's time lent an apparent confirmation to this view of the nature of ferments. Thénard had shown that yeast, when left to itself in a moist condition, rapidly fell into a state of putrefaction; Colin had found that yeast diminished in quantity during fermentation; Döbereiner had discovered that ammonia was produced during fermentation, and the formation of this substance is an almost invariable concomitant of the putrefaction of nitrogenous organic matter. In the two latter cases the facts observed were only true under certain conditions, and they never occur when a fermentation pursues a normal course. Liebig considered that fermentation, putrefaction, and decay were all caused in the same manner; fermentation was only a special case of putrefaction, in which the oxygen of the atmosphere co-operated, and in which the fermentable body was a nitrogen-free substance, just as the formation of acetic acid in the manufacture of vinegar might be regarded as a putrefaction of alcohol.

The fact that yeast was a living organism was to Liebig a matter of indifference; he considered it merely an accidental circumstance, and that its growth and other vital phenomena were in no way concerned with the process of fermentation. He adduced in support of this argument the production of acetic acid from alcohol, which could readily be effected without the intervention of a living organism body.

In later times, when the views of Pasteur began to meet with some acceptance in Germany, Liebig brought forward the following objections to them in a lecture delivered before the Bavarian Academy in 1868. He contended that, as all yeast contained a large amount of proteid matter, of which sulphur is an essential constituent, he could not conceive that, as had been asserted by Pasteur, yeast multiplication could take place in a solution consisting solely of water, sugar, ammonia, and "phosphates." With the increase of yeast new proteid matter must be formed, and this was impossible in the entire absence of sulphur. Pasteur had, however, when he spoke of "phosphates," meant the ash of yeast, and this contains a notable amount of sulphates. Objection was also taken to the reason assigned by Pasteur for the disappearance of the ammonia, since he regarded it as being utilised in the formation of new proteid substance; Liebig affirmed that it became converted into ammonio-magnesium phosphate, and as such was precipitated and withdrawn from observation. It has, however, been shown since then that yeast can obtain its supply of nitrogenous food from this salt.

This was followed, in 1870, by a series of articles in the *Annalen der Chemie und Pharmacie*, in which Liebig reiterated his former objections, and made a strenuous effort to overturn the theory propounded by Pasteur. At the same time considerable concessions were made; Liebig not only no longer denied that yeast development was a necessary concomitant of fermentation, but also completely renounced his previous theory of 1839. Still he attempted to explain the process in a way which should to some extent resemble his previous theory. In this he assumed that, during the life of the yeast, there was formed in the interior of the yeast-cell an enzyme, similar to diastase or invertase, which had the power of converting sugar into alcohol and carbon dioxide; and that the only correlation between the physiological act and the phenomena of fermentation is the production in the living cell of a substance which, acting as an enzyme, effects the decomposition of the sugar. The physiological act is only necessary for the production of this substance, and in no other way is it concerned with fermentation. This formed Liebig's last contribution to the subject of fermentation.[1] He died in 1873.

Traube's Views on Fermentation.—A new theory of fermentation was enunciated by Traube in 1858.[2] According to this, an enzymic body, having a definite chemical composition, being closely allied in its nature to the proteids, and which has a powerful affinity for oxygen, is formed in the interior of each fermentation organism. This he considered to be endowed with the peculiar property of taking up oxygen from one portion of the molecular group of a fermentable body and transferring it to another portion, but that it did not act in the manner suggested by previous observers, *i.e.*, by transferring the motion of their own particles to those of other bodies. He divided these supposed enzymes or oxygen-carriers into two groups, the first of which absorbed free oxygen and transferred it to another body, as in the conversion of alcohol into acetic acid; consequently, for the performance of their special function they required the presence of free oxygen. The second group, which he called "reducing enzymes," were supposed to transfer combined oxygen, and to this class belong the alcoholic ferments. In order to explain the action of the latter, he supposed that fermentable bodies consisted of two

[1] See Appendix D.   [2] *Theorie der Fermentwirkungen*, Berlin.

molecular groups, the former of which had an affinity for hydrogen, and as the enzyme had an affinity for oxygen, the two, acting conjointly, were able to tear asunder a molecule of water. The hydrogen of the water passed to this first molecular group, and the oxygen to the enzyme, and this constituted a true reducing action. The first group was the one which became alcohol in the alcoholic fermentation. The second molecular group possessed an affinity for oxygen, and this was satisfied by the oxygen that the enzyme had torn from the molecule of water, and which it now yielded up to this group; this in the alcoholic fermentation became carbon dioxide. During the course of this action the enzyme was not destroyed, but simply acted as an intermediary agent. As the oxygen, according to this theory, was entirely derived from water, this particular class of fermentation was able to proceed in the entire absence of free oxygen.

The Physiological or Vital Theory of Fermentation. —During the period we have just been discussing, the view that fermentation was intimately connected with the growth and development of the yeast organism had steadily been gaining ground amongst investigators; but owing in great measure to the vast authority of Liebig at that time, it was discredited, and even ridiculed.[1]

Yeast was examined for the first time microscopically, and its organised structure recognised, by Leeuwenhoek in 1680. He found that it consisted of spherical bodies, which he was, however, unable to distinguish from starch-cells; but we cannot wonder at this when the rudimentary form of the first microscopes is taken into account. Thénard, at the beginning of the present century, expressed the view that yeast was of an animal nature, but he only arrived at this conclusion on purely chemical grounds; he found it was extremely rich in nitrogen, and resembled in this respect the composition of animal bodies. In 1818 Erxleben, after examining yeast under the microscope, definitely pronounced an opinion that it was of a vegetable nature, and that it must be

---

[1] A curious specimen of this form of sarcastic ridicule, which undoubtedly came from the incisive pen of Liebig, was published in the *Annalen der Pharmacie* (vol. xxix. p. 100), one of the first and most serious scientific periodicals of that day. The writer professed that he had examined beer-yeast under the microscope, and found it to consist of an infinite number of minute spheres, which, when placed in a solution of sugar, were found to be eggs, for they rapidly developed into little animals having the appearance of an alembic without a receiver. That which resembled the tube of the helmet was a kind of trunk covered with long hairs. Though no teeth or eyes could be seen, a stomach, digestive canal, and urinary system were distinctly visible. These exceedingly voracious beasts greedily devoured the sugar, which, after undergoing digestion in the stomach, was excreted partly as alcohol from the digestive tract, and partly as carbonic acid by the urinary system. Such was the voracity of these little creatures, that they devoured and excreted in eighteen hours sixty-six times their own weight of sugar When all the sugar had been eaten up the animals perished, leaving behind them a plentiful supply of eggs, and these constituted yeast.

VITAL THEORY OF FERMENTATION.

regarded as the cause of fermentation. In 1822 Persoon microscopically investigated the film which occasionally appears on wine, and found it to consist of cells very similar to yeast-cells. Desmazières, in 1825, confirmed this observation, and considered the two sets of cells to be very nearly allied organisms; he, however, believed them to be of an animal nature. About this time it began to be generally conceded that yeast consisted of vegetable organisms. Cagniard de Latour, in 1835, made an exhaustive examination of yeast under the microscope, and found it to consist of minute spherical organisms, which reproduced themselves by means of buds. The appearance of yeast growing in a fermentable fluid is given in the accompanying figure (Fig. 45); in this, yeast-cells in various stages of growth and development may be seen.

As these organisms did not exhibit any power of movement, Latour considered that they were of a vegetable nature. He did not, however, completely abandon the old ideas, but held that in some cases oxygen might be the cause of fermentation. In 1838 Turpin formulated the vital theory of fermentation. He said: "Fermentation as effect, and vegetation as cause, are two things inseparable in the decomposition of sugar." About the same time Kützing and Mitscherlich, independently of one another and of other observers, came to the conclusion that beer-yeast consisted of organisms which were intimately connected with fermentation, though the latter considered that the organisms only acted by catalysis. It had now been decided by the majority of investigators that yeast was a vegetable organism, and the question arose as to what position should be assigned to it in that kingdom. To some it appeared to be a fungus without mycelium, to others it closely resembled the *Algæ*. While Turpin regarded the organisms as the spores of plants belonging to the genus *Torula* of Persoon, Meyen considered that yeast was a fungus, and created for it a new genus, the *Saccharomycetes;* this name was also adopted by Reess, Engel, &c. Kützing, who regarded the ferments as *Algæ*, arranged them in a separate genus of that class, which he named the *Cryptococci*. Later on we shall see that the yeast organism is capable of reproducing itself by spores, and now it is generally conceded that it belongs to the fungi.

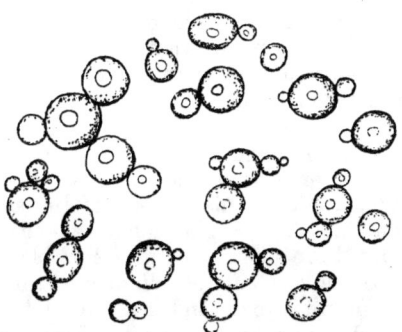

FIG. 45.—Yeast Growing in a Fermenting Fluid.

We now come to the important work of Schwann, carried out

during 1836–37, for with it commenced the real solution of the question. Though undoubtedly he arrived at his conclusions at a period subsequent to that of Erxleben, Latour, &c., yet it must be placed to his credit that he not only stated the proposition that fermentation was due to the presence of living organisms, but he also proved the correctness of his views by a number of the most ingenious experiments. Many years after the publication of the experiments of Schwann, we see them repeated in almost the same way, and with the same results, by Pasteur. It was, however, reserved for this latter investigator to obtain general acceptance for the vital theory of fermentation.

The Doctrine of Spontaneous Evolution.—At the time Schwann carried out his investigations, this doctrine, which assumed that it was possible for living organisms to be developed out of lifeless matter without the intermediate agency of eggs, seeds, or germs, was engaging the serious attention of the scientific world, which it divided into two great parties, the one affirming that living organisms could originate from lifeless matter, the other, that this never occurred. It was with a view to solving this problem that the experiments and observations of Schwann were undertaken. The well-known fact that an animal or vegetable infusion, when left to itself for a time, was invariably found to teem with living organisms, seemed to give some weight to the opinions of the upholders of the theory of spontaneous evolution; and as these organisms were of the simplest nature, an investigation into their origin and development seemed to be peculiarly adapted for the solution of the problem.

Some very curious opinions on this subject had been held from ancient times to the end of the Middle Ages; it was then an accepted fact that even large organised beings, such as mice, frogs, snakes, caterpillars, &c., were evolved spontaneously. These errors had been to a certain extent exploded by the criticism of the seventeenth century; but when, during the latter part of that century and the earlier part of the eighteenth, the microscope came to be largely employed, and by its means there was discovered a host of minute organisms in connection with which no sexual reproduction could be observed, then again the theory of spontaneous evolution became a prominent question. It was assumed that particles of matter, which at one time had entered into the composition of organised beings, after the death of these possessed the power, under favourable circumstances, of uniting into new forms of animated beings. On the other hand, there were always philosophers who maintained that this was not the case. The question of spontaneous evolution was brought prominently into notice by Needham, an English monk, who wrote in 1745, and also by Buffon in 1749. They showed that an aqueous extract of meat placed in closed vessels, after exposure to the heat of boiling water, was able to develop living organisms, which these observers conceived

DOCTRINE OF SPONTANEOUS EVOLUTION. 295

could have been formed in no other way than by spontaneous generation. Many other observers took an entirely opposite view on the subject, one of the most notable of whom was Spallanzani; he commenced in the year 1765 the publication of the results of a series of experiments, the results of which were in direct contradiction with those of Needham. Spallanzani exposed hermetically sealed flasks containing meat-extract solution for an hour to the temperature of boiling water, and found that in these no organisms made their appearance afterwards. Immediately, however, the seal was destroyed, and atmospheric air allowed to have access to the contents of the flasks, organisms quickly developed. From this he concluded the organisms did not arise spontaneously, but were derived from germs present in the air. In the former case these had been destroyed by the heating, but in the latter they obtained access to the flasks in an unaltered condition, and consequently were able to develop. To these experiments Needham and his adherents made the following objection: that, during the heating, the contents of the flasks had been so maltreated that they had lost their power of vegetative reproduction, and that the small quantity of air included in the flask had also been similarly corrupted.[1] Schwann now proved that air might be admitted in any quantity to solutions which had been boiled in flasks, without causing fermentation or putrefaction, provided the germs contained in such air were destroyed. This was effected in the following manner:— Each flask was provided with a stopper perforated with two apertures, into which tubes bent at an angle were inserted; one of these served for the introduction of air which had been previously passed through a red-hot tube. In this way he showed that though large quantities of air might in this manner be passed through the fluids in the flasks, in no case did fermentation or putrefaction take place in their contents. About the same time Schulze performed a series of experiments with flasks of a similar construction, but instead of heating the air, he passed it through strong sulphuric acid; he obtained results similar to those of Schwann. It was now objected by the opposite party that in all these experiments the air had been so altered in the treatment it was subjected to before being passed through the solution that it had lost its power of causing the dead particles to re-form themselves into living organisms. To demonstrate that this was not the case, a series of ingeniously contrived experiments were made in 1854 by Schröder and Dusch, in which the air, instead of being either heated or passed through strong acid, was simply caused to pass through a filter of cotton-wool.[2] Results similar to those of

---

[1] It must be remembered that at this time nothing was known definitely of the chemical constitution of the atmosphere.

[2] The use of cotton-wool for this purpose was suggested by some experiments of Löwel, who found that common air, after filtration through this material, was so far freed from solid particles as to be unable to cause the crystallisation of a super-saturated solution of sodium sulphate.

Schwann were obtained by this modification in all cases, with the exception of milk, so that even yet there was something that remained to be explained. It was, however, distinctly proved that such a fluid as beer-wort or grape-juice could be rendered completely sterile by boiling, and that by means of a simple filter of cotton-wool it was possible to remove those organisms from the air which caused fermentation or putrefaction. Undoubtedly the flasks of Schwann were the precursors of those so much employed afterwards by Pasteur and other investigators; and the cotton-wool filter invented by Schröder and Dusch is largely employed at the present time, both in the laboratory and in the apparatus for the cultivation of pure yeast.

Pasteur's Investigations on the Question of Spontaneous Generation.—In the pursuit of his studies on fermentation Pasteur was compelled to inquire into this subject, and this he did in such a masterly manner that the question may be considered as settled for ever.

He had previously observed that when solutions of sugar, to which small quantities of the magnesium and potassium phosphates and a little ammonium sulphate had been added, were kept warm for a time, fermentation, accompanied by the growth and development of certain bacteria, almost invariably set in, during which the sugar gradually became converted into lactic acid. When this stage had been reached, the mixture, if allowed to remain for some time longer, entered into another fermentation, accompanied by an entirely different set of bacteria, during which the lactic acid gradually became transformed into butyric acid. As such a solution, in which nothing but mineral substances were present, seemed peculiarly unfitted for the spontaneous generation of living organisms, Pasteur was led to conclude that the germs of these two different organisms could have obtained access to the fluid in no other way than by falling into it from the atmosphere. If this were the case, then it ought to be possible to find these organisms in the atmosphere.

Experiments showing that the germs of living organisms are present in the atmosphere.—With a view to ocularly demonstrating the presence of the germs of organisms in the atmosphere, Pasteur, after having inserted a plug of gun-cotton in a tube, drew a large quantity of air through it by means of an aspirator. He then dissolved the gun-cotton in a mixture of ether and alcohol, and examined the undissolved matter under the microscope. This was found to consist of minute particles partly of a mineral and partly of an organic nature; very frequently there were also present minute bodies which were undoubtedly the eggs of infusoria and the spores of fungi. The treatment with ether and alcohol unfortunately destroyed the vitality of any of the germs present, hence it was impossible to study their further development.

Experiments in which germs collected from the atmosphere

were actually grown.—Sweetened yeast-water is a fluid particularly prone to undergo change when exposed to the atmosphere; if boiled for a few minutes and allowed to remain in an uncovered vessel for a few days, it invariably falls into a state of change, and is invaded with bacteria, torulæ, mould fungi, &c. Pasteur demonstrated that this liquid could be preserved indefinitely without undergoing change if calcined air only were permitted to come in contact with it. In order to demonstrate this, the necks of a series of flasks were each drawn out to a narrow tube, and a quantity of sugared yeast-water introduced into each flask. The neck of each of these was now connected in turn, by means of indiarubber tubing, to a platinum tube maintained at a red heat. The contents of the flask were then boiled for two or three minutes, and, without being disconnected from the apparatus, its neck sealed up with the blowpipe. A series of flasks that had been treated in this way, and which were subsequently kept at a temperature of 30° C. (83° F.) for a considerable length of time,[1] did not show the slightest sign of change.

By an ingenious modification of the apparatus used in the foregoing experiments, a plug of ordinary cotton-wool, on which atmospheric dust had been collected in a similar manner to that in which it had been collected on gun-cotton, was introduced into one of these flasks charged with the sweetened yeast-water which had been boiled for a few minutes, precautions being taken that nothing but calcined air obtained access to the plug or to the interior of the flask.

In a few days numbers of organisms made their appearance in the liquid, similar to, but far greater in number than, those which would have appeared in the fluid had it been simply left exposed to the air. It is worthy of remark that the alcoholic fermentation was never met with in these experiments, though the fluid employed was one eminently adapted for undergoing this change. This series of experiments formed an effectual rejoinder to the objections which had been urged against the similar experiments of Schwann, viz., that the air could be so corrupted by incineration as to be incapable of starting life anew.

Liquids eminently liable to undergo change may be preserved for an indefinite period in free contact with atmospheric air, provided such air is freed from floating particles.—In order to demonstrate this, quantities of fluids liable to change, such as wort, grape-juice, &c., were introduced into flasks of convenient size (250 to 300 c.c.), the neck of each flask being afterwards drawn out to a fine tube and bent into the shape of a letter S. The contents of each flask were then boiled until steam freely issued from the orifice of the capillary tube, when the whole was allowed to cool.

---

[1] This treatment is now technically known as "forcing," and small samples of finished beer are frequently tested in this manner to ascertain if they are contaminated with bacterial organisms.

In this case, in addition to the liquid itself being sterilised during the boiling, the heated steam effectually sterilises the walls of the flask and the inside of the curved tube. As the temperature lowers during the subsequent cooling, air gradually enters to replace the steam, and any germs that it may happen to contain are caught and retained by the drop of condensation water which collects in the tube; they are in this way prevented from obtaining access to the contents of the flask. The tube eventually becomes dry, but the passage of air in and out of the flask is now so slow that its floating particles are all deposited on the side of the tube by gravitation. Organic fluids, such as beef-tea, grape-juice, &c., which are extremely liable under ordinary circumstances to enter into putrefaction or fermentation, and which are capable of being sterilised at a temperature of 100° C. (212° F.), may be preserved in such flasks for years. This form of flask was the precursor of the well-known Pasteur flask.

The Pasteur Flask.—In this, the form of the original flask is retained, but it has in addition a short straight tube inserted into its bulb, as shown in Fig. 46. This additional tube permits of additions to or withdrawals from the contents of the flask being made under suitable precautions, without running any risk of introducing foreign organisms. A suitable quantity of wort or other fluid having been introduced into the flask, a small piece of indiarubber tubing is attached to the side-tube, and the contents of the flask boiled. As the side-tube possesses the larger orifice, the steam rushes out of this first, effectually sterilising it and the indiarubber tube. When the boiling has proceeded for a few minutes the rubber tube is stopped by a piece of glass rod. The boiling is continued for a few minutes longer, and, as the steam is now compelled to pass through the fine curved tube, this is also sterilised. In order to prevent the chance of infection, which has occasionally been caused by insects crawling up the capillary tube and carrying germs with them, a small quantity of asbestos, previously ignited in the flame of a spirit-lamp or Bunsen burner, is inserted into the orifice of the fine tube just after the flask has been removed from the source of heat. When all these precautions have been fully carried out, the wort in such a flask will keep unaltered indefinitely. If there is any doubt that the sterilisation has not been perfect, the flasks so prepared may be easily tested by being "forced," that is, kept for a few days at a temperature of 24° to 28° C. (75° to 83° F.).

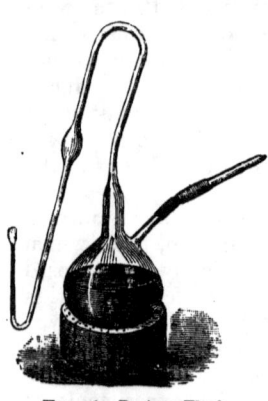

FIG. 46.—Pasteur Flask (*after Hansen*).

Such a temperature is exceedingly favourable for the growth and development of bacterial organisms, and in those extremely rare instances in which the sterilisation has not been perfectly attained, the presence of organisms is readily detected by the alteration in the appearance of the fluid which ensues.

Organic fluids, such as blood, serum, urine, &c., when taken from a healthy animal, may be preserved indefinitely without undergoing putrefaction if only allowed to come in contact with germ-free air.— The fluids contained in the various vessels and cavities of an animal when in a state of health are perfectly free from organisms, though, as is evinced by the extreme readiness with which they enter into a state of putrefaction under ordinary exposure to the air, they form excellent media for the growth and development of bacteria. It has also been conclusively shown that the blood of living animals does at times become invaded by organisms, and serious diseases are produced thereby. Pasteur demonstrated that such fluids as blood and various other secretions of the animal body could be preserved indefinitely in contact with air uncontaminated by germs in the following manner. A number of flasks, together with the air they contained, were thoroughly sterilised, they were then partially filled with blood taken directly from the blood-vessel of an animal—rigorous precautions being taken that the fluid suffered no contamination from the atmosphere during the operation—after which the flasks were hermetically sealed. Blood, which under ordinary circumstances exhibits the greatest proneness to putrefy, was kept in this way for several weeks at a temperature of 25° to 30° C. (77° to 80° F.) without showing the slightest tendency to putrefy. The same was shown to be true of many other organic secretions, and M. Gayon has since proved that it also holds good with reference to the contents of eggs.

These experiments form one of the greatest proofs of the fallacy of the theory of spontaneous generation, for they were made on substances which had been unaltered by heating; and they formed an effectual rejoinder to the arguments of some of the advocates of the theory of spontaneous generation, who had asserted, not without some show of reason, that materials which had been heated did not satisfy the conditions requisite for the spontaneous production of living from dead organic matter.

Tyndall's Experiments on the Floating Particles in the Air. —In 1881 Tyndall published the results of a remarkable series of experiments made at the Royal Institution, in which free exposure of highly putrescible fluids to the air was carried out in a still more extended manner.

Reasoning from the well-known fact that, when a powerful beam of light is passed through the air of a darkened room, its path is rendered distinctly visible because it illuminates the particles floating in that portion of the air which it traverses, he concluded that the path of such a ray would be invisible in an atmosphere which contained no floating particles. He argued that as all particles of matter, however small, must necessarily obey the law of gravitation and be precipitated in course of time, air if confined in a close space and kept perfectly free from disturbance, should, after the lapse of a sufficient length of time, be entirely freed from all the floating matter which it originally contained. In order to put this to the test, Tyndall had a box constructed, the front and sides of which were glazed. After the apparatus had been left at rest for some time, his expectations were fulfilled, for on passing the powerful beam from an electric lamp through the box he found its path invisible. He now reasoned that if the organisms which cause fermentation or putrefaction form a portion of the particles which float in the atmosphere, fluids liable to take on fermentative or putrefactive changes should, when placed in an atmosphere of this kind, remain unchanged for an indefinite time. In order to carry out this idea an exceedingly ingenious apparatus was constructed. It consisted of a four-sided box, in the front of which was fixed a plate of glass, the two sides of the box being similarly provided with smaller glass panes, while at the back of the box was a small door, so constructed that, when closed, the front, back, and sides of the apparatus were air-tight. The bottom of the box was perforated by several apertures into which fitted air-tight several large test-tubes. Passing through an aperture in the top of the box was a funnel with a long stem, fixed in such a manner that, whilst remaining completely air-tight, it could be moved about and its lower extremity placed in turn over the mouth of any individual test-tube. There were also inserted in the upper side of the box two tubes bent into the form of spirals. Before the apparatus was permanently closed the inside was smeared with a thin layer of glycerin, a liquid eminently adapted to retain any small particles coming in contact with it. The apparatus was now left at rest, until, on passing a beam of light from an electric lamp through it, no trace of floating matter could be observed; this was generally the case in about three days. The test-tubes were then partially filled by means of the funnel with various solutions, which under ordinary circumstances would have rapidly entered into a state of fermentative change; these were meat, game, and fish broths; decoctions of tea, vegetables, hay, &c. In order to destroy any germs which these liquids might contain, the lower portions of the test-tubes were immersed in a bath of heated brine and their contents boiled. The steam either condensed on the sides of the box or passed out of the bent tubes, and, after the heating, was replaced by atmospheric air, which entered by the spiral tubes. Treated in this way, liquids of the most putrefactive nature remained unchanged for months; when, however, the door at the back of the apparatus was opened for a few minutes and then closed, in a very short time some of the tubes began to exhibit signs of infection.

Here then was demonstrated the important fact that air, which resembled the outside atmosphere in every particular, with the single exception that it had been deprived of the floating particles which it originally contained, not by any violent means, such as passing it through a red-hot tube, or through sulphuric acid, or even filtering; but simply by natural gravitation, was unable to induce either fermentation or putrefaction in the most changeable liquids. Similar results were afterwards obtained by exposing solutions similar to the above to an atmosphere in which the organic particles had been burnt up by means of a spiral of platinum wire kept for some time in a state of incandescence by a current of electricity.

Organic Fluids of Varying Constitution require Different Temperatures to effect their Sterilisation.—It had been observed by investigators previous to Pasteur that, though most liquids were sterilised by exposure for a short time to a temperature of 100° C. (212° F.), yet there were exceptions to this rule, and of these milk formed a notable example. Pasteur made numerous investigations as to the reason for this, and found that while the nature of the solution itself exercised some influence, the resistance to sterilisation chiefly depended upon the reaction of the liquid. Thus, acidity of the liquid favoured sterilisation, alkalinity retarded it. For example, the highly acid fluid, vinegar, was deprived of its liability to change by being subjected to a heat of not more than 50° C. (122° F.); grape-juice, which is but slightly acid, required a higher temperature, but not such a high one as unhopped wort, this, being still less acid, required a temperature of 90° C. (194° F.). Fresh milk, which is a neutral fluid, needed a temperature of 110° C. (230° F.); but sour milk needed only to be heated to 80° C. (176° F.) to prevent its undergoing further change. The real cause seems to lie in the degree of permeability of the exterior coats of the germs to moisture, acidity of the solution apparently permitting the ready penetration of moisture into the interior of the cells, whilst alkalinity has the reverse effect. The germs of the *Bacillus subtilis*, found by Tyndall in hay, were able to withstand the temperature of boiling water for several hours; this resistance he also considered to be due to some peculiarity in the external coating of the germs.[1]

The state of dryness of an organism, or the reverse, is also of great influence in determining at what temperature it will succumb to heat; a mould fungus, which in a moist state cannot survive the temperature of boiling water, may, after thorough desiccation, be heated to a temperature of 120° C. (248° F.) with-

---

[1] Instances of the same nature are occasionally met with in the seeds of the more highly organised plants; for instance, Pouchet (*Bot. Zeit.*, 1878, p. 153) found that the seeds of a certain kind of lucerne, which adhere to the wool of sheep imported into France, were able to germinate after being boiled in water along with the wool for four hours.

out losing its fecundity. The same is true of yeast-cells, which, when perfectly dry, may be exposed to relatively high temperatures without losing their vitality.

Influence of the Acidity or Alkalinity of a Nutritive Fluid on the Nature of the Organisms which will Develop in it.—Pasteur has shown that neutral liquids, or those having a slightly alkaline reaction, form media particularly favourable to the growth of bacteria and infusoria; a slightly acid medium, on the contrary, is unfavourable to the ingrowth, but favours that of the mould-fungi and of the *Saccharomycetes*. A nutritive liquid when exposed to the air becomes contaminated with organisms of all kinds; much, therefore, depends upon its reaction as to what shall be the class of organisms that will flourish in it or be suppressed. When a liquid, such as grape-juice, which has a strongly acid reaction, is exposed to the air, the germs of the bacteria which fall into it are suppressed; while those of the moulds and *Saccharomycetes* flourish. If, however, the grape-juice is neutralised before being exposed to the air, then bacteria will flourish to the exclusion of the other organisms, for when once one particular set of these has taken possession of the field it is very difficult for others to obtain a footing. Under such circumstances, neutralised grape-juice, instead of undergoing the alcoholic fermentation, putrefies. The slight acidity of ordinary beer-wort has considerable influence not only in assisting its sterilisation in the copper, but also, to some extent, in guarding it against the invasion of foreign organisms during the processes subsequent to the boiling.

The difference which the nature of the organic fluid exercises in favouring the development of particular organisms lent considerable support to the theory of spontaneous generation. When an organic liquid, such as grape-juice, is exposed for a time to the atmosphere, it is invaded by the germs of all kinds of organisms, but owing to its slightly acid nature the growth and development of bacterial organisms are suppressed; *Saccharomycetes*, which thrive in a slightly acid medium, on the other hand, flourish. Consequently, yeast may be grown over and over again in a slightly acid medium, such as grape-juice, its reproduction not being interfered with by the acid of the medium; but if the yeast was originally impure, for instance, contaminated with any of those organisms which turn wine sour, the growth of these will be so checked in proportion to that of the yeast, that their number becomes relatively so small in comparison with that of the yeast-cells that they may easily escape detection by a microscopic examination. Under such circumstances, by assuming that such a yeast is pure, the mistake may readily be made that yeast can be transformed into the organisms which turn wine sour, if it is not tested by direct experiment for impurity.

This is the error into which M. Duval evidently fell when he imagined that he had demonstrated that yeast could be transformed into lactic ferment. By adding yeast (which he considered to be pure, since he had cultivated it many times in grape-juice) to a mixture of sour milk, glucose, chalk, and ammonium phosphate, he readily obtained the lactic ferment, and the product of its action, lactic acid, in the form of calcium lactate. As Pasteur points out, though Duval supposed that his yeast had been freed from foreign organisms by growing it repeatedly in grape-juice, yet he had not put it to the test of direct experiment, in order to ascertain with absolute certainty that it was pure.

The variety of the different organisms which are found to develop in fluids which have been sterilised by heat are comparatively few in number, since they can only be those which have obtained access to the fluid from the atmosphere after its sterilisation; obviously all the germs with which the fluid was previously contaminated, whether derived from the atmosphere, from the dust deposited on the containing vessels, or from the materials employed, are destroyed in the process of heating. The organisms which develop in such a fluid are not nearly so numerous as has been imagined, since air, unless in a state of violent agitation, can only hold in suspension the finest particles. All observations which have been made on spontaneous generation have been made in rooms or laboratories where the air is comparatively quiescent; consequently, the variety of organisms which would make their appearance in sterilised fluids after exposure to the atmosphere of such localities is limited in number. This is still further lessened because some of the germs which do find access to the fluid are suppressed by competition. That such is the case is shown by the fact that a much larger variety of organisms is found when a sterilised organic fluid is exposed for a short time to aërial contamination, and then immediately divided into several portions, each of which is closed up in a separate vessel. There is then the chance that only one or two species are enclosed in one vessel; and consequently there is not the same chance of suppression by competition.

The variety of the organisms which make their appearance in raw infusions of different organic substances, such as leaves, fruits, grass, &c., is much greater in number, since the substances carry with them the germs of larger organisms deposited from the air when it has been in a state of violent movement, and, in addition to these, plants are always more or less infested with minute parasitic organisms.

Competition amongst the Organisms Introduced into a Fermenting Fluid.—The fact that when a large number of organisms of a certain species are added to a fermentable fluid—such, for instance, as the addition of yeast, which consists principally of *Saccharomyces cerevisiæ*, in fairly large quantity to beer-wort—other organisms, such as the acetic and lactic bacteria, together with the germs of other *Saccharomyces* which may have accidentally obtained access to the wort, and which are comparatively few in number, are kept in subjection by the preponderance of the *Saccharomyces cerevisiæ;* they are, as it were, crowded out, or, in other words, lose in the struggle for existence, when the yeast is present in large quantity. The brewer seeks in this way, by adding sufficient vigorous yeast, to rapidly induce the alcoholic fermentation in his wort, and in this way keep in subjection those other organisms which would exert a prejudicial action on it. It must, however, be borne in mind that the deleterious organisms are not destroyed, they are merely suppressed, and when the alcoholic ferment has exhausted its course, then these inimical organisms are able to commence their functions. Thus, when the germs of the acetic ferment have obtained access to a wort, and though during the progress of the alcoholic ferment they have been held in check, yet, when that process is completed, they will then begin to manifest their presence by the souring of the beer. This peculiar property that one ferment possesses, of when in abundance suppressing by competition other ferments which occur in lesser numbers, has led to erroneous views being held by some observers, who, finding that as one ferment became exhausted and died off another of a different nature took its place, argued that one ferment had become transformed into the other.

Transformation of One Set of Organisms into Another.—Numerous observers, commencing with Turpin, and including Bail, Hoffmann, Hallier, Trecul, Robin, and Fremy, believed they had actually observed the transformation of one set of ferment organisms into another. One of the latter of these, M. Trecul, stated that he had seen this transformation take place in unhopped wort prepared at a temperature of from $60°$ to $70°$ C. ($140°$ to $160°$ F.). First, a number of fine granules appeared in the wort, these developed into active bacteria, which afterwards became quiescent and were then transformed into the lactic ferment. After a time larger granules made their appearance, which gradually increased in size, and were eventually transformed into globular and elliptical cells almost as large as yeast-cells; these, after a time, commenced to bud. He also asserts that he has sown the spores of the *Penicillium glaucum* in boiled wort and obtained from them the alcoholic ferment.

Pasteur pointed out that when wort is boiled in one of his flasks and allowed to stand for some time, though an abundance of granular deposit invariably forms, in no case is it transformed into a living organism. He then demonstrated that M. Trecul must, unintentionally, have introduced other germs along with those he thought he had sown in a pure state, for when a sterilised fluid is impregnated with a pure culture of any individual organism, nothing but that organism is ever reproduced.

Methods of Obtaining a Pure Culture.—For the purpose of securing a pure culture of an organism Pasteur employed the following two methods. In the first of these, a convenient-sized flask was partially filled with wort, its neck drawn out to a fine tube, and its contents boiled. While steam was issuing from its orifice, the tube was hermetically sealed up with the blowpipe. On cooling the flask the steam gradually condensed, and a partial vacuum was formed. When the point of the tube is broken off, air rushes in to fill the vacuum, carrying in with it any germs it may happen to contain. Sometimes the air admitted into a flask in this way contains the germs of several different organisms, as is shown by their subsequent development; at other times no germ is introduced, and the contents of the flask remain sterile. Occasionally only a single germ obtains entrance, and in this last case a pure culture is obtained, which may be used for impregnating the contents of an ordinary Pasteur flask containing sterilised fluid. In the second method, which Pasteur says is nearly as reliable as the first, the growth of a mould fungus is allowed to proceed on any object on which it may have adventitiously formed until it has produced a crop of spores. A piece of recently ignited platinum wire is then passed lightly over the fructifying heads of the fungus, when a few spores adhere to the wire. This with its adherent spores is then introduced into a Pasteur flask containing a sterilised solution. The spores germinate and are allowed to grow on the surface of the fluid until they have produced plants, from which some of the spores may be removed by means of a platinum wire and used in turn in a similar manner.

The Origin and Distribution of the Germs Present in the Atmosphere.—Now that it had been conclusively proved that fermentation and putrefaction were invariably caused by organisms present in the floating matter of the atmosphere, the origin and distribution of these in the air of different localities became a matter of considerable interest and importance. This subject has been ably investigated by Pasteur and Hansen, who, however, chiefly directed their attention to the organisms of fermentation found in the atmosphere; whilst other observers, such as Miquel, Petri, Frankland, and many others, interested themselves principally with the bacterial organisms that it contains.

Pasteur's Examinations of the Air of Different Localities for Organisms.—For this purpose Pasteur employed flasks prepared in exactly the same manner as those described at p. 304 for obtaining a pure culture of a mould fungus. A number of these were taken to the locality in which the nature of the organisms in the air was to be investigated, the points of the flasks were broken off in turn, and each immediately sealed up again. They were then brought back to the laboratory and kept at a forcing temperature for some time. The general appearance of the organisms which developed in them was noted, and a rough estimate of their number made.

In an experiment of this character performed at the Ecole

Normal, ten flasks containing sterilised grape-juice were opened at the bottom of the garden; another ten on the landing of the second floor of the building; and a third ten in the laboratory itself, where experiments on fermentation were being constantly carried on. Of the first ten flasks, only one was invaded by organisms; of the second ten, four were similarly affected; and organisms developed in all those belonging to the third set. The difference in the number of germs contained by the same amount of air in these three different localities was therefore considerable. The nature of the germs also differed; the first two series of flasks contained mould fungi only, whilst those taken in the laboratory contained *Torulæ* (p. 239) in addition to mould fungi.

Experiments of a similar nature were made upon the air of the most diverse localities, and it was found that germs were very abundant in the air of towns, but that they gradually decreased in number as the locality became more and more remote from human habitation. Of twenty-four flasks opened on Mount Poupet, at an elevation of 2800 feet above the level of the sea, only five were infected. Of another twenty, opened at Montanvert, near the Mer de Glace, 6560 feet above sea-level, when a strong wind was blowing from the glacier, only one was infected. In experiments made in the higher regions of the atmosphere by means of a balloon, the air was found to be absolutely free from germs.

Ferment Organisms or their Germs are Seldom Present in the Atmosphere.—But, said Pasteur's opponents, if it were true that fermentation was invariably caused by the germs present in the atmosphere, then they would be present in the air in such countless numbers that they would form a thick black fog as dense as iron; for it is a well-known fact that when grapes are crushed in any part of the world, even on the top of a high mountain, their juice invariably enters into fermentation. In reply to this, Pasteur demonstrated by a series of experiments that the germs of the fermentation organisms do not occur in extraordinarily great numbers in the air of the atmosphere.

It was a very rare occurrence to meet with fermentation organisms in the experiments just mentioned, or in those performed with plugs of cotton-wool on which the dust of the atmosphere had been collected. They were only found in appreciable quantities in the air of rooms where fermentations were being constantly carried on. For instance, out of seven flasks opened in Pasteur's principal laboratory, where this condition obtained, only one exhibited signs of fermentation. This fact was further demonstrated by the following experiments:—A number of bottles of wort that had been preserved by Appert's process were uncorked, all necessary precautions being taken to prevent their contents being contaminated by the dust deposited on the cork, neck of bottle, &c., and allowed to remain undisturbed in various places. Two of these were exposed in an underground room of the École

Normal, where experiments on fermentation were continually being carried on. The contents of the first were found to have fallen into a state of active fermentation in six days, and the second, which up to the sixteenth day had shown no sign of fermentation, was in active fermentation on the twenty-third. Of four bottles exposed for three months in a room only occasionally used for experiments in fermentation, the contents of none entered into fermentation. The contents of four bottles placed in an incubating chamber where fermentations were constantly in progress, and allowed to remain there five months, showed no sign of fermentation. Of ten bottles that were exposed in another room where fermentation experiments were continuously going on, the contents of one of them commenced to ferment in four days; the day after, those of another started; and in fourteen days two more. All the remainder had up to this time shown no signs of fermentation, and these, after each had been covered with a cap of sterilised paper, were kept under observation for three months. The contents of all these last failed to enter into fermentation.

Fermentation Organisms or their Germs are extremely Abundant on the Skins of Ripe Fruits.—Pasteur showed that ferment organisms were always present in great abundance on the exterior surfaces of ripe fruits. He picked a bunch of grapes from the vine, cut off each grape with a pair of sterilised scissors, washed it and the little stalk attached to it with a small quantity of water, by the aid of a perfectly clean badger-brush. A dozen grapes successively washed in this way with 3 c.c. of water yielded a turbid mixture, which, when examined under the microscope, was found to abound with minute organisms. The development of these was studied, and it was also proved that among them were some that were capable of inducing the alcoholic fermentation. This was also found to be the case with ripe gooseberries, plums, and pears. The woody portion of a bunch of grapes, from which the grapes had been detached, was similarly treated, and the number of organisms was found to be even greater on this than on the skins of the grapes themselves. In another series of experiments he proved that the organisms of fermentation were only present on the surface of the grapes and their stalks in the autumn, when the grapes were fully ripe; that they were not to be found on the immature fruit, and that grapes kept through the winter gradually lost their power of causing fermentation.

Pasteur also made a series of experiments in order to show that the organisms of fermentation were situated on the outside of the grape alone and did not exist in the juice in its interior, and also to explode the hypothesis of semi-organisation propounded by M. Fremy in 1864, and subsequently adopted and extended by M. Trecul, in an attempt to explain the formation of yeast from the albuminous portions of grape-juice. By this hypothesis it was assumed that organic bodies, such as albumin, casein, fibrin, &c.,

which have an extremely complex chemical constitution, were in reality semi-organised substances, that is to say, they held a position intermediate between dead and living matter. "These semi-organised bodies," says M. Fremy, "which contain all the elements of organs, have the power, like a dry seed-grain, of existing in a state of organic immobility, and of becoming active under circumstances which favour organic development. By reason of the *vital energy*[1] which they possess, they undergo a succession of decompositions, giving rise to new derivatives, and to the advent of ferments, not by any process of spontaneous generation, but by a process of vital energy which pre-exists in the semi-organised bodies, and is simply carried on, when this energy manifests itself, in these most varied organic changes."

Forty Pasteur flasks were taken and a portion of filtered grape-juice introduced into each of them. The contents of each flask were then boiled for a few minutes to effect sterilisation, after which the flasks were divided into four groups of ten each. The first group were kept without further treatment as a check on the experiments. Part of a bunch of grapes was washed with a few c.c. of water, and to each flask of the second group a few drops of this were added. To each flask of the third group a similar quantity of the water in which the grapes had been washed was added after it had been boiled. To each flask of the fourth group a single drop of grape-juice obtained from the interior of a grape was added, the experiment being so arranged that none of the organisms present on the exterior of the grape could obtain access to the drop. The whole series of flasks was then kept at the ordinary summer temperature. For some time the contents of the flasks belonging to the first, third, and fourth groups showed no change. All those of the second were found, at the end of forty-eight hours, to be in a state of fermentation, and in some of these the mycelia of mould fungi afterwards appeared; in others *Mycoderma vini*, but in no case bacteria, since these last would be suppressed by the fluid as it had an acid reaction. Obviously, if ferment could be formed from the albuminous substances present in grape-juice, as affirmed by Mons. Fremy and Trecul, the contents of every one of the flasks ought to have fermented. It was also shown that occasionally fermentation organisms were entirely absent on the surface of a grape; consequently, such a grape might be crushed without its contents fermenting—a fact that also militated against M. Fremy's theory.

The organisms which spontaneously cause changes in organic fluids are not, as a rule, derived directly from the atmosphere, but indirectly from it in the shape of dust which has been deposited on vessels and other objects with which the fluid comes in contact. That this is so is well exemplified in Pasteur's modification of the experiment of Gay-Lussac, mentioned on p. 288, which con-

---

[1] A purely gratuitous assumption, Pasteur remarked.

PRESERVATION OF FERMENTABLE FLUIDS.

sists in uncorking a bottle of grape-juice, preserved by Appert's method, and transferring its contents to another bottle, when the fluid so transferred invariably falls into a state of fermentation. This fact was adduced by Gay-Lussac as a proof of his theory that it was the action of the oxygen of the atmosphere on the albuminous constituents of the fluid which caused its fermentation. Pasteur repeated the experiment, and obtained the same result when no precautions were taken to ensure that the fluid was not contaminated by the dust on the cork and on the outside of the bottle; but when scrupulous precautions were taken, such as singeing[1] the cork and neck of the bottle, the corkscrew used, and taking care that the bottle into which the fluid was introduced was perfectly free from germs, no sign of fermentation made its appearance.

A Fermentable Fluid may be Preserved for Years in an ordinary Pasteur Flask, but if a little of the dust which has become deposited on its exterior be introduced into its liquid contents, change, indicated by turbidity, will manifest itself on the following day. The same object may be effected by tilting the flask so that its fluid contents come in contact with the lower portion of the capillary tube, where germs are invariably deposited; or even violently shaking the flask so as to disturb the deposited germs by a violent rush of air into the capillary tube is sufficient to infect the fluid.

Even the mercury sometimes employed in experiments of this nature may contain the sources of infection. This was exemplified in Pasteur's explanation of the fallacy involved in an ingenious experiment performed by M. Pouchet. In order to show that organisms could develop in a fluid which had never come into contact with ordinary air, M. Pouchet filled a bottle with boiling water, then plunged it into and uncorked it under mercury. When the water was cold, half a litre of oxygen gas was introduced, and a portion of hay which had been exposed for several hours to a temperature of 300° C. (572° F.) was passed under the mercury into the neck of the bottle. At the end of a week's time organisms appeared in the fluid. Pasteur showed that here the source of error lay in the particles of atmospheric dust with which the mercury was contaminated. By passing a beam of light through the room after being darkened, he was able to ocularly demonstrate to his audience the particles of dust which were continually floating about in the atmosphere, and, although these were apparently moving about in all directions, yet each one had the general tendency finally to settle down by gravitation. The surface of mercury which is frequently exposed to the atmosphere would undoubtedly receive its share of these particles; and it was

---

[1] This ingeniously simple method of sterilising small articles, such as forceps, scissors, &c., just before being used, is due to Pasteur. It is effected by passing the articles through the flame of a spirit-lamp or Bunsen burner; owing to the extreme readiness with which minute organic particles ignite, any germs deposited on such articles are immediately consumed.

clearly shown, by throwing a strong light on to the surface of perfectly clean mercury, spreading some dust on this, and plunging a glass rod into the metal, that the dust travelled to the spot where the rod was inserted, and was actually carried into the interior of the quicksilver.

Miquel's Experiments on the Floating Matter of the Air.—In these investigations, which were carried out at the observatory of Montsouris near Paris, Miquel employed, at first, an apparatus which he called an "aëroscope." In this the air to be examined was caused to impinge in a fine jet on a glass plate smeared with a mixture of gelatin and glucose, which, from its sticky nature, was eminently adapted to catch and retain the floating particles which the air contained. These particles were then evenly distributed over the glass plate, placed under the microscope, and counted. Since this method gave no information as to the number of living organisms contained in a measured quantity of air, he devised a peculiarly constructed flask in which a measured quantity of air was caused to bubble through a known quantity of water, certain precautionary measures being taken to avoid loss of organisms. In this way he obtained the dust of the atmosphere suspended in water, which, after being diluted with a further quantity of sterilised water, was added in small and equally measured portions to small quantities of sterilised broth contained in a number of flasks. When a large proportion of the contents of the flasks treated in this way exhibited no signs of infection, there was a considerable probability that each of the infected flasks contained only a single germ; consequently, by taking into account the quantity of air passed through the apparatus, and the number of flasks which became infected, it was possible to form an approximate estimate of the number of organisms that were present in a sample of air capable of development.

Koch's Method.—An entirely different form of apparatus was employed by this investigator. It consisted of a glass tube about a yard long and two inches wide. One end of the tube was closed with a perforated indiarubber cork, a glass tube plugged with sterilised cotton-wool being inserted in the aperture in the cork. A quantity of sterilised gelatin meat-broth was then introduced into the tube and its other end closed with an indiarubber cap. The whole was then sterilised, laid on its side, and allowed to cool. When about to be used, the glass tube was connected with an aspirator, the indiarubber cap removed, and air drawn slowly through the apparatus. In its slow passage through the tube any particles which the air might happen to contain would be gradually deposited on the gelatin. The indiarubber cap was then replaced on the end of the tube, and the whole transferred to an incubator, where it was allowed to remain until the organisms had developed and formed colonies. These were then counted and their general appearance noted. A curious fact was noticed in experiments conducted in this way; the bacteria which had developed were found in that portion of the tube nearest the end where the air entered, whilst the mould fungi were found much farther down the tube; this seems to show that the specific gravity of the germs of the former organisms is higher than that of the germs of the latter.

Hueppe's Method.—In this, the air under examination was caused to bubble slowly through warm sterilised gelatin meat-broth, which was afterwards spread upon glass plates. These were placed in an incubator, and the colonies that formed on them studied.

Method of Frankland, Miquel, and Petri.—In this, the floating particles of the air were collected by passing it through filters containing such substances as sterilised sand, powdered glass, glass wool, &c. The material of the filter was afterwards mixed with sterilised gelatin meat-broth and spread on glass plates; or the sand, &c., with the germs it contained, were placed in a Petri dish, intimately mixed with gelatin meat-broth, covered over, and placed in the incubator.

Objections to the Use of Gelatin Meat-broth in Experiments of this Nature.—As was shown by Miquel, there are several objections to the use of gelatin plates for this purpose. They cannot be used at a temperature higher than $24°$ C. ($75°$ F.), since above this the gelatin medium becomes fluid, and some bacteria require a higher temperature than this for their development. Many bacteria take a fortnight to form colonies at the usual temperature of the incubator, and by this time so much of the plate is liquefied by the action of certain bacteria, or the plate may be so covered over with the rapidly growing mould fungi as to be unfitted for further observation; consequently, those organisms which have not had time to develop entirely escape observation. Some species of organisms which readily develop in liquids will not do so on a gelatin plate. Finally, there is no absolute certainty that one colony only represents one species of organism. This was demonstrated by growing colonies on a plate in the ordinary way, then sowing each of them in a separate batch of meat decoction. A second plate culture was afterwards made with each batch of the infected broth; and, as in these latter colonies species were found which differed from those of the original colony, distinct evidence was afforded that the original colony by no means consisted of one species.

Hansen's Investigations of the Air.—These experiments, made between the years 1877 and 1882, were carried out with the object of ascertaining the nature and distribution of those organisms in the atmosphere which might exert a prejudicial effect on the operations of the brewer. For this reason, as in his method of examining water used for brewing purposes, he entirely rejected the use of gelatin, and used the two liquids with which the brewer is mostly concerned, *i.e.* wort and beer. Two forms of apparatus were employed in the course of these experiments:

first, the vacuum flasks (p. 304) introduced by Pasteur, only, instead of sealing up the ends of the tubes by fusing the glass, Hansen simply closed them with sealing-wax; the second consisted of ordinary boiling flasks partially filled with wort, which were sterilised after their necks had been covered with filter-paper. The flasks were taken to the place the air of which was to be examined, their caps removed, and allowed to stand open for various periods of time, in some cases as long as forty-eight hours. When using the vacuum flasks, it was found that the criterion for infection used by Pasteur, namely, a visible alteration in the contents recognisable by the naked eye, was not to be implicitly relied upon in all cases, for small quantities of various growths, such as those of *Aspergillus, Penicillium, Mucor, Cladosporium, Bacterium aceti*, and *Mycoderma cerevisiæ*, could only be detected by microscopic examination. He confirmed Pasteur's observation that it was often possible in this way to obtain a pure growth of one particular organism, a single germ only having obtained access to the flask; in fact, it was only in rare instances that several growths were obtained in the same flask. An observation made in this way necessitates the employment of nearly one flask for each germ present in the air examined, and although it gives the most certain information, it is exceedingly tedious and cumbrous in its application; it has the further disadvantage of only giving information as to the germs present in the air at the particular moment of opening each flask. Consequently, it was employed in conjunction with the second method, since this gave information as to the characters of the germs which appeared in the air during a longer or shorter period.

General Conclusions obtained from these Observations.— All these experiments tended to confirm the previous conclusions of Pasteur, that the air of places in immediate proximity to one another may contain at the same time not only the most diverse organisms, but also the most variable quantities of them. The germs of organisms seem to exist in the atmosphere in the form of clouds, the intervening spaces between which are comparatively germ-free. The number of bacterial organisms present in the air is much modified by the weather; they are most abundant in dry seasons, whilst after a spell of wet weather their number considerably diminishes. The reverse seems to be the case with the mould fungi; these occur in the greatest number in the atmosphere in damp weather, and in much smaller quantity in dry weather. *Saccharomycetes* are seldom met with; Hansen found that the yeasts increased in number from June to August, and are most abundant at the end of August and the beginning of September (the period of the ripening of fruit), after which their numbers again decrease. As there is strong evidence that the *Saccharomycetes* have their place of abode in the earth during the periods when they are not living on the surface of fruits, they are probably met with most frequently in the air when passing through it to change their respective resorts, viz., the ground and the fruit-trees. During this same period bacterial and mould fungi occur in the atmosphere in the greatest numbers, and hence it is at this season that the wort, when on the coolers and refrigerators, is likely to receive its greatest contamination from these organisms. This time of the year is proverbially known to be the worst for brewing operations.

Hansen's Investigations on the Air of Breweries.—Hansen now turned his attention to an examination of the air of the various rooms where the actual brewing, or the operations connected therewith, were carried on. The air of a passage leading to a room where barley was turned, as might readily be expected when the large number of organisms always present on its surface is taken into consideration, was found to be highly contaminated with organisms, especially those of a bacterial nature. The air in the neighbourhood of malting-floors was found to be extremely rich in mould fungi. The air of the fermenting-rooms of the old Carlsberg brewery was found to contain fewer organisms than that of any of the other localities hitherto examined; in that of the new Carlsberg fermenting-rooms organisms were comparatively abundant, from 55 to 100 per cent. of the flasks employed becoming infected. In the case of the old Carlsberg brewery the air is freed from germs by passing it through a bath of salt water, and by rigorous attention to the cleanliness of the room and all the utensils contained therein.

Morris[1] has recorded the results of a series of experiments made to ascertain the number of organisms that fell from the atmosphere into the wort during the time it was standing on the coolers. The season of the year is not given. These experiments were made at two different breweries; the cooler of the first had a superficial area of 70 square yards, that of the second an area of 426 square yards. It was found that rather more than 140,000 organisms fell on each square yard of cooler surface per hour, the numbers being pretty nearly equal in both breweries. He also determined the number of organisms with which the wort had been infected during its stay on the coolers in the following manner: wort was taken from the coolers when nearly the whole had run off and only a few barrels remained; this was added in measured portions, firstly, to 5 per cent. sterilised gelatin, secondly, to 5 per cent. sterilised meat-extract-gelatin. The number of colonies which had developed in three days were counted, and the quantity thus found calculated into wort per barrel. In the first brewery 163,548,000 colonies per barrel developed in gelatin, and 16,354,800 in meat-extract-gelatin. In the second brewery 1,406,512,800 developed in gelatin, and from 14,065,128,000 to 21,097,692,000 in meat-extract-gelatin. This great difference in

[1] *Trans. Lab. Club,* vol. iii. p. 36.

the respective numbers was attributed to the difference of the superficial area in the two coolers. Of the organisms which developed in gelatin alone, the majority consisted of *Saccharomycetes* and *Torulæ*, though bacteria and mould fungi were present in considerable numbers as well. The utmost cleanliness was observed in both breweries, and there can be no doubt that all the organisms found were derived solely from the air.

Yeast may be Dried and Reduced to a fine Powder without Losing its Vitality.—Pasteur made the following experiments,[1] to show that yeast may exist in a state of vitality in the form of dust floating about in the atmosphere. A few grammes of pressed yeast were rubbed up with five times their weight of sterilised plaster of Paris in a sterilised mortar. The mixture was then wrapped up in a piece of sterilised paper and dried at a temperature of 20° to 25° C. (68° to 77° F.). Two days afterwards a pinch of this powder was sown in sterilised wort, contained in a Pasteur flask, and the whole kept at a temperature of 20° C. (68° F.), when, in three days, signs of fermentation made their appearance in the wort. When the yeast was two and a half months old the experiment was repeated; this time fermentation commenced on the fourth day, hence the vitality of the yeast was not destroyed, but merely somewhat lowered. In a similar experiment made with the same yeast when seven months old, the vitality was not yet altogether lost, but was still more depressed, for it took eight days for signs of fermentation to appear. At the end of ten months the yeast, when again tested in this manner, was found to be completely dead; the wort impregnated with it, though observed for several months afterwards, gave no sign of fermentation.

Pasteur showed that it was quite possible to artificially impregnate the air of a room with yeast organisms. In order to effect this it was only necessary to drop a small quantity of the dried yeast powder from a height, when the presence of living yeast in the air of the room could be demonstrated by opening a series of vacuum flasks in it. It often happened in experimenting in this manner, that one flask received but one organism; and in this way a pure culture of a single species of yeast was obtained, since the whole of it must have sprung from one mother-cell. At the conclusion of the description of these experiments, Pasteur adds the following pregnant sentence : " Our preliminary observations, although incomplete, seem to favour the idea that numerous varieties of ferment are to be obtained by these means." Here is distinctly foreshadowed the important field of investigation which has been, in later years, so ably worked out by Hansen.

Pasteur's Experiments on Fermentation.—From what has gone before, it will be remembered that ever since the discovery had been made that yeast was a living organism, the opinion had been held by various observers that fermentation was inti-

---

[1] "Studies in Fermentation," p 81.

mately connected with the vital functions of the yeast. Pasteur was, however, the first to put this into a definite shape, and to attempt to give a really scientific explanation of the processes of fermentation. In studying the peculiar fermentation which impure calcium tartrate entered into in hot weather when allowed to remain under water, he was led to conclude that the fermentation was caused by a small living organism.

Struck by the part played by this organism in the fermentation of tartaric acid, he thought that it could not be an isolated fact, but that it must form a portion of some great law, the elucidation of which could not help throwing considerable light on the hitherto obscure problem of fermentation. He regarded fermentation and putrefaction as analogous processes, and recognised in them Nature's method of converting the dead refuse material of the animal and vegetable world into forms again available for the nutrition of plants.

Investigations into Lactic Fermentation.—When a mixture of milk, sugar, putrid cheese, and chalk[1] is kept in a warm place, it speedily falls into fermentation, and in course of time lactic acid is produced in abundance, according to the formula :—

$$C_6H_{12}O_6 \;=\; 2C_3H_6O_3$$
$$\text{Glucose.} \qquad \text{Lactic acid.}$$

The acid unites with the chalk to form calcium lactate. Hitherto no organisms had been detected in such a fermenting mass; but Pasteur, however, showed that they were there, and that they were merely hidden by the innumerable host of other particles present in such a mixture, since they could occasionally be detected in the grey scum which formed a zone on the surface of the mass. In order to obtain the ferment in a state of purity, a decoction of yeast, to which sugar and calcium carbonate had been added, was infected with a minute trace of this grey scum. Fermentation ensued, the calcium carbonate was gradually transformed into calcium lactate, and the solution gradually grew turbid from the formation of the lactic ferment. This, when examined under the microscope, was found to consist of minute bacteria.

Experiments where the Albuminous Material was Replaced by an Inorganic Nitrogenous Body.—In order to demonstrate the complete fallacy of the views held by Liebig and his followers, Pasteur repeated the foregoing experiment, employing, instead of the albuminous material derived from the yeast, an organic salt of the inorganic nitrogenous substance ammonia, together with a

---

[1] A good mixture for preparing lactic acid is made up as follows :—Skimmed milk, 2 gallons; raw sugar, 6 lbs.; water, 12 pints; putrid cheese, 8 ounces; chalk, 4 lbs. The mixture is kept at a temperature of 30° C. (86° F.) in an open vessel, and occasionally stirred. The lactic fermentation will be finished in about three weeks. The mass thus obtained is drained, pressed, and purified by recrystallisation in water, and oxalic acid added to remove the calcium from the calcium lactate. This leaves lactic acid in solution, which may be further purified by conversion into the zinc salt.

small quantity of the phosphates of potassium and magnesium as mineral food. The lactic ferment multiplied and caused fermentation just as in the previous case. An experiment was also conducted with a fluid of similar composition, but to which, instead of lactic ferment, a minute trace of beer-yeast was added. A similar growth and multiplication of the yeast-cells took place, accompanied with alcoholic fermentation. Evidently, then, this was a plain proof that yeast could assimilate the inorganic substance ammonia, and derive from it the nitrogen necessary for forming its albuminous constituents; the phosphates supplied the mineral food, and the carbonaceous portion was derived from the sugar. In these cases there was not the slightest possibility of fermentation being produced according to the theory of Liebig, for there was no albuminous matter present except that actually contained in the yeast itself; and to be certain that the albuminous material of the new yeast-cells could not be obtained from old and dying yeast-cells, only a minute trace of yeast was added at the outset. Here, too, it was demonstrated that fermentation is a phenomenon of nutrition, during which the yeast is able to build up the complex bodies which enter into its composition from sugar and purely mineral substances.

The Ferment of Butyric Acid.—If, in the process mentioned above for the production of lactic acid, the action is allowed to proceed for a considerable length of time, the calcium lactate is gradually transformed into calcium butyrate, carbon dioxide and hydrogen being evolved, according to the formula:—

$$2C_3H_6O_3 = C_4H_8O_2 + 2CO_2 + 2H_2$$
Lactic acid.　　Butyric acid.　　Carbon dioxide.

This has led to the idea that the lactic acid ferment was gradually transformed into the butyric acid one. Pasteur demonstrated that the two ferments exhibited important distinctions; the lactic ferment is motionless, the butyric is in a constant state of lively motion. The latter ferment is also distinguished by another extraordinary property; it is able to live and perform its functions in the entire absence of oxygen; in fact, its vitality is destroyed by that gas. If a current of air be passed for a few hours through a liquid undergoing the butyric fermentation, the process is completely arrested, the bacteria cease all active movement and fall to the bottom of the vessel. When sown in a fluid capable of undergoing the butyric acid fermentation, the ferment increases sensibly in weight, though, as with the rest of the fermentation organisms, this is exceedingly small when compared with the amount of substance decomposed.

This important discovery led Pasteur to divide organisms into two classes, the first, which are only able to live when supplied with free oxygen, he designated "aërobic;" the second, which are only able to exist in its entire absence, he styled "anaërobic."

Fermentation a Phenomenon of Nutrition.—Pasteur having decided in his own mind that fermentation was a phenomenon of nutrition, the next question for elucidation was wherein it differed from the ordinary process of nutrition. As a rule, an organism assimilates nearly the whole of the food which it absorbs, but in the case of a fermentation organism the amount of substance absorbed and decomposed is enormously greater than the small portion which it appropriates for its own nutrition. The fact that the butyric acid bacteria lived and could perform their functions in the absence of free oxygen struck him as probably furnishing the key to the mysterious problem of fermentation. Perhaps it was only the anaërobic organisms which were able to functionate as ferments, the aërobic ones being incapable of doing so. He conceived that there might yet be a third class of organisms represented by the yeast fungi, which had two phases of existence, the one aërobic, the other anaërobic. In their aërobic state of existence they would live and carry on their nutritive functions after the manner of ordinary fungi; in the anaërobic, they would act as ferments. The following observations and experiments seemed to confirm this view.

Several of the Mould Fungi are able to Act as true Ferments. —In 1857 M. Bail had discovered that the mould fungus, *Mucor mucedo*, was able to excite alcoholic fermentation; this he wrongly attributed to the transformation of the fungus into yeast. Pasteur afterwards found that several of the mould fungi, when entirely submerged in a saccharine fluid, that is, when they were deprived of free oxygen, were able to produce alcohol and evolve carbon dioxide; in other words, to act as true ferments. His first experiments were made with *Aspergillus glaucus*.[1] A pure culture of this, obtained by the method described on p. 304, was sown on the surface of wort contained in a Pasteur flask, the experiment being so arranged that the little plant was supplied with an abundance of air. It then grew after the ordinary manner of a mould fungus, that is, absorbed oxygen and exhaled carbon dioxide, and decomposed no more sugar than was necessary to supply the requirements of its own nutrition. When the solution on which it had been growing for some time was distilled, it was not found to contain the slightest trace of alcohol.[2] When, however, in a similar experiment, the plant was deprived of air by being submerged in the fluid, its general appearance and method of existence suffered

---

[1] Afterwards he expressed a doubt as to whether the plant he had in hand was *Aspergillus glaucus*, or a particular bluish species of *Penicillium* which bore large round spores. This, however, in no way affects the argument.

[2] Pasteur determined the presence of alcohol in minute traces in a fluid by distilling it from a long-necked retort connected with a Liebig's condenser. When alcohol is present, the first few drops distilled "present the form of little drops or striæ, or, better still, oily tears." In this way the presence of 0.001 part of a volume of alcohol can be detected in a fluid, and with care and practice as little as a tenth part of this quantity. By collecting the first third of the distillate, and redistilling this, even smaller quantities may be detected.

great alteration, the hyphæ became swollen to some six times their natural thickness, and assumed a singular contorted appearance. The plant still continued to evolve carbon dioxide, and when the fluid was distilled it yielded a small quantity of alcohol. The same phenomena manifested themselves when *Mycoderma vini* was similarly treated. But much more marked was the fermentative change when another mould fungus, *Mucor racemosus*, was submerged and grown in a saccharine fluid, as much as 3.5 per cent. (by weight) of alcohol being produced.

Relation between the Weight of the Fungus Formed and the Amount of Sugar Decomposed.—Pasteur found that when 120 c.c. of wort were impregnated with a minute trace of the spores of *Mucor racemosus*, and the growth resulting from these was kept submerged in the fluid, at the end of the fermentation, the plant weighed 0.37 gramme, and the alcohol formed 4 grammes; consequently, an amount of alcohol nearly eleven times as great as the weight of the fungus had been produced, and this quantity would require for its formation nearly twenty times the weight of the fungus in sugar. In another similar experiment the weight of *Mucedo* formed was 0.25 gramme, and that of the alcohol produced 4.1 grammes, or nearly thirty-three times the weight of the fungus, and this quantity of alcohol would necessitate for its formation the destruction of nearly sixty-six times the weight of the plant in sugar. Obviously, then, when a mould fungus is compelled to live under these forced conditions, it does not, as might be expected, perish for lack of oxygen—in fact, drown—but seems to be able to adapt itself to its new conditions and still maintain its existence. That these conditions are entirely abnormal is shown by the signs of decrepitude which the cells soon begin to show when the plant is compelled to live in this way without a supply of free oxygen. They put on an old, shrivelled, and worn-out appearance, their contours become irregular and their contents become exceedingly granular. Coincidently with these appearances the fermentation ceases, and with it the evolution of carbon dioxide. The cells under these conditions are not dead, for, when supplied with a trace of free oxygen, they immediately revive, and fermentation and evolution of gas again commence; when the cells are examined, they are found to have once more put on their pristine appearances of activity. Thus, in order to produce the amounts of alcohol yielded in the experiments just mentioned, it was found necessary every few days to decant the contents of the flask into another one, the slight contact with the air during the decantation being sufficient to revivify the cells.

Confirmation of this Theory by a Consideration of the Conditions under which Yeast Functionates in actual Practice.— In the Jura wine is manufactured by a peculiar method. The grapes are gathered and placed in large receptacles, which would be entirely closed were it not for an aperture at the top of about four inches in diameter, which is left open for the escape of the carbon dioxide. Each of these receptacles is filled three parts full with grapes, some whole, others crushed, and all more or less moistened with the juice of the injured fruits. After a short time the whole mass falls into a state of active fermentation. Obviously, under these circumstances, fermentation with the concomitant formation of a considerable quantity of new yeast must proceed in the absence of oxygen, for the small quantity of that gas originally contained in the vessel will soon be driven out and replaced by carbon dioxide. In a large vat of fermenting wort the process must likewise be carried on in the entire absence of oxygen, for the small amount of the gas originally dissolved in the wort will be soon absorbed by the yeast; and access of air to the liquid is afterwards effectually prevented by the layer of carbon dioxide which forms on its surface. Nevertheless, in every such case the yeast-cells not only live, but rapidly multiply and actively exercise their functions for several days; and the immense disparity between the amount of matter which they have actually assimilated, as shown by the weight of the newly-formed yeast, and the amount of sugar decomposed and converted into alcohol, carbon dioxide, and other products, evinces the highly fermentative character which yeast displays when compelled to live under these conditions.

Yeast Grown under the more or less Perfect Exclusion of Oxygen.—In order to test the truth of this view, Pasteur performed the following series of experiments. A large flask holding three litres, the neck of which was drawn out and bent, as in a Pasteur flask, was completely filled with a five per cent. solution of sugar-candy in yeast-water,[1] and the open end of the neck immersed in a small dish filled with mercury.

Experiment A.—The fluid contents of the flask were impregnated with the small quantity of yeast obtained by fermenting 15 to 20 c.c. of a solution similar to that contained in the flask, and to which a minute trace of yeast had been added in the body of a small tapped funnel which had been fused on to the short side tube of the flask. In less than twelve days all the sugar of the fluid in the flask was completely fermented, and the yeast produced, after being dried, was found to weigh 2.25 grammes, the proportion of yeast to sugar fermented was $\frac{2.25}{150} = \frac{1}{67}$. As such a quantity of fluid would not absorb more than 16.5 c.c. of oxygen, this is the maximum amount of the free gas which could have been utilised in the fermentation of 150 grammes of sugar. A similar experiment (C) was performed, but in this case the sugar solution was almost entirely deprived of air (it contained less than one milli-

[1] By yeast-water is meant water in which yeast had been boiled, and which, consequently, contained nitrogenous matter and salts extracted from the yeast.

gramme of oxygen) by boiling, and precautions were taken to ensure that the yeast added by the funnel was entirely free from extraneous germs. Simultaneously a parallel experiment (B) was made in an ordinary Pasteur flask double the size of the former flask, the three litres of fluid occupying about one-half the flask; consequently a large surface was exposed to the air. The contents of the flask were boiled to ensure sterilisation, and, after cooling, impregnated with a small quantity of yeast from the funnel of the former flask. In the case of the flask (C) fermentation commenced on the second day, and on the following days fine and actively budding yeast-cells were observed in the yeast driven out from the end of the tube. The fermentation gradually became more and more languid, and at the end of nineteen days, when the experiment was brought to a close, 4.6 grammes of sugar still remained unfermented. The yeast seemed on examination to be very pure in both cases, and from the sharply-defined appearance of the borders of the cells of that from the first flask, capable of easy revival by aëration; after drying it weighed 1.638 grammes. The proportion of yeast to sugar fermented was, therefore, $\frac{1.638}{145.4} = \frac{1}{89}$ In the flask (B), where free exposure to air took place, the fluid was in active fermentation on the second day, and all the sugar had completely disappeared in eleven days; the yeast from this, after being dried, weighed 1.97 grammes, giving a proportion of yeast to sugar fermented $\frac{1.97}{150} = \frac{1}{76}$.

In a further experiment (D) conducted on the same lines as C, still greater precautions were taken to ensure the entire absence of air from the fluid which was impregnated, as before, with a small quantity of yeast from the funnel; but instead of using the yeast whilst the contents of the funnel were in an active state of fermentation, at which time the yeast would be in a young and vigorous condition, it was added when fermentation was entirely at an end in the fluid in the funnel. Under these circumstances, fermentation was still going on in the flask at the end of three months, when the experiment was brought to a close. The weight of sugar fermented in this case was only 45 grammes, and the weight of yeast formed was 0.255 gramme, the proportion being $\frac{0.255}{45} = \frac{1}{176}$. The fermentation had evidently proceeded with great difficulty, and the cells varied much in appearance, some being large and elongated, some very old and granular, others were more transparent, but all might be considered abnormal. In Experiment A, where, at the commencement, the fluid was saturated with air, the proportion of sugar to yeast was $\frac{1}{67}$; in B, where the liquid had almost been deprived of air by boiling, but where it was afterwards permitted to have free access to the surface of the fluid, the proportion was $\frac{1}{70}$; in Experiment C, where there was almost an entire absence of air, the proportion was $\frac{1}{89}$; and in the last (D), where air was entirely absent in the fluid and old yeast was employed to start the fermentation, the proportion was $\frac{1}{76}$. From these results Pasteur concluded that the less oxygen the yeast has at its command the larger the quantity of sugar which the same amount of yeast is able to ferment when the length of time required for the fermentation is not taken into consideration.

Difference in Fermenting Power of Yeast of Various Ages. —Pasteur points out that it is only young and vigorous yeast which will ferment a solution entirely deprived of oxygen. If a fermentation be commenced, and, as it progresses, small portions of yeast (as nearly as possible equal in quantity) are successively taken from the yeast which subsides to the bottom of the vessel, and used for starting other fermentations (all of which are, as far as possible, conducted under the same conditions as regards exclusion of air, temperature, &c.), and this continued for some time after the original fermentation has come to an end, it is found with each successive portion of yeast taken, that the first visible signs of fermentation appear later and later the more distant the removal of the portions was, in point of time, from the commencement of the original fermentation.

Microscopic Appearance of Vigorous and of Worn-out Yeast. —If each of these portions of yeast are examined microscopically, those taken at the commencement of the original fermentation will be found to present those appearances which characterise young and vigorous yeast, that is, the cells are somewhat swollen in size, have thin walls, and their contents are plastic and almost fluid, exhibiting scarcely any signs of granulation. As time progresses, the cells gradually take on the appearance of weakness and exhaustion, the cell-walls become thickened, their protoplasm becomes denser and more and more distinctly granulated. "These progressive changes in the cells," as Pasteur observes, "after they have acquired their normal form and volume, clearly demonstrate the existence of a chemical work of a remarkable intensity, during which their weight increases, although in volume they undergo no sensible change, a fact which we have often characterised as 'the continued life of cells already formed.' We may call this work a process of maturation on the part of the cells, almost the same that we see going on in the case of adult beings in general, who continue to live for a long time, even after they have become incapable of reproduction, and long after their volume has become permanently fixed."

Yeast Cultivated under Conditions of Free Exposure to Air. —The opposite conditions in which the yeast was allowed to ferment in the presence of more or less free oxygen, were then studied. This was effected, in the first instance, by conducting a fermentation in a flask of 4.7 litres capacity, only 200 c.c. of a 5 per cent. solution of sugar in yeast-water being used. The fermentation was complete in forty-eight hours, and the yeast formed weighed 0.44 gramme; the ratio was therefore $\frac{0.44}{10} = \frac{1}{23}$. Another experiment was conducted in a flat glass dish, with a layer of fluid that was less than a quarter of an inch in depth, and which consisted of 1.72 grammes of sugar dissolved in 200 c.c. of yeast-water. Fermentation was complete in forty-eight hours, and the yeast weighed 0.127 gramme; the ratio was $\frac{0.127}{1.04} = \frac{1}{8}$. In another experiment, conducted in a similar receptacle, where fermentation was only allowed to proceed until sufficient yeast had been formed to permit of its being weighed, the proportion was found to be $\frac{0.024}{0.098} = \frac{1}{4}$. This was the highest ratio attained, and Pasteur considered that under these circumstances the yeast more closely approached the manner in which an ordinary mould fungus lives, the eliminated carbon dioxide in this case having been derived from respiration and not from fermentation, which latter process now proceeded very languidly; and that, if it had been possible to keep each yeast-cell entirely surrounded with air, yeast would then have lived just as an ordinary mould fungus does, and vegetated without causing fermentation.

Importance of the Aëration of Yeast and Worts.—These observations and the deductions to be drawn from them are of the greatest importance to the brewer, and their publication threw an entirely new light on the cause of the deterioration which so often takes place in the yeast used in the brewery, and which necessitates its renewal. As pointed out by Pasteur, if it were not for the oxygen the yeast meets with dissolved in the wort, and also that which it seizes upon when manipulated in contact with air (for yeast which has been deprived for some time of free oxygen absorbs this gas with the greatest avidity when the opportunity is afforded it) it would soon cease to act as a ferment. The enunciation of these principles led to much care being given to the due aëration of wort after it had left the copper; closed or horizontal refrigerators were abandoned in favour of vertical ones, and in more recent times various kinds of apparatus have been contrived for converting the wort into the form of a spray in its passage from the copper to the cooler. The air supplied in this way enables the yeast to preserve its vitality and fermentative faculties.

The Amount of Air which Yeast is capable of Absorbing during its Growth.—A number of investigations were made by Pasteur with the object of ascertaining the amount of oxygen which yeast absorbed in carrying out its functions. An endeavour was made to ascertain how much of this element yeast absorbed when caused to live under conditions approaching as nearly as possible those of the ordinary manner of life of a mould fungus. For this purpose a flat-bottomed flask was employed, the neck of which was drawn out and curved as in a Pasteur flask. Into this was introduced 60 c.c. of yeast-water containing in solution 2 per cent. of sugar and the whole impregnated with a minute trace of yeast. The orifice of the fine tube was then sealed up and the flask and its contents kept at a temperature of 25° C. (77° F.) for fifteen hours. The end of the fine tube was then broken off under a jar of mercury, and the gas thus obtained submitted to analysis; and from the amount of oxygen which was found to have been absorbed in this, that of the whole vessel was calculated. In this way it was found that the yeast, which, after being dried, weighed 35 milligrammes, had absorbed 14.5 c.c. of oxygen; consequently, at this rate 1 gramme of yeast would absorb 414 c.c. of oxygen.

Growth of Yeast without a Supply of Free Oxygen.—In Experiment A, described on p. 319, 2.25 grammes of yeast were produced in three litres of a saccharine solution under conditions where the liquid could not obtain more than 16.5 c.c. of free oxygen. But if the original cells from which this amount of yeast was derived could not have multiplied without the amount of free oxygen absorbed by the yeast in the previous experiment, that is, 414 c.c. for each gramme of yeast produced, $2.25 \times 414 = 931.5$ c.c. of this gas must have been necessary. The larger portion of the yeast formed in the experiment just mentioned must, therefore, have been an anaërobic growth. Similarly, a mould fungus requires, when grown in its ordinary aërobic condition, a large quantity of oxygen. In the case of a growth of a mould fungus, which had been obtained from a single spore by means of a vacuum flask, the air contained in the flask, when analysed after the lapse of a year, showed that the minute plant, the weight of which, when dried, only amounted to 8 milligrammes, had absorbed no less than 43 c.c. of oxygen, that is, in the proportion of 5375 c.c. of this gas for each gramme of plant-substance formed.

Pasteur's Theory of Fermentation.—From a consideration of the deductions to be drawn from these observations and experiments, Pasteur formulated the following theory of fermentation, which we give in his own words.[1] "Fermentation by yeast, that is to say, by the type of ferments properly so called, is presented to us, in a word, as the direct consequences of the processes of nutrition, assimilation, and life, when these are carried on without the agency of free oxygen. The heat required in the accomplishment of that work must have necessarily been borrowed from the decomposition of the fermentable matter, that is, from the saccharine substance, which, like other unstable substances, liberates heat in undergoing decomposition. Fermentation by means of

---

[1] "Studies in Fermentation," p. 259.

yeast appears, therefore, to be essentially connected with the property possessed by this minute cellular plant of performing its respiratory functions, somehow or other, with oxygen existing combined with sugar. Its fermentative power—which power must not be confounded with the fermentative activity or the intensity of decomposition in a given time—varies considerably between two limits, fixed by the greatest and least possible access of free oxygen which the plant has in the process of nutrition. If we supply it with a sufficient quantity of free oxygen for the necessities of its life, nutrition, and respiratory combustions; in other words, if we cause it to live after the manner of a mould, properly so called, it ceases to be a ferment; that is, the ratio between the weight of the plant developed and that of the sugar decomposed, which forms its principal food, is similar in amount to that in the case of fungi. On the other hand, if we deprive the yeast of air entirely, or cause it to develop in a saccharine medium deprived of free oxygen, it will multiply just as if air were present, although with less activity, and under these circumstances its fermentative character will be most marked; under these circumstances, moreover, we shall find the greatest disproportion, all other conditions being the same, between the weight of yeast formed and the weight of sugar decomposed. Lastly, if free oxygen occurs in varying quantities, the ferment power of the yeast may pass through all the degrees comprehended between the two extreme limits of which we have just spoken. It seems to us that we could not have a better proof of the direct relation that fermentation bears to life, carried on in the absence of free oxygen, or with a quantity of that gas insufficient for all the acts of nutrition and assimilation."

"Another equally striking proof of the truth of this theory is the fact that the ordinary moulds assume the character of a ferment when compelled to live without air, or with quantities of air too scanty to permit of their organs having around them as much of that element as is necessary for their life as aërobic plants. Ferments, therefore, only possess in a higher degree a character which belongs to many common moulds, if not to all, and which they share, probably, more or less with all living cells, namely, the power of living either an aërobic or anërobic life, according to the conditions under which they are placed."

Yeast may be grown in an Albuminous Fluid such as Yeast-Water when an Unfermentable Carbohydrate is present.—Pasteur found that when a trace of pure yeast was sown in 150 c.c. of sterilised yeast-water containing $2\frac{1}{2}$ per cent. of milk-sugar (which is unfermentable by ordinary yeast), the yeast increased, and in three months' time, after being dried, weighed 50 milligrammes. Not the slightest trace of alcohol could be detected in the fluid. In another experiment, yeast which had been grown under precisely similar conditions was found to readily excite fermentation in sterilised wort. These results prove that, under certain circumstances, yeast may live as an ordinary plant, and that fermentation is not an invariable phenomenon of its life, but one which depends on the conditions under which the plant is compelled to live. It is capable of carrying on its existence either as a ferment or not, and even after being grown under conditions where it is unable to exercise its fermentative power, it is able to again manifest this when it is again allowed to vegetate under other conditions. It must be noted that yeast when grown in the former manner will only reproduce itself when freely supplied with air. Pasteur concludes that the power of exciting fermentation is not a property enjoyed by cells of a special nature, nor is it a permanent character of any particular structure, but that it is mainly dependent on the conditions of environment under which living cells are placed.

Extension of the Property of Exciting Fermentation to every Living Cell.—If, then, as has been stated, the ability to excite fermentation is not a power peculiar to certain cells, there is a strong probability that all living cells, when placed under suitable conditions, will act as ferments. Pasteur felt so certain on this point that one day he said to M. Dumas, "We would lay a wager that if we were to plunge a bunch of grapes into carbon dioxide gas, alcohol and carbon dioxide would immediately be produced as the consequence of a renewed action starting in the interior cells of the grapes, in such a way that these cells would assume the function of yeast-cells." The experiment was made, and on the following day alcohol was detected in the juice of the grapes. No yeast-cells were to be found in the interior of the grapes, although a careful examination was made to see if any were present. Fresh experiments made by immersing grapes, a melon, plums, oranges, and rhubarb-leaves gave similar results. Twenty-four plums were placed under a glass cylinder, which was immediately afterwards filled with carbon dioxide, and were allowed to remain for eight days. During this period they evolved a considerable amount of carbon dioxide. They were then crushed and submitted to distillation, when they yielded 6.5 grammes of alcohol, or rather more than 1 per cent. on the weight of the plums. In all these cases there was a liberation of heat, which could be detected by first placing the hand against that portion of the cylinder with which the plums were in contact, and afterwards against another portion where they were not. A somewhat similar series of facts had been demonstrated in 1869 by Lechartier and Bellamy,[1] who found, as had been previously shown by Berard in 1821, that fruits absorb oxygen and liberate almost an equal volume of carbon dioxide. They found that, in addition to the fruits mentioned above, chestnuts, potatoes, grains of wheat, linseed, &c., when any of these were placed in test-tubes and the mouths of the tubes plunged into mercury so as to exclude air, the

[1] *Comptes Rendus*, vol. xix. pp. 336–466.

oxygen of the air in the tube was quickly absorbed and replaced by carbon dioxide. The exhalation of this gas continued for several months, and, when a considerable period had elapsed, notable quantities of alcohol were found. Two pears which weighed respectively 157 and 125 grammes were kept under these conditions for seven months; they yielded at the end of that time 2.62 grammes, or about 1.4 per cent. of their weight of alcohol. Beetroots, when kept excluded from the air for a length of time, exhale a strongly acid liquid, and this Pasteur ascribes to the commencement of the lactic and other fermentations.

From these and other facts of a similar nature Pasteur infers, if perhaps only in an indirect way, that the cells of an organism can give rise to several kinds of fermentation. Reasoning from the fact observed by Lechartier and Bellamy, that the proportion of carbon dioxide to alcohol yielded in the case of their experiments with fruits was entirely different to that obtained in the fermentation of wort with beer-yeast, he states: "In the present day it must be borne in mind that the equation of a fermentation varies essentially with the conditions under which that fermentation is accomplished, and that a statement of this equation is a problem no less complicated than that in the case of a living being. Moreover, it is quite erroneous to suppose that the presence of a single one of the products of a fermentation implies the co-existence of a particular ferment. If, for example, we find alcohol among the products of a fermentation, or even alcohol and carbonic acid gas together, this does not prove that the ferment must be an alcoholic ferment, belonging to alcoholic ferments in the strict sense of the term. Nor, again, does the presence of lactic acid necessarily imply the presence of lactic ferment. As a matter of fact, different ferments may give rise to one, or even several identical products. We could not say with certainty, from a purely chemical point of view, that we are dealing, for example, with an alcoholic fermentation, properly so called, and that the yeast of beer must be present in it, if we had not first determined the presence of all the numerous products of that particular fermentation, and that they were present in those proportions characteristic of that fermentation under conditions similar to those under which the fermentation in question had occurred." These extracts show how far-reaching Pasteur's views on the subject of fermentation had become; they were, in fact, not only a denial that there were any special fermentation organisms, but also an affirmation "that every cell may cause every fermentation, alcoholic, lactic, acetic, &c., when deprived of its supply of free oxygen; and the kind of fermentation and of the products formed thereby are merely due to external conditions, the way in which the cells are nourished, and the temperature of the fermentation." "In fact, fermentation is a general phenomenon; it is life without air."

Meyer's Theory of Fermentation.—This philosopher[1] sought to explain the phenomena of fermentation by what he called "intramolecular respiration." The cell of every living plant absorbs oxygen, which, by the agency of its protoplasm, is combined with carbon, and again evolved as carbon dioxide. It is, in fact, a process of combustion in which a certain amount of heat is produced, or, in other words, energy liberated which is expended in performing the work necessary to carry on the vital functions of the cell, such as growth, &c. But when a living cell is deprived of the access of free oxygen, it does not cease at once to evolve carbon dioxide, carbon is still consumed in its interior, and the oxygen necessary for effecting this is derived from some of the constituents of the cell. This then is "inner" or "intramolecular respiration." But when oxygen is obtained under these forced conditions, the compounds contained in the cell which furnish this oxygen undergo decomposition and are converted into other bodies, e.g. sugar into alcohol and carbon dioxide. The explosion of gunpowder or nitro-glycerine is adduced as a somewhat analogous process; in both these cases combustion ensues, and a large amount of energy is liberated in the entire absence of free oxygen, this element being obtained from a portion of the materials which constitute the explosive. The increase of yeast in the absence of free oxygen is explained in a somewhat similar manner; the ferment organisms under such a condition derive the force necessary for the performance of their vital functions from the oxidation of carbon by intramolecular respiration. Though the growth of yeast is greater when it is supplied with free oxygen, yet it can continue to reproduce itself for a considerable length of time in the absence of this element in the free state. Respiration is then supposed to take place at the expense of the sugar molecule, which is split up in that peculiar way which we characterise as fermentation. Meyer considers that though yeast may cease to grow, it is still able to excite fermentation, because intramolecular respiration can still proceed in yeast though it may have ceased to increase, just as is found to be the case with the cells of the higher plants when they are deprived of the presence of free oxygen.

Though in this latter case the production of alcohol and carbon dioxide may be unquestionably attributed to intramolecular respiration, it is doubtful if it affords a true explanation of the production of these bodies in the ordinary course of fermentation.

Brefeld's Views on Fermentation.—In 1874 and 1875 Brefeld published the results of a number of experiments undertaken with the view of further elucidating the subject of fermentation.[2] His interpretation of the results of these, and the

---

[1] *Lehrbuch d. Gährungs-Chemie*, Heidelberg, 1879.
[2] *Berichte*, 1874 and 1875.

propositions he founded on this interpretation, gave rise to considerable discussion and much experimenting at the time, which ended in throwing considerable light on some obscure points in the process. He assumed that the cells of yeast, like all other living cells, must, in order to carry on their vital functions, have a supply of free oxygen. That this was the case with yeast he demonstrated by showing that when placed in an atmosphere entirely devoid of oxygen it very soon ceased to grow, and then a series of characteristic changes commenced in it which ultimately ended in its death. He also proved that it was possible for yeast, when very freely supplied with oxygen, to grow for a time in a saccharine medium without exciting fermentation. From these facts he drew the erroneous conclusion that it was only when all the free oxygen in a fermentable fluid had been completely exhausted, and the yeast-cells it contained began to take on the peculiar condition mentioned above, that they commenced to excite fermentation. This they do, according to Brefeld, by secreting an albuminous substance, which, acting on the sugar as an enzyme, causes it to split up into alcohol and carbon dioxide. If the yeast be removed from the saccharine fluid, the secretion of this substance ceases; hence the impossibility of isolating it. Fermentation is, then, according to Brefeld, a pathological process, which only proceeds when the yeast is in a drowning state that ends with its death.

Traube and Pasteur successfully controverted this view by proving experimentally that young and vigorous yeast can increase in a fluid entirely devoid of oxygen. Meyer showed that yeast can simultaneously increase and excite fermentation in a fluid containing an abundance of oxygen; that fermentation without growth can take place when old cells are immersed in a saccharine fluid which contains no oxygen; and that young cells, when supplied with a superabundance of this element, can grow without producing fermentation. But, as Meyer points out, between these two extremes there are probably many intermediate stages in which both yeast growth and fermentation can proceed simultaneously. In reality, some of the observations of Brefeld, which were undertaken in support of the theory of Liebig, go to confirm those of Pasteur, such as those where he finds that there is a sort of correlation between the respiration of yeast and its fermentative action, and that the more the former is suppressed the more actively the latter proceeds.

Schützenberger's Views on the Theory of Fermentation.[1]—This observer disputes the validity of Pasteur's doctrine that fermentation is "life without air." With reference to the experiments mentioned on p. 318, where *Mycoderma vini*, when grown immersed in a saccharine medium, was able to act as an alcoholic ferment, Schützenberger is disposed to think that though *Mycoderma*, when grown on the surface of such a fluid, and supplied with an abundance of free oxygen, completely oxidises

---

[1] "On Fermentation," 1886.

the sugar to water and carbon dioxide, it may still produce alcohol, which, at the moment of its formation, suffers further decomposition, and hence eludes observation. The volume of the carbon dioxide evolved in these experiments was found to be nearly equal to that of the oxygen which disappeared, and this strongly points to the carbon dioxide being obtained either from carbon or from a hydrocarbon, which would coincide with the view that alcohol is primarily formed, and then immediately afterwards broken down to carbon dioxide and water. He also thinks that the phenomena observed by Lechartier and Bellamy (p. 325) may be explained in a similar manner. They prove conclusively that the cells of fruits, under certain circumstances, are capable of transforming sugar into alcohol and carbon dioxide, just as yeast does, only in a much less energetic manner. Schützenberger thinks that under the ordinary circumstances of normal plant life alcohol may be the first formed product. He states that neither Pasteur nor others have definitely proved that the commencement of this cellular fermentation coincides with the moment when the cells are partially or wholly deprived of free oxygen. Here also alcohol may be the normal production of the cell, which is, under the influence of an abundant supply of oxygen, as in the case of *Mycoderma*, immediately transformed into carbon dioxide and water.

He then draws attention to Pasteur's own statements, that fermentation, when conducted in the entire absence of oxygen, is very slow; but when it is carried on under circumstances of free access of oxygen, such as when a large surface of the fermenting fluid is exposed to the air, fermentation proceeds much more rapidly. Pasteur also states that the great activity exhibited at the commencement of a fermentation is due to the air dissolved in the fermenting fluid. These statements seem to be in obvious contradiction to the view that the cause that incites a yeast-cell to ferment is its deprivation of free oxygen. Pasteur, however, explains this apparent contradiction by showing that yeast multiplies with much greater rapidity in the presence of free oxygen, and though the fermentative power of the yeast is lowered, yet the quantity of yeast formed under these circumstances is so much greater, that fermentation, in spite of the lowered fermentative activity of the yeast, proceeds more rapidly. Schützenberger questions the validity of measuring the fermentative power of yeast by the proportion between the weight of the new yeast formed and that of the sugar decomposed, and suggests that a more rational method of measurement would be the quantity of sugar decomposed by the unit weight of yeast in the unit of time. Quoting his own experiments on the absorption of oxygen by yeast, he points out that though yeast absorbs a certain amount of oxygen when it is diffused in water, yet it absorbs much more when placed in an aërated saccharine liquid which also contains nutritive proteid matter. Moreover, he measured the amount of alcohol produced in two fermentations in which all the conditions were identical, with the single exception that in one the fermentation was carried on in the entire absence of oxygen, in the other the fluid was kept saturated with this gas. In this latter case, strictly in accordance with Pasteur's experiments, the amount of alcohol produced during the same period was sensibly greater, but it was found that during the whole fermentation there had been a continuous absorption of oxygen. If Pasteur's theory be true, under these circumstances the fermentation should have been either stopped or considerably lessened in intensity, because, as there was always a supply of free oxygen at hand, there was no necessity for the yeast to obtain its supply of this gas by decomposing the sugar molecule. According to this, yeast, when placed in a fermentative fluid kept continuously saturated with oxygen—that is, under conditions where it is able to fully satisfy its respiratory needs with the oxygen,—is still able to produce a very active fermentation.

Schützenberger finally concludes that the respiratory and fermentative functions of yeast are two entirely separate and independent powers inherent in the yeast-cell; that they go hand-in-hand; as the one becomes heightened so does the other; and that they do not, as Pasteur assumes, stand in inverse proportion to one another—that is, as the one function is lowered the other is heightened. To more fully demonstrate the correctness of this view, he performed a number of experiments which show that all those cases which tend to lower the vitality of the yeast coincidentally lower both its respiratory and fermentative powers.

Nägeli's Moleculo-Physical Theory of Fermentation.—This, which differs essentially from all the other theories of fermentation hitherto propounded, was published in 1879.[1] Nägeli considers that the theories so far proposed fail to account for the whole of the observed phenomena, and seeks to establish one which shall more perfectly include them all. By the extension, in a certain sense, of his views on enzymic action he assumes that the numerous and complicated arrangement of atoms and atomic groups which constitute the protoplasm of a ferment organism are in a state of molecular vibration, each having its own particular rhythm. These vibrations the protoplasm is able to communicate to compounds, which, by diffusion, have actually entered into its substance, or which, dissolved in the liquid, are in close proximity to it. These vibrations acting upon the atoms or atomic groups of the fermentable body cause them to rearrange themselves into new compounds. Since, in this case, we have a number of substances present in the protoplasm, each of which has its own particular rate of vibration and is consequently capable of effecting a different transformation in the

---

[1] *Theorie der Gährung*, München.

fermentable substance, we meet with a great variety of products in fermentation, such as alcohol, glycerine, succinic acid, &c. This also explains why a substance capable of producing fermentation can never be isolated from the living cell: it does not exist. Nägeli considers that there is no necessity for an organism to excrete an enzyme, unless the action has to be performed at a distance from the organism itself. When an organism has to effect a chemical transformation under such circumstances, it then secretes an enzyme, which being able to diffuse itself through a fluid, can travel to and act at a point more or less removed from the organism itself. This we see in the secretion of diastase, &c., by the growing germ of the barleycorn, which, in order to secure its own nourishment, must dissolve tissues situated at a distance from itself. Nägeli considers it very questionable if the organism ever secretes ferments which shall be employed within itself; he thinks that much more powerful energetic means for chemical action are at its disposal in the molecular forces of its own protoplasm. Again, the action of all enzymes is to convert a substance either totally or partially unfitted for the nutrition of an organism into compounds better fitted, as, for example, the conversion of starch into sugar, albumin into peptone, &c. On the other hand, the substances produced in fermentation are of no further use whatever to the organism; in fact, they are always prejudicial to it. Thus, yeast in a saccharine medium produces alcohol, which, in sufficient quantity, is absolutely poisonous to it. In enzymic action the chemical compound acted upon is transformed in a simple manner; each of its molecules suffers identically the same transformation. In fermentation, since the living protoplasm is constituted of many compounds, each of which has its own particular molecular movement, the molecules of the fermentable substance do not each undergo the same identical change, but are transformed into a variety of bodies, such as alcohol, carbon dioxide, succinic acid, glycerin, &c.; and, moreover, these bodies do not occur in the same proportions in every fermentation; they vary considerably according as the particular conditions under which each individual fermentation takes place differ. Fermentation is invariably accompanied by the evolution of carbon dioxide, enzymic action never. Enzymes can be replaced by purely chemical agents, such as acids, alkalies, or even heat alone; fermentation has never been produced by such agents. Such profound differences in the two classes of reactions show that they must be produced by very different agencies. With reference to Pasteur's theory, Nägeli found that the presence of free oxygen favoured fermentation; consequently, fermentative action could not be caused by the ferment being compelled to derive its supply of oxygen from the sugar molecule. He found that access of free oxygen to the yeast-cells was advantageous, and he assumes that its invigorating action heightens the molecular movements of their protoplasm. Nägeli then endeavours to decide whether fermentation takes place inside or outside the yeast-cell. He shows from the results of his own experiments that the yeast-cell, when acting under favourable conditions, decomposes 1.67 times its own weight of sugar and produces 0.85 of its weight of alcohol in an hour. If the fermentation takes place entirely within the cell, there must pass into each of the most vigorous cells 3.34 times its weight of sugar in an hour, and out of the cell in the same time 1.7 time its weight of alcohol. Though these quantities seem enormous, they were found to be quite within the osmotic capacity of the yeast-cell; hence no conclusion can be drawn from this as to where fermentation takes place. He then seeks for analogous phenomena in other living cells. No plant-cell is able to effect a transformation outside itself or to withdraw anything from an insoluble body so situated. For instance, though yeast-cells are able to withdraw oxygen from the corpuscles of the blood, they only do this indirectly; they withdraw it by osmosis from the serum in which the corpuscles are bathed, and the corpuscles in turn yield up the gas by osmosis to the serum. Living bacteria have the power of decolorising, by a process of deoxidisation, litmus dissolved in a nutrient medium. As the litmus molecules are entirely outside the investing membrane of such organisms, here is a case in which chemical action produced by a living cell takes place outside itself, and this is perhaps the only known instance of this kind. Nägeli found that when the thinnest membrane was interposed between the yeast and the fermentable fluid no fermentation took place. He considers that this fact not only indisputably proves that ferment action can take place in the closest proximity to the yeast-cell, but also that it effectually disposes of those theories which assume that fermentation is caused by an enzyme secreted by the yeast-cell; such enzyme must necessarily pass through a membrane by osmosis. But this also does not settle the question. The following facts seem to point to the action taking place outside the cell. Nägeli found in 1853 that when certain fruits were immersed, in their entire condition, in grape-juice which had been submitted to the action of the fumes of burning sulphur, fermentation took place in their interior, though no trace of fermentation could be detected in the grape-juice in which they were immersed. But if the fruits, before being treated in this manner, were first carefully deprived of their skins, then no fermentation took place at all. Since yeast-cells or their spores are always to be found on the skins of ripe fruits, he considers that the fermentation in the former case was caused by the vibrations of the protoplasm of the yeast-cells being transmitted through the skin of the fruit into its interior. Nägeli strongly impresses the point that this fermentation is not to be confused with the entirely intramolecular one discovered by Lechartier and Bellamy, described on p. 325, because the fleshy portion of a fruit, when deprived of its skin and adherent yeast-cells, caused no fermentation. Another circumstance which strongly favours the theory that ferment action occurs outside the cell is the formation of acetic ether when yeast and acetic bacteria are present in the same fermentable fluid. The yeast-cells are supposed to produce alcohol in their immediate proximity, and the acetic bacteria similarly acetic acid; these two bodies unite in their nascent state and produce acetic ether. The prejudicial influence which one energetic ferment exercises upon the nutrition and growth of another ferment seems also to be readily explicable on the assumption that ferment action takes place outside the cell. From experimental data Nägeli calculated that the greatest distance from itself at which an organism could excite fermentation was the one-fortieth to one-thirtieth of a millimetre. He considers that the molecular vibrations of the protoplasm of a ferment organism are transmitted in the same way as those of sound, light, or heat. The vibrations of one molecule set up vibrations in another, and these again in others, and so on. The vibrations of the protoplasm are first communicated to the sugar molecules which have penetrated into the interior of the yeast-cell by osmosis; these molecules transmit their vibrations to the sugar molecules contained in the membrane surrounding the yeast-cell, which is saturated with the saccharine solution, and these, in turn, to the sugar molecules dissolved in the fluid outside the cell. Finally, Nägeli concludes that the cause of fermentation is the living protoplasm of the ferment organism, and therefore resides in its interior, but that it is able to extend its sphere of action to a distance of at least one-fiftieth of a millimetre beyond the investing membrane of such organism. The actual fermentation action is supposed to take place to a small extent inside the cell, but to a much greater extent outside.

Nägeli considers that this is true only with reference to the principal products of fermentation, *i.e.* alcohol and carbon dioxide; the subsidiary products, glycerin, succinic acid, &c., are most probably entirely formed within the cell.

He thinks that the same theory may be extended to all other fermentations—lactic, butyric, putrefactive, &c. The acetic fermentation, in which alcohol becomes oxidised to acetic acid, may be explained as follows: The molecular vibrations of the protoplasm of the acetic bacterium are communicated to the alcohol and oxygen molecules absorbed by the organism, and by them transmitted to the alcohol and oxygen molecules outside the cell. When the disturbance of the equilibrium of these molecules has reached a certain degree of intensity, a molecular transformation, *i.e.* a chemical change, ensues. A part of the process takes place in the interior of the cell, but the greater part outside.

He considers that the action of a ferment may be divided into two stages; in the first of these the equilibrium of the molecules is disturbed, and to effect this a small amount of energy is expended by the yeast-cell; in the second stage, by virtue of the new attractions and repulsions brought into play amongst the molecules, new phases of equilibrium become established among them, which result in the formation of new compounds, and this process is accompanied by the liberation of energy, which appears as heat.

Hoppe-Seyler's Theory of Fermentation.—This physiologist[1] considers that fermentation is caused by an enzyme secreted by the yeast, and that the division of ferments into organised and unorganised is entirely wrong. If the whole organism is to be included under the denomination "ferment," then no division is made between it as a whole and its several constituents, nor between its whole life and the individual processes of its life. The idea of an organised ferment, in Pasteur's sense, might be applied to every organism, even to man himself; the ferment (enzyme) which converts glucose into alcohol and carbon dioxide is identified with the whole organism of the yeast-cell, and the production of glycerin and succinic acid are assumed to be connected with the same process. But the yeast-cell produces fat, cellulose, invertase, &c., and Hoppe-Seyler considers that to assume that one series of vital processes is inseparably connected with the whole life of the cell, and that another series is not, is quite an arbitrary assumption.

It has been hitherto supposed that oxidation in the living organism was purely a process of addition, similar to that which is effected by ozone or potassium permanganate. It was thought that oxygen combined directly with organic substances and converted them into compounds becoming progressively richer and richer in oxygen, until the final stages of water and carbon dioxide were reached. This view is, however, contradicted by recent observations, and it seems much more probable that the transformation process depends essentially upon the action of water, which, though itself a comparatively inert substance, yet, when under certain conditions co-operating with those organic bodies which we call enzymes, can then cause profound changes. All organisms contain and produce enzymes, and the higher organisms produce a large number of different varieties. The enzymes themselves are extremely unstable bodies which have the power of decomposing hydrogen peroxide $(HO)_2$, and, with the assistance of water, can transform certain organic compounds into others which possess less heat of combination. The organisms which are concerned with fermentation are, according to Hoppe-Seyler, able to produce and to be the bearers of ferments. These are just as inseparable from the organism as the contractile substance is from the muscle, or the protoplasm from the organism itself; with the death of the organism the ferments cease to act.

Water, according to Hoppe-Seyler, is essentially necessary for

[1] *Physiolog. Chemie.*

fermentation. In ferment action, according to his theory, the oxygen atoms split up from the hydrogen atoms and attach themselves to the carbon atoms; the liberated hydrogen, being in the nascent state, converts any oxygen which may be present into active oxygen.

In the alcoholic fermentation the sugar molecule is split up by the ferment into two atomic groups, and two molecules of water are split up into two hydrogen atoms (H_2) and one hydroxyl $(HO)_2$ molecule. The oxygen of the hydroxyl exercises an oxidising action on the two molecular groups into which the sugar is split up, the hydrogen simultaneously exercising a reducing action. Through this combined action two molecules of carbonic acid, $CO(OH)_2$, and two molecules of alcohol are yielded, thus:—

$$\underset{\text{Glucose.}}{C_6H_{12}O_6} + \underset{\text{Water.}}{2H_2O} = \underset{\text{Carbonic acid.}}{2CO(OH)_2} + \underset{\text{Alcohol.}}{2C_2H_5(OH)}$$

The carbonic acid then breaks up into water and carbon dioxide. The production of glycerin, succinic acid, and the higher alcohols (fusel oil) are not considered to be connected with this process, which is essentially concerned in the decomposition of the sugar.

Experiments of Adrian Brown on Fermentation.— Though the theory of fermentation propounded by Pasteur is perhaps the one which has met with the widest acceptance, yet in later times numerous observations have been made which tend to throw considerable doubt upon its validity. We have seen how both Schützenberger and Nägeli found that fermentation by yeast was aided rather than hindered by the presence of oxygen. Blankenhorn observed the same thing in the process of wine-making; at the end of the fermentation of the grape-juice a certain amount of sugar is usually left unfermented, and if at this stage the young wine be aërated, fermentation commences anew, and the last portions of sugar disappear. More striking still are the experiments of Adrian Brown performed with a view to the further elucidation of the phenomena of fermentation. In these, following the example of Pedersen, Hansen, and others, he employed a method of counting the yeast-cells formed at the various stages of the process by means of an instrument which, as it was originally invented to count the red corpuscles of the blood, is called a "a hæmatimeter."

The Hæmatimeter and the Method of Using it.—This instrument consists of an ordinary glass slip, to which is cemented a cover glass of a definite thickness, say the one-tenth of a millimetre, in which a circular aperture has been cut out. This forms a shallow trough one-tenth of a millimetre in depth, consequently, when a drop of fluid is placed in the trough and covered with a thin cover-glass, it is evident that we have a layer of fluid one-tenth millimetre in thickness. On the bottom of the trough a series of squares are engraved, the sides of which are one-twentieth of a millimetre in length; consequently, the area of each square will be $\frac{1}{20} \times \frac{1}{20} = \frac{1}{400}$ of a square millimetre, and the amount of fluid represented by one of these squares is $\frac{1}{400} \times \frac{1}{10} = \frac{1}{4000}$ of a cubic millimetre. If a drop of fluid containing yeast-cells is placed in the trough, taking care that the drop is of such a size as to nearly fill the trough, but not so large as to exude when the cover-glass is put in its place, and the whole allowed to remain at rest for a short time, the yeast-cells will settle to the bottom of the trough. The slide is then placed under the microscope, and the number of yeast-cells counted gives the average number of cells contained in each $\frac{1}{4000}$ of a cubic millimetre of the liquid. For instance, if, on counting 64 squares, 168 cells are found, then this number divided by 64 will give 2.625 as the number of cells per $\frac{1}{4000}$ cubic millimetre. Whatever the volume per square of the instrument may be, this is taken as the standard volume; and as all calculations are made to this one volume, some little latitude in its size is permissible. Before placing the liquid in the instrument, it is necessary to agitate it for some time, so that the yeast-cells may be evenly distributed and an average sample obtained; also to see that the yeast-cells are not too numerous to be conveniently counted; this may always be secured by sufficiently diluting the liquid and making a subsequent allowance for the amount of dilution. The fluid used should have a specific gravity so high that the yeast-cells do not sink in it too rapidly, and yet not so high as to cause them to take an inconvenient length of time in subsiding. Wort of the strength usually used in brewing operations offers no difficulties in either of these directions. In those cases where the yeast-cells exhibit a great tendency to cling together, and also for the purpose of stopping the growth of the cells, Hansen recommends the addition of one or two volumes of 10 per cent. sulphuric acid.

The Amount of Increase of the Yeast-cells in a Fermenting Fluid is Determined by the Volume of the Fluid.—Adrian Brown[1] found that when equal quantities of the same fermentable fluid were seeded, the one portion with 0.93 yeast-cells per standard volume,[2] and the other with 7.44, at the conclusion of the fermentation the number of yeast-cells found in each of the two solutions were 25.14 and 27.08 respectively—that is, practically the same number. From this and other similar experiments it was conclusively shown that the number of cells found at the end of a fermentation was little influenced by the number added at its commencement, provided they did not exceed a certain number; in fact, that the number of yeast-cells found at the end of a fermentation was dependent, to a great extent, on the volume of the fluid; in other words, that with a certain volume of fluid there was a certain maximum beyond which no further increase

---

[1] *Journal of the Chemical Society*, 1892, p. 369.
[2] The standard volume used in all these experiments was $\frac{1}{4000}$ of a cubic millimetre.

of yeast-cells appeared possible. By adding more yeast-cells at the commencement of a fermentation than was the maximum for the particular volume of liquid being used, the further increase of yeast-cells was almost entirely suppressed; and when two or three times this maximum number was added, then no further increase, as evinced by signs of budding, was visible. By working in this way, it was possible to eliminate from a fermentation all the influences due to the multiplication of the yeast.

In two parallel experiments conducted under conditions where this large excess of yeast was employed, air was caused to continuously bubble through the fluid in the one case, while the other was conducted in the ordinary manner. It was invariably found in those experiments which were liberally supplied with air that the most sugar had disappeared in a given interval of time. In order to ascertain if this increase of fermenting power was caused by the agitation of the liquid caused by the air bubbling through it, parallel experiments were made in which either carbon dioxide or hydrogen was passed through the liquid in a similar manner. Fermentation, however, always took place with greater rapidity in those cases where the fluid was liberally supplied with air. These results, therefore, stand in direct contradiction with the hypothesis of Pasteur, that under the influence of free aëration the fermentative activity of yeast is diminished.

In a more recent paper,[1] A. Brown brings forward further arguments and fresh experiments in support of his contention. The discussion is confined to fermentation produced by yeast alone, and no reference is made to those experiments with mould fungi, bacteria, &c., by which Pasteur sought to extend to every living cell the property of being able to excite fermentation, since these latter could but strengthen the theory by analogy after it had been conclusively proved to be true with regard to yeast. Brown considers that Pasteur's experiments may be divided into two classes, the first of which suggests the probability of the correctness of his theory, the second those by which he attempts to prove its truth. To the first class belongs the experiment on p. 322, where a large quantity of oxygen was absorbed by yeast when grown in a thin layer of fermentable fluid freely exposed to the air; this, however, merely proves that yeast is able to grow under aërobic conditions, and gives no information concerning its fermentative functions. So the experiments on pp. 319 and 320, which show that yeast can grow and carry on a brisk fermentation in a fluid when entirely cut off from access of air, only demonstrate the fact that yeast can also live under anaërobic conditions, and that yeast, when existing under the latter conditions, can cause fermentation.

To the second class belong those experiments which tend to

[1] *Journal of the Chemical Society*, 1894, p. 911.

show that the more and more yeast is caused to act under anaërobic conditions, the greater becomes its fermentative power; and on these conclusions is based the theory that fermentation is "life without air." But as A. Brown says, if Pasteur's experiments are to conclusively prove his case, "two points are essential—first, that the ratios of the weight of the yeast formed to sugar fermented should be numerical expressions of the fermentative powers of the yeast; and, secondly, that these ratios should be comparable with each other. For by comparisons between these ratios Pasteur determines the relative fermentative powers of the yeast-cells under varying conditions of experiment, and established his theory upon the conclusions thus arrived at. If, then, the ratios of yeast to fermented sugar do not express the true fermentative powers of the yeasts, or if these ratios are not comparable, Pasteur's deductions fail, and his theory remains without experimental proof." According as fermentations were carried out under different conditions with reference to air supply, the proportions of yeast formed to sugar fermented varied from $1:176$ to $1:4$; that is, the greater the supply of air the less the yeast acted as a ferment, and the nearer it approached the manner of living of an ordinary fungus. But, as Brown remarks, the ratios given are assumed to express fermentation powers that are comparable with each other; and in order to do this correctly, the numbers given must either express the total or some known fraction of the fermentative power of the yeast; though, undoubtedly, Pasteur assumes that these numbers expressed the total fermentative power which the yeast was capable of exerting. In the experiment in flask C (p. 120), as sugar was present in the solution when the experiment was terminated, the yeast had been able to exercise its fermentative powers during its whole existence, and here, perhaps, its total power was represented as nearly as possible, though a very languid fermentation was still proceeding. In the experiment in flask B, as the whole of the sugar was fermented, the fermentation must necessarily have come to an end, and no evidence is afforded that in this case the whole of the fermentative power of the yeast was obtained, since, had the yeast had more sugar at its command, it would undoubtedly have still continued to exercise its fermentative functions. Still, Pasteur assumes that the ratios obtained in these two instances were comparable. This would be true if the ratio of yeast produced to sugar fermented was a constant one, but such is by no means the case, as the following experiments show. Five fermentations were conducted, each with 100 c.c. of sterilised yeast water, to which weights of cane sugar varying from 2.5 to 30 grammes were added, each being impregnated with a trace of a pure culture of high fermentation yeast. The results are appended in the following table:—

No. of Experiment.	Time of Fermentation.	Grammes of Sugar in Solution.	Grammes of Sugar Fermented.	Weight of Yeast Formed, in Grammes.	No. of Cells per Standard Vol.	Ratio of Yeast to Sugar Fermented.
1	5 days	2.5	2.5	0.1240	8.51	1 : 20.2
2	5 days	5.0	5.0	0.1550	9.94	1 : 32.3
3	7 days	10.0	10.0	0.1775	10.44	1 : 56.3
4	12 days	20.0	20.0	0.1400	11.17	1 : 142.9
5	20 days	30.0	25.17	0.1380	12.26	1 : 182.4

These results show how little the weight of yeast formed is influenced by the strength of the solution. The yeast increases in weight in solutions up to 10 per cent. strength, after which there is a diminution as the solution becomes stronger.[1] These differences in the weights are very slight, and distinctly prove that the weight of the yeast formed bears no definite relation to the weight of the sugar fermented. According to Pasteur's views these ratios represent the fermentative powers of the yeasts, and should therefore be comparable with one another. If this were so, then the yeast in experiment 4 had about four and a half times the fermentative power of the yeast in experiment 2, and in both these cases the sugar was entirely fermented. The only difference in the conditions of the two experiments was the amount of sugar present, and probably the reason why the yeast exhibited this increased fermenting power was because it had more sugar to act upon. Nothing in either case shows that the yeasts had exerted more than an unknown fraction of their fermenting power. In experiment 5 sugar was still present at the termination of the fermentation, consequently the yeast here decomposed as much sugar as it was capable of doing. In this case it is notable that the ratio of yeast formed to sugar decomposed was 1 : 182, a lower result than the lowest obtained by Pasteur when working under the most perfect anaërobic conditions; yet air had access to the fermenting fluid all the time. Brown further concludes that though there are many other factors concerned in fermentation, the amount of sugar present is the principal one influencing the ratio of the weight of yeast formed to that of the sugar fermented, and not the relative condition of aëration or non-aëration. The following experiments show clearly that the true ratio is not given by experiments of such a nature as that of B (p. 320). Three flasks numbered 1, 2, and 3, each containing 200 c.c. of a 5 per cent. solution of cane sugar in yeast water, had their mouths closed with cotton-wool plugs, and were sterilised. One of these, No. 1, was so arranged that when the fermentation in it was ended, the solution could be partially withdrawn and replaced by fresh. Each was impregnated with a trace of a pure culture of high

---

[1] The yeast-cells still continue to increase in number, but become smaller in size.

fermentation yeast and kept at a temperature of 26° C. (79° F.). In four days fermentation had ceased, and in flask 2 the ratio of yeast formed to sugar fermented was found to be 1 : 29. The same day 120 c.c. of the fermented fluid were removed from flask 1, and 50 c.c.[1] of a solution containing 10 grammes of sugar were passed into it. A rapid fermentation ensued, which ceased in three days, when a further portion of the fermented liquid was again removed, and replaced by another 50 c.c. of a solution containing 10 grammes of sugar. Fermentation was over in another six days, and the yeast-cells were then counted and weighed. The yeast in flask 3, which had remained at rest the whole time, was similarly treated. The results of the three experiments are appended:—

No. of Experiment.	Duration of Experiment.	No. of Cells Found per Standard Vol.	Total Weight of Yeast Found, in Grammes.	Total Sugar Fermented, in Grammes.	Ratio of Yeast to Sugar Fermented.
1	13 days	9.38	0.2990	30	1 : 100.4
2	18 days	9.02	0.3435	10	1 : 29.1
3	13 days	9.40	0.2435	10	1 : 41.0

Though the number of yeast-cells in each experiment was practically the same, three times as much sugar was fermented in experiment 1 as in 2 and 3. The weight of the yeast formed in the three experiments somewhat differs, and has decreased in experiments 1 and 3. In the latter the decrease is probably owing to the fact that it had remained in an exhausted solution for some time, and it is known that yeast does lose weight under these conditions. The yeast in experiment 1 had actually been employed in fermenting all the time, and the fact that it had lost weight distinctly proves that the fermentative functions of the yeast-cell are not in any way connected with its growth and development. These experiments, then, distinctly show that the ratio of yeast formed to sugar fermented do not express the true fermentative powers of the yeast, or any known fraction of that power, when the yeast has only a limited amount of sugar at its command, and such ratios are therefore not comparable. With reference to Pasteur's experiment on p. 322, where by growing yeast under conditions of very free aëration, and where the ratio of yeast formed to sugar fermented was 1 : 4, Brown objects that here the element of time was introduced; fermentation was arrested when there was sufficient yeast formed to be weighable, its action was limited in point of time, and there was nothing to show that if the experiment had been continued the yeast would

[1] The volume of this solution was kept small in order to prevent any multiplication of the yeast-cells, as this would have necessarily vitiated the result of the experiment.

not have gone on fermenting. Hence, according to Pasteur's own views, the ratio thus obtained does not represent the fermentative power of the yeast, but its fermentative energy.

But there is still a further objection to this experiment: Pasteur employed cane sugar. It has been distinctly shown by J. O'Sullivan[1] that the fermentation of cane sugar is invariably preceded by its inversion, and Brown proves, by the following experiments, that during the time that Pasteur allowed his experiment to proceed, the yeast was really occupied in inverting the sugar. Fermentations were started in six flasks, with, in each case, a trace of yeast. Three of the flasks contained a 10 per cent. solution of cane sugar in yeast water, the remaining three a similar solution of glucose. In forty-eight hours, when sufficient yeast had formed to be weighed, the experiment was stopped in one flask of each series, and the ratio of yeast formed to sugar fermented determined. The same operation was performed in two other flasks at the end of sixty-six hours, and with the remaining two at the end of ninety hours. The results are appended:—

Cane Sugar Experiments.				Glucose Experiments.		
Time from Inoculation.	Yeast Found, in Grammes.	Grammes of Sugar Fermented.	Ratio of Yeast to Sugar Fermented.	Yeast Found, in Grammes.	Grammes of Sugar Fermented.	Ratio of Yeast to Sugar Fermented.
42 hours	0.0985	0.531	1 : 5.4	0.1040	2.334	1 : 22.4
66 hours	0.1785	4.026	1 : 22.5	0.1692	4.364	1 : 25.8
90 hours	0.1900	6.812	1 : 35.8	0.2090	6.920	1 : 33.1

These distinctly show that when yeast is added to a solution of cane sugar, the yeast is employed for a considerable length of time in inverting the sugar before it commences to exercise its fermentative functions; but when a considerable quantity of sugar has become inverted, fermentation proceeds with much the same rapidity as it does in a solution of glucose.

Brown finally concludes that as Pasteur has not correctly determined the fermentative powers of the yeasts, and that as these are not comparable with each other, his theory remains without proof: "If, in the place of M. Pasteur's theory that fermentation is 'life without air,' is substituted the consideration that yeast-cells can use oxygen in the manner of ordinary aërobic fungi, and probably do require it for the full completion of their life history, but that the exhibition of their fermentative functions is independent of their environment with regard to free oxygen, it will be found that there is nothing in the results of any of M. Pasteur's experiments contradictory to such a hypothesis."

[1] *Journal of the Chemical Society*, 1892, p. 593.

Investigations of Dr. Emil. C. Hansen.—We now come to the epoch-making work of the Danish physiologist, Dr. E. C. Hansen, on the organisms of fermentation, which has had the effect of placing our knowledge of these bodies on an entirely new and much more definite basis. His researches have led to the discovery of properties and peculiarities possessed by these organisms hitherto unheard or even undreamed of, the importance of which to the brewer it is hardly possible to estimate.

The happy idea occurred to this philosopher that, in order to investigate with any degree of exactitude the life conditions or the specific action of any one set of these organisms, it was first of all necessary to secure a supply of them in a state of absolute purity. Fully recognising the fact that the only way to effect this must be to start such a culture from a single organism, he set about to devise a method by means of which this could be accomplished.

The Dilution Method.—In his earlier experiments pure cultures from a single yeast-cell were obtained by a method similar to that which had been employed by Lister in 1878 for securing pure cultivations of lactic acid bacteria, and which is known as "the dilution method." In this a number of organisms are distributed in a quantity of a culture liquid, and the number present in a measured drop ascertained by means of the hæmatimeter. Another drop of the same mixture is then added to a quantity of sterilised water sufficient to ensure that each drop of this diluted mixture shall contain, on an average, less than one organism. If then a number of flasks containing sterilised culture fluid are taken, and a drop of this mixture added to each of them, and if, after a sufficient time has been allowed to elapse for the organisms to develop, some of the flasks remain sterile, it is extremely probable that the remainder will have only received one organism each.

Instead of employing the hæmatimeter for counting the yeast-cells, Hansen made use of a microscopic cover-glass, the central portion of which was ruled in a series of squares. A small drop of the fluid containing the yeast was placed on this ruled portion, the glass turned over and placed on a moist chamber, and the number of yeast-cells in the drop counted under the microscope. If such a drop, containing twenty cells, were mixed with 40 c.c. of sterilised water, 2 c.c. of this mixture would contain, on the average, one cell. If, then, to the sterilised contents of forty flasks 1 c.c. of this mixture was added to each, presumably one-half of the flasks would, in course of time, show signs of infection, and the other half would remain sterile. It was found in practice, however, that this did not always happen, for sometimes, in spite of all precautions, two organisms would gain entrance to one flask. This difficulty was obviated by the discovery that, if the flasks were vigorously shaken, in those to which two yeast-cells had gained access the cells were almost invariably widely separated by the shaking, and the colonies which they subsequently formed made their appearance in two different parts of the flask. In this way an almost absolute degree of certainty was introduced into the dilution method. It was by this modification that Hansen's earlier observations were made, and upon these his principles were grounded.

Plate Cultivation.—In the year 1872 a method for securing pure cultures of bacteria was proposed by Schröter in which a semi-solid substratum was employed, and this, as improved by Koch, now forms the process chiefly used for the isolation of bacteria. The method does not, however, provide that an absolute pure culture is obtained with certainty; for, as shown by Miquel, 100 colonies obtained on a gelatin meat-broth plate may contain 108 different varieties; and Holm, working with yeast cultures on gelatin-wort plates, found that 100 colonies contained, on the average, 108 varieties and in extreme cases as many as 135.

Hansen's Modification of Koch's Plate Method.— Hansen, however, so modified the plate-cultivation method that it yielded

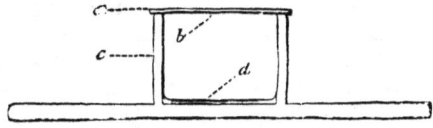

FIG. 47.—Bottcher Moist Chamber (*Hansen*).

absolutely certain results; it was so arranged that the individual yeast-cell was kept under observation during the whole time occupied in forming a colony.

Preparation of the Pure Culture from a Single Cell.—In Hansen's first method a dilute mixture of yeast and sterilised water is made in a sterilised flask. A drop of this mixture is examined under the microscope, and if the field contains more than twenty or thirty cells, more water is added until the proper degree of dilution is obtained. A test-tube containing sterilised wort-gelatin (prepared with clear bright wort of about 1056 specific gravity, and containing 5 or 6 per cent. of gelatin) is warmed until its contents are just fluid. A small portion of the yeast mixture is taken up on the platinum wire or loop and stirred into the gelatin, and, after the tube has been closed with a cotton-wool plug, it is turned and twisted about until the yeast-cells are evenly distributed throughout the gelatin. A drop of the mixture is then examined under the microscope to see if the dilution has been carried to the right extent—that is, that the individual yeast-cells are so far apart that there is no chance of the colonies they subsequently form touching one another. If this has not been secured more wort-gelatin is added. The right degree of dilution having been attained, a small drop of the gelatin is spread out in a thin layer on a microscope cover-glass and placed under a bell-glass until the gelatin has set; this takes place in about a quarter of an hour. It is then placed, gelatin side downwards, on one of the Bottcher moist-chambers shown in Fig. 47.

The chamber consists of a piece of glass tube about ¾-inch in diameter and about ½-inch in length, ground perfectly true at both ends. One of these is cemented to an ordinary microscopic slip; the other end is slightly smeared with vaseline, so that when the cover-glass is placed in position an air-tight joint is formed. A drop of water is placed in the chamber previous to its being used, in order to prevent evaporation from the surface of the gelatin. The whole is then placed under the microscope, and the miniature cultivation plate examined with a power of about 250 diameters. Those cells which are so far apart from one another that there is no chance of the colonies they will subsequently form touching, are marked. This may be done by making small circles round the cells with an ordinary pen and ink, or with a small camel-hair pencil and a little Chinese white; or the object-marker of Klonne and Müller may be used. This last is an arrangement which screws on to the nose of the microscope in place of the objective, and which, when the body of the instrument is pushed down, prints a little circle on the cover-glass. A cover-glass which has been previously marked out into small squares by means of hydrofluoric acid may also be employed, the columns of squares being numbered on two adjacent sides of the marked-out portion. It is well to make a sketch of the field, as this will materially assist in the subsequent identification of the cells. After the individual yeast-cells have been marked in one of these ways, and a number appended by the side of each circle to indicate it, or the two columns which intersect to form the square noted, the moist chamber and its contents may either be left at the ordinary room temperature or placed in an incubator and kept at a temperature of 24° to 25° C. (75° to 77° F.); in the former case the development of the colonies will be somewhat slow. The growth of the yeast-cells is watched day by day to see that the colonies do not encroach upon one another. In a few days they will have become so large that they can be seen by the naked eye. When they have arrived at this stage, portions of them are removed from the gelatin film and placed in wort in the following manner:—A piece of thin platinum or iron wire about half an inch long is taken up with a pair of tweezers and heated to redness in a Bunsen burner, and, after being allowed to grow cold, its end is inserted into one of the colonies and gently stirred about, so that some of the yeast-cells adhere to it. The wire is then dropped into a Pasteur flask containing sterilised wort. The flask, after being labelled, is placed in an incubation-chamber and kept at a temperature of from 25° to 28° C. (77° to 80° F.). In two or three days the contents of the flask will be found to be in a state of lively fermentation, and in a week or so a fair quantity of yeast will have formed.

Some of the Diseases of Beer may be Caused by the Saccharomycetes.—When pure cultures of yeast, obtained in this way from a single cell, came to be investigated, it was found that the ordinary yeast used in breweries never consisted of one single species, but of a number of varieties exhibiting markedly different properties. This led to the important discovery being made that some of those deteriorations which beer is liable to undergo in the course of its manufacture, and which up to that time had been solely attributed to the presence of those bacterial organisms commonly known as "disease ferments," could be produced by some varieties of the yeast itself.

Distinction of the Different Varieties of Yeast.—Hitherto yeasts had been solely differentiated by the various forms they presented when observed under the microscope, but it was soon discovered, by working with pure cultures of one single species, that this criterion could not be relied on. Although a circular or slightly oval shape is to a certain extent characteristic of the *Saccharomyces cerevisiæ*, an oval one of the *S. ellipsoideus*, and a sausage-shape of *S. Pastorianus*, yet under certain circumstances any one of these varieties may temporarily assume the forms of the others. Hence it became necessary to discover some test or tests by which the different varieties of yeast could be distinguished with certainty. This problem was also solved by Hansen, who, by making observations on the occurrence of spore formation at different temperatures, found that each species only developed spores between certain definite temperatures, and that when the whole of the six races were cultivated at a definite temperature at which they were all able to form spores, much difference was exhibited in the time which elapsed before spore formation commenced in the various species. By taking the times of the development of spores as abscissæ, and the temperature as ordinates, it was possible to construct a curve for each yeast, and these curves differed materially from one another.

A number of organisms which had hitherto been included amongst the *Saccharomycetes* were found not to yield spores, such as *S. apiculatus, S. exiguus, S. mycoderma*. This pointed to a great difference in the nature of these organisms and that of the true *Saccharomycetes*. In testing yeasts from various sources Hansen found that there were three distinct species of spore-bearing *Saccharomycetes*, and that each of these species could be further subdivided into varieties. The variety was indicated by placing a Roman numeral after the name of the species: thus we have *S. cerevisiæ I., S. cerevisiæ II.*, and so on. Their general forms are illustrated in the accompanying figures (48 to 53), and their temperature limits in the tables appended.

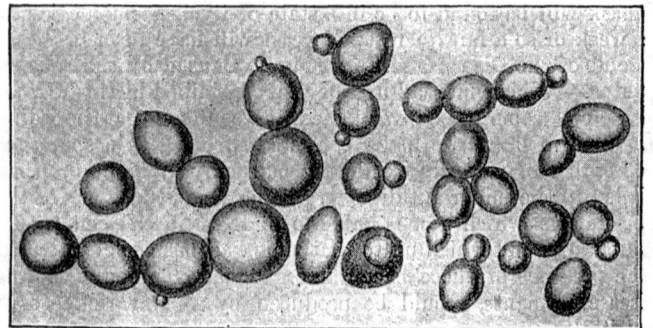

FIG. 48.—Saccharomyces cerevisiæ I (*Hansen*).

At 37.5° C. no ascospores are developed.
,, 36–37° C. the first indications are seen after 29 hours.
,, 35° C.   ,,   ,,   ,,   25 ,,
,, 33.5° C.   ,,   ,,   ,,   23 ,,
,, 30° C.   ,,   ,,   ,,   20 ,,
,, 25° C.   ,,   ,,   ,,   23 ,,
,, 23° C.   ,,   ,,   ,,   27 ,,
,, 17.5° C.   ,,   ,,   ,,   50 ,,
,, 16.5° C.   ,,   ,,   ,,   65 ,,
,, 11–12° C.   ,,   ,,   ,,   10 days.
,, 9° C. no ascospores are developed.
Size of the spores, 2.5–6 μ.

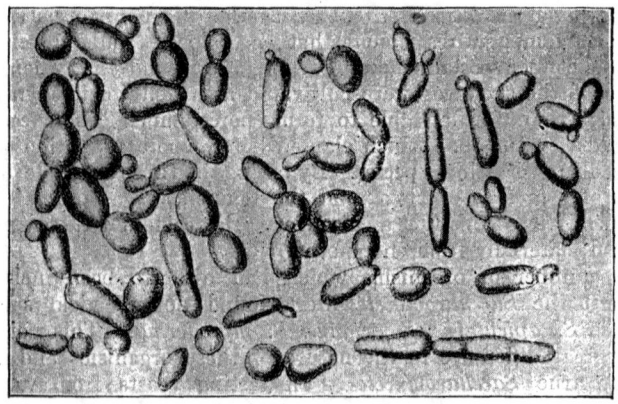

FIG. 49.—Saccharomyces Pastorianus I. (*Hansen*).

At 31.5° C. no ascospores are developed.
,, 29.5°–30.5° C. the first indications are seen after 30 hours.
,, 29° C.   ,,   ,,   ,,   27 ,,
,, 27.5° C.   ,,   ,,   ,,   24 ,,
,, 23.5° C.   ,,   ,,   ,,   26 ,,
,, 18° C.   ,,   ,,   ,,   35 ,,
,, 15° C.   ,,   ,,   ,,   50 ,,
,, 10° C.   ,,   ,,   ,,   89 ,,
,, 8.5° C.   ,,   ,,   ,,   5 days.
,, 7° C.   ,,   ,,   ,,   7 ,,
,, 3°–4° C.   ,,   ,,   ,,   14 ,,
,, 0.5° C. no ascospores are developed.
Size of the spores, 1.5–5 μ.

DIFFERENT VARIETIES OF YEAST.

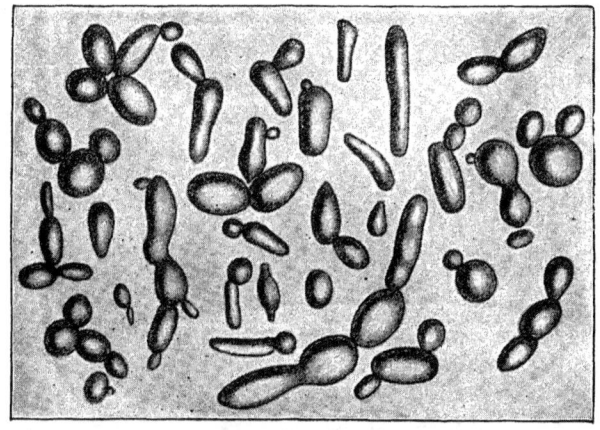

FIG. 50.—Saccharomyces Pastorianus II. (*Hansen*).
At 29° C. no ascospores are developed.
,, 27°–28° C. the first indications are seen after 34 hours.
,, 25° C.   ,,   ,,   ,,   25 ,,
,, 23° C.   ,,   ,,   ,,   27 ,,
,, 17° C.   ,,   ,,   ,,   36 ,,
,, 15° C.   ,,   ,,   ,,   48 ,,
,, 11.5° C.   ,,   ,,   ,,   77 ,,
,, 7° C.   ,,   ,,   ,,   7 days.
,, 3°–4° C.   ,,   ,,   ,,   17 ,,
,, 0.5° C. no ascospores are developed.
Size of the spores, 2–5 μ.

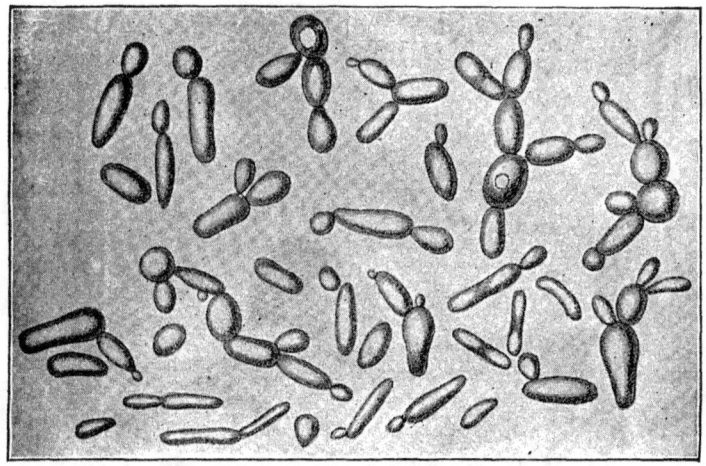

FIG. 51.—Saccharomyces Pastorianus III. (*Hansen*).
At 29° C. no ascospores are developed.
,, 27°–28° C. the first indications are seen after 35 hours.
,, 26.5° C.   ,,   ,,   ,,   30 ,,
,, 25° C.   ,,   ,,   ,,   28 ,,
,, 22° C.   ,,   ,,   ,,   29 ,,
,, 17° C.   ,,   ,,   ,,   44 ,,
,, 16° C.   ,,   ,,   ,,   53 ,,
,, 10.5 C.   ,,   ,,   ,,   7 days.
,, 8.5 C.   ,,   ,,   ,,   9 ,,
,, 4° C. no ascospores are developed.
Size of the spores, 2–5 μ.

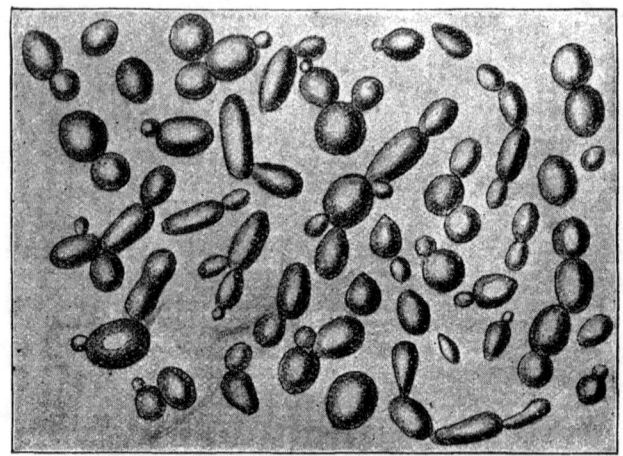

FIG. 52.—Saccharomyces ellipsoideus I. (*Hansen*).

At 32.5° C. no ascospores are developed.
,, 30.5°–31.5° C. the first indications are seen after 36 ours.
,, 29.5° C. ,, ,, ,, 23 ,,
,, 25° C. ,, ,, ,, 21 ,,
,, 18° C. ,, ,, ,, 33 ,,
,, 15° C. ,, ,, ,, 45 ,,
,, 10.5° C. ,, ,, ,, 4½ days.
,, 7.5° C. ,, ,, ,, 11 ,,
,, 4° C. no ascospores are developed.
Size of the spores, 2–4 μ.

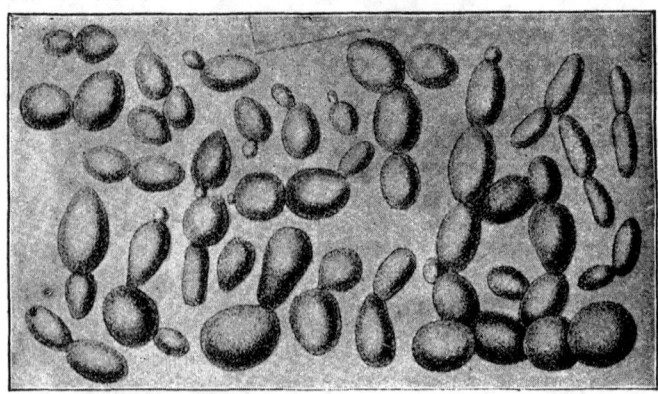

FIG. 53.—Saccharomyces ellipsoideus II. (*Hansen*).

At 35° C. no ascospores are developed.
,, 33°–34° C. the first indications are seen after 31 hours.
,, 33° C ,, ,, ,, 27 ,,
,, 31.5 C. ,, ,, ,, 23 ,,
,, 29° C. ,, ,, ,, 22 ,,
,, 25° C. ,, ,, ,, 27 ,,
,, 18° C. ,, ,, ,, 42 ,,
,, 11° C. ,, ,, ,, 5½ days.
,, 8° C. ,, ,, ,, 9 ,,
,, 4° C. no ascospores are developed.
Size of the spores, 2–5 μ.

INTRODUCTION OF PURE CULTURES. 349

The temperature at which the divergences are widest with these six varieties is about 52° F., at which ten days are required by *S. cerevisiæ* to form spores, less than four by *S. Pastorianus I.* and *II.*, less than seven by *S. Pastorianus III.*, less than four and a half by *S. ellipsoideus I.*, and less than four and a half by *S. ellipsoideus II.* In order to find out if a given sample of yeast is contaminated with any of the last five varieties—or, as they are termed, wild yeasts—it is only necessary to cultivate a portion of it on a plaster block at a temperature of 52° F., and, if no spores make their appearance before ten days, the sample may be pronounced free from wild yeasts. In making such a culture it is necessary that Hansen's directions be rigidly adhered to, or the spore formation will be irregular; he has shown that if the young cells of a particular sample of yeast are cultivated in the same wort for two days instead of one before being transferred to the plaster block, the time of spore formation will not be the same.

Of the varieties of yeast capable of producing diseases in beer, *S. Pastorianus I.* communicates an intensely nauseous bitter flavour to beer, and *S. Pastorianus III.* and *S. ellipsoideus II.* cause persistent cloudiness. *S. Pastorianus II.* produces no ill effects in beer, neither does *S. ellipsoideus I.*, which is the principal yeast occurring in the fermentation of grape-juice.

Introduction of Pure Cultures of Yeast into the Brewery.— In 1883 Dr. Hansen published a paper in which he described the removal of an irregularity which had for two years shown itself in the products of the Tuborg Brewery, near Copenhagen; the beers of which were affected with persistent turbidity. This was effected by introducing into the brewery yeast grown from a single cell. At about the same time the beers of the old Carlsberg Brewery became affected with a nauseous bitter taste and unpleasant odour. The yeast in use there was examined and found to consist of four different varieties. After experimenting with these singly and in combination, it was found that only one of the four was capable of yielding a satisfactory beer. Another of them was found to be *S. Pastorianus I.*, and it was this which had given to the beer its objectionable flavour and odour. The state of matters was rectified by the introduction of a pure culture of that yeast which gave the normal beer, and this same variety, which is now known all over the world as Carlsberg bottom yeast No. 1, has been employed there ever since. This introduction into the fermentation industries of yeasts consisting entirely of one race or species, obtained by starting the culture from a single cell, constitutes Hansen's great reform; it is undoubtedly one of the greatest reformations of modern times. The benefits to be derived therefrom are obvious. Before, brewers were completely in the dark as to the nature or properties of the organisms they were cultivating in their worts. Now, it is comparatively easy to find the particular race or species of yeast best fitted for producing each class of beer, and having once obtained it, it is possible to secure ever afterwards a constant supply of yeast possessing the same identical characters and properties. Thus, where all was problematical before, a definite result is now assured by determinate means.

Difference in the Properties of Various Yeasts.—When the method of making pure cultures of yeasts from a single cell began to be extensively practised, it was soon discovered that yeasts differed remarkably in their properties. Thus some were found to attenuate wort much more rapidly than others, some to yield beers of a much more stable nature than others. The flavours of beers, produced from a wort similar in all respects, were found to vary considerably when different races or varieties of yeasts were employed. Considerable difference was found in the amounts of the dextrins fermented away; in other words, some yeasts attenuated to a much higher degree than others. Such is the effect of the yeast upon the flavour of the fermented product that it has been found possible to obtain from malt wort, by employing for its fermentation one of the wine yeasts, a beverage scarcely distinguishable from wine.

The Practical Application of Pure Cultures of Yeast in the Brewery.—In introducing a culture of single-cell yeast into a brewery, the sample of yeast being used at the time is taken and from it a number of single-cell cultures made. It is best for this purpose to take the yeast from the upper strata of the wort at the beginning of a fermentation, for it has been found that when the pitching yeast contains wild yeasts, the proportion of these as compared with the cultivated yeasts is then proportionally small. In those cases where the actual isolation of the yeast-cells has to be performed at some distance from the brewery, a sample of the ordinary yeast is forwarded. The operator begins by starting a fermentation with a little of the yeast in a Pasteur flask, and, after fermentation has fairly set in, a number of single-cell cultures are made with a little of the wort, which will now contain an abundance of young and vigorous yeast-cells. The contents of a number of large Pasteur flasks containing ordinary sterilised wort are then infected with the various colonies thus obtained, and after fermentation is completed in each, the resulting beer is examined as to taste, odour, clearness, &c. The yeast which has yielded the most promising beer is then developed until a sufficient quantity has been obtained to start an ordinary fermentation in the brewery. As this sample will have to be tried in actual practice before a definite opinion can be formed as to its suitability, the remainder of the flasks should be kept, so that, in case the sample taken into use does not prove satisfactory, the other yeasts may be tried in turn. With this object in view, they should be placed in a dark cupboard, where they may be kept alive for a considerable time. The best preservative for yeast, however, is a 10 per cent. solution of cane sugar; in this, yeast will preserve its vitality for seven or eight years.

PRODUCTION OF SINGLE-CELL YEAST.

Small quantities of yeast may also be preserved in sterilised filter-paper. For this purpose a speck of yeast is placed on a piece of filter-paper which has been sterilised in a hot-air oven, and which is, immediately before being used, passed through the flame of a Bunsen burner. It is then wrapped up, and enveloped with a few more layers of sterilised filter-paper and placed in a drawer. According to Hansen, samples of yeast may be preserved in this way for five months, and often for much longer. This is an exceedingly convenient method for transmitting samples by post.

The Production of Single-Cell Yeast in Quantities sufficient for the Brewery.—For this purpose five Pasteur flasks are required, each of the capacity of a quart. Their contents are pitched with the selected yeast, after which the flasks are allowed to stand for a week at the ordinary temperature. Four metal vessels, each holding about two gallons, of the form shown in Fig. 54,[1] known as "Carlsberg vessels," are also necessary. These latter are made of tinned sheet copper, and are something like an ordinary oil-can in shape. A short tube is provided at a for emptying the vessel of its contents; it is fitted with a piece of indiarubber tubing and a glass stopper, similar to those used in connection with the side tube of a Pasteur flask, and is, in addition, provided with an ordinary spring burette-clip. Another tube, which serves for the introduction of fluids into the vessel, is inserted at b; it is also provided with an indiarubber tube and a glass stopper. A short tube is fixed on the mouth of the can

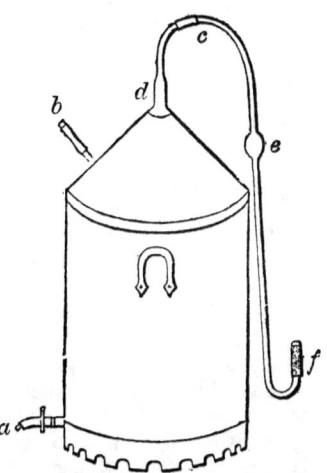

FIG. 54.—Carlsberg Flask (*Hansen*).
a. outlet tube; b. inlet tube; c. bent tube; d. mouth of can; e. chamber filled with cotton-wool; f. air-filter.

d, to which, by means of a piece of indiarubber tubing, the curved side tube c is attached. In a more recent modification, the tube c is attached to the neck by a screw union joint, so contrived that an absolutely air-tight joint is secured. To the other extremity of the bent tube c a small brass cylinder is attached which can be screwed on and off; this is filled with cotton-wool well pressed together, or with asbestos fibre; it is provided with a loosely fitting cover. The oval expansion at e acts as a safeguard against infection, especially just after the sterilisation has been effected, when the condensing steam is liable to cause a sudden inrush of

---

[1] This and the following figures are taken from Dr. Hansen's "Practical Studies in Fermentation," London, 1896, to which the reader is referred for fuller information on the subject of this chapter.

air. It is necessary that the vessel and all its various parts should be absolutely air-tight, so that no air is able to enter except that which passes through the air-filter.

Sterilisation of the Carlsberg Vessels.—In order to effect this, one of the vessels, after being thoroughly cleansed, is about half filled with water. A flame is then applied to its under surface, the water brought to the boil, and kept in a state of vigorous ebullition for an hour, the air-filter f and the stopper of the tube b being meanwhile removed. At the end of the hour's boiling, the stopper, after having been passed through a Bunsen burner, is replaced in the tube b. The boiling is kept up for another quarter of an hour, during which steam rushes through the bent tube c. In order to secure the sterilisation of the tube a, shortly before the flame is extinguished its stopper is removed and about 100 c.c. of hot water allowed to flow away; it is then closed with the glass stopper, which has previously been made fairly hot in a Bunsen flame, and the clip replaced. The source of heat is now removed and the air-filter f screwed on, the tube a opened, and the bulk of the water allowed to run off; the little which remains in the vessel is of no consequence. One and a half gallons of wort are now introduced through tube b, the air-filter is removed, heat applied, and the boiling conducted in exactly the same manner as with the water; the only extra precaution to be taken is that while the 100 c.c. of wort are being run off from tube a, and during the next quarter of an hour in which the wort is being boiled, the bent tube c is strongly heated with a Bunsen flame. The sources of heat are then removed, and the air-filter, which has in the meantime been sterilised in the hot-air oven, is screwed tightly into its place. Before one of these vessels is taken into actual use it must be kept for several days, and a sample of its contents then drawn off from tube a and examined. During this operation the bent tube c is heated with a Bunsen flame.

Since it has been found, that when yeast is grown in one of these vessels, in which the wort is not saturated with air, the yeast produced does not give normal clarification for some time when subsequently employed in the brewery. It is necessary therefore that the wort should be aërated if a normal yeast is to be obtained; and this aëration may be effected by allowing the vessel to stand for some time before being used. But the aëration is much more quickly attained by removing the stopper from tube a, connecting with it a piece of sterilised glass tube, and forcing in air which has been passed through an air-filter. A much more effective aëration is, however, secured if air is passed through the wort while it is still hot; consequently, the better plan is to commence the passing of air through the wort as soon as the flame is removed from the vessel, and continue the aëration until the contents of the vessel have fallen to 30° or 35° C. (87° to 95° F.).

STERILISATION OF THE CARLSBERG VESSELS.

During all this time a flame is kept playing on the bent tube C, and this is not removed until the wort has acquired the above temperature. The air-filter is then screwed on. According to Nielsen's experiments, 60 litres (2.1 cubic feet) of air passed through the wort, at such a rate that the time occupied by its passage shall be about five or six hours, suffices to effectually aërate the contents of one of these vessels.

Of the five large Pasteur flasks, which should now contain a vigorous cultivation of the selected yeast, four are used for the subsequent pitching of the wort in four Carlsberg vessels, the fifth being kept as a reserve in case of accident. One of the flasks is then taken, the bulk of the liquid poured off from the deposited yeast, and the flask gently shaken so as to bring the yeast into suspension in the remainder of the fermented wort. This mixture of yeast and wort is then introduced into one of the Carlsberg vessels, all necessary precautions being taken to avoid contamination from the external atmosphere. The remaining three flasks and three Carlsberg vessels are treated in the same way. Fermentation will have thoroughly commenced in all of these by the next day, when the air-filter may be removed from each. The fermentation may be hastened by gently shaking the vessel occasionally, so as to remove a portion of the carbon dioxide gas, but, before doing this, the bent tube should be heated in a Bunsen flame. In the course of about seven days as much yeast will have been formed as can be produced under these conditions, and the four vessels, which will collectively contain sufficient yeast to pitch a hectolitre (22 gallons) of wort, may then be taken to the brewery. Here a small vat provided with a loosely fitting cover, and having a capacity of about 36 gallons, is thoroughly cleansed, and about 22 gallons of the ordinary hopped wort being used in the brewery are run into it. After the exterior surfaces of the Carlsberg flasks have been singed with a Bunsen flame, they are gently shaken up so as to bring the yeast into suspension, and their contents poured one after the other into the wort. When the contents of the vat are in a lively state of fermentation and yeast begins to appear on the surface, the whole may be added to three or four hectolitres of wort (66 to 88 gallons) contained in a larger vat; or a larger vat may be used at the commencement, 22 gallons of wort being run into it and pitched with the contents of the four vessels, and after its contents are in an active state of fermentation, another 70 or 80 gallons of wort are added. The yeast from this is collected, weighed, and used for pitching an ordinary brew. In working with the smaller quantities of wort it is advisable to carry on the fermentation at a temperature slightly higher than that usually employed in the brewery.

Pure Yeast Cultures occasionally Work Irregularly at first.—Hansen directs attention to the fact that the yeast produced in this way occasionally gives, at first, a higher degree of attenuation

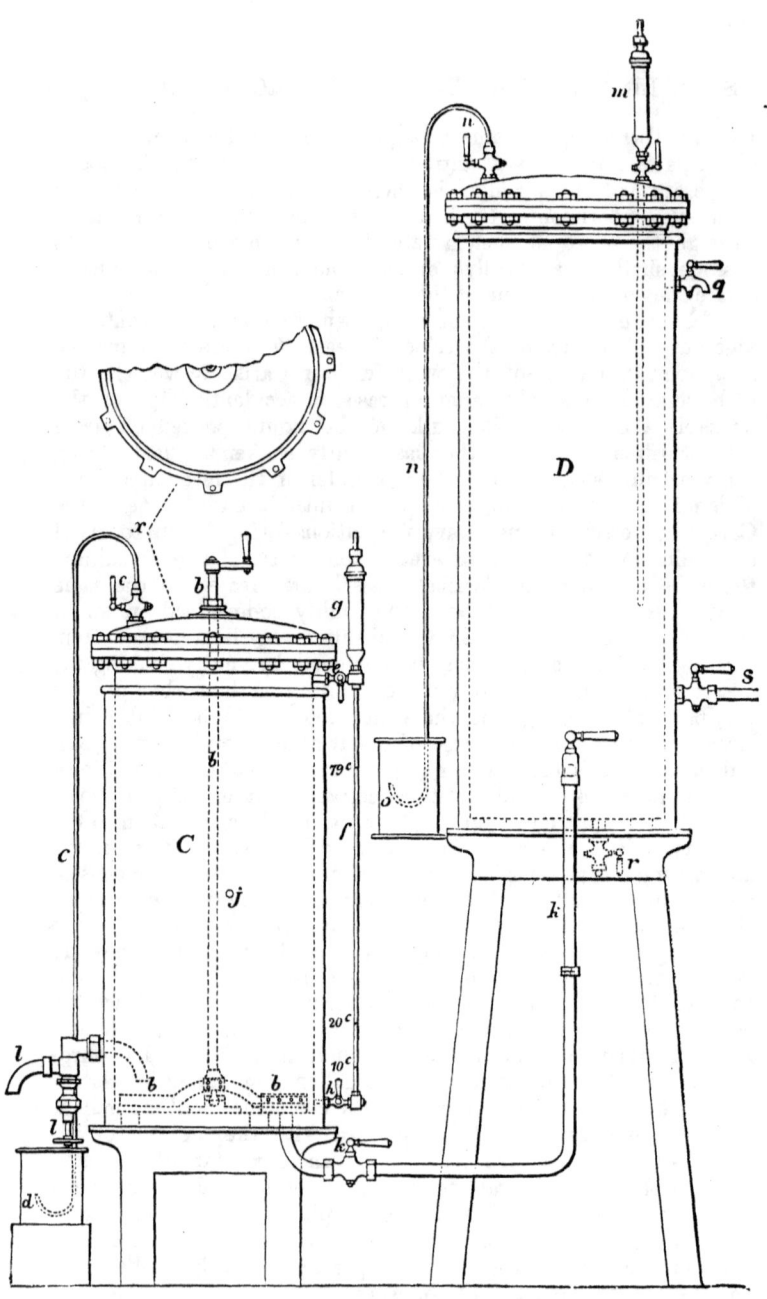

FIG. 55.—Hansen's Pure Yeast Culture Apparatus.

C, fermenting vessel; D, wort sterilising vessel; b, rouser; c, bent tube ending in vessel d, which is partly filled with water; e, upper tap of gauge; f, gauge; g, air-filter; h, inlet tap for air; j, tube for introduction of pitching yeast; k, pipe and tap connecting steriliser and fermenter; l, tap for drawing off yeast and wort; m, air-filter; n, bent tube connecting steriliser and vessel o, partly filled with water; q, tap for indicating height of wort; r, tap for drawing off wort; s tap connected with the ordinary wort-main.

THE PURE CULTIVATION APPARATUS.

than the normal, and that it does not yield such a brilliant beer as it will subsequently. He cautions brewers not to be alarmed if these appearances show themselves, as they will disappear after the yeast has passed through two or three fermentations.

The Pure Cultivation Apparatus.—When pure yeast is introduced into a brewery it is exposed to a number of contaminating influences, in consequence of which it becomes sooner or later impure. The length of time which yeast will remain in a comparative state of purity differs much with the various species and races of yeast, but the most resistant succumb after a time.

FIG. 56.—General View of Hansen's Apparatus.

It is, therefore, far better to be provided with some means of keeping up a constant and regular supply of pure yeast, than to be obliged to go through the laborious process just described each time a change of yeast is required, and for this purpose various forms of apparatus have been contrived. The first of these, which was devised by Hansen and Kühle, and introduced to public notice in 1887, is now employed in hundreds of breweries, and has proved itself a thoroughly reliable apparatus; all the other forms may be simply regarded as modifications of this. The original apparatus is shown in the accompanying illustration (Fig. 55). It consists of two principal portions, a closed fermenting vessel, C (Fig. 55), and a closed wort-sterilising vessel, *D*. Besides these, there is a large metal cylinder, capable of withstanding a pressure of 100 lbs. to the square inch, furnished with a pressure-gauge and a safety-valve, and also an air-pump, shown to the right in Fig. 56. By means of the latter, air drawn through an air-filter is forced into the cylinder until a pressure of three or four atmospheres is obtained; this supply of compressed air serves, in the first instance, for the aëration of the wort while in the sterilising vessel, and afterwards for aërating it when in the fermenting vessel. The wort cylinder, *D* (Fig. 55), has a cover which can be screwed down air-tight, and into this two taps are inserted. One of these is furnished with a tube reaching about three-fourths of the way down the vessel, as shown by the dotted lines, to this tap the air-filter *m* is attached, and to the other the bent tube *n*, the extremity of which dips into the vessel *o*, which is nearly filled with water. This tube should not be less than half an inch in diameter, or it is liable to be blocked up by pieces of hop. Four other taps are inserted into the body of the vessel; the one communicating with the pipe *k* serves for conducting the sterilised wort to the fermenting vessel. The tap at *s* is connected with the ordinary wort main of the brewery which leads from the copper to the cooler. The tap *q* serves to indicate when sufficient wort is in the cylinder, and the tap *r*, inserted in the bottom of the cylinder, is used for drawing off the first portions of the wort introduced, which are diluted with the condensation water from the steam used in sterilising. The cylinder may be provided either with a circular pipe immediately under its lid, having a number of small holes on its inner side through which cold water is passed, and which, after streaming down the outside of the vessel and cooling its contents, is caught in a trough at its base; or the cylinder may, as shown in the figure, be supplied with a mantle, through which a supply of cold water is passed. The whole arrangement is fixed at a higher level than the fermenting vessel, so that the wort will flow from the one to the other by gravitation.

The fermenting cylinder *C* is similar in shape to the wort cylinder, and is also provided with a removable air-tight cover. To the centre of this lid a stuffing-box is attached, through which passes the axle of the rouser *b*, one end of which is provided with a handle, while the other is supported in a socket. To its lower extremity the blades of the rouser are attached, one of which is provided with a piece of sheet indiarubber, which sweeps the bottom of the vessel when the rouser is put in action. A tap is also inserted in the lid, on which the bent tube *c* screws. The lower extremity of this tube is bent into a half circle, and is immersed in the vessel *d*, which is partially filled with water. To the right of the cylinder will be seen the gauge *f*, which serves to indicate the height of the fluid in the vessel; it is marked at several points. The topmost mark indicates 31.6 inches from the floor of the cylinder, and the vessel, when filled up to this, contains about thirty-seven gallons. The mark below this is at a distance of eight inches, and the lowermost four inches above the floor of the vessel. The pipe kk leading from the wort cylinder enters at the bottom of the fermenting cylinder; in addition to the taps shown in the figure, there are two others connected with the pipe kk (shown in Fig. 56), one for admitting high-pressure steam, the other for allowing the escape of air or steam. In the middle of the fermenting cylinder at j there projects a small tube about three-quarters of an inch long, provided with an indiarubber tube and glass stopper; it serves for the introduction of the pitching yeast. The tap l, seen to the left, serves for drawing off the yeast and wort, its screw valve is raised or lowered by the handle underneath, and the tap is so contrived as to render contamination from the outside atmosphere impossible when the vessel is being emptied. Communicating with this tap there is a pipe which projects into the interior of the vessel and bends downwards, terminating at a distance of one and a half inches from the bottom of the cylinder. This arrangement prevents the entrance of air into the cylinder through the tap l when it is being emptied. The exterior of the vessel may be clothed with wood, or, as shown in the figure, surrounded with a mantle through which cold water for attemperating may be passed.

Air-Filters.—These, which are represented in the engraving at m and g, consist of metal tubes, one and a quarter inches in diameter and nine inches long, and are closely packed with cotton-wool, at least one and a quarter ounces being used for this purpose. They are provided with metal lids which screw on, each lid carrying a union joint, by means of which it may be placed in communication with the pipe from the air-vessel. The air-filters, before being used, are sterilised in the hot-air oven, where they are allowed to remain for two hours at a temperature of 150° C. (302° F.).

Method of Working Hansen's Apparatus.—Before being used, the apparatus must be carefully tested to ascertain that it is perfectly air-tight; this may be done either with steam or water under pressure. The next thing is to sterilise the two cylinders and all the communicating pipes, taps, &c., with steam. The fermenting cylinder (Fig. 55) is first sterilised by admitting steam to the pipe kk, and opening the tap which admits steam to the cylinder C, the air filter g having been previously removed. The various other taps, e, h, l, c, are opened in turn, one after the other, so that they and the bent tube cc are all effectually sterilised; this takes about half-an-hour. The taps e and h are then closed, the filter g screwed on to its place, and all the other taps, with the exception of c, shut off. The tap from the air vessel to the filter g is now opened and also the tap h, so that a current of sterilised air may flow into the cylinder through h. The steam is then very gradually shut off, care being taken that steam mixed with air escapes from the end of the bent tube c for some time. As the admission of steam becomes less and less, the cylinder gradually cools, and when steam has nearly ceased escaping from c, the vessel d is partially filled with water and the tap k closed. If the apparatus is cooled too suddenly, air will be drawn in through the bent tube by the vacuum formed by the condensation of the steam, or a pressure may be developed sufficient to cause the vessel to collapse. The cooling should occupy about two hours. The water in the vessels d and o, in which the bent tubes are inserted, serves to show that a slight air pressure is being maintained in the vessels. The wort cylinder is sterilised in a similar manner, but here there is no occasion for the slow cooling; as soon as the vessel is sterilised the tap s is opened, and hot wort allowed to flow into the cylinder until it begins to run out at the tap q. The air and steam above the wort escape partly by q and partly by the bent tube n. It is advisable to remove the first portions of the wort which enter the cylinder by the tap r, as these are diluted by the water that has condensed during the steaming-out process. When the cylinder is filled to the proper height the taps q and s are closed, and air allowed to bubble through the wort by means of the pipe communicating with the air-filter m for an hour (hot aëration). The wort is then cooled by passing cold water through the mantle, and when its temperature has been reduced to 60° F. it is admitted to the fermenting cylinder C through the pipe k. The fermenting cylinder is, at first, only filled so far that the level of the wort is a little below the aperture j, through which the pure culture of yeast from a large Pasteur flask is then introduced, all the usual precautions against infection from the outside being taken, the flame of a spirit-lamp being used for singeing, if a Bunsen flame is not at hand. The rouser is then set in motion and the yeast well mixed in, after which wort is run in from vessel D, until its level reaches the highest mark on the gauge. During the fermentation it is advisable that the wort should be aërated; the supply of air being so regulated that a small bubble escapes every few minutes from the end of the tube c. In a week or ten days, according to the temperature at which the fermentation has been conducted, the newly-formed yeast may be withdrawn from the apparatus. In order to do this the tap l is opened and the beer allowed to run off until foam begins to appear. The tap is then closed, and sterilised wort from the wort cylinder run in to the lowest mark but one on the gauge. The rouser is set in action, and the mixed wort and yeast drawn off into a clean vessel until the level of the liquid has sunk to the lowest mark on the gauge. This operation is repeated, and in this way eighty-eight gallons of wort, containing sufficient yeast to pitch five barrels of wort, are obtained. Enough yeast is left in the apparatus to start the next fermentation after it has been again filled with wort. Care must be taken during all the time the drawing-off of the contents of the vessel proceeds that air passes in through the air-filter g, in quantity sufficient to prevent any chance of the liquid in the trough d, or of air, being drawn through the bent tube c. Sterilised wort is again admitted into the cylinder until the topmost mark on the gauge is reached, the whole well roused, and a fresh fermentation started.

Apparatus of Bergh and Jörgensen.—This is a very efficient arrangement, and occupies less space than the apparatus just described. It is so constructed that the wort is sterilised in the same cylinder that the fermentation afterwards takes place in. The apparatus, which is shown in Fig. 57, consists of two cylinders, A and B, one placed above the other, and an air-vessel (not shown) communicating with the tap Y, situated immediately above the air-filter D. A pipe is continued down the side of the apparatus, and in its descent gives off three branches, each of which is commanded by a three-way tap, A, B, C. The topmost cylinder, A, holds about ten gallons; it is provided with a rouser, E, an air supply through the tap A, a tap F and bent tube, and a small opening, a, for the introduction of yeast and the withdrawal of samples. The upper and lower vessels are connected by the tube P, on which is fixed the tap G. The lower cylinder B has a closely fitting lid, through which passes the air-pipe from the tap B, and the axle of the rouser I, which is actuated by the bevel-wheels; the float L serves to show by the pointer outside the level of the fluid inside. The lid also carries the tap K, to which is attached a bent tube. A circular pipe, M, perforated with fine holes, surrounds the lower chamber immediately under its cover, and supplies the refrigerating water. The upper portion of B is enveloped in a mantle for cold

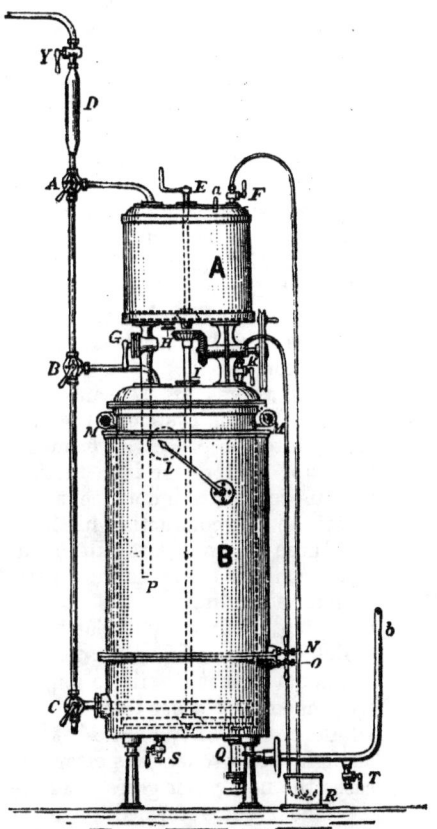

FIG. 57.—Propagating Apparatus of Bergh and Jörgensen.

water, the lower portion with another mantle for high-pressure steam, by means of which the contents of the vessel are boiled and sterilised. In the floor of the lower cylinder is the tap Q, communicating by means of the pipe b with the wort main of the brewery from which the vessel is filled.

When the apparatus is about to be used, the lower chamber B is filled to the proper height with wort, steam is admitted to the under-mantle, and the wort vigorously boiled, the taps B and A being placed in such a position that the steam passing away from the wort sterilises the upper chamber as well. The wort is then cooled and aërated, the yeast added through a, and a little wort forced into the upper chamber by means of the compressed air, roused well with the added yeast, and allowed to run down into the lower cylinder. When the yeast is all in the lower chamber the tap G is shut and fermentation allowed to proceed. A good crop of yeast having formed, the contents of vessel B are well roused, and a portion of the fermented wort with the yeast in suspension is forced up into the vessel A; this serves for starting the next fermentation. The remainder of the wort and yeast are then run off and used for pitching in the brewery. In this way the vessel B serves in turn for a wort steriliser and for a fermenting vessel.

Advantages of Hansen's System.—The principal advantage afforded by this system is the introduction of an element of certainty into the process of fermentation which was never attainable hitherto. Before the introduction of this great reform, the brewer was in complete ignorance as to the nature of the yeast with which he was conducting his fermentations. In all cases, excepting those where a pure cultivated yeast is employed, the pitching yeast consists of an unknown mixture of various types and races, which, according to the laws of competition, are continually changing with every variation of the surrounding conditions in which the yeast is cultivated. At times the yeast becomes so degenerated that it will no longer fulfil its functions properly, and a change of yeast has then to be obtained from some other brewery, where probably little or no better conditions reign. Consequently, though such a change is often attended with benefit, yet in some cases the new yeast gives worse results than the old.

Under the reformed system the brewer makes choice of the one individual race or species of yeast which he finds best adapted to his requirements, and having once secured this, he is afterwards able to provide himself with a continuous supply of absolutely the same yeast. When the influence which the yeast exercises on the flavour of the beer, on its clarification, and on its stability, is taken into account, it is impossible not to recognise the importance of this great reform. It affords absolute protection against the diseases of beer and their attendant pecuniary losses, and for the necessity of obtaining periodical changes of yeast from other breweries, with its attendant uncertainty. The Continental brewers, who employ low fermentation yeast, were the first to avail themselves of Hansen's method, and, as a consequence, the system is now, and has been for a good many years, employed with the most satisfactory results in hundreds of Continental and American breweries. Its introduction into breweries which work on the high fermentation system has not been nearly so rapid, principally on account of the ill-founded prejudices and preconceived ideas that yeast of one species or variety was unable to effect a satisfactory secondary fermentation in the cask. That all such ideas are chimerical has been abundantly proved by Jörgensen, who has been engaged for a number of years in the selection of types of single-cell yeast fitted for breweries working on the high fermentation system, and who states that he has introduced pure yeasts into numerous breweries of this class with the most marked success. Vuylsteke, Bau, and others have all testified to the same purport, and, to the author's personal knowledge, pure single-cell yeast has now been used for several years in two large breweries in this country with the most successful results.

Analysis of Yeast by Hansen's Method.—It is highly necessary, where pure cultivated yeast is employed in a brewery, to have some means of determining when the yeast in actual use is becoming so contaminated that a new supply from the propagating apparatus should be introduced, and also of ascertaining that the apparatus itself is fulfilling its duties properly. This is effected by making a spore cultivation on a block of plaster of Paris at a definite temperature, and noting the time when spore-formation makes its appearance. This varies with every species of yeast, and, according to the experiments of Holm and Poulsen, the two best temperatures for conducting experiments of this nature with low fermentation yeasts are $15°$ and $25°$ C. ($59°$ to $77°$ F.). They state that low fermentation yeasts may be divided into two classes, one of which at the latter temperature yields spores later than the wild yeasts. At this temperature the other class sporulates at about the same time that the wild yeasts do, but at the lower temperature it forms spores some time after the wild yeasts.

The Yeast which is to be Examined in this Way should be taken at the End of the Primary Fermentation.—The sample is removed from the fermenting-room in a sterilised flask, and, after being allowed to settle for a few hours, is spread on the plaster blocks. The cultures carried out at $25°$ C. are examined at the end of forty hours, those at $15°$ C. at the end of three days. The method is said by Hansen to be so sensitive that an admixture of one two-hundredth of a part of wild yeast can be readily detected in the Carlsberg bottom-yeast. This is much beyond the degree of delicacy actually required in practice, for Hansen has shown that, when working under normal conditions and storage, one part of *S. Pastorianus III.* or of *S. ellipsoideus II.*, both of which in large quantities produce beer turbidity, are, in these small proportions, incapable of causing any ill effects, and that *S. Pastorianus I.* is only able to impart its disagreeable flavour and odour when it exists in a larger proportion than 1 part in 22 of an otherwise pure yeast.

Analysis of the Yeast of the Propagating Apparatus.—The purity of the yeast contained in this apparatus should be tested at the end of every fermentation. For this purpose a small portion is drawn off into a Pasteur flask, and cultivated for a few days at a temperature of 70° to 80° F. in yeast-water, when bacterial contamination, if present, will show itself. After being allowed to settle, the beer is decanted from the yeast in the Pasteur flask and the yeast passed through several cultivations in a solution of cane sugar acidified with tartaric acid, then a few times in beer-wort, after which it is tested by spore-formation on plaster blocks. This treatment with tartaric acid is exceedingly favourable to the development of wild yeasts and to the suppression of the cultivated varieties; consequently the former, if present in the sample, are brought into prominence.

Differences in the Actions of the Various Species of Yeast.—In addition to the broader differences between the members of the several races and varieties of the *Saccharomycetes* which have been already mentioned, many minor diversities are observable in the action even of varieties of the same species.

Differences in Fermenting Power.—The remarkable differences in the fermenting power of the various yeasts has attracted considerable attention lately. It was prominently brought into notice by the difference in the behaviour of two varieties of yeast, one of which had been isolated from the yeast of a brewery at Saaz, and is now known as Saaz yeast; the other from the yeast used in the brewery of Herr Frohberg at Grimma, and which now bears the name of Frohberg yeast. In conducting experiments with the same wort, it was found that the Frohberg yeast invariably carried the attenuation to a considerably lower point than the Saaz yeast did; the former attenuated a wort of sp. gr. 1049 to 1018, the latter was only able to reduce it down to 1023. It has now become customary to distinguish yeasts which possess high attenuating powers as Frohberg types, and those which have more feeble powers in this direction as Saaz types; between these extremes there are many yeasts which hold various intermediate positions.

Since then another yeast, one of the *Schizo-saccharomycetes*, the *Saccharomyces pombe*, has been discovered, which possesses an even greater power of attenuation than the Frohberg yeast. It was found that it could reduce the original gravity of a wort 84 per cent., while the Frohberg could only reduce it by 70 per cent.

This divergence is not confined to these yeasts, but is exhibited in a greater or less degree by all which have been examined so far. A number of experiments which were conducted by Dr. Lindner[1] on different yeasts demonstrate this very clearly. In each of these, 150 c.c. of sterilised wort were seeded with a trace of a pure culture of a particular variety of yeast, and this, when it had entered into a state of active fermentation, was added to a larger quantity of sterilised wort, generally about 1200 c.c. In each series of experiments the same wort was employed; some of the flasks employed were plugged with cotton-wool, others were closed with a bulb tube containing sulphuric acid. Consequently, in the former a slow interchange between the air of the flask and that of the outside atmosphere took place; in the latter, this was effectually prevented. High and low fermentation, and also wild yeasts, were experimented with. Of the high fermentation yeasts examined, the one which exhibited the highest attenuative power reduced the gravity of a wort of 1049° to 1018°; the weakest in this respect only reduced the same wort to a gravity of 1024°. The remainder exhibited various attenuative powers intermediate between these extremes. Similar differences in attenuative power were observable with the low fermentation and wild yeasts.

The Differences in the Attenuative Powers of the Various Saccharomycetes on Wort are dependent upon the Nature of the Enzymes which they are able to Secrete.—While the vast majority of the yeasts are able to ferment the two sugars glucose and fructose, which require no previous inversion, it is not by any means the case with the disaccharide sugars maltose, cane sugar, and milk sugar. Certain yeasts, such as the *S. apiculatus*, do not secrete an inverting enzyme, hence they are unable to ferment any of the disaccharides. Others, such as the *S. albicans*, Rees, are able to invert and ferment maltose, but not cane sugar. The *S. kefir*, Beyerink, can invert and ferment cane sugar and milk sugar, but not maltose. Some yeasts, such as the *S. Ludwigii*, are unable to transform any of the sugars into alcohol, but appear to produce oxalic acid instead. These distinctions are also found in the mould fungi, and their ability to ferment the various disaccharides, or the reverse, depend upon the same causes; thus *Monilia candida* has recently been shown to secrete invertase.

The Yeasts commonly used in Brewing appear to be endowed with the Power of Secreting a Number of Different Enzymes. —In addition to producing the well-known enzyme, invertase, it has been recently found that the cultivated yeasts secrete another enzyme, glucase, which, previous to its fermentation, converts maltose into glucose. Evidently others are produced which have the power of degrading a greater or less number of those intermediary dextrins which occur in wort, and to this is to be attributed the differences which are observable in the attenuative

---

[1] *Mikroskopische Betriebskontrolle*, p. 112.

powers of the various yeasts. The power of inverting and fermenting milk sugar seems to be entirely wanting in these yeasts.

Differences in the Behaviour of Various Yeasts to the Trisaccharide Raffinose.—It was shown by Berthelot, in 1889, that the various yeasts exhibited marked differences in their behaviour to this complex sugar; the ordinary yeast of the French breweries (low fermentation) fermented it completely, whilst bakers' yeast (top fermentation) only fermented a third part of it, and Loiseau found that this was true with respect to all bottom and top fermentation yeasts. In consequence of doubts which had arisen as to the validity of these observations, Bau reinvestigated the subject in 1894, and obtained the following results. It was found that top fermentation yeasts of the Frohberg type inverted raffinose into melibiose and fructose, but were only able to ferment the latter sugar. They were also able to ferment completely two bodies occurring in wort, which he termed α and β isomaltose.[1] Top yeasts of the Saaz type behaved in a similar manner towards raffinose; they could only ferment α isomaltose, but not the β variety. The Frohberg low yeasts completely inverted raffinose into glucose, galactose, and fructose, all of which sugars they are able to ferment; and this is also the case with α and β isomaltose. Saaz low yeasts behaved similarly with raffinose, but could only ferment β isomaltose. In this way he founded a classification of the various yeasts according to their action on these various sugars.

Variations in the Quantity of Alcohol Produced.—In connection with the series of experiments performed by Dr. Lindner, and which have already been mentioned on p. 363, portions of a 29 per cent. solution of maltose were pitched with various yeasts, in order to find the amounts of alcohol they were capable of yielding. The top fermentation yeasts were found to produce from 9 to 15 per cent. of alcohol by volume, and the low fermentation yeasts slightly less quantities. Saaz yeast yielded a little over 8 per cent., and Frohberg not quite 11 per cent.

Quantities of Acid Produced by Different Yeasts.—Five top fermentation yeasts respectively yielded amounts of acid which, reckoned as lactic acid, varied from 0.117 to 0.135 per cent. of the wort. The Pombe yeast produces a much larger quantity of acid than this, and consequently has considerable power in keeping bacterial organisms in check.

Amounts of Nitrogenous Matter Removed from Wort.—This, with the various top fermentation yeasts, was found to range from 9.4 oz. to 13.8 oz. per barrel in a wort having a gravity of 1048.6. Low fermentation yeasts did not act so vigorously; they only removed from 5.64 to 11.52 oz. per barrel from the same wort. Saaz yeast removed 7.03 oz., and Frohberg 10.66 oz.

Reproductive Powers.—Yeasts vary much in the crops of

---

[1] The existence of these two bodies is somewhat problematical.

yeast which they yield. In the same series of experiments the amounts of the new yeast formed varied from 9.25 to 23 grammes, the fermentation being started with a mere trace. In all but two instances the larger yeast crop was associated with a higher attenuative power. In those cases where parallel experiments were made, the one set in flasks closed with cotton-wool, the other in flasks sealed with a sulphuric acid tube, the yeast crop and the attenuation were always greatest in the former, and this demonstrates the influence of even a slight amount of aëration.

Energy of Fermentation.—Leaving out of consideration the different physiological condition of the yeast, the various races show marked differences in their fermentative activity. This can be estimated by observing the amount of gas evolved in a given time, or by the time taken for a given amount of gas to be evolved. The latter method was that employed in Dr. Lindner's series of experiments, the measurement being indicated by the number of days taken to evolve 36 grammes of carbon dioxide. The number of days required by a number of the yeasts to do this varied from six to fifteen, but some of them did not evolve this quantity of gas during the whole period of the fermentation.

Hayduck's Apparatus for Ascertaining the Fermentative Energy of Yeast.—This consists of a flask, Fig. 58, a, for containing the yeast and a solution of cane sugar; it is placed in a water-bath provided with a perforated false bottom, fixed about half an inch above the real bottom, the

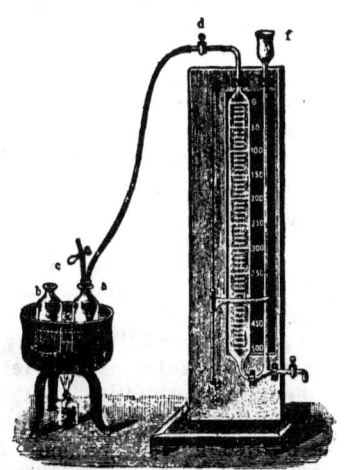

FIG. 58.—Hayduck's Apparatus for Ascertaining the Fermentative Energy of Yeast.

contents of the bath being kept at a constant temperature of 30° C. (86° F.). The flask is connected by means of an indiarubber tube with an apparatus for measuring the carbon dioxide evolved. The flask a is furnished with an indiarubber cork having two perforations, in each of which short pieces of glass tube are inserted. To one of these the tube communicating with the measuring apparatus is connected; to the other a short piece of indiarubber tube, which can be closed by an ordinary burette-clip. The measuring apparatus consists of a U-shaped tube, one limb, which is considerably wider than the other, is constructed to hold 500 c.c.; it is graduated into 10 c.c. divisions. The extremity of the narrower tube carries a funnel-like expansion f, and is provided at its lower end with a tap e. The upper extremity of the wider limb carries a narrow tube bent at a right angle, and also a tap d. When a determination is made, 10 grammes of the yeast under examination are intimately mixed with 400 c.c. of a 10 per cent. solution of cane sugar and introduced into the flask a. This is then placed in the water-bath, which has been previously heated to, and is now constantly maintained at, a temperature of 86° F., and is allowed to remain uncorked for an hour. In the meantime the measuring apparatus is made ready. Five c.c. of paraffin oil are first poured into the funnel f, and afterwards water. This latter drives the oil before it, and this, when it enters the measuring chamber, floats on the surface of the water, its object being to form a protective layer on the surface of the water, and prevent the absorption of carbon dioxide. Sufficient water is then introduced to bring the surface of the paraffin layer to the zero mark on the scale. After the flask has stood for an hour in the water-bath the apparatus is connected up, the tube c being left open while this is being done. This is now closed with the clip, the tap d opened, and the time noted. As fermentation proceeds the disengaged gas gradually displaces the liquid in the wider limb, and the tap e is opened just sufficiently to allow the water to escape at the same rate as the gas enters. At the expiration of half an hour the taps d and e are closed, the clip c is removed, and, with the assistance of the pointer, the level of the liquid in the two limbs adjusted at the same height, either by pouring water into the funnel or drawing it off by the tap e, and the quantity of the collected gas read off.

The fermentative activity of a yeast may be expressed by the number of cubic centimetres of gas it evolves in half-an-hour under these conditions; but the better and more usual method is to express it in terms of the number of grammes of sugar fermented by 100 grammes of yeast in half-an-hour. This is done by multiplying the number of c.c. of carbon dioxide collected during the experiment by the factor 0.03841 (which is obtained by taking into account the ratio of gas evolved to the amount of sugar fermented, the ratio of the weight of carbon dioxide to its volume), and multiplying the result by ten.

PART III.

CHAPTER VI.

WATER.

Chemical Composition.—Water was, until comparatively recent times, regarded as an element, its compound nature being discovered by Cavendish in 1781, who showed that when hydrogen gas was burned in air, a quantity of water was formed equal in weight to that of the two gaseous elements which disappeared. Shortly after this Lavoisier proved that water was a compound of oxygen and hydrogen; and in 1805 Humboldt and Gay-Lussac determined its composition by volume. Chemically pure water consists of two parts by weight (two atoms) of hydrogen and sixteen parts by weight (one atom) of oxygen, or two volumes of the former gas and one volume of the latter. Of all known fluids water possesses the most extensive solvent powers; an immense number of different substances, liquid, solid, and gaseous, are readily dissolved by it. In consequence of this property, water is never met with in nature in a state of purity, the nearest approximation to this is the distilled water of the chemist's laboratory. Even this is not absolutely pure, for it contains a variable amount of dissolved gases derived from the air; indeed, the removal of the last traces of nitrogen is so difficult that it is doubtful if absolutely pure water has ever yet been prepared.

Occurrence.—Water is met with in nature in great abundance, where it plays an important part in most terrestrial phenomena. It is found in all the three physical states it is capable of assuming; in the solid state it occurs permanently in the polar regions and on the tops of high mountains as ice and snow, and in this state it appears periodically in temperate regions. In the liquid state it is met with in the ocean, lakes, seas, and rivers, &c., which constitute more than three-fourths of the surface of the earth; also under the earth's surface, where it forms the source of the supply of springs, wells, &c. The atmosphere is always more or less charged with water in the shape of vapour, the amount which it contains at any one time being governed by local and general meteorological conditions.

In the economy of nature water is always in a state of constant circulation; its tendency, from whatever source it may be derived, is to flow back to the ocean, which may be regarded as an immense reservoir provided by nature for its storage. From this store it is evaporated by the heat of the sun's rays and diffused in the form of vapour through the atmosphere, from whence it is again condensed, and appears in the form of rain, snow, hail, dew, &c. When precipitated to the earth in the form of rain, snow, &c., one portion flows off the surface of the land, giving rise to the various brooks, streams, lakes, rivers; a second portion evaporates, whilst a third portion sinks into the earth. This latter portion descends through the more porous portions of the soil until it reaches some impervious stratum by which it is retained to form the subterranean lake known as the "ground water," from which the water of surface wells is derived. The level of the surface of the ground water[1] varies considerably; in some districts it almost touches the surface of the soil, whilst in other places it is situated at a considerable distance below the surface; hence the difference in the depths of wells in various districts. In hilly countries, where the strata are inclined, the underground water, obeying the natural laws of gravitation, forms subterranean watercourses, and these,

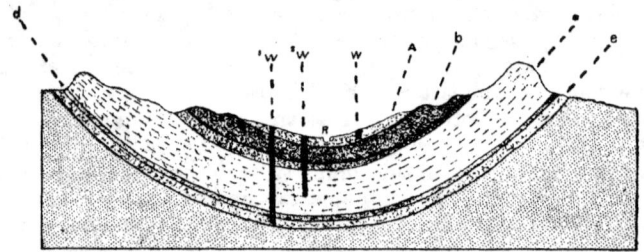

FIG. 59.—Diagrammatic Section of the Strata forming the London Basin.[2]

when they find outlets in districts at a lower level, appear as springs. The accompanying diagram (Fig. 59) will help to explain these various conditions. The stratum A, in the middle of the basin, is sand, gravel, or some layer pervious to water. The rain, snow, &c., falling on this portion would be partially evaporated, a part would drain off the surface and fall into the river, R, and the remaining portion would sink into the soil until it reached the stratum marked b, which is clay, a substance impervious to water; here it would be retained and form the ground water. A well, W, sunk into this bed would consequently derive its supply from this ground water. The stratum c immediately below the clay is the chalk, and e greensand, whilst at d there is again an impervious stratum.

[1] The surface of the ground water is very far from being horizontal; as a general rule it conforms to the inequalities in height of the surface of the ground.

[2] As with all geological sections, the height is greatly exaggerated in comparison with the width.

As a consequence of this arrangement of the strata, a certain amount of the water falling on the exposed portions of c would sink into and be retained in this stratum, since, as b is impervious, no water can escape upwards, and as d is also impervious, no water can escape downwards. A well, W_2, driven into this stratum, would derive its supply from this pent-up body of water, and would consequently have the character of a chalk water. The water derived from a well, W_3, driven through the impervious layer d, would have the characters of a water coming from the greensand. Should the level of the water contained in the beds c and d be much higher than the surface of A, the water from these two wells would of itself rise above the level of the mouths of the wells; should it be lower, the water would only rise a proportional distance in the well and have to be pumped the remainder. From this it will be seen that the water obtained from a well may be collected at a great distance from the well itself.

Rain Water.—This, since it is derived directly from the condensation of aqueous vapour, is, when collected in the open country at a distance from human dwellings, the nearest approach to pure water met with in nature. It contains, however, many impurities derived from the air, for, in traversing the atmosphere, water dissolves an amount of oxygen, nitrogen, and carbon dioxide gases equal to about 0.05 of its volume; small quantities of ammonia, nitric acid, and sodium chloride are similarly taken up. In towns, rain water takes up sulphuric and sulphurous acid, the presence of which in the atmosphere is due to the burning of coal; sometimes hydrogen sulphide is also absorbed.

Besides these, rain water takes up small traces of the solid impurities which exist in the atmosphere in the form of dust. These are partly of an inorganic nature, such as calcium and magnesium compounds, or, more rarely, those of potassium, iron, and manganese; partly of an organic, such as portions of decaying vegetable tissue, the spores of bacteria and mould fungi, occasionally microscopic plants of the lowest orders (*Protococcus pluvialis*, &c.).

Rain water collected from roofs is generally very impure, since it contains much organic matter derived from the surfaces on which it has been collected, these being, as a rule, far from clean. When such water is stored in cisterns or other closed vessels it is apt to become putrid.

Waters from Wells and Springs.—Of the water which is precipitated on the surface of the earth in the form of rain, that portion which sinks into the ground varies very much, and is regulated by the nature, configuration, &c., of the ground on which it falls. In magnesian limestone districts about 26 per cent. finds its way into the subsoil; in the new red sandstone (Triassic), about 25 per cent.; in the chalk, about 42 per cent.; in the loose tertiary sand, as much as 90 to 96 per cent. After penetrating the soil, water absorbs a large amount of carbon dioxide from the air contained in the soil (the "ground air"), which is especially rich in this gas, containing, as it does, some 250 times more carbon dioxide than the air of the atmosphere. In descending farther it dissolves as much of the substances with which it comes in contact as it is capable of taking up, and it is much aided in this direction by the carbon dioxide which it holds in solution; thus when water saturated with this gas meets with chalk (calcium carbonate), the latter is converted into acid calcium carbonate, which is a soluble salt. But the decompositions which take place between water charged with carbon dioxide and the constituents of the soil are often of a much more profound nature than this; for instance, many silicates undergo partial decomposition, small quantities of soluble alkaline silicates, alkalies, &c., being liberated. Double decompositions also often take place between the constituents of the soil, as when calcium sulphate and magnesium carbonate mutually decompose one another to form calcium carbonate and magnesium sulphate.

General Constituents of Well and Spring Waters.— Of the inorganic matters in solution, the most commonly met with are the following:—Bases: sodium, potassium, calcium, magnesium, and iron; acids and halogen: sulphuric, nitric, nitrous, and phosphoric acids; silica, carbon dioxide, and chlorine. Of the organic matters taken into solution, some entirely consist of carbon, hydrogen, and oxygen; others contain, in addition to these three elements, nitrogen, and often sulphur, but we possess no very definite knowledge as to the exact nature of these bodies. Substances are sometimes met with in suspension, such as clay, &c. The respective bases and acids of the inorganic constituents are estimated directly by chemical analysis, but the organic substances present in solution can only be estimated indirectly.

Statements of Results of Water Analyses.—In these statements it is now customary to tabulate the metals found in terms of their oxides: thus calcium is expressed as lime (CaO), and the acids in terms of their anhydrides, sulphuric acid being expressed as SO_3. A second statement then follows, in which the various bases and acids are given in the state in which they are supposed to be combined in the water. Unfortunately, in tabulating the amount of the various constituents of a water, chemists employ very different methods of statement. Some express their results in grains per gallon, and probably for trade purposes this form of statement is the best, as every one has some notion of the weight of a grain and the quantity of a gallon. This has been objected to on the ground that a decimal form of statement is a more scientific one, and this view of the subject has led to the results of water analyses being stated in parts per million, parts per 100,000, and in some few cases in parts per 1000. Of these different methods, the first, which corresponds to milligrammes per litre, is frequently used on the Continent, and is also often employed in this country

RESULTS OF WATER ANALYSES.

in expressing the amounts of such substances as ammonia, &c., which occur in waters in infinitely small amounts, though the remainder of the results are expressed in grains per gallon. The second method—that is, parts per 100,000—is the one which was used by the Rivers Pollution Commissioners, and has consequently to some extent official sanction. The third method was formerly much used on the Continent, but seems to be now nearly obsolete. These different methods of statement are easily convertible into one another, if it be borne in mind that the gallon of water contains 70,000 grains; it is obvious that a sum in simple proportion will convert any one form of expression into any of the other forms. For instance, if we have a water which contains 14 grains per gallon of calcium oxide, this amount expressed in parts per 100,000 may be obtained as follows :—

$$70{,}000 \; : \; 100{,}000 \; :: \; 14 \; : \; 20$$

or, more shortly, by dividing the number of grains per gallon by 0.7 : thus, $14 \div 0.7 = 20$. Similarly, parts per 100,000 are converted into grains per gallon thus—

$$100{,}000 \; : \; 70{,}000 \; :: \; 20 \; : \; 14$$

or, shortly, by multiplying by 0.7 : thus, $20 \times 0.7 = 14$.

Grains per gallon are converted into parts per million by dividing by 0.07 : thus, 14 grains per gallon divided by 0.07 equal 200 parts per million; for the converse process multiply the parts per million by 0.07 : thus, 200 parts per million $\times 0.07 = 14$ grains per gallon.

Attempts have been made at various times to put an end to the statement of results in so many different ways, and to induce chemists to employ one uniform plan, but so far these have been unsuccessful.

As has been observed before, the bases and acids are determined separately in every water analysis, but it is customary to append to each report, in addition to the bare statement of the results obtained, another which represents the way in which the bases and acids are probably combined as salts in the water. This statement is, however, of an entirely arbitrary nature, because the way in which the bases and acids are combined in a water is not known with absolute exactitude.

It has long been recognised that it is impossible to imitate artificially a mineral water; such a water may be analysed with all possible care, and its ingredients exactly ascertained. But if these are added to distilled water in the exact proportions found, the artificial water produced in this way will differ, as Professor Liebrich points out, "from the natural one in taste and value; and this difference is not easy to explain." As being in strict accordance with this view, the results of experiments conducted at a brewery in Berlin may be cited, where it was attempted to produce beer having the qualities of the renowned Munich beer. The water was artificially prepared so as to resemble the Munich water as closely as possible, and all the other materials, with the exception of the water, were imported from the Bavarian capital. The quality of the beer produced, however, did not reach the Munich standard.

The statement of the results of a water analysis, in which the constituents are given in their probable state of combination, is convenient, and enables an opinion to be more easily formed as to the general characteristics of the water, than one which gives a mere statement of the various bases and acids it contains.

Waters considered in respect to the Geological Formations in which they occur.—The composition of the various rocks, strata, &c., which water meets with, either in percolating through the soil or in flowing over its surface, differs widely. The salts found in a particular water depend upon the composition of the various strata with which it comes in contact, and these will differ as the geological formation of the district varies. It is impossible to say with absolute certainty what these strata will be in any particular district, because, as a rule, the information we possess is not sufficiently minute as to what the geological formation of any particular district is, for only the broad features have been laid down by geologists, and these are usually insufficient for this purpose. Thus a hard selenitic (calcium sulphate) water may be found occasionally in the middle of a sandy district which yields in other places a soft water.

Hardness of Water.—This term, which is in everyday use, is in reality a rough expression of the comparative soap-destroying power which a water exerts before it permits the soap to exercise its well-known detergent action. Every one is aware of the fact that, when soap is used with a very hard water for the purpose of ablution, a certain amount of soap has to be dissolved before a lather can be formed; whereas, when a very soft water is employed, a lather is obtained at once: the formation of a lather serves to indicate the point at which the cleansing action of the soap commences. Dr. Clark very ingeniously founded a process for the comparative estimation of the hardness of waters on this principle. In this an alcoholic solution of soap of a definite strength is added little by little to a measured quantity of the water placed in a stoppered bottle. After each addition of soap solution the water is vigorously shaken, and so soon as a lather forms on its surface, which is permanent for two minutes, the quantity of soap solution is read off, and from the amount used in this way the hardness of the water is expressed in degrees. Soap is a compound of an alkali, such as soda, with one or more of the fatty acids, and when its solution is added to a water containing calcium or magnesium salts, decomposition ensues, and the earthy bases combine with the fatty acids to form insoluble compounds; it is only after the whole of these bases are precipitated that a lather can be formed. The formation of the first lather indicates that the point has been reached at which all the earthy bases have been removed from solution; consequently, the amount of soap solution used to reach this point forms the basis for an estimation of their quantity.

A sample of water is tested in this way both before and after boiling. Obviously a water containing the acid carbonates of lime and magnesia will show a less degree of hardness after boiling (p. 382). The amount removed in this way is termed the "temporary hardness," in contradistinction to that which depends upon the presence of the sulphates and chlorides of the earthy bases, and which, not being removable by boiling, is termed "permanent hardness."

Since the term "hardness of a water," when applied according to its commonly accepted meaning to a class of waters very often used for brewing purposes, conveys an entirely erroneous impression as to the character of such waters, this term, if applied at all to brewing waters, should be strictly limited to permanent hardness. As Clark's test is but a rough one, it is much better to form an opinion of the hardness of a water from the results of an actual analysis, by means of which it is possible to learn exactly what proportion of the calcium and magnesium salts will be precipitated on boiling and what proportion will remain in solution.

Consideration of Brewing Waters with reference to their Mineral Constituents.—Since water is one of the most important substances used in brewing, it is one of the first and most indispensable conditions that every brewery should possess an abundant and never-failing supply of suitable water. The most usual sources of brewing water in this country are wells and springs, but occasionally lake or river water is employed; and rain water, if collected under proper conditions, may also be employed. As a proper supply of water is of such primary importance in brewing operations, before the erection of a brewery is commenced in any particular locality it is absolutely necessary to ascertain whether the district is capable of affording a sufficient amount of water of the requisite quality. Though this proposition may appear to be self-evident, yet instances are not unknown where breweries have been erected without an investigation of this kind having been made, and where considerable difficulties have subsequently arisen from the oversight.

The waters obtained from the different sources just mentioned differ considerably as to the amount of organic and inorganic constituents which they contain. Some of them are eminently adapted in their natural state for the production of one particular class of ale, and can be modified by artificial treatment so as to become fitted for the production of ales of all classes; others do not lend themselves readily to such treatment; whilst others, either from the amount of organic pollution which they have been subjected to, or from the quantity or quality of the inorganic salts which they contain, are entirely unfitted to enter into the composition of beer, though they may be employed for refrigerating purposes.

It has been found that the waters best adapted for brewing purposes are those which contain inorganic salts in quantities which should not vary beyond certain limits. In all public water supplies for domestic purposes, the quantity of inorganic constituents demanded by hygienic considerations is much less than that permissible or, as a rule, advisable for brewing purposes; consequently, waters of this class are not generally fitted for the production of all the classes of ales without further modification; but in all cases these waters may be so treated artificially as to render them suitable for use in the brewery. The waters chiefly used for brewing purposes are derived from wells or springs, though rain, lake, or river water may also be used for this purpose, provided it possesses the requisite purity, organically and inorganically. The waters derived from springs, wells, &c., as met with in various parts of the country, differ widely in the quantity and nature of the inorganic constituents which they contain. Of the various salts generally met with in well and spring water, some, such as the calcium and magnesium sulphates, the sodium, magnesium, and calcium chlorides, exercise a beneficial influence when not present in too large quantities; whilst others, such as sodium sulphate and carbonate, when occurring in any considerable quantity, are prejudicial. Naturally, water for brewing purposes must not contain any of the salts of the poisonous metals; and the salts of iron, when present in anything but mere traces, are detrimental.

Experience has shown that the waters of some districts are particularly adapted for the production of ales of one particular character, whilst they are unsuitable for the production of those of another class. Thus the waters found at Burton-on-Trent have been long renowned for yielding excellent pale ales, whilst those of Dublin have been equally famed for producing black beers. In other districts waters are found which are admirably adapted for the production of mild ales. When investigations were made into the chemical constitution of these various waters, it was found that the difference in their behaviour in brewing operations bore a constant relation to their chemical composition. Thus the waters of Burton were found to contain large quantities of the sulphate of lime, and to the presence of this salt their excellence in the production of that particular class of ales for which this district has been so long famed is undoubtedly to be attributed. The

WATERS ADAPTED FOR PALE ALES.

Dublin waters, on the other hand, were found to contain calcium sulphate in small quantity, but fairly large amounts of the acid calcium and magnesium carbonates; and, since these latter salts are almost completely precipitated on boiling, the water as it is actually used in brewing contains but two or three grains of solid matter per gallon. The waters which experience has shown to yield the best mild ales also show a marked difference in composition from the other two. In these, calcium sulphate only occurs in quantities of from 5 to 7 grains per gallon, whilst sodium chloride predominates. These important facts having been established, it occurred to chemists that a water inorganically unsuitable to the production of a particular class of ale might be so modified in its chemical composition as to be rendered fit for the production of another class of ale. It was found that some waters could be so treated, and thus arose the various systems for treating water chemically which have been practised to such a large extent during recent years, and which have been followed with very advantageous results. Thus the London waters have been long noted for their adaptability for the production of black beers, and for their unfitness for yielding pale ales; so much was this the case, that formerly it was the custom for London firms either to obtain their pale ales from Burton or to have a branch brewery there. By chemically treating the London waters, it is now found possible to brew pale ales which are able to compete with those produced at Burton.

Waters Adapted for Producing Pale Ales.—As has just been observed, the waters most suitable for the production of pale ales are those which contain calcium sulphate in fairly large quantity. Of these the Burton waters may be taken as typical examples. The following are the results of an analysis[1] of the water from a deep well situated in that town:—

Type I.

	Grains per Gallon.		Grains per Gallon.
Silica	0.49	Sodium chloride	3.90
Alumina	0.49	Potassium sulphate	1.59
Iron oxide	trace	Sodium nitrate	1.97
Lime	36.33	Sodium sulphate	10.21
Magnesia	10.15	Calcium sulphate	77.87
Soda	7.25	Calcium carbonate	7.62
Potash	0.86	Magnesium carbonate	21.31
Chlorine	2.37	Silica and alumina	0.98
Sulphuric acid	52.29		
Nitric acid	1.25		

In this and similar waters the amount of calcium sulphate is exceedingly high, whilst a fair amount of calcium and magnesium

[1] E. Brown, "The Geology and the Mineral Waters of Burton-on-Trent," p. 23.

carbonates, which are precipitated on boiling, are also present; the chlorides are very small in quantity. The amount of calcium sulphate in this particular water is undoubtedly very large, and most probably all the possible beneficial effect to be derived from the presence of calcium sulphate in a brewing water may be obtained with from 40 to 50 grains per gallon of that salt.

Waters Suitable for Black Beers.—As a contrast to this class of waters, an analysis of one of the Dublin well waters is appended:—

Type II.

	Grains per Gallon.		Grains per Gallon.
Silica	0.26	Sodium chloride	1.83
Iron oxide and alumina	0.24	Calcium sulphate	4.45
Lime	9.79	Calcium carbonate	14.21
Magnesia	0.43	Magnesium carbonate	0.90
Soda	0.97	Iron oxide and alumina	0.24
Chlorine	1.11	Silica	0.26
Sulphuric acid	2.62		

Waters of this class are distinguished by the small quantity of calcium sulphate and all the other constituents, with the exception of calcium carbonate, which they contain. This last salt is almost entirely removed on boiling such a water.

Waters Fitted for Mild Ales.—The following is the analysis of a water adapted for the production of mild ales:—

Type III.

	Grains per Gallon.		Grains per Gallon.
Silica	0.22	Sodium chloride	35.14
Iron oxide and alumina	0.24	Calcium chloride	3.88
Lime	13.13	Calcium sulphate	6.23
Magnesia	1.91	Calcium carbonate	16.37
Soda	18.62	Magnesium carbonate	4.01
Chlorine	23.81	Iron oxide and alumina	0.24
Sulphuric acid	3.67	Silica	0.22

The essential characteristic of this class of waters is the high amount of chlorides and the comparatively small amount of calcium sulphate which they contain.

Artificial Treatment of Brewing Waters.—From the above generalisations of the inorganic constitution of those waters which have been found by experience to be the best fitted for the production of the different classes of ales, it is obvious that no single water possesses the qualifications necessary for producing every class of beer. Fortunately, the knowledge acquired during the past few years has shown that it is possible so to modify the inorganic constitution of many waters that this important result

ARTIFICIAL TREATMENT OF BREWING WATERS. 377

may be attained. Some waters do not lend themselves so readily to this treatment, and there still remain others which it is absolutely impossible to convert into good brewing waters. As an example of the first of these may be adduced the waters from the chalk, which are very frequently met with, and which are highly valued for brewing purposes.

The following is the analysis of one of these:—

	Grains per Gallon.		Grains per Gallon.
Calcium carbonate	17.92	Calcium chloride	0.21
Magnesium carbonate	0.49	Sodium chloride	0.84
Calcium sulphate	0.07	Magnesium nitrate	1.05
Potassium sulphate	0.56	Silica	1.12

Such a water as this, without any other treatment than boiling, is eminently fitted for the production of black beers, since, when boiled, the carbonates are almost completely precipitated, and very little solid matter of any kind remains in solution.

To convert such a water into one suitable for the production of pale ales, an addition of those salts in which the water is deficient must be made, and its inorganic constitution brought more into agreement with that of a water of Type I., of which an analysis has just been given. This will be secured by the addition of the following salts, in quantities equivalent to the following amounts of the respective anhydrous salts in grains per gallon:—

Calcium sulphate	50 grains.
Magnesium sulphate	6 grains.
Calcium chloride	10 grains.

These salts, as met with in commerce, always contain definite amounts of crystalline water, hence in all calculations this must be allowed for. The amounts of the anhydrous salts given above, sufficient for one barrel of water, are worked out in the following statement, in which their water of crystallisation is taken into account. It is arranged in a convenient manner for everyday use:—

Calcium sulphate (gypsum, $CaSO_4 + 2H_2O$)	$5\frac{1}{4}$ ounces.
Magnesium sulphate (Epsom salts, $MgSO_4 + 7H_2O$)	1 ounce.
Calcium chloride (saturated solution)	$\frac{1}{20}$th of a pint.

The gypsum, which should be in the state of a fine powder, is intimately stirred into sufficient water to yield a mixture of the consistence of thin whitewash; the remainder of the ingredients are added to this, and the whole mixed with the water of the hot-liquor tank, when it is in a state of lively ebullition. If the water in the tank is not boiling, the contents of the hot-liquor back should be well roused from time to time, in order to secure the solution of the calcium sulphate, which is by no means a very soluble salt. Obviously the whole of the water used for the brew—mash and sparge liquor—should be similarly treated. Calcium chloride ($CaCl_2 + 6H_2O$) is such an extremely deliquescent salt that it is difficult to handle it in the solid state. Consequently, it is much better to obtain this substance in the form of a saturated solution, which has a specific gravity of 1380.

In order to convert the original water (marked A) into one suitable for producing mild ales, an addition should be made per gallon equivalent to the amounts of the following salts in their anhydrous condition:—

Calcium sulphate	5 grains.
Magnesium chloride	5 grains.
Sodium chloride	30 grains.

This result will be obtained by adding the following quantities of the commercial salts to each barrel of brewing liquor:—

Calcium sulphate	¾ ounce.
Calcium chloride (saturated solution)	$\frac{1}{40}$th of a pint.
Magnesium chloride (45 per cent. solution)	$\frac{1}{10}$th of a pint.
Sodium chloride (common salt)	2½ ounces.

As magnesium chloride, like calcium chloride, is an exceedingly deliquescent salt, it is best kept in the form of a 45 per cent. solution, made by placing 4½ lbs. of the chloride in a vessel, adding four pints of water, and, when the salt is dissolved, making the quantity up to one gallon. The specific gravity of such a solution will be about 1160.

Another method of securing a very similar result is the addition of three-quarters of an ounce of calcium sulphate and three ounces of Kainit[1] to every barrel of mash liquor. Kainit is a mineral substance which in its pure state contains one molecule of potassium sulphate, one of magnesium sulphate, and one of magnesium chloride, together with six molecules of water ($K_2SO_4 + MgSO_4 + MgCl_2 + 6H_2O$). But as met with in com-

[1] Kainit is obtained from mines sunk in a very peculiar geological formation which exists at Stassfurt in Germany, and which is supposed to have been formed by the evaporation of such an inland sea as the Dead Sea. The lowest bed consists of an enormous mass of rock salt, interspersed with thin veins of gypsum. Next comes a layer of Polyhallite ($2CaSO_4 + MgSO_4 + K_2SO_4 + H_2O$), then one of Kieserite (KCl 11.8 per cent., $MgSO_4$ 21.5 per cent., $MgCl_2$ 17.2 per cent., NaCl 26.7 per cent., $CaSO_4$ 8 per cent., H_2O 1.3 per cent.), then Carnallite (KCl 15.5 per cent., $MgCl_2$ 12.4 per cent., $MgSO_4$ 14.5 per cent., NaCl 22.5 per cent., $CaSO_4$ 1.9 per cent., H_2O 26.1 per cent.). On the top of this last bed is found the Kainit, above which there is found relatively small quantities of Sylvinite (KCl 30.55 per cent., K_2SO_4 6.97 per cent., $MgSO_4$ 4.80 per cent., $MgCl_2$ 2.54 per cent., NaCl 46.05 per cent., $CaSO_4$ 1.80 per cent., H_2O, &c., 7.29 per cent.). The way in which these deposits occur points strongly to their various components having been separated from an aqueous solution by slow evaporation, the less soluble salts being the first to be deposited. Most of these beds, with the exception of the rock salt, are now largely mined for manurial purposes, the large quantities of potassium salts which they contain rendering them very valuable manures.

ARTIFICIAL TREATMENT OF BREWING WATERS. 379

merce it is never pure, being largely contaminated with sodium chloride. Its average composition, as given by Maercker,[1] is as follows :—

	Per Cent.
Potassium sulphate	21.3
Magnesium sulphate	14.5
Magnesium chloride	12.4
Potassium chloride	2.0
Sodium chloride	34.6
Calcium sulphate	1.7
Clay	0.8
Water	12.7
	100.0

We now come to another and numerous class of waters, which require the addition of various salts not only to supply the natural deficiency of the waters in certain saline constituents, but also to remove by double decomposition salts which are in themselves detrimental. For instance, suppose we have a water containing 10 to 15 grains of calcium sulphate and 5 or 6 grains of magnesium sulphate per gallon, which is in other respects of fairly pure quality, and that we wish to produce black beers with such a water. To do this successfully the calcium and magnesium of the sulphates must be removed, and this can be effected by adding sodium or potassium carbonate, or a mixture of both, in proportions equivalent to the respective amounts of the sulphates we wish to displace. On boiling such a water, after this addition, double decomposition ensues, sodium or potassium sulphate being formed, together with calcium and magnesium carbonates, and these latter, since they are nearly insoluble, are precipitated and removed. Obviously, by such a method we produce a considerable amount of sodium or potassium sulphate, or of a mixture of the two salts, and these salts, especially the sodium one, are not very desirable ingredients in a brewing water, as they are apt to extract a somewhat coarse flavour from the hops. Still, in the brewing of black beers this is not a matter of such primary consideration as would be the case in the production of a pale or mild ale. The equivalent of anhydrous calcium sulphate is 136; that of crystallised sodium carbonate ($Na_2CO_3 + 10H_2O$), 286; and that of crystallised potassium carbonate ($K_2CO_3 + 2H_2O$), 174. Consequently, each grain of anhydrous calcium sulphate present in a water will require for its decomposition 2.13 grains of sodium carbonate, or 1.28 grains of potassium carbonate; similarly, each grain of anhydrous magnesium sulphate (equivalent 120) will require 2.38 grains of sodium carbonate, or 1.45 grains of potassium carbonate.

Obviously, the calcium sulphate of such a water as that of Type I. might be removed in this manner; but, in such a case, the

[1] *Die Kalidungung*, p. 5.

large amount of sodium or potassium sulphate formed would render the water unfitted for brewing any class of ale.

Many waters are met with that contain small quantities of calcium and magnesium in the form of carbonates only, but which contain the carbonates and sulphates of the alkalies in considerable proportion. Of such a nature are the waters found in the strata beneath London, which are derived from the greensand below the London clay. It is from these waters that the black beers for which London has so long been celebrated are brewed. The following are the results of an analysis of one of these :—

	Grains per Gallon.
Sodium carbonate	11.13
Sodium chloride	9.88
Sodium sulphate	1.34
Calcium carbonate	6.75
Magnesium carbonate	1.85
Silica and alumina (including traces of calcium sulphate)	0.39
Iron oxide	0.49

Such waters as these in their natural state are eminently fitted for the production of black beers, but they require considerable treatment before being used for brewing ales. This treatment consists in the addition of substances which convert the salts prejudicial to ale-brewing (the sodium carbonate and sulphate) into compounds of a harmless nature. This object is best effected by the addition of calcium chloride in proper proportion. Double decomposition then ensues on boiling the water: the carbon dioxide of the sodium carbonate goes over to the calcium of the calcium chloride to form insoluble calcium carbonate; whilst the chlorine goes over to the sodium and forms sodium chloride, which remains in solution. The sulphuric acid of the sodium salt goes over to the calcium of the calcium chloride to form calcium sulphate, whilst its sodium unites with the chlorine to form sodium chloride, both which salts remain in solution. Such waters as these may be easily converted into waters well adapted for the production of mild ales, and this would be effected by the addition per barrel of $1\frac{3}{4}$ ounce of saturated solution of calcium chloride. If, however, pale ales are to be produced from one of these waters, a further addition will have to be made; this will be the addition of sufficient calcium sulphate to bring the water up to the standard of Type I., and in this case the quantity to be added would be about $5\frac{1}{4}$ ounces per barrel. Obviously a water treated in this manner would contain an amount of sodium chloride much larger than the typical class of waters best fitted for the production of pale ales; hence an ale produced from such a water after artificial treatment is never equal to the real Burton water. Owing to the large amount of sodium chloride which it would contain after treatment, an ale produced from such a water could never have the cleanness and delicacy of flavour possessed by one brewed with the natural water. Much of the "running ales" brewed in London and its neighbourhood, which are of an exceedingly satisfactory nature, are produced with this class of water in its natural state.

Waters Entirely Unfitted for Brewing Purposes.—Waters are occasionally met with which, owing to the excessive amount of either one or several of the mineral constituents that they contain, are entirely unfitted for brewing purposes. For instance, waters are sometimes found containing over 100 grains per gallon of calcium sulphate. Such waters interfere so seriously with the course of fermentation that it is impossible to produce a satisfactory article with them. In the case of a water of this kind, which is in other respects a suitable one for brewing purposes, the only way is to dilute it with a purer water, such as that supplied for domestic purposes. Waters are also occasionally met with which contain enormous quantities of sodium chloride, amounting to as much as 160 grains per gallon. These, likewise, are entirely unfitted for brewing purposes, since they seriously interfere with the fermentations, and cause rapid weakening of the yeast. Perhaps 50 grains per gallon may be put down as the largest amount of sodium chloride permissible in a brewing water; the only way out of the difficulty with such a water is dilution with one from a purer source.

Removal of Iron from Brewing Waters.—Occasionally salts of iron occur in water in such quantities as to demand their removal. In a water which contains a fair amount of the calcium and magnesium carbonates, simply boiling the water effects the removal of the iron. The calcium or magnesium combines with the acid of the iron salt, iron carbonate being formed, which, at the temperature of boiling water, is quickly converted into insoluble iron oxide, and is precipitated along with the calcium and magnesium carbonates. Where these bodies do not exist in a water, the removal of the iron is best effected by the addition of sodium carbonate in quantity rather more than is sufficient to decompose the iron salt present, and after the water has been well boiled, to add calcium chloride in a rather larger quantity than is sufficient to decompose the excess of sodium carbonate. In the report of a water analysis the iron is always returned as ferric oxide. An iron salt containing one grain of ferric oxide (Fe_2O_3) requires for its decomposition 1.79 grain of crystallised sodium carbonate; and for the decomposition of one grain of crystallised sodium carbonate 0.4 grain of calcium chloride is necessary. Thus, in a water shown by analysis to contain 2.26 grains per gallon of iron oxide, 4.05 grains per gallon of crystallised sodium carbonate will be required to destroy the iron salt present. To effect this object in actual practice, it is best to add a little more than the theoretical quantity—say, in this case, 6 grains per gallon, which would leave an excess per gallon of 1.95 grain of sodium carbonate. For the decomposition of this quantity of sodium carbonate 1.95 × 0.4 = 0.78 grain of anhydrous calcium chloride would be required, and as the addition of $\frac{1}{10}$th of a fluid ounce (20 fluid ounces = 1 pint) of a saturated solution of this salt to each barrel of water would be equal to the addition of about one grain per gallon, the object in view would be effected, and a slight excess of calcium chloride left.

Influence of Boiling upon Certain Waters.—The chemical composition of many waters is considerably modified when they are boiled. Waters which contain the acid carbonates of lime and magnesia suffer great alteration on boiling. The second molecule of carbonic acid, which is somewhat loosely attached, becomes liberated, and, as a consequence, the calcium and magnesium carbonates, being insoluble bodies, are precipitated thus:—

$$H_2Ca(CO_3)_2 = H_2CO_3 + CaCO_3.$$
Acid calcium carbonate.   Carbonic acid.   Calcium carbonate.

This reaction takes place gradually, and about half-an-hour's vigorous boiling is required to ensure the fairly complete elimination of the carbonates.

Many natural waters are met with which contain calcium and magnesium sulphates in solution along with acid sodium carbonate. The composition of these is considerably affected by boiling; double decomposition ensues, one molecule of the carbonic acid of the sodium being liberated, the other passing over to the earthy bases, while the sulphuric acid from these combines with the sodium, thus:—

$$2(HNaCO_3) + CaSO_4 = H_2CO_3 + CaCO_3 + Na_2SO_4$$
Acid sodium carbonate.   Calcium sulphate.   Carbonic acid.   Calcium carbonate.   Sodium sulphate.

Consideration of Water with reference to its Organic Constitution.

—Experience has long shown that the presence of any considerable quantity of organic matter in a water used for brewing purposes is highly undesirable; beers produced from such contaminated waters show a great tendency, especially marked in hot weather, to become "fretty," go turbid, turn sour, and refuse to be clarified in the process of fining. The sources of the organic matter in water are manifold; rivers are often polluted with the refuse of manufactures, such as tanneries, paper mills, &c., &c.; frequently they are defiled by the sewage of towns. Surface water is often contaminated by flowing over arable ground on which manure is spread regularly. Wells in and about towns are often rendered organically impure by the drainage from cesspools, liquid portions of manure from stables, &c. Waters collected from uncultivated land frequently contain large quantities of organic matter of vegetable origin (peaty waters).

ORGANIC CONSTITUTION OF WATER.

The organic bodies which contaminate a water cannot, like the inorganic ones, be directly isolated; consequently their quantity cannot be determined directly. We are therefore obliged to be content with a determination of one or more of the constituents of the organic matter, and to deduce from these the amount of contamination which any particular sample of water has been subjected to. Several of the processes that have been proposed for this purpose will be afterwards described.

In forming a conclusion from the data afforded by the results of a water analysis, it may be useful to bear in mind the following facts. Urine, which contains nearly 1 per cent. of sodium chloride, is one of the most frequent sources of pollution; consequently, a water which contains a much larger amount of this salt than other waters occurring in the same district, and which are known to be uncontaminated by sewage, is to be regarded with suspicion. A fairly large amount of a highly nitrogenous substance called "urea" is also present in urine, and this, under the influence of certain bacteria, is rapidly transformed into ammonium carbonate; consequently, waters directly polluted by urine contain considerable amounts of free ammonia as well as chlorides. The drainage from sewers, middens, and cesspools are rich in both chlorine and ammonia. Ammonia is also one of the chief compounds derived from the decomposition of animal organic matter; hence its presence in the water of a shallow well often points to recent pollution with organic matter of animal origin. Rain water as it descends through the atmosphere dissolves a fairly large amount of ammonia; in districts remote from towns the average amount of this substance found in rain water is 0.49 part per million parts of water; in towns the quantity may reach 2.10 parts per million. Nitric acid is also found in a small quantity in rain water, in country places about 0.12 grain per gallon. Much chlorine is often present in rain water falling near the sea; as much as 3 grains per gallon of chlorine has been found in such a water. The washings of solid excreta contain little chlorine, but much organic matter, and also phosphoric acid.

Effect of Filtration Through, and Drainage Over, the Soil on the Organic Constituents of Water.—Filtration through the soil is the great natural purifier of water, and its beneficial effects are most marked in soils of a loose, porous nature, especially when these are well aërated. In passing through them, the ammonia which is present in the water, whether derived from the atmosphere or from the putrefaction of organic matter, is rapidly oxidised into nitrous and nitric acids by certain bacteria which exist in every soil; and these acids, immediately they are formed, combine with one or other of the alkalies or alkaline earths present in the soil to form the corresponding salts. It has been found that 97 per cent. of the nitrogen of the London sewage is converted in this way into nitric acid by slowly traversing a stratum of gravelly soil 5 feet thick. A water containing a large quantity of nitric acid has almost certainly been contaminated, at some portion of its history, with organic matter of animal origin. Vegetable matters, under like circumstances, yield either no nitric acid or mere traces. As the presence of nitric acid (combined as nitrates) in any quantity less than 10 grains per gallon may be permitted in a water used for brewing purposes, the presence of nitrates to this amount in the water of a deep well is of no importance, since such a water must have passed through an immense thickness of soil, rock, &c., before being finally collected. When, however, nitric acid is present in large quantity in the water of a shallow well, the matter stands very differently, and especially when this acid is accompanied by nitrous acid. Here the filtering strata are of a comparatively small thickness, and though they may be sufficient at most times to convert into nitric acid and render innocuous all the organic matter originally present in the water, yet there is always the danger that the filtering power of such a soil may become overtaxed, and then organic matter in a state of putrefaction will obtain access to the well. The soil itself has also the remarkable property of absorbing and retaining several substances, such as ammonia, potash, phosphoric acid, &c., which the plants growing upon its surface remove for their nourishment; hence, water which is obtained by surface drainage from uncultivated tracts of land, such as moors, is found to be almost destitute of these bodies, though ammonia and nitric acid are invariably contained in the rain which falls on the soil. When the water is collected from cultivated tracts of land, the influence of manuring becomes evident by the appearance of ammonia and nitric acid in the water.

Ammonia is often contained in fairly large amounts in the water of deep wells, especially in those which are sunk through the London clay into the chalk. In this case it is derived from the nitrates present in the water, and is of even less importance than the nitrates from which it is derived, since it is still more remote from the original organic contamination. Waters derived from deep wells often contain large quantities of nitric acid as nitrates, these may occasionally amount to 4 or 5 grains per gallon.

The Combustion Process of Professor Frankland.—In this process a quantity of water is first boiled with sulphurous acid, to remove nitric and nitrous acids and carbon dioxide, a trace of ferric chloride being also added to ensure the destruction of the nitrates. The water is then evaporated to dryness, certain precautions being taken so as to ensure that no impurity is taken up by the water during this part of the process. The dry residue is then removed from the dish, mixed with cupric oxide, and placed in a combustion tube. The tube is then partially filled with cupric oxide, a cylinder of copper gauze being placed in front, and afterwards another layer of cupric oxide. The tube and its contents are now placed in the combustion furnace, connected with a Sprengel air-pump, and the air completely exhausted. Heat is gradually applied to the tube from before backwards. The organic matter in the water-residue is burnt up by the combined agency of heat and the oxygen liberated from the cupric oxide, its carbon being given off as carbon dioxide, its hydrogen as water, and its nitrogen as nitrogen and nitric oxide. The gases as they escape are collected in a jar over mercury, and, when the combustion is finished, are transferred to a graduated tube, and their respective quantities measured. From the measurements thus obtained the respective weights of the nitrogen and the carbon dioxide are calculated. These are then calculated into parts per 100,000 parts of water, and expressed in these terms in the statement of the results of the analysis. In addition to this, the free ammonia in the water, the chlorine, the nitrogen of the nitrates and nitrites, as well as the total solids of the water, are estimated and recorded.[1] The following are the results of the analyses of two waters by this process; the first of these is a very pure water, the second is one which has undergone considerable organic contamination:—

Statement of Analysis in Parts per 100,000.

Source of Water.	Total Solids.	Organic Carbon.	Organic Nitrogen.	Ratio of C to N.	Ammonia.	Nitrogen as Nitrates and Nitrites.	Total Combined Nitrogen.	Previous Sewage Contamination.	Chlorine.
Loch Katrine .	2.40	0.185	0.022	8.4:1	0.001	0.000	0.023	0.000	0.85
Polluted well .	77.44	0.294	0.186	1.6:1	0.026	2.130	2.337	21.190	12.40

In forming an opinion of the organic purity or otherwise of a water from the results of an analysis in this form, the absolute amounts of organic carbon and nitrogen are first considered. In a water fitted for domestic purposes the organic carbon should not exceed 0.2, nor the organic nitrogen 0.02 part of water; but in a water employed exclusively for brewing purposes considerably more latitude may be allowed, say four times these amounts.

The ratio existing between the *organic* carbon and nitrogen is next taken into account, and this is supposed to afford considerable evidence as to the kind of pollution the water has been subjected to. When the pollution has been of vegetable origin the ratio of organic carbon to nitrogen is high, about 8 : 1, or even higher, as may be seen in the analysis of the pure lake water just given. When, on the other hand, the organic matter has been derived

---

[1] For further details of the process, "Water Analysis," by E. Frankland, Ph.D., may be consulted.

from animal sources, the ratio of carbon to nitrogen is relatively low, 3 : 1, or even less, as is exemplified in the case of the polluted well water given in the table.

The original ratio existing between the organic carbon and nitrogen of a water is often considerably modified by processes which the water is subjected to after having received the actual contamination; such, for instance, as exposure to the air, in the case of the water of lakes, streams, or rivers. The organic matter then suffers gradual oxidation, but as its carbon is oxidised away at a more rapid rate than its nitrogen, the ratio previously existing between the two is gradually altered. When an impure water filters through a porous soil, a considerable quantity of the nitrogen of its organic matter is invariably converted into nitrates by the agency of the nitrifying bacteria present in the soil; and in this case, as the carbon is little affected, its quantity remains practically the same. Consequently a disturbance of the C to N ratio takes place, but in the opposite direction to that just described.

In the case of waters from springs and deep wells, owing to the thick strata which these waters pass through, the greater part of the inorganic matter is fully oxidised; the amount of organic carbon remaining ranges from 0.02 to 0.1 per 100,000 parts, the average for deep wells being about 0.6 part carbon and 0.02 part nitrogen. In such waters the C to N ratio may vary from 2 : 1 to 6 : 1, and probably it is here determined more by the extent to which oxidation has taken place than by the original nature of the organic matter.

Previous Sewage or Animal Contamination.—This expression was invented by Frankland as a means of forming some conclusion as to the amount of the organic contamination to which a water had been subjected during some portion of its history, when the contaminating substances had been converted into harmless bodies. It is obtained by adding together the total nitrogen of the nitrites, nitrates, and the ammonia, and deducting from the quantity so obtained an amount of nitrogen equivalent to the average quantity of ammonia found in rain water. The remainder is then converted into parts of sewage, on the assumption that sewage contains 10 parts of nitrogen per 100,000 parts. Thus, if the total nitrogen of the nitrites, nitrates, and ammonia of a polluted water was found to be 2.151 parts per 100,000, from this is deducted 0.032 part nitrogen (the average amount found in rain water), when 2.119 is left, and the "previous sewage contamination" is found by the following proportion sum:—

$$2.119 : 10 :: 100,000 : 21,190$$

N found.　　N in sewage.　　Parts of sewage.　　Previous sewage contamination.

Objections Raised to the Frankland Process.—Many exceptions have been taken to this process, with regard to its want of accuracy, its tediousness, and to the cumbersome and costly apparatus required for its performance.

The Albuminoid Ammonia Process of Wanklyn.—This process is the one most frequently used at present for the determination of the organic matter in water; it is easily performed, and does not require the use of any special apparatus beyond that found in every laboratory. In the determination a certain amount of water is taken (generally half a litre), placed in a retort connected with a Liebig's condenser, a portion distilled off, and the amount of ammonia contained in the distillate ascertained by Nessler's reagent; this quantity is registered as free ammonia, as in Frankland's process. A strongly caustic alkaline solution of potassium permanganate (previously well boiled) is then added to the water remaining in the retort, the whole again submitted to distillation, and the ammonia estimated in successive portions (50 cubic centimetres each) of the distillate. The amounts of ammonia found in the several portions of the distillate are added together, and the quantity thus found returned as albuminoid ammonia. Thus, Wanklyn endeavours to determine the amount of organic nitrogen contained in a water by converting it into ammonia and estimating it as such. In judging of a water by this process, Wanklyn states that a water for domestic purposes should not contain more than 0.002 part of free ammonia and 0.005 part of albuminoid ammonia per 100,000 parts. For brewing waters these limits may be considerably exceeded, but, as a general rule, neither the free nor the albuminoid ammonia should, either of them, exceed 0.01 part per 100,000.

Some idea as to the nature of the organic contamination of a water—that is, as to whether it is of animal or vegetable origin—is obtained by noting the rapidity with which the nitrogen is converted into ammonia under the influence of the alkaline permanganate. In those cases where the organic matter is of animal origin this takes place rapidly, and the first 50 c.c. of the distillate contain nearly the whole of the ammonia; in those where it is of vegetable origin, the successive portions of the distillate contain ammonia in proportions more nearly alike, the second portion of the distillate containing an amount of ammonia but slightly less than the first one, and so on with the third and following portions.

Objections to the Wanklyn Process.—It has been urged against this process that the total amount of the nitrogen of the organic matter of the water is never obtained as ammonia, and this is undoubtedly true. In some comparative experiments made on the same waters by the two processes, Dr. Hill found that the ratio of the nitrogen found by the Frankland method to that of the ammonia yielded by the Wanklyn method varied from 15 : 1 to 4 : 1. In a series of experiments where various nitrogenous organic bodies were submitted to the Wanklyn process, it was found that some of them gave up the whole of their nitrogen as ammonia, others two-thirds, some one-fourth, some very little, and others, amongst which urea is a notable example, yielded none whatever. In spite of this, Dr. Cornelius Fox [1] states that on comparing the details of the results of ninety-nine analyses obtained by both processes, only in one instance was there a distinct conflict of opinion as to the quality of the water, and this divergence was readily explicable. Experience has taught that the indications of the Wanklyn process, when taken in conjunction with the other details of a water analysis, are capable of yielding much useful information, and the extensive practical application which the process has met with also points to the same conclusion.

Determination of the Organic Matter in a Water by the Quantity of Oxygen Required to Completely Oxidise it.—This process, which was originally invented by Forchammer, has received various modifications at the hands of other analysts, amongst these the late Dr. Tidy, who proposed the modification that is generally used in this country. A measured quantity of the water is first acidified with sulphuric acid, and a known quantity of potassium permanganate in solution added. The mixture is allowed to stand for several hours (generally three), at the end of which time the amount of undecomposed potassium permanganate remaining in the water is estimated. In most cases a certain amount will have disappeared, having been destroyed in oxidising the organic matter, and this amount, stated in terms of the quantity of oxygen which it would yield up under the conditions of the experiment, is returned in the results of the analysis as "oxygen required to oxidise the organic matter." Its quantity in a water for domestic purposes should not exceed 0.05 part per 100,000 parts, but may probably be allowed to rise as high as 0.25 in a water used for brewing purposes.

Nitric Acid Existing as Nitrates.—There are several methods of estimating nitric acid in water, all of which are equally reliable. There are two ways of stating the quantity of nitric acid found in analysis, as "nitric acid" (really nitric anhydride N_2O_5), or as "nitrogen existing as nitrates." It must be borne in mind that when stated in the former method the same amount will appear much larger than when stated in the latter way; the ratio between the two is 14 to 54; consequently, nitrogen as nitrates $\times$ 3.86 = nitric acid, and nitric acid $\div$ 3.86 = nitrogen as nitrates. Pure waters from different sources contain various amounts of nitrogen as nitrates; spring water contains on an average 0.2 grain per gallon, river water about 0.15, while that from wells in the chalk often contain from 0.6 to 0.7 grain per gallon. In a water for domestic purposes the nitrogen thus combined should not exceed 0.5 grain per gallon; but for brewing waters it may be allowed to rise to 1.5 grain per gallon, and waters containing as

---

[1] "Sanitary Examinations of Water, Air, and Food," p. 63.

much as 2.5 grains per gallon are known which produce excellent beers, but with these latter amounts the water must be organically pure in other respects. Nitrates in excessive amount cause rapid weakening of the yeast.

Nitrogen Existing in a Water as Nitrites.—The nitrogen existing in a water in this condition can be accurately estimated by several processes. Nitrites, when present, should always be looked upon with the greatest suspicion, as, in the vast majority of cases, they indicate that organic matter is obtaining access to the water of a well in quantities beyond those which the adjacent soil, acting in its capacity as a natural filtering bed, can convert into the more harmless nitrates. If the filtering powers of a soil under such conditions should become a little more taxed, then organic matter in its most objectionable form will obtain entrance to the well. Moreover, it has been shown by Laurent that nitrous acid, when set free from any of its salts by an acid in the fermenting wort, exerts a very prejudicial effect on the yeast, and as all worts contain a certain amount of free acid, there is a great probability that this result will take place in the fermenting vat when a water containing nitrites is employed.

Chlorine Viewed from an Organic Standpoint.—As has been stated before, the presence of chlorides in fairly large amount are, in many cases, decidedly beneficial in a brewing water. When, however, the other indications of the water point to the chlorides being derived from sewage, an excess of these salts must be regarded from an entirely different point of view. Good natural waters contain from 0.7 to 1.5 grain of chlorine per gallon; those from the greensand often contain as much as 5 or 6 grains per gallon; while in the neighbourhood of the sea, well waters may contain still larger amounts of chlorine without its being an indication of organic pollution. One of the best guides as to the significance of the amount of chlorine in a water is to compare it with the quantities contained by waters in the same district which are known to be unpolluted.

Phosphoric Acid.—This acid, which exists in combination with some one or other of the bases present in nearly every water, is, as a rule, met with only in traces. Its amount is seldom estimated quantitatively, unless for some special reason; the precipitate produced on the addition of a small quantity of ammonium molybdate and nitric acid to a water residue is noted; and the quantity estimated in this way is returned as "minute traces," "traces," or "heavy traces." The presence of phosphoric acid, in anything but the smallest traces, may generally be regarded as an indication of pollution with organic matter of animal origin, since all animal excreta contain an abundance of phosphoric acid.

Behaviour of Water Residues on Ignition.—In every water analysis it is usual to estimate the total solids of the water, in order to obtain a check on the other results of the analysis. For this purpose a measured quantity of the water is evaporated to dryness in a platinum dish over the water-bath, the residue dried in a hot air-bath at 180° C., and weighed. A perfect agreement between the amount thus obtained and the sum of the respective quantities of the bases and acids, determined separately in the analysis, is never obtained; because, during the evaporation of the water, the magnesium chloride partially loses chlorine and becomes converted into a basic salt; the silica displaces some of the carbon dioxide of the carbonates; the last traces of moisture are expelled from the magnesium carbonate with great difficulty; and this salt, though it is recorded in the results of the analysis as the neutral carbonate, is really left in the residue in the form of basic magnesium carbonate. More or less organic matter is always present in a water residue, and the weight of this is never subsequently determined with any approach to accuracy; nor is any notice taken of the effect which the ammonia compounds may exercise on the other constituents of the water residue. In order to overcome this difficulty, Fresenius[1] directs that all the salts in the water residue shall be converted into sulphates. During the process of ignition after the addition of the sulphuric acid, all the organic matter is destroyed and removed. The separate bases, as obtained in the detailed analysis, when calculated into their respective sulphates, ought to add up to an amount closely agreeing with the quantity obtained by treating the water residue with sulphuric acid in this manner.

It was customary in the olden days to ignite a water residue, cool, and again weigh, and as the loss in weight was then considered to be entirely due to the organic matter destroyed, it was taken as the measure of it. Afterwards it was found that the carbonates lost more or less of their carbon dioxide during the ignition, and this was replaced by treating the residue with ammonium carbonate solution, evaporating to dryness, and then igniting again very gently. But, even when these precautions are taken, no reliable result can be obtained in this way, for we do not know the exact state in which the magnesia salts are left behind in a water residue after evaporation and ignition; also, a further quantity of carbon dioxide is driven out from some of the carbonates by the silica, and as this latter acid combines with the base, that portion of the base so united is not re-carbonated on treatment with ammonium carbonate solution. Though no information of value can be obtained by estimating the loss which a water residue suffers on ignition, something may be learned from its behaviour under these circumstances. In those cases in which a water is free from, or contains mere traces of, organic matter, the residue when ignited suffers no discoloration; when the organic matter is in a somewhat higher amount, a transient browning takes place; this soon passes away, and leaves the residue either white or

[1] *Anleitung z. Quant. Chem. Anal.*, 6th edition, p. 165.

slightly yellow. When a still higher amount of organic matter is present the residue blackens, and the application of a much higher degree of heat is required to remove the discoloration than in the previous case. When the residue contains much organic matter, intense blackening takes place on the application of heat, which is removed only on prolonged ignition, and even then with the greatest difficulty. Under such circumstances, and when the organic matter is of animal origin, an odour like that of burning feathers may often be perceived. In the residues of very bad waters which contain an abundance of nitrates, small sparks are sometimes seen, owing to minute deflagrations taking place between the nitrates and the organic matter present; occasionally red fumes are evolved. At times a smell of sulphurous acid is detected, and this points to the presence of sulphur compounds in the water. According to the late Professor Parkes, "three grains per gallon of either vegetable or animal organic matter cause some blackening, six grains per gallon a good deal, and ten grains per gallon a great amount," in water residues on ignition. The residues of peaty waters often blacken considerably on ignition, but this by itself is an indication of no importance, as the organic matter here is of a vegetable nature.

Heisch's Sugar Test.—This test was originally proposed by the late Mr. Heisch as a means of ascertaining the presence of sewage in water. It is performed by filling a perfectly clean stoppered bottle of about five ounces capacity with the water under examination, and adding thereto one gramme of the purest cane sugar. The stopper having been accurately adjusted so as to exclude atmospheric air, the whole is kept, exposed to the light, at a temperature of 70° F. for several days, the sample being inspected at intervals. In the event of the water being impure, small bodies will be seen floating about in it at the end of twenty-four hours, which, when examined under the microscope, are seen to consist of cells with very brilliant nuclei. These cells afterwards unite into groups which resemble bunches of grapes. Bad waters turn opalescent and milky, and in very bad cases the odour of butyric acid is developed. A similar experiment is conducted with another portion of the water after it has been well boiled, as some waters, especially those which contain carbonates, lose the power of forming these organisms on boiling; and this is the most important point to the brewer, for he is always able to boil his water before using it. Should no change take place in three days in a water subjected to this test after being boiled, though it may have responded to the test in the unboiled state, such a water may be used with impunity, providing it is in other ways organically pure; but if a marked action shows itself at or before the end of three days, the water should be regarded with suspicion. Much difference of opinion has been expressed by different observers as to the value of this test. Dr. Frankland considers that the formation of these peculiar bodies is merely an indication of the presence of phosphates in a water; Messrs. Matthews and Lott view the matter somewhat differently. Whilst confirming to a certain extent the views of Frankland, they state that butyric fermentation takes place only in those waters which contain an abundance of phosphates, and in such cases the rest of the results of the analyses invariably show that such waters have been contaminated. Matthews[1] considers that the reason why some waters do not respond to this test after boiling, which they did before, is to be attributed to the precipitation of the phosphates caused by the expulsion of carbon dioxide which takes place on boiling; for the addition of a mere trace of phosphates to such waters again starts the process, and this serves to show that the organisms are not destroyed on boiling. He found that such a substance as egg-albumin, which contains phosphorus in a state of organic combination, mixed with sugar, forms a better food for bacteria than one composed of mineral phosphates and sugar; and that a water to which albumin has been added, and which has been subsequently boiled and filtered, is capable of supporting the growth of bacterial organisms. He considers, therefore, that if a water containing calcium and magnesium carbonates responds to the Heisch test after being boiled, it may be presumed that such a water contains organic matter of an albuminoid nature.

Microscopic Examination of the Sediment of a Water.—An examination of this kind, taken in conjunction with the chemical analysis of a water, is often of assistance in forming an opinion as to its value from an organic point of view. For this purpose a fairly large quantity of the water should be allowed to stand for some time (say twenty-four hours) in a covered conical vessel; a portion of the sediment is then removed by means of a pipette,[2] transferred to a glass slip, and examined under the microscope. The deposit generally consists of earthy particles, such as portions of sand, clay, &c., vegetable and animal organisms, portions of decaying vegetable matter, animal matters in a process of decay, such as the portions of the dead bodies of insects. These substances are found in most waters. Amongst the vegetable matters which afford indications of pollution are such substances as the fibres of linen, hemp, cotton, &c.; amongst animal matters such objects as fibres of silk, wool, human hair, hair of other animals, epithelial scales, decaying muscular fibre, fragments of feathers, &c.

The mineral portion of the sediment may be neglected, as it affords no evidence of contamination. Many vegetable and animal organisms are met with in pure waters, some in running streams, others in still water; and their presence, when existing in comparatively small numbers, is of no serious import, though it is

---

[1] *Brewers' Guardian*, p. 509.
[2] Or the ingenious micro-filter of W. J. Dibdin may be employed. For description, see *The Analyst*, vol. xxi. p. 3.

MICROSCOPIC EXAMINATION OF SEDIMENT.

always an indication of the existence of more or less organic matter in the water, as evidently such organisms cannot exist without food. Amongst the harmless species may be placed the *Algæ* and *Diatomaceæ, Euglenæ, Hydrozoa,* and the larger animal forms, such as the water-flea, &c. Bacteria, when occurring in large quantities, and especially when these belong to many different species, are of ill-omen, and this applies also to such forms as the *Rhizopoda, Amœbæ,* &c. *Vibriones* and *Spirilla* also indicate the presence of organic matter in a state of putrefaction. The presence of fungi is also undesirable. But the most objectionable matters are those which point most conclusively to a water having been subjected to direct pollution, such as cells of the potato, spiral tissue from cabbage and other vegetables used for food, epithelial scales, portions of muscular fibre, and other débris of vegetable and animal origin.[1]

The detection of these various objects found in water requires a considerable amount of skill in the use of the microscope, which can only be acquired by long practice. Their examination is, however, an extremely interesting and fascinating study, and one which can be recommended as an excellent means of educating the eye of the observer, and enabling him to form a keen discrimination in the form and size of objects when viewed through the microscope. The water from almost any pond or ditch will be found, especially during the hotter months of the year, to contain microscopic organisms in abundance.

---

[1] To those interested in this study, "A Guide to the Microscopical Examination of Drinking Water," by Dr. Macdonald (Churchill), which contains engravings of a large number of the organisms and substances usually met with in water, will be found extremely useful.

CHAPTER VII.

BARLEY AND MALTING.

Barley.—Though a number of the seeds of the various species of the *Gramineæ* or grasses have been at times used for the purpose of malting, barley has been found, from an experience of many thousands of years, to be the grain best adapted for yielding a malt possessing to the fullest extent those properties which are necessary for the production of good beer. It seems, also, the grain most naturally adapted for malting, for it suffers less injury in the process than any other corn. Of barley there are a great number of varieties, some of which have proved themselves by long use to be particularly adapted for brewing purposes, as, for instance, the variety known as "Chevalier," so called from the clergyman of that name who originated it by careful selection and cultivation. Barleys are often spoken of as six-rowed, four-rowed, or two-rowed; in the first of these varieties there are three rows of spikelets on each side of the ear, each spikelet of which bears a fertile flower. Consequently, when ripe, there are six rows of barley-corns on each ear. In the four-rowed variety the middle rows of spikelets on each side of the ear are abortive—that is, produce no seeds; consequently in an ear of this variety there are four rows of barley-corns. In the two-rowed kind only the central rows on each side are fertile; consequently only two rows of corns are produced. The "Chevalier" belongs to this last type, and, since there are considerably fewer corns on each ear than in the other varieties, a much better chance for luxurious development is afforded them. The corns from the six- and four-rowed varieties may be distinguished by a peculiar curving of the furrow, which characterises all the corns of the four-rowed variety, and also those of the outside rows of the six-rowed barley. The best malting barleys are those grown on light, warm, friable soils; but fine barleys can also be produced on rich, loamy, well-drained soils. On the other hand, good malting barley is rarely, if ever, yielded by stiff, heavy, cold, clayey land.

Barley is much influenced by the season. Continued wet weather, especially during the period of ripening, has a most prejudicial effect upon the barley crop; so much so that, in extreme cases, it is impossible to convert the barley grown in such seasons into good malt. Until comparatively late years all the malt used in this country for brewing purposes was made from home-grown barley; now much is imported from foreign

CHOICE OF BARLEY FOR MALTING.

countries, as, owing to the more regular climatic conditions, the barleys there grown are more uniform in character.

Choice of Barley for Malting.—As barley is almost invariably bought in the open market, the purchaser has to rely on certain external signs, the knowledge of which can only be acquired by actual experience in the handling of samples. The following characteristics may be of some little use in starting, but long experience alone will make a person a good judge as to the quality of barley.

Vitality.—One of the most essential conditions of good barley is vitality; obviously, if barley will not grow, it is impossible to convert it into malt. Should, in a particular sample, anything but an extremely small percentage of the grains be incapable of germinating, good malt cannot be made from it, for all ungerminated corns represent merely so much raw barley in the finished malt, and they, moreover, form breeding-grounds for mould during the malting process. Defective vitality may arise from under- or over-ripeness, incipient germination in the ear owing to a wet harvest, death of the germ through heating in the stack, or improper storage, the attacks of vermin and insects, or damage to the grain in the threshing. Barley rapidly loses its vitality in storage, and especially so if stored in a damp condition. Under the best circumstances it should not be used for malting purposes when more than two years old. Barley, especially that which has never been stacked, improves in vitality when stored under favourable conditions for a few months. This is, doubtless, owing to the gradual drying which takes place, for it has been shown by Brown and Morris that barley will not germinate until it has attained a certain degree of dryness.

Tests for Vitality.—The most certain test for vitality is actually to grow a sample of the barley, but as this takes time, it is of no assistance to the purchaser in the open market. The germination test may be applied in several ways. A hundred to two hundred corns, taken promiscuously from the sample without any picking or choosing, are placed between two layers of flannel on a plate or dish, and the whole covered with water for twenty-four hours. The water is then poured off, the dish and its contents being kept at a temperature of 50° F., when, in from forty to fifty hours, those corns which are capable of germinating will begin to chit—that is, a small protuberance will show itself at that end of the grain where the germ is situated. Contrivances have been invented for effecting experimental germination, of which Coldewe's germination apparatus is a good example, and which is illustrated in Fig. 60. It consists of a glass vessel with a constriction about $1\frac{1}{2}$ inch from the top, on which rests a porcelain plate provided with 100 perforations. Into each of these a corn of barley is inserted, with the germ end downwards. The apparatus is nearly filled with water, the plate inserted, a small quantity of moist sand placed around the barley, and a wooden cover, which has its under side covered with a layer of felt, placed over the sand. A small thermometer attached to the lid serves to indicate the temperature at which the experiment is being conducted. The number of sluggish and dead corns can be seen at a glance, and if the germination be allowed to proceed for a few days, the relative degrees of vitality of the different corns can be readily estimated.

When a large number of samples of barley have to be tested, the apparatus of Aubry[1] is extremely convenient. It consists of a cubical chamber about 10 inches each way, constructed of sheet metal, and provided with doors in front and behind. On either side of its interior are attached a number of strips of sheet metal about half an inch wide, and fixed at a similar distance from each other; these serve as supports to the glass plates on which the barley is germinated. The top and bottom of the apparatus are perforated with a few small holes, to admit the quantity of air necessary for the growth of the barley, but not more than this, or too free evaporation will take place. Each of the glass plates is loaded with 500 corns, which are either simply counted or computed with such an apparatus as that of Westfelt. These are steeped in water for six hours at the ordinary temperature. A piece of damp blotting-paper the size of the plate is first attached to it, then the barley is spread evenly on the paper in an even layer. It is next covered with two similar sheets of blotting-paper, which are afterwards moistened with a damp sponge, when the plate and its contents are placed in the apparatus. The glass plates, which form a series of shelves, are withdrawn periodically for examination, and the manner in which the barley grows noted. The figures thus obtained are converted into percentages by dividing by five.

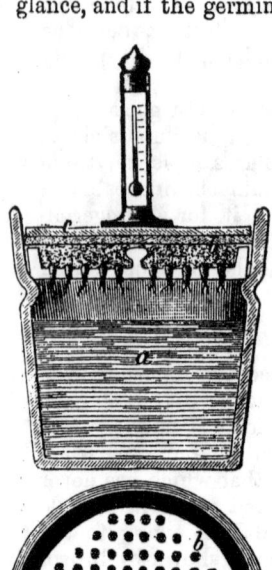

FIG. 60.—Coldewe's Germinating Apparatus.

a. apparatus with water; *b*. perforated plate; *c*. felt cover over sand.

In examining a sample of barley in the open market, many of the conditions which produce a want of vitality can be determined by an experienced hand, though some of them are so extremely difficult to

[1] *Zeit. f. Brau.*, 1885, p. 77.

detect, that they can hardly be discovered in any other way than by the germination test. Corns damaged by vermin or insects can be readily detected, as can also the signs of incipient germination; the acrospire will be slightly developed, and the germ end of the sprouted grain have become soft and friable. In extreme cases rootlets protrude. On dividing a corn with a sharp knife through the furrow, and examining the divided germ with the aid of a magnifying glass, this should have, according to Slopes, "a juicy, fairly firm, yellow appearance, very closely resembling in consistency and colour good, firm, freshly churned grass butter, or have a greenish-yellow colour very like wax." If grey, its vitality is low; if reddish-brown, dark, or dried and shrivelled, it is dead. Haberland has proposed a test for vitality which consists in placing a drop of strong sulphuric acid on the germ and observing its action. Healthy, vigorous germs quickly assume a fair rosy tint, feeble germs become brown, dead ones black.

Vegetative Energy and Vegetative Capacity.—The former of these expressions is used to denote the percentage of grains in a sample of barley which, when placed under favourable conditions for germination, vegetate within a definite time; this, at the ordinary temperature, is generally taken as three days. By the latter term is expressed the percentage number which are found capable of germinating irrespective of time. In a good sample of barley the vegetative energy should be not less than 90 per cent., and the capacity not below 95 per cent. The more closely these two characters correspond the better, since any considerable variation indicates that such a sample would grow irregularly on the floor.

Age.—The vitality of barley seems to improve for several months after it has been harvested, and then to slowly decline; so much is this the case, that it is unsafe to malt barleys which are more than two seasons old. Lots which have been stored in a damp condition suffer most in this respect.

Condition of Endosperm.—When a grain of barley is bitten or cut in two, the contents of the endosperms, in samples which are capable of being converted into good malt, are of a white and friable or mealy nature. Excessive hardness or any variation in colour from white are indications that such samples will require special care in malting. Grain which is very resistant to the bite, and the fracture of which has a vitreous appearance, can never be converted into superior malt. As nearly all malts contain a certain proportion of corns of these two latter classes, it is well to ascertain the respective numbers of each variety in a sample and value it accordingly. This can readily be done with one of the "farinatomes," which cut through 50 or 100 corns with one stroke of the knife. It has been found that light will pass through glassy but not through mealy barley-corns, and upon this fact there has been founded a method of discriminating between the two. The distinction is so sharp that the mealy portions can be detected in corns which are partially glassy. Several appliances have been constructed for carrying out this test, one of the simplest of which is that of Vogel (Fig. 61). It consists of a quadrangular box made of sheet metal, in the interior of which there is a petroleum lamp. The roof of the apparatus is provided with a number of slits, into each of which a barley-corn is placed. The quality of their contents is adjudged from their relative transparency. In an examination of barley, by either the farinatome or the transparency test, it is usual to operate on 500 corns. These are divided into three classes—floury, vitreous, and semi-vitreous. The numbers thus obtained divided by five give the respective percentages of the different classes.

Coldness.—A sample of barley when handled should not feel cold to the hand, for this is a sign that it is either damp, or immature, or has not been sufficiently sweated in stack. Damp barley stores badly; it is liable to become mouldy, and to rapidly lose its vitality. In addition to this, in buying barley which contains an excess of water by weight, the price of barley is paid for this excess of water. Dampness in the sample which is good in other respects may be removed by sweating—that is, drying it for several hours on the kiln at a low temperature.

FIG. 61.—Vogel's Apparatus for Testing Barley for Transparency.

Maturity.—Barley should neither be under- nor over-ripe, for in either of these conditions it grows badly. Immature corn feels cold to the hand, has a sickly greenish or greenish-yellow colour, and, when dry, a starved, wrinkled appearance. Over-ripeness in barley is distinguished with difficulty; the skin of such corn has often a dead white appearance, but little difference can be detected in the condition of the endosperm. The degree of ease with which a number of corns slip through the fingers affords some slight insight into the maturity of a sample; mature and dry grain slips through more readily than unripe or damp.

Mouldiness.—It is hardly possible to obtain barley absolutely free from mould, but it should be as nearly so as possible. The more it is contaminated in this way, the more liable it is to become excessively mouldy during the malting process.

Damaged Corn.—Barley may be so damaged as to have its vitality destroyed in several ways. When in store it may be attacked by vermin and insects which partially and totally destroy a number of corns. In careless threshing or dressing many corns may be damaged by being cut in pieces, having their skins partially stripped off, or portions of the corn removed along with the awn. All corns damaged in this way, as well as dead corns, invariably become mouldy during the malting process, and form centres of contagion from which mouldiness spreads to the rest of the barley on the floor.

Odour.—Only sweet barley, that is, barley possessing no objectionable odour, will yield first-class malt. Corn having a mouldy, fusty, or other objectionable smell is a highly undesirable article. Barley which is free from the defects previously enumerated, and which possesses the good qualities indicated, is fitted for the production of first-class malt; but in addition to these qualities, there are others which, if not absolutely indispensable, are highly desirable.

Size.—This may be considered from two points of view—that of absolute size of the individual grains, and the relative size of the grains to one another. Obviously, if each corn were a perfect sphere, such a form would allow of the largest proportion of contents to husk; consequently, the fatter and plumper the grains are, or the nearer they approach this figure in form, the larger the proportion of contents to husk. As it is the contents alone which yield extract, the plumper the grains, other conditions being equal, the more extract will a given weight yield. The more uniform the individual grains are in size, the more regularly they grow; hence the value of grading where considerable difference exists in the size of the individual corns.

Weight.—Barleys vary much in weight per bushel, heavy samples weighing as much as 56 lbs. per bushel, light ones not more than 50 lbs. Much importance was formerly attached to the weight per bushel; but, providing the more essential characters are present, viz., vitality, friability, &c., light barleys will yield equally good malt. This is especially the case with the thin foreign barleys, the malt from which yields sound worts, especially distinguished by rapid clarification of the resulting beers. The weight per bushel of barley is generally ascertained by some form of instrument called a "chondrometer," which consists of a small balance provided with a small measure, a funnel for filling, and a strike. The measure is filled by means of the funnel with a fair average sample of the bulk, from which the coarser impurities, such as stones, have been removed. The surface is levelled with the strike, and the weight ascertained. The weights bear the same proportion to the measure as pounds do to the bushel, therefore the weight in pounds per bushel is read off directly.

Uniformity.—As different-sized corns, when steeped, are not evenly wetted, they consequently germinate irregularly; in a firstrate sample of barley the individual corns should be pretty much of the same size. This is a difficulty which can be readily got over by grading the barley with one of the numerous machines constructed for this purpose; and as no sample of barley is perfectly uniform in size, it is advisable to grade all barley before proceeding to work it.

Colour.—Barley should be of a light straw-yellow colour; a greenish tint indicates unripeness, too pale a colour over-ripeness. The tips should have the same colour as the skin; dark or black tips are a sign of heating in the stack, or that the sample has been harvested during wet weather, or that it is old. Such corns grow irregularly at best, and have often entirely lost their vitality. Sometimes the barley is dried on the kiln, and sulphured to improve its colour and odour. This treatment with sulphur is very objectionable; if the light colour of a sample leads to the suspicion that it has been bleached in this way, it should be submitted to a chemical examination, which will reveal the presence of traces of sulphuric acid on it.

Skin.—Good barley should have a thin skin, smooth, and slightly wrinkled. Thick, harsh, rough, crinkled skins indicate that the barley has been grown on cold heavy land.

MALTING.

The malthouse consists of a main building two or three storeys high, and of a drying kiln, together with stores for barley and malt. Malthouses are built in many shapes and sizes; the preferable form is that in which the plan of the house is a long parallelogram, with the cistern placed at one of its short ends, and the kiln at the other. The cistern, which may be of brick lined with cement, of wood, of lead, or of cast-iron, is generally placed in the lowest storey. It is provided with a false bottom perforated with holes; this, acting as a sieve, allows the water to run away whilst retaining the grain. The cistern should have a capacity of about twelve cubic feet for each quarter of barley to be steeped. In the construction of the floors many materials have been used; the principal requisite being that the floor has a smooth, even surface, which can be readily cleansed and which will not absorb moisture. Probably well-faced Portland cement makes one of the best floors. One hundred and eighty to two hundred square feet of floor space should be provided for every quarter of barley to be steeped. The uppermost storey is generally used as a store-room for the barley, from which it is shot down into the cistern at each steeping. After being steeped (and couched) it is either worked on the lowest floor or removed to the upper ones in baskets. In more modern houses the uppermost storey, in addition to forming the barley store, also contains the machines for cleaning and grading the barley, and the steeping cistern or cisterns as well. These are in this case made of iron, the lower portion being constructed in the form of an inverted cone, which has an outlet valve at its apex. This form of cistern secures considerably better drainage, and as the barley when steeped is simply shot out to the place where it is wanted, all damage from shovelling is avoided. An abundant supply of water should be provided for each cistern, and also a drain to take away the steep-water. Every cistern should be provided with some arrangement by means of which the water can be warmed to a slight degree in very cold weather. Malting is the first and one of the most important processes antecedent to the actual operation of brewing, and much afterwards depends on the way in which it is carried out; unless performed with skill and care, however good the barley may be, if not properly treated in this process, it will never yield malt capable of producing first-class beer. As a rule, it is considered much the better plan for the brewer to be his own maltster, for then he can conduct the malting process without hurry; whereas the maltster who only makes for sale is rather inclined to hurry the process on, since the more malt he can produce during the season with his plant the greater his profits will be. This hurried method of working has of late years given rise to the manufacture of much malt of an inferior quality, which has often caused serious trouble in the brewery. It is not the privilege of every brewer to be able to choose his own malt, but in all cases he ought to be allowed to do so, otherwise he cannot be expected to be held responsible for the results of his brewing operations.

Screening.—Barley, before being steeped, should be screened by means of one of the numerous machines which are in the market for this purpose. These not only remove the dust and dirt, but also separate the small and broken corns, and grade the barley into different sizes. This last is a matter of considerable importance when a quantity of barley irregular in size has to be dealt with, for unless the individual corns are pretty equal in size, some absorb more water than others, and, as a consequence, irregularity of growth ensues.

Steeping.—The barley, after being screened, &c., is immersed in water in the cistern, in order that it may absorb an amount of that liquid sufficient to arouse its dormant vitality and set in action the train of physiological processess which we collectively include under the term "germination." The amount of water which different samples of barley will absorb in a given time varies much with their size, condition of skin, &c.; consequently, a certain amount of judgment, only to be arrived at by experience, has to be exercised by the maltster in this operation. About fifty hours may be taken as the average length of time of steeping for English barleys, but, as has just been stated, this will vary much according to the nature of the barley and the temperature of the steep-water. The light Danubian and Odessa barleys require about sixty hours' steep, and the Smyrnas about seventy-two hours. The proper quantity of water which should be absorbed is about 50 per cent. In steeping, the barley increases considerably in bulk, one quarter expanding to a volume of from 1.18 to 1.20 quarters. In conducting the operation of steeping, it is best to place the proper quantity of water into the cistern, and then to run in the barley in a slow stream, stirring vigorously all the time. In this way the contact of every grain with the water is assured, and any dust or refuse which floats on the surface of the water can be removed.

Change of Steep-Water.—As water dissolves a certain amount of organic and mineral matters from the barley, and this affords a supply of food for the numerous bacterial organisms which are always adherent to the barley, the steep-water sooner or later becomes putrid, and requires changing. It is recommended by a very good authority to change the water twice during the first day and once each subsequent day. Barley should never be oversteeped or sodden, as, in such a case, the vitality of some grains is considerably weakened and that of others completely destroyed. Deficiency of water can be afterwards remedied by sprinkling on the floor, but it is almost impossible to remedy the ill effects of over-steeping. Some maltsters find it advantageous for preventing mould to use bisulphite of lime in the steep-water, at the rate of about one gallon of the commercial solution to every thirty quarters of barley.

Signs of Proper Steeping.—When a corn of properly-steeped barley is taken by its ends between the finger and thumb, the exertion of a certain degree of pressure in squeezing it together should cause the skin to leave the grain. If it yields to pressure too easily, it is over-steeped; if too resistant, it is insufficiently steeped. When a grain is cut through with a sharp knife, the whole contents should appear evenly wetted, but should not be soft and milky. It should be possible to pass an ordinary sewing needle easily through the grain.

Temperature of Steep-Water.—This is a point of considerable importance, and one which is very often overlooked. The subsequent growth of barley is much prejudiced by being submitted to too low a temperature in the cistern. The temperature should be between 50° and 55° F., and should never fall below the first of these figures. On the other hand, a temperature above 56° F. favours bacterial action, as evinced by the steep-liquor rapidly becoming sour or putrid.

Quality of Steep-Water.—It has been a much-debated question as to whether steep-water should be soft or hard; the former extracts more potassium phosphate, the latter more organic substance. An excessive quantity of sodium chloride (above 10 grains per gallon) is said by Märcker to retard germination and to cause an imperfect development of the rootlets, the malt when finished being deficient in extract and containing a high percentage of nitrogen; so also do excessive amounts of calcium chloride or magnesium chloride. Ferruginous waters are also unsuitable. The presence of nitrates in moderate quantities appears to be favourable to germination. Prior[1] considers that any fairly pure water which is free from the above-mentioned salts may be used for steeping—preference being given to soft water, or to a moderately hard water in which the hardness is due to calcium sulphate.

The Changes which Take Place during Steeping.—During this process the grain absorbs a large amount of water; its quantity has been given by Ford as an average of 47 per cent. on the original weight of the grain—that is, 100 lbs. of barley, after being steeped, and with the surface moisture removed, would weigh 147 lbs. Thausing gives a somewhat higher figure than this—from 50 to 55 per cent. with a barley containing originally 12 per cent. of moisture; but the amount varies with each sample of barley, and its condition as to dryness. Stopes[2] states that 20 per cent. of imbibed water is sufficient to start germination, and that 30 per cent. is sufficient to carry some barleys through the whole process of malting, though others require 40 per cent., and some even more than 50 per cent. The grain also considerably increases in bulk, 100 bushels of barley measuring from 118 to 120 bushels after being steeped. According to Holtzner, the water finds its way into the barley partly through the husk, but chiefly through the apertures at each end of the grain, between the husk and the grain; it thence passes by diffusion from the periphery of the grain to its centre. A certain amount of various substances, organic and inorganic, such as cane sugar, gum, diastase, colouring matters, &c., as well as a portion of the mineral constituents, are dissolved out of the barley by the steep-water. The total amount of the organic matters removed in this way have been estimated by various observers at from 0.6 to 1.5 per cent., and about one-tenth of the mineral substance is similarly lost. The quantity so dissolved will vary according to the temperature of steeping, the condition of the barley, and the quality of the water employed. The amount of acid originally present in the barley is increased by the formation of fixed organic acids. But trifling quantities of phosphoric acid are extracted in this way; so small are they that the percentage of this acid may, owing to the comparatively larger amount of the other substances dissolved out, be greater in the steeped barley than it was in the original corn. About one-half of the soluble nitrogenous constituents of the barley pass over into the steep-water. Many of the spores of bacterial organisms and of the mould fungi, &c., which are always present in abundance on the husk of the grain, are washed away.

Germination.—In this portion of the operation the two chief objects of the maltster are to secure what is known as the "modification" of the contents of the endosperm, and to generate

[1] *Chem. u. Physiol. Malzes u. d. Bieres.*, p. 98.
[2] "Malt and Malting," p. 334.

the enzyme diastase in sufficient amount. This modification of the endosperm constitutes the chief difference between barley and malt; and on its perfect accomplishment depends the amount of substance which will subsequently enter into solution in the mash-tun; in other words, it determines in a great measure what shall be the amount of extract yielded. It also to a considerable extent determines the quality of the dissolved matters. As we have seen before (p. 275), these changes are due to the effects of the physiological processes which take place during germination; an enzyme is secreted by the secretory layer and gradually invades the endosperm. This gradually permeates the substance of the endosperm, dissolves up the cellulose walls of the starch-containing cells, and leaves them, as it were, in a naked condition and directly accessible to the diastase dissolved out by the mash-liquor. It is upon the action of the enzyme that the "modification" of the contents of the endosperm mainly depends. But there is much difference in the degree to which its action extends in different barleys; the cellulose of some samples appears to be peculiarly resistant to its influence, and, consequently, such specimens can never be made to yield tender, friable malt. Sometimes, or perhaps it may be said often, through deficient germination, the enzyme does not reach the extreme ends of the corns, in which case a malt is produced, many of the corns of which have hard ends, and all such unmodified masses represent so much raw barley. We therefore desire that the fullest action of the diastase should be exerted on the cellulose throughout the whole endosperm. As regards the diastase, though we wish to have it secreted in large quantities, yet we endeavour to limit its action on the starch as much as possible, to keep this down, as nearly as we can, to the extreme limit capable of furnishing food to the plant during the period necessary for inducing the requisite changes in the constituents of the barley-corn. If the action be greater than this, an amount of starch is unnecessarily destroyed. The respiration of the embryo must, therefore, be kept within due limits in order to avoid waste, and this is chiefly effected by never permitting the heat of the germinating barley to exceed a temperature of 60° F. All the carbon which is expended in keeping the heap of germinating barley at a temperature beyond this represents so much lost material.

During germination there is a gradual increase of the diastase; Kjeldahl found this, as estimated by the diastatic power of the malt, to be :—

	Diastatic Power.
First day	70
Second day	73
Third day	80
Fourth day	105
Fifth day	150
Sixth day	190
Seventh day	220
Eighth day	226

As diastase is able to act upon starch in the cold, it naturally follows that its presence will be attended with the products of its action on the starch of the endosperm. The first sugar produced is, according to Brown and Morris, maltose, and this becomes partially transformed into cane sugar in the embryo, most probably by means of an enzyme. Kjeldahl found that when a barley which originally contained 1.5 per cent. of cane sugar, had arrived at the stage of green malt, it then contained 4.7 per cent. of this sugar. The former observers were able to prove that the largest amount of cane sugar was contained in the germ, while comparatively small amounts existed in the endosperm. Thus, of the total amount present in barley, the embryo contained 5.4 per cent., the endosperm but 0.3 per cent.; in the same barley, after ten days' germination, 24.2 per cent. were found in the embryo and 2.2 per cent. in the endosperm. Simultaneously with the formation of cane sugar, a small amount of invert sugar seems to be formed by the hydrolysis of the former, and this accounts for the presence of glucose and lævulose in the finished malt.

In germinating barley we are absolutely obliged to produce a certain quantity of rootlets; these, after the malt has been dried, are removed, and as they are a comparatively valueless material, they represent a loss, for the substance from which they are formed is derived from the endosperm. By different methods of germination very different quantities and weights of rootlets may be produced, and it is therefore the maltster's aim to keep root-production to as low a limit as is consistent with securing the proper modification of the barley. The comparative state of wetness or dryness in which the barley is germinated has great influence upon the amount of root-production; if the barley is worked too wet, and especially at too high a temperature as well, root-production proceeds at a high rate. In such a case the growth of the acrospire and the modification of the contents of the endosperm do not keep pace with the root-formation.

The Barley during the Whole Process of Germination Respires.—During the period of germination barley absorbs oxygen and exhales carbon dioxide, and any undue accumulation of this latter gas either stops growth or seriously impedes it. Provision has, therefore, to be made for removing the carbon dioxide and supplying fresh oxygen. This necessary aëration of the barley is secured by the continual turning it undergoes, and by securing the frequent renewal of the air in the malthouse by proper ventilation. At the same time the amount of ventilation must not be too great, for, in that case, moisture will be removed from the barley. Schütt[1] found that, in a period of nine days' germination, about 196 cubic feet or 36 lbs. of carbon dioxide were given off by a quarter of barley, which is equivalent to a loss of 22.5 lbs. of starch; during the same period 11.5 lbs. of water were formed.

[1] *Woch. f. Brau.*, 1887, p. 673.

The amount of the carbon dioxide evolved was found to be slightly greater than that of the oxygen inhaled; this shows that intramolecular respiration takes place to a limited extent. According to Day,[1] a definite relation exists between the amount of carbon dioxide evolved and the water formed.

As germination proceeds, a considerable portion of the hitherto insoluble proteid bodies is converted into soluble forms; thus Hilger and Van Becke found that, after five days' germination, the soluble proteid matter of a sample of barley was three times as large as that of the original grain, and was nearly four times as large in the finished malt. The amount of acidity of the barley also augments during germination; Belohoubek found an increase of from 0.34 per cent. (reckoned as lactic acid) in the original barley to 0.60 per cent. in the green malt. Prior[2] determined the respective amounts of the free and fixed organic acids, of the primary phosphates, and of the total phosphoric acid, in two samples of barley during germination. After two days' germination the volatile acids had slightly increased in one sample, while in the other they remained the same. The fixed acids had shown a not inconsiderable diminution, and the primary phosphates had increased in an almost exactly proportionate amount. This he considered to be due to the action of the fixed acids on the secondary and tertiary phosphates contained in the grain, which were converted into primary phosphates according to the action expressed in the following equation:—

$$K_2HPO_4 + C_2H_4(OH)COOH = KH_2PO_4 + C_2H_4(OH)COOK.$$
Secondary potassium phosphate.   Lactic acid.   Primary potassium phosphate.   Potassium lactate.

And similarly with the other phosphates of magnesium and calcium. This took place chiefly between the second and fourth day, and caused the rise in total acidity noticeable at this period. An apparently slight diminution in the total amount of phosphoric acid was also observed, and this was thought to be due to an increase in the total weight of the grain, caused by the combination of a small amount of the absorbed oxygen with some of the constituents of the grain. From the second to the fourth day there was a rise in the total acidity, chiefly caused by the formation of primary phosphates; meanwhile the volatile acids remained about the same, but there was a slight decrease in the amount of the fixed acids. A slight increase in the total phosphoric acid also occurred, but this was probably only apparent, being due to some of the combined oxygen being dissociated. From the fourth to the sixth day the total acidity of one malt had augmented, while that of the other had slightly diminished. The

---

[1] *Journal of the Chemical Society*, 1880, p. 645.
[2] *Chem. u. Phys Malz. u. Bier.*, p. 112.

augmentation was not due to the volatile acids, for these were the same in amount, but to the fixed acids, which appeared to have been released to some extent from the bases which they had before converted into primary phosphates; apparently, then, a reversal of the original action had taken place. From the sixth to the eighth day there was a very slight decrease in the total acidity. The volatile acids were not increased in amount, but the fixed acids had slightly decreased, and a corresponding increase had again taken place in the primary phosphates. The total amount of phosphoric acid appeared to be slightly higher, probably owing to a diminution in the weight of the other constituents of the grain. The total acidity of the rootlets was twice as great as that of the green malt itself.

Growth of Mould.—Another point of the greatest importance is to prevent the growth of mould while the barley is on the floor. It is impossible to grow barley entirely free from mould, but its growth should be kept down to the lowest possible limit. This is effected by the removal of damaged corns, and by exercising care in steeping, so that the vitality of the corns is neither depressed nor destroyed, by securing an abundant supply of fresh air in the malt-house, and by working the barley without an undue excess of water.

Couching.—We now proceed to the actual operation of germination on the malthouse floor, taking for example what may be considered as the average course of a steeping of barley. The first operation, after the barley leaves the cistern, is known as "couching," so called because, in the days when the duty was collected on the malt, the barley, on leaving the cistern, was placed in the couch-frame, levelled, and there gauged by the Excise officers. The steeped barley is levelled up into a heap of twelve or fourteen inches in depth, either on the couch, where such still exists, or, where it does not, on the floor. Here it is left for a period of about twenty-four hours, during which time it begins to exhibit the first signs of vitality. Now that it is removed from the water in which it was immersed, and is surrounded by air, it is able to breathe, and the evidence of respiration is soon afforded by the rise in temperature of the heap. The barley during this period sweats, and gives off a distinctly pleasant odour. The barley while on the couch is turned, as a rule, every five or six hours, in order to secure proper aëration, equability of temperature, and an even distribution of moisture. When the heat of the heap has risen to from 60° to 63° F. it is broken down, that is, extended in superficial area and reduced in depth. At this period it will be found that many of the corns show the first visible signs of growth, little white protuberances having made their appearance at the ends of the grains where the germ is situated. When this occurs the barley is said to "chit."

Flooring.—This portion of the process may be said to commence with the breaking down of the couch, at which time the piece is called a "one-day floor;" the next day it becomes a "two-day floor," and so on. The parcels of malt are called "pieces," and each of these is turned, as a rule, twice a day, but this will vary according to circumstances. The turning is effected by means of large wooden shovels, and is an operation requiring a considerable amount of dexterity, which can only be obtained by long practice. When properly effected, the mass of grain is entirely turned over, that portion which formed the surface of the original piece becoming the undermost layer in the fresh piece. In this way a certain uniformity in moistness and temperature is secured. Each succeeding piece is turned on to a portion of the floor farther away from the cistern; consequently, it gradually travels towards its destination, the kiln. The thickness of each succeeding piece is determined by the temperature of the previous one; if the temperature of that was too high, the next is spread out into a thinner layer, or *vice versâ*.

Sprinkling.—Probably by about the fifth day the rootlets will begin to lose their turgid appearance, as evidenced by flaccidity, a loss of their bright colour, and an inclination to turn yellow. When this appearance presents itself it is a sign that the piece is becoming deficient in moisture. This deficiency is supplied by sprinkling water over the piece at the rate of from four to six gallons of water per quarter of barley, according to the condition of the piece. The sprinkling is usually effected by means of a can, provided with a long spout perforated with a number of fine holes; or, where water is laid on for the purpose, by means of a hose and fine rose. Probably one or more sprinklings may be found necessary, but this is a point which must be left entirely to the judgment of the maltster. Bisulphite of lime is often added to the sprinkling water at the rate of two pints per barrel in cold weather, and four pints in warm; it serves to check the growth of mould.

Growth.—As before stated, chitting will have commenced in an average steep at the end of twenty-four hours from the removal of the barley from the cistern. On the following day many of the corns will show from one to three distinct rootlets; by the third day nearly the whole of the grains will have from three to four roots; and on the fourth, rootlets to the number of four or five will have developed. It is the object of the maltster to obtain a short, abundant, bushy root consisting of five or six rootlets. Too long rootlets, with too short a growth of acrospire, are a source of waste, for they show that the modification of the endosperm is not keeping pace with the growth of the rootlets.

By the evening of the third day the acrospire will have commenced to move up the back of the grain, and will be distinctly visible; in another three or four days it will be half-way up, at which stage sprinkling is generally commenced; in another two or three days it will have reached a length of two-thirds of the grain. When this point is reached the withering is commenced, during which the acrospire either remains entirely stationary or grows very slightly.

Withering.—By means of this, the maltster endeavours to get rid of as much water as possible before placing the green malt on the kiln. In this stage the piece is allowed to remain for twenty-four hours unturned, by which treatment it collects heat and loses moisture, as is shown by the withered appearance which the rootlets gradually assume. Some maltsters heap up the malt into a thicker piece during the withering, in order to increase its temperature. This is not advisable if the piece exhibits signs of mouldiness; indeed, in such cases it is advisable to get the malt on the kiln as soon as possible, even at the risk of its containing a little too much moisture. Since, during the withering process, the piece is not so freely ventilated, the carbon dioxide exhaled by the corn collects, and, by arresting respiration, checks growth. Under these conditions the action of the enzyme still appears to proceed, it continues to dissolve the walls of the starch-cells, and to form sugars from the starch. The sugars formed at this time, not being oxidised so rapidly as they would under conditions of freer respiration, accumulate to some extent during the withering process.

Pneumatic Malting.—The malthouse of the present day has pretty nearly the same primitive construction which it has had from times immemorial; the improvement in the appliances employed in malting cannot be said to have kept pace with that of the appliances employed in other manufactures—compare, for instance, the machine-shop of to-day with that of five hundred years ago. During the last twenty years the spirit of reform has extended itself in a small degree to the malthouse, and some little improvement in the hitherto primitive condition of things has manifested itself. This has consisted in the introduction of the conical steeping cistern, the various improved forms of apparatus for cleaning and grading the barley, machinery for the removal of broken corns, &c. But the first really radical reform was introduced by Galland in 1874, who designed an apparatus which was intended to obviate the large amount of hand-labour expended in the constant turning of the barley. The steeped barley was placed in a thick layer on a floor, that consisted of perforated iron plates, and through which moist, cool air was continually drawn and compelled to traverse the barley. In this way the grain was abundantly supplied with oxygen, and at the same time the carbon dioxide was effectually removed. As the air was saturated with moisture, it was expected that it would exercise no drying action on the barley; but, owing to the friction generated, some loss of moisture ensued. This principle has been preserved in all the pneumatic methods of malting which have been since proposed; and obviously, if such a method were found practicable, it would be possible to make malt all the year round, because the

air used in the process could be cooled down at any season of the year to the temperature required for malting.

Galland's idea was to do away with the turning altogether, but this was found impracticable, for the mass of germinating barley gradually became matted and felted together, and more and more impervious to the air-current; the rootlets also penetrated and filled up the perforations in the floor, thus offering another obstacle to the air-current. In consequence of this a considerable degree of force was required to drive the air through the mass of grain, considerable friction ensued, and the heat developed in this way tended to remove water from the barley, even though the air employed was saturated with moisture. Some means had therefore to be provided for turning the barley and keeping it in a loose condition, and as the pieces in the apparatus were from thirty to forty inches in thickness, it was impossible to do this by hand. To remedy this defect Saladin invented a mechanical arrangement for turning the barley. It consisted of a number of vertical screws, which, in addition to having a revolving motion on their own axes, travelled from one end of the apparatus to the other. This turning apparatus was put in motion every eight or twelve hours, when it effectually turned the grain and left its surface perfectly level. The apparatus in this improved form has proved fairly successful, and is employed to a limited extent in France and Germany.

Drum Pneumatic Malting.—Galland fully recognised the imperfections of his original method, and, in order to remedy them, proposed to vegetate the grain in a slowly revolving drum, provided with various air-passages, by means of which a constant current of moist air at a suitable temperature could be continuously passed through the germinating grain. Since the barley was being continuously turned, it was kept in a loose condition and not permitted to felt together. This system was further improved by Hemming, and now bids fair to become, at no very distant period, the universal method of malting. The general appearance of a pneumatic malthouse is shown in Fig. 62. One of the drums is shown in section in Fig. 63, together with the arrangements for providing for the circulation of the moistened air. The passage L, from which the air is drawn, communicates with a tower nearly filled with coke, which is kept moist by a small sparging apparatus continuously delivering a spray of water; this may be either cold or warm, according to the general temperature at the time. In passing through this layer the air is either cooled or warmed, and at the same time saturated with moisture. From this point it passes up the tube past the valve D, which can be opened or closed at will, to a number of passages, c, on the inner surface of the drum, best seen in Fig. 64. (The letters indicate the same portions of the apparatus in both diagrams.) The drum T is a cylinder of sheet metal which revolves on its axis, and is supported on the wheels $b\,b$. The rotation is effected by a worm acting upon a series of cogs which surround the circumference of the drum. The drum is provided with an aperture, which can be closed by a lid, for the introduction of the steeped barley and the removal of the green malt. A tube m is fixed along the axis of the drum; this communicates with the valve D_1, and with the passage S_1. From this last the air is drawn by means of the exhaust fan Z. The steeping cistern is shown at W_1, one cistern sufficing for two drums; the shoot, as seen in

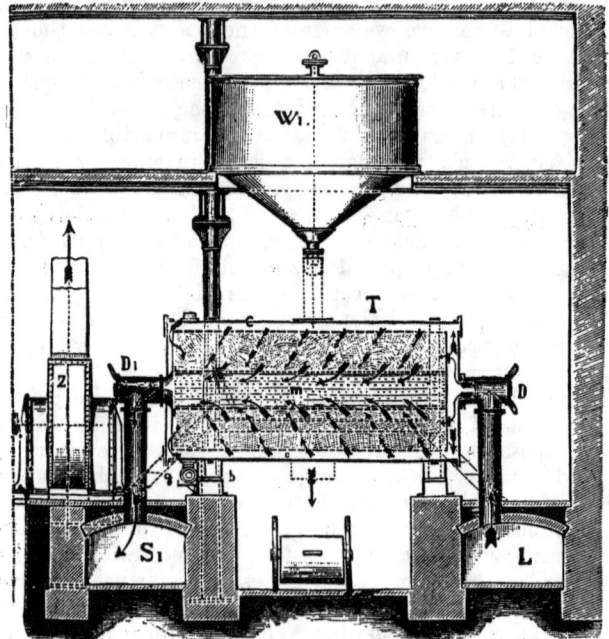

FIG. 63.—Section of Pneumatic Malting Drum.

Fig. 64, being movable. The direction in which the air passes through the germinating barley is shown by the arrows. The slow rotation of the drum (once in forty to forty-five minutes) keeps the grain in a loose condition and readily permeable to air. An air arrangement is also provided by which the air supply can be shut off from the passage L and drawn from the outside atmosphere, either directly or after filtration through an air filter.

Advantages of the System.—The greatest of these is the command which it gives to the maltster in completely controlling the temperature of his growing barley, in adjusting the degree of aëration to which it shall be subjected, and in regulating the amount of moisture which the air supply shall contain. All damage to the corn by turning with a shovel, or by the feet of the workmen, is avoided. Owing to the abundant supply of fresh air the malt is produced absolutely free from mould, and this is no doubt considerably aided by the purification (removal of germs) which the air undergoes in its passage through the coke tower. There is also an enormous saving in space—only one-sixth of that required for the ordinary growing floors. Hand-labour too is reduced to a

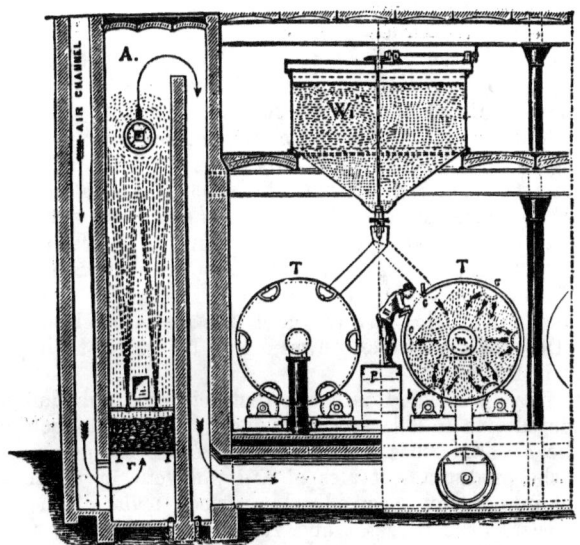

FIG. 64.—Section of Pneumatic Malting Drum.

minimum. Against this must be placed the expenditure for power in driving the exhaust-fans and drums, together with the pumping of the water for sparging the coke.

The Drying Kiln.—This consists of a building having a roof in the shape of a cone, which terminates in a short chimney, usually surmounted by a cowl. The building is divided into two chambers by the drying floor, which may consist of wire gauze, perforated iron plates, or perforated tiles. In the space beneath the drying floor is the arrangement for supplying heat, and this, in its simplest form, is merely an iron cage in which the fuel is burnt. In order to equally diffuse the heated air, a large plate, generally made of iron, and covered over with bricks or some other non-conducting material, is placed over the fire. Openings are provided in the walls of the room under the drying floor; by the opening and closing of these, together with the manipulation of the fire, the temperature of the air passing through the green malt is regulated. Occasionally drying kilns have two floors, one above the other. The green malt is first placed on the uppermost, and, after having attained a certain degree of dryness, is then shot down to the lower floor, where the operation of drying is completed. In this case provision has to be made for the entrance of cold air into the space between the two floors, or some arrangement provided by which the bulk of the heated air coming from the lower floor can escape without actually passing through the grain on the top floor. In such kilns there is undoubtedly a saving of fuel.

In order to increase the volume of air passing through the green malt, especially in the earlier stages of the drying process, it has been proposed by Mr. Free to place a Blackman fan at the apex of the roof. This plan, he says, has been carried out with a considerable degree of success in his own maltings.

Drying.—The objects of this process are to effectually arrest further growth in the green malt by the withdrawal of water, which has the effect of rendering the malt stable, so that it will keep on storage, and also to confer flavour on it. In no other department of malting is so much care required as in the kilning, for the character of the beer produced by a malt is most essentially influenced by the way in which the malt has been dried. Many of the troubles of the brewer are to be traced to improper treatment in this stage.

Malts may be divided into two distinct types, pale and high-dried; they derive to a large extent their different characteristics during the kilning process. The malts of the former type are used for the production of ales of a dry flavour, or which have to be kept in store for a considerable period; malts of the latter type for beers of a sweeter and fuller nature, which, as they are rapidly consumed, are technically known as "running ales." Between these two different classes of malt there are a number of intermediate varieties, partaking more or less of the characters of the one type or of the other.

In all cases the drying of malt is commenced at a temperature not exceeding 100° F., and this should not be exceeded on the first day, care being taken that a large quantity of air passes through the malt. The heat is gradually raised during the second day, until, at its close, the temperature of the malt has risen to 110° or 120° F. During the next twelve hours the heat is gradually raised to the temperature at which it is desired to finish the malt at. This will vary from 190° to 230° F., according to the character of the malt being manufactured; it is kept at this point for six hours before being taken off the kiln. This, which is generally known as the "curing stage," is a most important portion of the operation, and should never be neglected.

The object of the very slow application of heat and abundant air supply during the earlier stages of the drying process is to remove the superfluous water, and to effect this object without either destroying too much of the diastase or rendering the malt hard and steely. Diastase in the moist condition is extremely sensitive to heat, but when perfectly dry it will bear exposure to the comparatively high temperatures necessary for conferring those desirable qualities which are imparted to the malt during the curing stage. If the green malt is exposed to too high a temperature at first, the starch in the external layers of the endosperm becomes gelatinised, and dries into a horny mass, which effectually prevents the proper escape of moisture from the interior of the grain.

Changes Effected during the Kilning Process.—During the first stages of the drying process the green malt loses from 30 to 35 per cent. of water; it then becomes what is known as "hand dry." This is accompanied with a considerable shrinking in volume. During this period changes similar to those which are brought about by germination still continue to a limited extent. Thus Prior found that the diastatic power in a malt had increased from 88.9 to 134, and that the reducing sugars had similarly increased from 12.58 to 13.74 per cent. The total acidity also increased, in one sample from 0.60 to 0.74 per cent., in another from 0.56 to 0.64 per cent. (reckoned as lactic acid), owing to an increase in the fixed organic acids and in the primary phosphates. As the drying proceeds the remainder of the water is lost, but a series of changes take place which are largely governed by the amount of water which the malt contains at the earlier part of this period. When, as in pale malt, this is small, all those changes of a vegetative nature which had hitherto been taking place are arrested, and the malt gradually loses its raw flavour, but never acquires that highly aromatic flavour characteristic of the highly dried malts. As a very considerable portion has been removed at a comparatively low temperature, the diastase is not seriously impaired; and since during the subsequent stages little moisture is present in the malt, and as dry diastase is not affected by heat to nearly such a great extent as it is when in a moist condition, the diastatic powers of such malts are fairly high at the finish. But when, on the other hand, the heat is permitted to rise to temperatures towards 150° F. while the malt still contains from 12 to 15 per cent. of water, then the formation of those products on which the aroma of malt depends commences, and the more this stage is prolonged the greater is their amount. According to Thausing,[1] the formation of the aromatic principles commences at a temperature of about 150° to 170° F., and it is impossible to produce aroma in a malt when its

[1] *Malzber. u. Bierfabrik.*, p. 380.

moisture has fallen below 7 or 8 per cent. Consequently, the brewer, by keeping these principles in mind, has the power of varying the character of his malt, as regards flavour, as he chooses. The contents of the corns become browner in colour according to the degree of heat to which the malt has been subjected in this middle stage. This deepening in colour has been attributed by various observers to the alteration of very different constituents of the malt, but its formation depends, to a certain extent, upon the alteration in the proteid constituents, as was first suggested by Habich. This is also borne out by the author's own experiments, who found that the albumoses derived from high-dried malt were always considerably darker in colour than those derived from pale malt. Caramel is evidently produced, to some slight extent, probably from the lævulose, the amount of which may be increased during the drying process by the hydrolysis of cane sugar by the acids of the malt. During the drying the diastase loses much of its activity, and this is advantageous, for, if such were not the case, it would be impossible to brew full-bodied beers. The diastatic power of green malt diminishes from 100° to 120° on Lintner's scale to the average of 45° to 30° under the ordinary conditions of drying; but with careless manipulation it may sink much lower than the latter figure, or even be nearly destroyed.

Heaping Up.—After the malt has passed through the curing stage it is generally heaped up for a few hours; this is also said to increase its aroma.

Removal of Rootlets.—After the completion of the drying process, the rootlets, technically known as the "coombs," are removed from the malt. This is effected by workmen, who wear heavy boots for that purpose, treading the malt. In this way the brittle rootlets are easily broken off; they are then removed by passing the malt over a screen. It is the custom of some maltsters to store malt with the rootlets still attached, but this is an objectionable practice, since malt coombs have a much more powerful attraction for water than the malt itself.

Storage.—In no other department of malting or brewing has such short-sighted policy been pursued as in the storage of malt. The maltster or the brewer, after all the trouble he has taken to remove, as far as possible, the last traces of water from the malt, and well knowing its extremely hygroscopic nature, often leaves it simply lying in heaps on the floor of the store-room; or, at best, merely covered with a layer of coombs. Malt, to preserve its good qualities intact, should be stored in bins, which are made as nearly air-tight as possible; these may be constructed of wood, and lined with sheet iron or zinc, or air-tight rooms may be built of brick, and lined with cement, as the malt silos so frequently seen in Continental malthouses are. According to Brand, malt should never be stored while hot, especially if in large masses; for, under such circumstances, it gathers heat, its good qualities become depreciated, and it is liable to become rancid. A slight oxidation process, which leads to the formation of carbon dioxide and water, takes place in malt when first stored, and in this way the malt acquires a slightly higher percentage of water. This oxidation has much to do with the "mellowing" which is brought about in the malt during the first few weeks of its storage; it is well known that malt does not yield good results in the mash-tun before it has undergone this mellowing process. Malt should be stored at an even temperature; if one portion of the mass is warm and another portion cold, the moisture distils from the warmer portion and condenses in the colder; consequently there is a tendency in the latter portion to become slack, and in extreme cases even to become mouldy. In spite of all the best conditions of storage, malt depreciates in quality after a time; it is probably at its best two or three months after being made. This is a point which tells strongly in favour of pneumatic maltings, in which malt can be made all the year round.

Chemical Examination of Barley.—In addition to a physical examination, barley is often subjected to a chemical analysis as well. In this way the amounts of moisture, of starch, of nitrogenous matters, soluble and insoluble, and of ash which a given sample contains are determined, and also its degree of acidity.

Moisture.—The amount of water contained in barley generally ranges from 12 to 18 per cent., the average being 15 per cent. Any sample which contains more than 20 per cent. would be considered damp.

Starch.—This is by far the most important constituent of barley, for on the amount of starch which a given sample contains chiefly depends the amount of extract it will yield. Barley contains, as a rule, from 60 to 66 per cent. of its weight of starch.

Nitrogenous Matters.—Barley contains from 8 to 15 per cent. of nitrogenous matter, the average being about 10 per cent. Part of this is soluble in water, but by far the larger part (80 to 85 per cent.) is insoluble. Some of the soluble nitrogenous constituents are coagulable by heat; these should, according to Kukla, amount to about one-half of the total soluble nitrogenous matter. Barleys which contain lower proportions of coagulable proteids than this require to be worked slowly on the floor in order to secure perfect modification. Samples containing a large total amount of nitrogenous matter were considered at one time to be unfitted for malting, it being thought that such grain invariably yielded hard, steely malts. It is now pretty generally conceded that it is not the absolute amount of nitrogenous matter, but the state in which it exists in the barley, which is of paramount importance. The amide bodies are contained in very small quantities in barley; Märker and Farsky found the amounts contained in two barleys to be 0.275 and 0.169 per cent. respectively. The ammonia amounted to 0.054 and

0.012 per cent. in the same samples; this latter body is therefore contained in still smaller quantities.

Acidity.—An aqueous extract of barley is always more or less acid, and this acidity was formerly attributed solely to the presence of lactic acid, hence the custom which still prevails of stating the amount of acid contained in barley, malt, or wort in percentages of lactic acid. According to Prior,[1] the acidity of barley mainly depends on the presence of the primary magnesium, calcium, and potassium phosphates, and to a lesser extent on that of fixed and volatile organic acids. He states that the respective proportions of these bodies in two samples of barley, expressed in percentages of lactic acid on the water-free substance, were as follows:—

	I.	II.
Volatile organic acids	0.07	0.05
Fixed organic acids	0.06	0.05
Primary phosphates	0.29	0.25
Total	0.42	0.35

Mineral Constituents.—Barley, on incineration, leaves behind from 2.5 to 3.0 per cent. of ash, reckoned on the dry substance. According to Wolff, the ash of barley consists of the following substances, expressed as percentages on the ash:—

Potassium	20.92
Sodium	2.39
Lime	2.64
Magnesia	8.83
Iron oxide	1.19
Phosphoric acid	35.10
Sulphuric acid	1.80
Silica	25.91
Chlorine	1.02
Total	99.80

The preponderating bases are potash and magnesia, and the predominating acids phosphoric and silicic. The potassium and magnesium phosphates are the most important constituents of the ash, since they serve as mineral food for the yeast.

Malt Substitutes.—In former times malt was invariably produced by germination, but within the last few years many preparations—so-called "malts"—made from rice and maize, have been placed on the market. In these the more or less complete gelatinisation of the starchy contents of the grain has been brought about by the action of moisture and heat. It has been recently proposed to make malt by growing on barley or other grain a mould fungus, in the same way as the Japanese manufacture koji from

[1] *Chem. u. Phys. d. Malz. u. d. Bier.*, p. 40.

rice. Several of the mould fungi secrete diastatic enzymes in great abundance.

Germinated Barley Malt.—Malt is divided into several classes according to its colour; thus we have white, pale, high-dried and amber malt; crystal and imperial; brown and black. In the first four of these the requisite modification is produced by variations in the method of drying. In the manufacture of white malt, comparatively little of which is now made, the best and palest barleys are employed; these are worked with great care and dried at a low temperature. Pale malt is also made from superior barleys, but dried at a higher temperature than white malt. For high-dried malt, inferior or even slightly damaged barley is often employed; the curing takes place at a considerably higher temperature than with the foregoing malt. Amber malt is made from similar corn; it is somewhat more highly dried than the last, and a quantity of wood is thrown on the fire towards the end of the process. Crystal malt is dried in an entirely different manner. After being grown and withered it is placed in a wire cylinder which is rotated over a fire, consequently the grain is dried at a very high temperature. Imperial malt is made in a similar manner to amber, the heat of the kiln being allowed at the finish to rise to from 240° to 270° F., oak or beech wood being used as fuel. The use of brown or blown malt, formerly much in demand for making stout, is dying out. The barley for this was germinated in the usual manner, but not allowed the usual length of time on the withering floor. It was then placed on the drying kiln in a layer not exceeding an inch and a half in thickness. Moderate heat was applied at first until much of the moisture had been driven out, when the heat, which was obtained from burning wood, either oak or beech, was suddenly and intensely augmented. Under the sudden application of the heat the grains swelled some 25 per cent., and the malt became embued with a strong empyreumatic odour, derived from the products of combustion. Since in this method corns which have not vegetated do not swell, only barleys which possess good vegetative powers should be employed in the manufacture of brown malt.

Black or Roasted Malt.—This, which now almost entirely replaces the brown or amber malt formerly used in the production of black beers, is prepared by roasting malt in a cylinder, constructed something after the fashion of those in which coffee is roasted. Though well-made malt yields a black malt of a superior character both in point of colour and flavour, yet inferior malts are often used for this purpose, and at times even raw barley. A good sample of black malt is of a uniform chocolate colour, and should not be black. The individual grains should be distinct and clean; if matted and run together and the corn burst, neither the colour nor flavour will be good. Black malt is of an exceedingly hygroscopic nature; no large stock of it, therefore, should be kept in hand. It is best prepared the day before being used, but this entails the necessity of having a roasting apparatus on the premises.

Quality of Malt.—In estimating the quality of a malt we depend upon certain physical and external indications, combined, if possible, with a chemical examination. The first of these enables us to determine the characteristics which it has derived from the barley employed in its manufacture, and also several other conditions induced during the process of malting.

Growth.—This, which affords a valuable indication as to the amount of modification which has taken place in the endosperm, is ascertained by noting the length to which the acrospire has grown up the back of the corn. Those corns in which the acrospire has not started at all were evidently dead; they are termed "idlers." Growth should be regular; in a given sample all the acrospires should be grown pretty nearly to the same length. The degree of growth of the acrospire is stated in terms of the length of corn; thus, if it is half the length of the corn, it is said to be half-way up; if three-quarters the length, three-quarters up; and if it extend beyond the end of the corn, it is said to be grown out. The length usually found to coincide with the proper modification of the endosperm is when the acrospire extends from two-thirds to three-quarters up the back of the grain, but now-a-days the maltster relies more upon the condition of the endosperm itself than upon the length of the acrospire. To detect regularity in growth a number of corns are taken promiscuously from a sample, say several lots of 200 each, and the number of corns which have grown to the various different lengths of one-eighth, one-quarter, three-eighths, one-half, five-eighths, three-fourths, or grown out are determined by inspection. The different groups of figures so obtained divided by two, give the percentages.

Thus in 200 corns from a sample the numbers at the various degrees of growth were as follows :—

Degrees of growth	0	$\tfrac{1}{8}$	$\tfrac{1}{4}$	$\tfrac{3}{8}$	$\tfrac{1}{2}$	$\tfrac{5}{8}$	$\tfrac{3}{4}$	$\tfrac{7}{8}$	grown out
No. of corns	4	6	14	20	38	78	34	4	2
Percentage	2	3	7	10	19	39	17	2	1

Such a sample as this exhibits extreme irregularity in growth. A good sample of malt should be evenly grown, the acrospire should be from two-thirds to three-quarters up the back, and it should not contain more than 2 per cent. of idlers.

Condition of Contents.—The condition of the contents of the endosperms is of immense importance; they should be tender and friable; the corns should be crisp, and, when bitten, should crumble between the teeth; a broken corn, when drawn across a board, should leave a mark such as a piece of soft chalk would do. Hard glassy corns indicate that either a bad sample of barley has been malted, or that there has been mismanagement in the drying process. The number of friable, glassy, and medium corns may be ascertained by cutting a number through with the farinatome used in examining barley, and the percentage of the different classes determined in the same way.

Percentage of Moisture.—Malt on leaving the kiln should be practically free from moisture—at all events, it should contain less than 1 per cent.; it is, however, an extremely hygroscopic substance, and consequently absorbs water with great avidity, or, in other words, it soon becomes "slack." When malt has absorbed even small quantities of water it rapidly deteriorates; it takes on a series of unexplained changes which render it in a short time quite unfit for the production of first-class beers; in fact, slack malt is probably one of the most frequent causes of trouble in the brewery. Though it may be improved to a slight extent by re-drying, it never regains its original good qualities. The amount of water contained in a malt is best ascertained by a chemist, who has all the apparatus necessary for such a determination at hand; but considerable information can be gained by the way in which a malt "bites." In slack malt all the crispness is gone, the contents of the corn are soft, and convey a general impression of staleness, difficult to describe, but easily appreciated by practice. As it is almost certain that the objectionable changes in malt caused by excess of moisture begin before slackness can be determined in the way just described, the only certain method of ascertaining when a malt contains a detrimental quantity of moisture is by an actual moisture determination in the laboratory.

Broken Corns.—These, when they occur in the sample of malt in anything beyond the smallest proportions, exercise a very prejudicial effect upon the beer brewed from it: their presence is wholly undesirable, and in a good malt they should not exceed 2 per cent.

Age.—Malt should not be used until it is about six weeks old. If used when too new, it causes trouble in the clarification of the resulting beer. Malt over four months old, when kept under ordinary storage conditions, and even before then if kept under bad ones, will become slack; it will then exhibit all the ill effects due to this condition. When kept in air-tight rooms or bins it can be preserved without deterioration for considerably longer periods. Much more attention has been paid in recent times to the provision of proper air-tight chambers for the storage of malt, and this has been followed by a marked improvement in the quality of samples which have been stored for a considerable time. The introduction of pneumatic malting, which is gradually taking place, will remove in course of time the necessity for long storage and its attendant evils.

Odour and Flavour.—Good malt has a pleasant aromatic odour; if it possesses a faint and unpleasant, or a fusty, mouldy, sour, or rancid smell, it cannot be used with any degree of confidence. The flavour, which depends very materially on the way in which the drying process has been carried out, is best detected by chewing a few grains; it should not be raw, for that denotes insufficient heat during the curing process, but sweet and aromatic, the aroma being developed to the greatest extent in those samples which have been submitted to a high temperature for several hours at the end of the kilning process. The aroma is not so marked in pale malts, which are finished off at a lower temperature. A burnt, bitter flavour points to mismanagement in the kilning process.

Size.—It is possible to make equally good malt from either large or small corn, provided it possesses the other requisite qualities; but of two samples, equal in other respects, the one with the larger corns will yield the more extract, since there will be a smaller proportion of husk to contents.

Weight.—Much importance used to be formerly attached to the weight of malt. From 40 to 44 lbs. a bushel is the usual weight met with; and undoubtedly heavy malt, provided it possesses the other necessary qualities, yields the best extract. Samples should be carefully examined, in order to ascertain that the malt does not derive this quality from some imperfection, such as steeliness.

Cleanliness.—Malt should not contain gross impurities, such as stones, dirt, fragments of iron, or foreign seeds in any quantity. All these should be removed from the barley before steeping. The surface of the malt is always more or less contaminated with bacterial organisms, mould, and yeast fungi, which were deposited on the barley when growing in the field, and have survived the heat of the drying process. Machinery is now obtainable which, by a combination of screening and brushing, removes a considerable proportion of the adherent dirt, both of an inorganic and organic nature, and this cleansing is exceedingly advisable. By placing a small quantity of malt in a test-tube containing sterilised water, plugging up the tube with sterilised cotton-wool, and keeping the whole at the ordinary temperature for a few days, a microscopic examination of the liquid will show what a number of the germs of different organisms are deposited on malt. Malt should be free from rootlets;[1] their presence indicates defective screening.

Chemical Examination of Malt.—This affords exceedingly valuable information, not only as to the way in which the malting process has been conducted, but also as to the way in which the malt will comport itself in the brewery. The report of an analysis of a malt usually gives in percentage terms the moisture contained in the sample, which is the index of its slackness; the amount of

[1] A concentrated extract of malt rootlets is frequently added to wort as "yeast food."

extract the malt is capable of yielding, together with the general character of the extract yielded; the amount and condition of the nitrogenous matters of the malt; its diastatic power; the amount of ash it contains; and its acidity.

Moisture.—In samples directly off the kiln the moisture should not exceed 1 per cent., and malt by the time it contains 5 per cent., or even less, will be considerably deteriorated in quality.

Extract.—The amount of extract yielded by different malts varies considerably; fine, plump malt, when of good quality, will naturally yield more extract than thin samples. The amount of extract yielded by malt in this country is always stated in pounds of brewer's extract per quarter, hence the cost of each pound of extract can be obtained by dividing the price per quarter by the number of pounds of extract yielded. Thus, if a malt yield eighty pounds extract, and costs thirty shillings a quarter, each pound of brewer's extract costs fourpence halfpenny per pound. But the quality of the extract must also be taken into account, for it is of quite as much importance as its quantity. Malt yields from seventy-five to ninety-five pounds extract per quarter, but this yield varies considerably in different seasons. The report of a chemical analysis of a malt, in addition to the amount of extract yielded, contains particulars as to the colour of the wort yielded, its odour, the way it filters—bright or cloudy—the way in which it breaks on boiling, and the proportion of matter fermentable during the primary fermentation.

Colour of Wort.—This naturally depends upon the kind of malt, whether it is pale or high-dried, and is generally stated as pale, medium, dark, &c. As it is now possible to measure shades of colour with the utmost degree of nicety by means of the ingenious tintometer of Mr. Lovibond, the tint of the wort of a given strength from a sample of malt should be given in degrees of this instrument.

Nitrogenous Constituents.—The total amount of the nitrogen contained in the malt is first ascertained, and the figure thus found multiplied by a factor, generally 6.25, and the quotient given as the total nitrogenous matter of the malt. The factor 6.25 is employed under the erroneous assumption that all the nitrogenous matters of malt contain 16 per cent. of nitrogen, but it is now well known that the amide constituents contain considerably larger percentages of nitrogen. Figures obtained in this way, however, serve for purposes of comparison. The amount of nitrogenous matter soluble in water is then found in a similar manner, and deducted from the total nitrogenous matter found before; this gives the percentage of the insoluble nitrogenous substance. A malt contains a slightly less amount of nitrogenous matter than the barley from which it was made, since a portion of the nitrogenous substances passes into the rootlets, which are removed from the finished malt. During the malting process a large quantity of the insoluble nitrogenous matters present in the barley become converted into soluble modifications. Attempts have been made to subdivide the nitrogenous constituents of malt into various groups by Ullik, Bungener and Fries, and others. The following is a comparative analysis of malt and the barley from which it was made by the two last-mentioned chemists :—

Barley.

	Nitrogen. Per Cent.		Factor. Per Cent.		Per Cent.
Nitrogen of total nitrogenous matter	1.69	×	6.25	=	10.6
Nitrogen of total soluble nitrogenous matter	0.407	×	6.25	=	2.54
Consisting of albumen	0.161	×	6.25	=	1.00
peptones	0.040	×	6.25	=	0.25
amides	0.026	×	6.25	=	1.29

Malt.

	Nitrogen. Per Cent.		Factor. Per Cent.		Per Cent.
Nitrogen of total nitrogenous matter	1.58	×	6.25	=	9.87
Nitrogen of total soluble nitrogenous matter	0.824	×	6.25	=	5.16
Consisting of albumen	0.230	×	6.25	=	1.44
peptones	0.060	×	6.25	=	0.37
amides	0.534	×	6.25	=	3.34

The nitrogen figures have been multiplied by the factor 6.25 for purposes of comparison.

The following figures were obtained by the author some years since from samples of malt from different sources, all of which were yielding good results in the brewery :—

	Per Cent.	Per Cent.	Per Cent.	Per Cent.	Per Cent.	Per Cent.
Total nitrogenous matter	8.1	9.2	9.4	10.0	10.1	11.8
Soluble ,, ,,	3.8	3.6	3.9	4.5	4.5	4.1
Coagulable ,, ,,	0.9	0.7	0.8	0.6	1.7	1.4
Proteoses and peptone	0.7	0.8	0.5	0.9	0.7	0.6
Amide	2.2	2.1	2.6	3.0	2.1	2.1

The amount of the soluble nitrogenous matters was obtained by estimating the nitrogen of a cold-water extract of each malt; this was then reduced to a percentage of the weight of the original malt, and multiplied by the factor 6.25. Another portion of the same extract was boiled, and the amount of matter coagulable on boiling ascertained. The proteoses and peptones were precipitated in the filtrate from the coagulable matters by means of phosphotungstic acid, and the nitrogen in the precipitate estimated. After deducting the amount of the coagulable proteids, proteoses, and peptones from the total nitrogenous matter, the remainder was put down as amides. In a number of subsequent experiments, where the proteoses were precipitated by saturating the malt-extract with ammonium sulphate, only the barest traces of peptone could be detected.

Diastatic Power.—The method for ascertaining the diastatic power of malt generally employed is that of C. J. Lintner, and the results are given as degrees on Lintner's scale. Malts vary much in diastatic power, according to the way in which they have been grown, and especially according to that in which they have been dried. As a rule, the longer the acrospire is grown, the more diastatic will be the green malt. Diastase, when in a moist state, is exceedingly sensitive to heat; consequently, if the green malt is submitted to too high a temperature at the earlier stages of the drying process, when it still contains a large quantity of water, the diastase may be almost completely destroyed. During the drying process a considerable portion of the diastatic power of malt is invariably lost, and the higher the heat to which the malt is exposed the greater will be this loss. Consequently pale malts have a higher diastatic power than high-dried ones; in chocolate or black malts the diastase is completely destroyed. Fortunately this partial destruction of the diastatic power is advantageous to the brewer, for malts as ordinarily manufactured contain, as a rule, more diastatic power than is required, and consequently this has to be still further reduced in the mash-tun. It is this excess of diastatic power which enables the brewer to use a considerable proportion of unmalted material, such as rice, maize, &c., in the mash-tun. The distiller and the vinegar maker, who require malts that possess the highest diastatic powers it is possible to obtain, employ such as would be, from a brewer's point of view, excessively grown; they are also dried off at as low a heat as possible. Green malt taken straight off the floor or out of the cylinder, and therefore possessing the maximum of diastatic power, is frequently employed in these special branches of the fermentation industries.

The diastatic power of English brewing malt ranges between 30° and 45°; higher or lower figures than these generally point to some defect in the malting process. The diastatic power of green malt ranges from 110° to 125°, consequently an enormous destruction of diastase takes place during the kilning process.

Modification of the Endosperm.—Malts differ considerably in the degree of modification which has taken place in the endosperm, and this may be due either to the nature of the barley from which the malt was made, or to faults in the malting process. Very often the enzyme has not penetrated to the extreme end of the barley-corns, and there exerted its action; in this way we get malt with "hard ends." An ingenious method has been recently proposed by Moritz for ascertaining the amount of modification which has taken place during the malting process. Two experimental mashes are made; in the first the malt is treated for several hours with a cold-water extract of green malt (which will, of course, contain the cellulose dissolving enzyme); in the second the malt is similarly treated with the same amount of the same malt extract which has been boiled (in order to destroy the enzyme), and allowed to grow cold. After heating both to mashing temperature, the worts are filtered off and their respective gravities ascertained. The difference between the gravities affords an indication as to the amount of modification.

Acidity.—The normal proportion of free acid in malt is from 0.2 to 0.3 per cent., expressed in terms of lactic acid, and it should not exceed 0.4 per cent. This method of expression is purely conventional, the acidity is never entirely due to lactic acid, for there are other organic acids present in malt, and part of the acidity is also due to the presence of acid phosphates; however, it suffices for purposes of comparison.

The Ready-Formed Sugars of Malt.—It seems conclusively established that maltose and invert sugar are invariably present in malt, but opinions differ as to the presence of dextrin, maltose, and raffinose. Kühnemann seems to have been the first to detect the presence of cane sugar in malt.[1] O'Sullivan[2] found the ready-formed sugars of two barleys and malts to be as follows:—

	Barley I.		Barley II.	
	Before Germination.	After.	Before Germination.	After.
	Per Cent.	Per Cent.	Per Cent.	Per Cent.
Cane sugar	0.9	4.5	1.39	4.50
Maltose	...	1.2	...	1.98
Glucose	1.1	3.1	0.62	1.57
Lævulose	...	0.2	...	0.71
Total	2.0	9.0	2.01	8.76

Brown and Morris[3] found that the ready-formed sugars of barley and malt were distributed in the following manner:—

	Barley after 48 hours' steep.		Barley after 10 days' germination.	
	Embryos.	Endosperms.	Embryos.	Endosperms.
	Per Cent.	Per Cent.	Per Cent.	Per Cent.
Cane sugar	5.4	0.3	24.2	2.2
Invert sugar	1.8	0.2	1.2	2.2
Maltose	...	...	...	4.5
Total	7.2	0.5	25.4	8.9

Jalowetz found in malt, in addition to the fore-mentioned sugars, a carbohydrate, which had the same opticity and reducing

[1] *Berichte*, viii. p. 202.
[2] *Journal of the Chemical Society*, 1886, p. 58.
[3] "Text-Book of the Science of Brewing," p. 74.

power as maltose, but which was unfermentable; he also considered that he had been able to detect the presence of dextrin, but the results of other investigators render this extremely doubtful.

Morris[1] describes a body which he had found in the unfermentable residues of wort and beer, and which reduces Fehling's solution. This he thinks may probably be a product of enzymic action on the cellulose walls of the starch-containing cells of the embryo of the barley.

Import of the Ready-Formed Sugars.—Moritz and Morris[2] consider that an excess of the ready-formed sugars, or, what is a better expression, of the ready-formed carbohydrates, point to errors in the malting process, such as forced growth, and also to wet loading on the kiln. According to their views, these bodies should not, except in the case of very highly dried samples, exceed 16 per cent.; malts containing a higher percentage than this are stated to "give bad results in brewing." Abnormally low percentages (under 10 per cent.) point to insufficient germination. In this way they propose to use the estimation of the ready-formed carbohydrates as a test for the way in which the malting process has been carried out, but the utility of the method was strongly controverted by Chapman and Heron in a discussion which ensued on the reading of a paper on this subject by Morris.[3]

Maltol.—This substance, which has been found in caramelised malts, is chiefly interesting because it gives with ferric chloride a reaction very similar to that yielded by salicylic acid, and this might lead to a mistake being made in the analysis of beer or porter for that acid. Maltol does not, however, give the intense red coloration with Millon's reagent which salicylic acid does; hence the two can be readily differentiated. Maltol is a crystalline, colourless, odourless, volatile substance which melts at 159° C.; it is readily soluble in hot, difficultly soluble in cold water; very soluble in chloroform, fairly so in alcohol and ether, but nearly insoluble in petroleum spirit. It has the formula $C_6H_6O_3$, and exhibits the characters of a weak acid, forming crystalline salts with several of the alkalies and heavy metals.

[1] *Journal of the Federated Institute of Brewing*, 1895, p. 125.
[2] "Text-Book of the Science of Brewing," p. 165.
[3] *Journal of the Federated Institute of Brewing*, pp. 239, 241.

CHAPTER VIII.

ARRANGEMENT OF THE PLANT IN A BREWERY.

IN giving a general survey of the arrangement of the various vessels used in brewing, the type of brewery built upon what is termed the "tower system" has been chosen, not that it is necessarily the best arrangement possible, but because it is the simplest and most readily understood. It also occupies the smallest superficial area, and this is a matter of importance when the site is limited.

A "tower or gravitation brewery" is so called because one portion of the building is built at a greater elevation than the remainder. The liquor and goods used for the brew are raised to the highest storey of this elevated portion or "tower," and as the operation of brewing proceeds, they gradually descend by gravitation. In this arrangement each vessel is said to "command" the next in order of use, and all intermediate elevation or pumping is avoided.

The accompanying figure 65 illustrates the arrangement of a modern brewery on this system. In the roof of the topmost room of the tower is the cold-liquor back A, and on the floor of the same the hot-liquor back B, also the upper portion of the grist-case C, by the side of which is seen the highest part of the elevator. In the room immediately below are placed the lower part of the grist-case, the mashing-machine, the mash-tun D. On a platform immediately under its floor is the under-back E, and in a room to the side the copper F and the hop-back G. Coming back into the tower proper, we find the cooler H, and immediately below, the refrigerator, and the malt-case J. In the lowest room there is the malt-mill L, the lowest portion of the elevator, the engine, and the pumps N, situated over the well. Slightly to the left are the steam-boilers O, which furnish the steam for heating the hot-liquor back, the copper, and for driving the engine. In the uppermost storey of the portion of the brewery to the right is the storeroom for the malts and hops, on the floor, below the fermenting rounds. The lowermost storey forms the racking-room, and is furnished with the racking tanks PP.

The tower system, though it is admirable so far as the actual process of brewing is concerned, has several disadvantages. It does not permit of extension as the business of the brewery increases; there is, as a rule, no room for further enlargement of plant; and, when this is required, the erection of an additional tower is necessitated. It is also inconvenient for the brewer when engaged in the upper portion of the tower, as his attention is frequently required in other parts of the brewery.

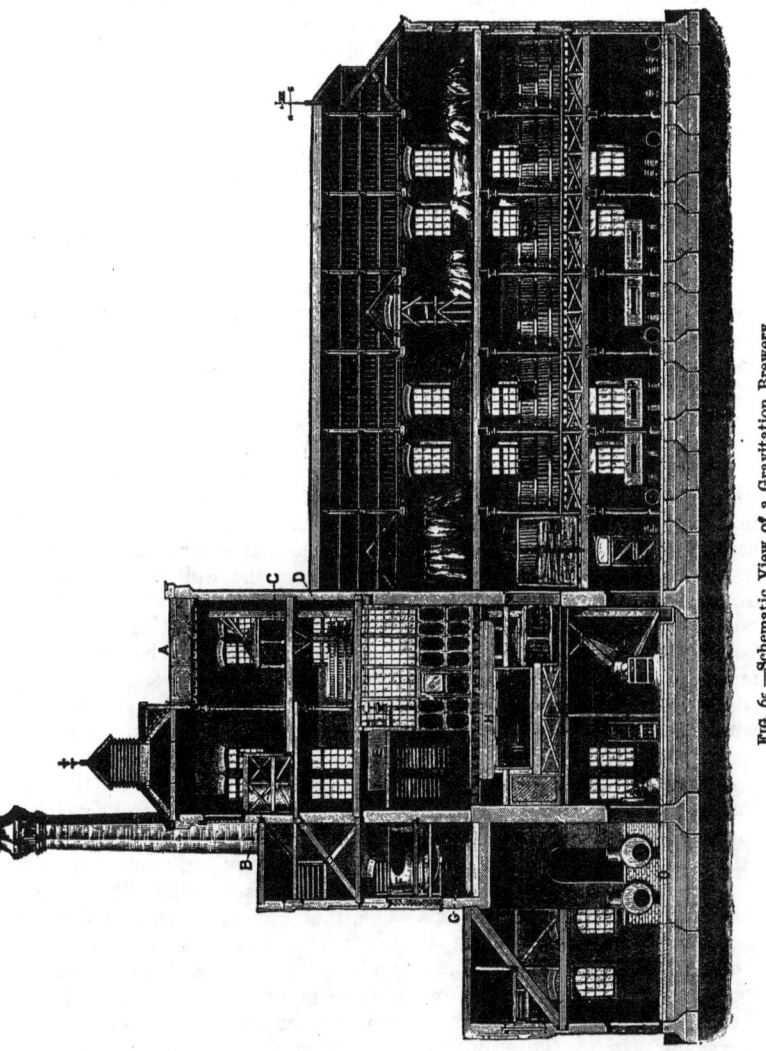

FIG. 65.—Schematic View of a Gravitation Brewery.

Other plans have consequently been devised to reduce the height of a brewery, though this increases the extent of surface occupied. In these the wort is elevated at one or more points;

it may, for instance, be pumped into the copper as it leaves the mash-tun, the copper being placed at the same or a higher level than the mash-tun; or the wort, as it leaves the hop-back, may be pumped into the cooler, which in this case may be situated in the topmost room of the brewery. This latter is the more preferable plan, when the wort is only pumped once.

The height of the brewery may be still further reduced, and the superficial area on which it stands increased, by pumping the wort at both these points, and this is the plan most frequently found in old breweries. In times gone by it was thought that any building could be transformed into a brewery, and the result of commencing with inappropriate structures, and afterwards making alterations in them when increase of business has necessitated such a course, is the many complicated and ill-contrived breweries so often met with, where it is impossible to apply those rules of rigid cleanliness which science has taught to be so indispensable in the successful practice of brewing. Fortunately at the present time, from the efforts of a number of architects who have devoted much attention to the construction of breweries and malt-houses, their general arrangement is now being placed on a much more scientific and rational basis.

Cold-Liquor Back.—This is a large rectangular tank, sometimes made of wood, oftener of flanged cast-iron plates, the planed flanges of which are bolted together. It is, as a rule, placed in the highest storey of the brewery, and is occasionally so arranged as to form a portion of the roof. The water from the well is pumped directly into this vessel, which very often acts as a sort of general reservoir for the brewery, both brewing and attemperating water being drawn from it. Such an arrangement can scarcely be recommended; the refrigerating water is much better drawn directly from the well, the water of which is much cooler in summer than that which may have been standing for some hours in the back. The size of the back will depend upon whether its contents are to be used for brewing alone or for both brewing and refrigerating. In the former case, this may be fixed at the rate of $2\frac{1}{2}$ barrels per quarter for the largest number of quarters ever brewed at one time; in the latter, five barrels per quarter will not be too ample an allowance.

Hot-Liquor Back.—This is a covered vessel, generally constructed of iron, and which may be either circular or rectangular in shape, and is situated at a higher level than the mash-tun. To prevent loss of heat by radiation, the back should be either lagged with wood or covered with some non-conducting material. Arrangements are provided for heating its contents, and these are of various kinds. The simplest plan is an arrangement of perforated pipes that allow the admission of steam into the liquor, but this, though very effective, has several drawbacks. For instance, impurities may be carried over with the steam if the boiler is fed with dirty water, or if chemicals are used to prevent boiler incrustation forming. If the engine used be a condensing one, oil and grease of various descriptions find their way into the boiler along with the feed water, and these may be transported by the steam into the hot-liquor back. Moreover, the steam when entering the liquor, especially when the latter is cold, makes much noise and causes a considerable amount of vibration; contrivances, however, are now made which obviate these latter inconveniences. The most common method of heating the contents of a hot-liquor back is by a coil of copper or iron pipes placed along the bottom of the back, into which steam is admitted. The steam in condensing gives up its heat to the cold liquor, and escapes as water at the end of the coil. The coil has a steam-trap attached to its farthest extremity, which, while keeping up the steam pressure, permits the escape of the condensed water. Instead of a coil, high-pressure heaters may be used. These are cylindrically shaped vessels, closed in top and bottom, through which a series of tubes about two inches in diameter pass; into the space between these and the sides of the cylinder steam under pressure is admitted. These are said to cause a greater circulation in the back during the heating process than the coil does. When the latter is employed, the liquor underneath it is apt to escape the heating process, so that when the outlet pipe is attached to the bottom of the back the first portions of the mash liquor drawn off are not at the required temperature. A thermometer bent at right angles is inserted into the back, into which the bulb projects for a short distance, whilst that portion which carries the scale stands outside and is protected with a case. A chain running over two pulleys, with a float attached to one end and a pointer to the other, extends from the hot-liquor back to a scale in the mashing-room, upon which is indicated by the pointer the quantity of liquor in the back. The cover of the back is either removable or is provided with a manhole; in this way access is afforded to the interior of the back for cleansing purposes, or for the introduction of the materials for hardening the liquor where such are employed.

The hot-liquor back should be large enough to contain the whole of the liquor required for one brew, and this is generally reckoned as six barrels per quarter. It is often made only large enough to contain sufficient liquor for the mash, the remainder of the liquor for the brew being heated whilst the mash is standing on. The former is by far the more preferable arrangement, since when the brewing water contains calcium and magnesium carbonates in solution, there is often not sufficient time to effect their removal from the sparge liquor.

The Malt-Mill.—This apparatus consists essentially of a pair of smooth rollers, either of steel or chilled cast-iron, mounted on axles. Motion is communicated by a strap running over a pulley fixed to the end of one of the axles; at the opposite end of the same axle a toothed wheel is fixed, and this gears into a similar toothed wheel fixed on the axle of the other roll, by means of which motion is conveyed to it. The distance between the rolls can be regulated with the utmost precision by two set screws which act on the axle bearings of one of the rolls. Grooved rollers are occasionally employed. These, which cut rather than crush the malt grain, are very effective in their action, and do not powder the malt so much as ordinary rolls. They are, however, easily damaged by stones, nails, &c., in the malt, and as they require a special machine to sharpen them, this necessitates their removal from the brewery to be sharpened, whilst the ordinary rolls can often be refaced on the premises.

A large rectangular receptacle of wood or iron, the lower portion of which has the shape of an inverted pyramid, is fixed immediately above the malt rolls, and this serves as a receptacle for the malt which is about to be crushed. The lower portion of the hopper terminates in a narrow opening situated some inches above the rolls, in which an iron roller, deeply grooved in the direction of its length, is placed; this slowly rotates when the mill is in action, its object being to secure an even feed to the mill and to distribute the malt equally over the rolls. One side of the hopper is provided with a slide, worked by a rack-and-pinion movement, and by varying the distance between the edge of this slide and the feed roller, the quantity of malt passing to the rolls in a given time can be regulated. The whole of the moving parts are cased in to prevent loss of malt-dust, suitable doors being provided for obtaining access to the various parts when necessary for cleansing or other purposes. The best position for the malt-mill is directly over the receptacle known as the "grist-case," which holds the ground malt or "grist."

When, owing to the arrangement of the brewery, the grist-case is situated at a higher level than the malt-mill, a hopper is attached to the under side of the malt-mill, the shoot of which communicates with the lower portion of an elevating apparatus commonly known as a "Jacob's ladder." This machine consists of an endless leather band, working over two pulleys, one of which is driven by power. A number of light metallic buckets are fixed on the band at regular intervals; these scoop up the grist, elevate it to the required height, and in turning over the top pulley discharge their contents into a shoot which leads to the grist-case. The whole is closed in to prevent the loss of malt-dust. Should it be required to move the malt in a horizontal direction, a trough, in which an endless screw revolves, is used to effect this object. Since both these latter arrangements are apt to separate the more finely ground portions of the malt from the husk, and as this is decidedly disadvantageous, these appliances should only be used where it is absolutely impossible to have the malt-mill placed directly over the grist-case. For the same reason a long drop from the malt-mill to the grist-case is to be avoided.

Malt-Mills in Pairs.—A pair of malt-mills fixed side by side is a very common and recommendable arrangement, as the rolls in one of the mills can be set a little closer than those of the other; by passing the smaller malt through the more closely set mill, a more perfect grinding is secured. In this case the malt is passed through a screen, which generally consists of a revolving cylinder provided with three sets of differently sized apertures. Through the smallest of these the dust, dirt, &c., find their way, and fall into an appropriate hopper and retaining-box; the next set of holes are just large enough for the smaller malt to pass through, and the remainder of the malt passes through a series of larger holes; any large foreign bodies, such as stones, flints, nails, &c., fall out at the end of the cylinder. Sometimes, in addition to this, a set of magnets are so arranged as to remove pieces of iron, such as nails, which are frequently found in malt. All these arrangements are decidedly advantageous, since they prevent injury to the malt-rolls and lessen the danger from explosions. When hard substances such as stones, flints, &c., pass through the rolls, sparks are often struck off, which are apt to ignite the finely divided malt-dust suspended in the air of the mill and the various receptacles. This, when ignited, explodes much as a mixture of an inflammable gas and air does. Various effective contrivances are now made which cut off all connection between the mill and the various passages and receptacles if the malt-dust becomes ignited.

Speed.—Malt-mills should not be driven at too high a speed, or their action becomes too much of a grinding instead of a crushing nature. A speed of about 180 feet per minute for the periphery of the rolls is the outside limit. This, with rolls 8 inches in diameter, would be ninety revolutions per minute.

Care in Grinding.—As much of the success of the process of brewing depends on the way in which malt is ground, considerable care should be exercised in this operation. The more finely malt is ground the more extract it will yield; but, unfortunately, if it is too finely ground, serious difficulties arise in connection with the drainage from the mash-tun. The aim of the brewer should be just to hit the happy medium, grinding to such a degree of fineness as shall secure him a good yield from his malt, and yet not so fine as to involve himself in subsequent drainage difficulties. Every corn should be crushed, or waste will take place, since corns which escape crushing yield up little or nothing of their constituents in the subsequent mashing. Tender, friable malt, especially when recently dried, pulverises very readily. Such malt when coarsely ground yields up its soluble constituents much more perfectly than inferior material.

As a matter of convenience, malt is generally ground the day before it is to be used, and this length of time should not be exceeded. Grist absorbs moisture with extreme avidity, and hence rapidly deteriorates if kept long on hand.

Regular Cleaning.—Malt-mills, hoppers, and all other adjuncts to the grinding of malt, should be regularly cleansed to remove the malt-dust which collects in them. If this is not done regularly, the dust becomes damp, and gradually undergoes changes which render its presence most undesirable in the wort.

The Grist-Case.—This receptacle is situated immediately over the mash-tun, its upper portion is rectangular in form and its lower portion shaped like an inverted pyramid, the sides of which should slope at an angle of about $45°$, this inclination being found best for the regular delivery of the grist. It is made either of well-seasoned wood, smoothly planed and with well-fitting joints, or of iron well painted and varnished. Wood is the preferable material, since, being a worse conductor of heat than iron, the steam rising from the mash-tun is not so liable to condense upon the grist-case, and form little rills of water which often find their way into the inside of the case and its appendages. The position of this receptacle will vary according as the mashing is entirely performed by rakes or is effected by a mashing-machine; in the former case, its apex will be a short distance above the level of the top of the mash-tun, and a little way inside it; in the latter case, its apex will be a short distance from the side of the tun. A slide fixed at its apex allows, when opened, the contents of the grist-case to fall either directly into the mash-tun or into the mashing-machine. Closely fitting doors should be provided to afford free access to every part of the grist-case for cleansing purposes. It is especially necessary that those portions adjoining the slide should be carefully attended to, since malt-dust moistened by the condensed vapour from the mash-tun is very apt to lodge on these parts.

The Mash-Tun.—This is a cylindrical vessel (Fig. 66) of either wood, copper, or iron; in rare cases, of wood lined with copper, closed at the bottom. When constructed of wood, it tapers slightly towards the top, so that the staves may be tightened by driving down the hoops; when made of iron, its sides are parallel. Iron mash-tuns are lagged with wood round the sides and at the bottom, a space being left between the iron and the lagging, which is filled with some non-conducting material, such as slag-wool, asbestos, &c. Mash-tuns are generally provided with thick wooden covers, which for convenience in handling are generally made in two or more portions. Large tuns are often provided with a hood-shaped cover, hung on chains, working over pulleys and counterbalanced by weights, so that the hood can be readily raised or lowered bodily. In every case the cover should be constructed of non-conducting material, so as to retain the heat in the mash-tun as much as possible. This vessel should be placed in a situation as free from draughts as possible, since these tend to carry off its heat, and nothing is worse in practice than what is known as a "cold mash-tun," that is, one which loses heat rapidly.

THE MASH-TUN.

False Bottom.—At a distance of one and a half to two inches above the real bottom of the mash-tun is situated the false bottom, which may be of copper, or of iron, or of gun-metal, constructed in sections for easy removal. These sections are technically spoken of as "the plates," and, when in position, should fit closely together, no interspaces being left open. Iron plates are perforated with holes of from one-sixteenth to one-eighth of an inch in diameter, situated at a distance of about one inch from each other, their under sides being countersunk. Copper and gun-metal plates are similarly perforated, or they are slotted, *i.e.* have fine slits cut through them. Cave's patent plates are either made

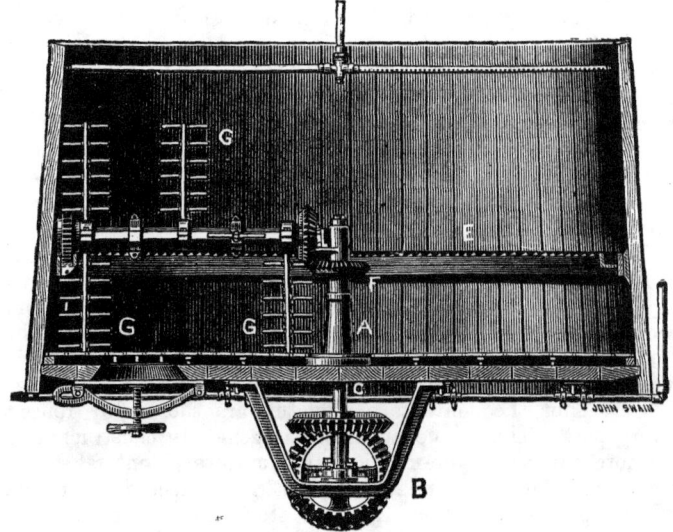

FIG. 66.—Mash-Tun.

of copper one-eighth of an inch in thickness or of cast-iron three-eighths inch thick. Holes of three eighths of an inch in diameter are bored at a distance of two inches from each other, and in each of these is inserted a thin copper thimble, the top of which is on a level with the upper surface of the plate, and is perforated with nine small holes one-thirty-second of an inch in diameter. They give excellent results in practice, and the contact of the dissimilar metals in the iron plates does not seem to give rise to any difficulties arising from galvanic action.

The plates, when in position, rest upon small wooden or metallic supports fixed to the upper side of the real bottom of the mash-tun, and are firmly secured in their places by bolts and nuts.

Rakes, or Internal Mashing-Machine.—The rakes were invented by Matterface in 1807, and have with little alteration survived to the present day. They consist of a vertical shaft A (Fig. 66), which revolves in the axis of the mash-tun, and may be driven by a bevel wheel either from above or below. When driven from above, the shaft works in a footstep fixed to the centre of the mash-tun bottom; but when driven from below, the shaft passes through a water-tight stuffing-box in the bottom of the mash-tun. About half-way up this shaft is a collar which revolves round the shaft, and which carries the gun-metal bearing that supports one end of a horizontal shaft that carries the rakes, GGG. A small toothed wheel is fixed to the other end of this shaft, and this engages with the circular toothed rack E, which is bolted to the inside of the mash-tun. A toothed bevel wheel, F, fixed on the vertical shaft A, transmits motion to a similar wheel placed on the horizontal shaft. The rakes GGG are fixed at right angles to the shaft D, and are arranged around it in a spiral, so that if the shaft be looked at endways, the rakes appear like the spokes of a wheel. The pair of rakes at the extreme end of each shaft are fixed in a vertical position and do not revolve.

When in action the motion is a somewhat complicated one. As the vertical shaft A turns round, the bevel wheel F moves along with it, and rotates the wheel attached to the end of the horizontal shaft which carries the rakes. The wheel at the other end of the shaft engages with the rack E, and causes the horizontal shaft and the rakes to slowly revolve round the vertical shaft A. Thus the contents of the tun are thoroughly roused in every part. The whole apparatus is generally constructed of iron, more rarely of gun-metal. In Conron's patent rakes the end of each rake is expanded into a stirrup-shaped form, to the horizontal portion of which a piece of stout sheet indiarubber is attached. When in action these elastic flaps sweep the surface of the plates. Their utility, except in special cases, is extremely doubtful.

External Mashing-Machines.—These were invented to ensure a more equable admixture of the hot liquor and grist than was possible by the methods already mentioned. The first machine of this class was invented by Steel in 1853; it bears his name and is the one most frequently used at the present time. The machine (Fig. 67) consists of a horizontal cylinder of iron or copper, A, from 3 to 6 feet in length and from $8\frac{1}{2}$ to 22 inches in diameter. One end of the cylinder is closed with a cap, having in its centre a stuffing-box, through which passes the shaft B. At the other end is a cap, C, which closes half the aperture and carries the bearing in which the end of the shaft turns. The shaft is revolved by means of the pulley D, and is provided with a series of arms, fixed at right angles. The screw blade, H, is

EXTERNAL MASHING MACHINES.

a slight improvement made to the original machine with a view to adding to its efficiency. The grist is admitted to the machine by an opening, F, the supply being regulated by a slide which is opened or closed by a rack and pinion actuated by the wheel G. The hot liquor is admitted through the aperture E, a cock being placed immediately in front of it to regulate the supply. The streams of grist and liquor meet at the end of the cylinder, and during their passage down it are thoroughly and intimately mixed by the revolving screw and arms. The gearing should be so arranged that the shaft makes from 150 to 180 revolutions per minute. The machine is extremely rapid in its action; with one of the larger machines it is possible to mash 200 quarters of malt in twenty minutes.

A number of self-acting mashing-machines are also in use. The first of these was invented by Maitland in 1863, and consists

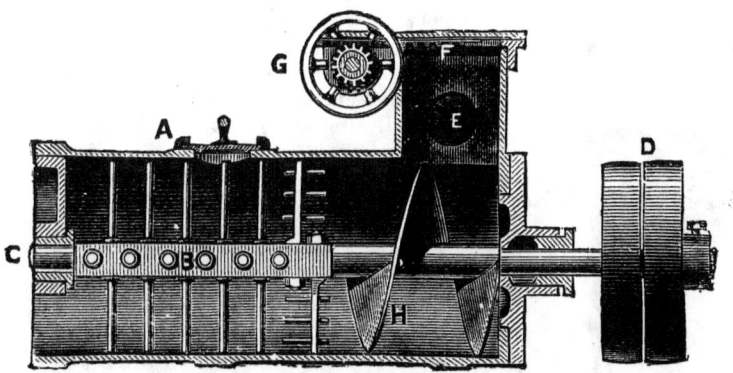

FIG. 67.—External Mashing-Machine (Improved Steel).

of an external copper cylinder which encloses an inner copper cylinder of smaller diameter, the space between these being closed in at the top and bottom. An aperture at the side admits the hot liquor to this space, and the grist falls through the internal cylinder. This is perforated with a number of fine holes through which the water rushes, and forms a series of jets which impinge against the stream of falling malt. A cross tube situated at the lower end of the cylinder carries a fairly large jet which throws a stream of liquor directly upwards, forming as it were a small fountain. The grist in its passage through the machine is met by numerous streams of hot liquor which impinge upon it in every direction and a thorough admixture ensues. Maitland's machine is rapid and efficient in its action, and has no moving parts to get out of order. Southby also invented a self-acting machine, in which the grist falls in a thin layer between two horizontal sets of fine jets of hot liquor. This form requires a less head of liquor to work it than the others.

Appliances for Raising the Temperature of the Mash-Tun.—The simplest of these is a pipe from the hot-liquor back terminating in the bottom of the mash-tun, and known as the "underlet." Hot liquor of any required temperature can be admitted by its agency underneath the plates, whence, through the perforations in the false bottom, it finds its way into the tun, with the contents of which it can be mixed by the rakes. Liquor introduced in this way is technically called "piece liquor."

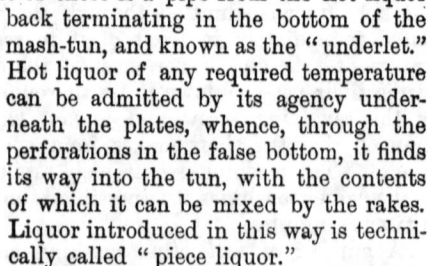

Fig. 68.—The Sparge.

Coil Under False Bottom.—A copper or iron coil is sometimes introduced between the plates and the bottom of the mash-tun, by means of which steam under pressure can be introduced. This has the advantage over the underlet of not diluting the goods, but it is somewhat slow in its action. In order to remedy this defect, the coil may be perforated at its sides with fine holes, to admit steam directly into the mash. The steam, owing to the large amount of latent heat which it contains, acts very quickly, and it is possible in this way to raise the contents of the mash-tun to the temperature of boiling water. Obviously this method can only be employed where the steam is uncontaminated with impurities.

Steam-Plough.—The same object can be effected by the steam-plough invented by Cave. It consists of a couple of hollow vessels, each of which is somewhat similar in shape to a ploughshare, attached to, and revolving with, the rake-shaft. It is placed close to the false bottom and just clears the rakes as it revolves. Steam or hot or cold liquor may be circulated at will through the ploughs; hence the mash can be either heated or cooled down to any desired temperature.

The Sparge.—This apparatus, shown in the accompanying illustration (Fig. 68), consists of a cylindrical copper vessel, technically called "the basin," so arranged as to revolve easily round its axis. This is effected either by mounting the basin on a pivot,

THE SPARGE.

or, as shown in the engraving, by means of friction wheels, bearing on a small platform attached to the shaft which revolves the rakes, or it may be attached directly to the end of the sparge liquor-pipe by means of a water-tight gland. A pipe communicating with the hot-liquor tank, and provided with a tap, terminates just over the basin of the sparge; this supplies the necessary sparge-liquor. The sparge arms (generally two, sometimes three in number) are attached near the bottom of the basin. These are perforated on one side by a series of fine holes, which should be so arranged that the sparge liquor is distributed equally over the superficial area of the mash. The arms can be removed, and are also provided with removable caps at their ends, for the purpose of cleansing. As the perforations are very liable to become stopped, the sparge requires frequent attention; a sparge which works inefficiently is often the cause of loss of extract.

In order to ascertain that the sparge is delivering the liquor uniformly, a wedge-shaped trough, a few inches deep and made of sheet metal, is employed, its length being half the diameter of the mash-tun, and its shape a sector of a circle of a similar diameter. The large end may be about two feet wide. At equal distances of about six inches each from the pointed end, a series of partitions are fixed; these conform in shape with the circumferences of a series of circles having for their radii the distance from the pointed end of the trough to each partition. The spaces between these partitions constitute so many separate compartments. The apparatus having been placed on the plates in the mash-tun, the sparge is set in action for a short time. It is then stopped, and the level of the water in the different compartments observed. If the sparge is working uniformly, the water in all these will be at the same level; if not, the holes in the sparge immediately above any compartment in which it is at a lower level should be slightly opened, the operation being repeated until all compartments are equally filled in the same time.

The principle of the action of the sparge is the same as that of the well-known Barker's mill; the liquid in the sparge arms tends to exert an equal pressure in all directions, but as it is permitted to escape at one side of the arm, a reaction is set up at the opposite side, and this drives the sparge round.

The Copper.—Coppers are of two principal kinds—"fire-coppers," *i.e.* those which are heated directly by fire, and "steam-coppers," which are heated by steam. Of the former there are several varieties, the simplest being a copper or iron pan placed over a furnace, the flues from which are carried round the copper nearly to its top. An outlet is provided near the bottom, to which a tube is attached which carries the cock for emptying the copper. This pattern is now nearly obsolete, and is only met with in small or old-fashioned breweries.

Fire-Copper.—The most common form of open fire-copper is that shown in the annexed figure (69), known as the "bench copper." The lower portion is of the shape of an ordinary pan; at about half-way up it suddenly widens out a few inches, to form what is termed the "bench." From this the copper is continued upwards with parallel sides. The flues which surround the copper are only carried as far as the bench; consequently the portion above the bench, not being heated, has, to some extent, a cooling action on the wort, which tends to prevent boiling over.

The dome-copper is, as its name implies, covered in with a dome, round which there is a sort of tray, which has an outlet into the body of the copper. At the summit of the dome is a large opening, to which is attached a wide tube $1\frac{1}{2}$ to 2 feet in length.

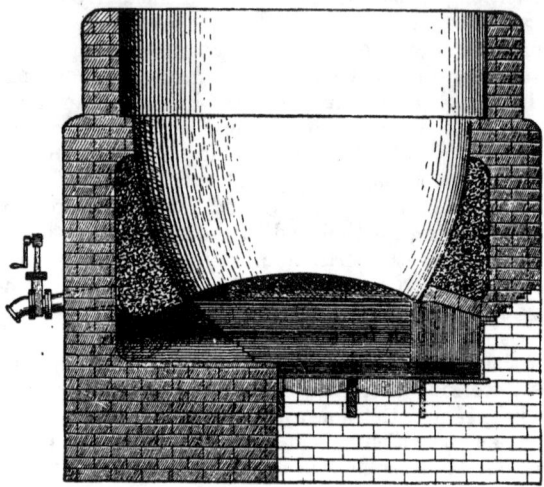

FIG. 69.—Fire-Copper.

When the copper is boiling, the communication is left open between the tray and the copper, and through this the wort, as it boils out at the wide tube, after pouring over the sides of the dome and falling into the tray, finds its way back into the interior of the copper. When the wort has finished boiling the heat is slackened, the plug fixed in its place, and a second batch of wort may be placed in the tray, which will be heated to some extent before being allowed to pass into the copper. This heated wort is also ready for immediately running into the copper as soon as the other wort is out, to "save the copper," that is, to prevent the bottom from being burnt.

Coppers are sometimes provided with a strainer for keeping

STEAM-COPPERS.

back the hops when the vessel is emptied; but this is not an arrangement which can be recommended, as the strainer soon becomes more or less blocked, and consequently the copper empties itself very slowly.

Steam-Coppers.—These differ considerably in shape from fire-coppers. The bottom is bellied outwards, and is made of considerable thickness to withstand the pressure of the steam (Fig. 69a). The sides may, as shown in the figure, be carried straight up or they may slightly taper outwards. Attached to and surrounding the bottom of the copper there is an iron dome, which is somewhat larger in size, so that a cavity of from three

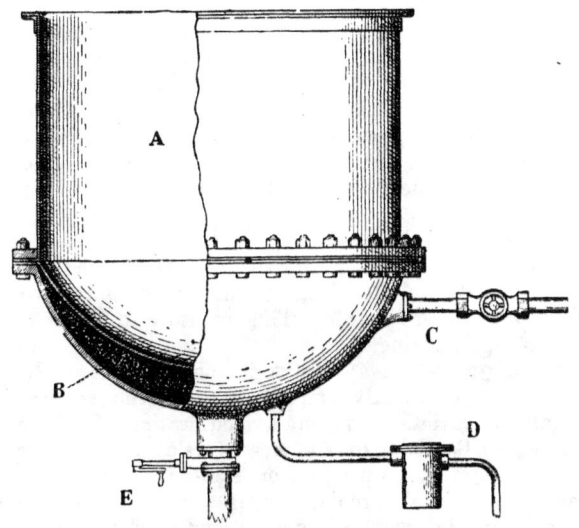

FIG. 69a.—Steam-Copper.

A, copper; *B*, steam-jacket; *C*, steam-pipe; *D*, steam-trap; *E*, outlet-valve.

or four inches *B*, is left between its interior and the bottom of the copper. Into this high-pressure steam is admitted by a cock, *C*, an air-tap being provided to allow the imprisoned air to escape when steam is turned on. This is a very necessary precaution, since, as air is so highly expansible by heat, if its exit were not amply provided for an explosion might occur. A pipe for carrying off the condensed water passes from the bottom of the steam-jacket to the steam-trap, *D*, which latter permits the water to escape but holds back the steam. The wort-pipe, *E*, passes through the centre of the copper bottom. The steam-copper possesses several distinct advantages over the older fire-copper.

It can be heated up before the introduction of the wort, and the wort quickly raised to a temperature which puts an end to all further diastatic action. The readiness with which the heat can be applied or arrested is a point very much in its favour. It is also vastly more economical; a fire-copper, according to Professor Schwackhöfer,[1] only utilises about 30 per cent. of the available heat of the fuel consumed, while a steam-copper utilises 70 per cent., and thus a saving of more than half the fuel employed is effected.

Hop-Back.—This vessel is required in all cases where the copper is not provided with a strainer. It may be constructed of wood or of iron, and its shape be either rectangular or circular. At a short distance from the bottom a false bottom of perforated plates is provided to keep back the hops. The hop-back is placed at a lower level than the copper. A series of drainage pipes proceed from holes distributed at equal distances in the bottom of the back to the cooler, or to the pumps which raise the wort into the cooler. Circular hop-backs are often provided with a sparge something similar to the one in the mash-tun, through which hot liquor is sparged on the hops to remove the wort held back by them.

Cooler.—This is a large, shallow, rectangular vessel of wood, iron, or copper. It is placed in a room in the brewery to which air can be admitted from all sides, for which purpose the windows are provided with louvre boards. This exposure to a free current of air ensures rapid cooling.

Refrigerator.—This apparatus is made in many different forms, that most commonly employed being the vertical one, which, in all its variations, is but a modification of the refrigerator invented by Baudelot over twenty years ago. In its original shape it consists of two upright iron standards, between which a number of horizontal circular copper tubes are fixed, having attached to their under sides narrow strips of sheet copper, the lower borders of which are cut zigzag something like the teeth of a saw. The cold liquor is admitted into the lowest tube, from this it passes to the tube immediately above, and so on, until at last it emerges from the uppermost tube. The wort is introduced into a trough placed at the top of the apparatus and perforated with a series of fine holes. Through these the wort flows on to the first tube in a thin layer; after passing over this tube it falls on to the next, and so on until at last it reaches the trough in which the apparatus stands, from whence a pipe conveys it to the fermenting vessel. The toothed strips attached to the lower side of each of the tubes secure an equal distribution of the wort.

In the more modern varieties of this apparatus the tubes are made oval instead of circular in section, thus providing a larger cooling surface, and are soldered into openings in the uprights,

[1] *All. Anzeiger f. Brau.*, 1896, p. 1209.

which are in this case made of gun-metal. On the exterior of the uprights are fixed a series of removable caps, which can be removed so that the tubes may be cleaned out periodically, an operation which is frequently necessary when the water used for cooling purposes is at all dirty or contains much acid carbonate of lime in solution. In another variety (the Laurence) the cooling liquor passes between two sheets of copper worked into a zigzag form. A series of removable caps give access to the interior of the apparatus for cleaning, which is effected by a peculiar-shaped brush. The whole machine is fixed on trunnions, and can be turned bodily over. In another (Shear's) the tubes are arranged in the form of a circle, the whole apparatus having the appearance of a vertical cylinder.

Horizontal Refrigerators.—When height is a consideration, a refrigerator of the horizontal pattern must be used. This has the form of a shallow rectangular trough, divided at intervals by a number of partitions, through each of which passes a horizontal tube of rectangular section with rounded corners, through which the cold liquor passes. Each tube occupies the centre of a division, a small space being left between the underside of the tube and the bottom of the trough. The wort flows into the first division, passes underneath the horizontal tube, and, on reaching the level of the top of the partition, passes over into the next division, and so on, until, having passed the last division, it escapes into the wort main. The refrigerator is set at a slight inclination, being highest at the end where the wort runs in. Horizontal refrigerators are much more difficult to clean than vertical ones, since those parts through which the wort passes are so much more inaccessible.

The quantity of refrigerating liquor required to cool each barrel of wort varies with the temperature of the wort to be cooled, the temperature of the liquor, &c., from two to six barrels, or even more; horizontal refrigerators are said to consume about twice as much cold water as vertical. In breweries provided with an ice-machine, cold brine is passed through the lowermost half-dozen or so of the tubes of the refrigerator, which are provided with a separate inlet and outlet.

Collecting Vessel.—In some breweries a separate vessel is provided in which the wort, after being cooled in the refrigerator, is collected, but more often the fermenting vats themselves are used as the collecting vessels. Where separate collecting vessels are employed, they are constructed in the same form and of the same materials as the fermenting vats. Where the collecting vessel is provided, the wort must, according to the excise regulation, remain in it for at least twelve hours after collection.

Vessels for Primings, &c.—In breweries where it is the custom to add syrup or wort to the ale when in cask, or in other words to "prime," a special vessel must be provided for such fluids. In these the excise officer takes the gravity and dip when filled, and afterwards the dip from day to day. Where a solution of caramel is used for colouring purposes, a similar vessel must also be provided for it.

Fermenting Vessels.—Many materials have been employed from time to time in the construction of fermenting vessels, such as glazed bricks and tiles, brick lined with glass plates, wooden vessels lined with lead or copper, enamelled iron plates, &c. None of these have stood the test of experience, and now-a-days the material chiefly used is wood, though in some parts of the country stone is extensively employed, and slate is also used to a limited extent. One of the principal points with reference to the material used in the construction of a fermenting vessel is that it should be capable of being worked to a fine smooth surface, and retain this during usage, as such a surface permits of the vessel being readily cleansed. The material used in construction should also be able to bear the heat of boiling water without being liable to fracture, since the destructive action of heat is one of the chief agencies relied on in the brewery for keeping micro-organisms in check.

Fermenting vats are generally made either round or square; occasionally oval. The first of these are technically known as "rounds," the second as "squares." Rounds are built of staves held together by iron hoops, and are, as a rule, made of oak. They are open at the top, and are carried up some two or three feet higher than the level of the surface of the wort fermented in them, this extra space being provided for the yeast. On one side, at a distance of a few inches above the wort level, an opening of about thirty inches square is made, so as to permit of access to the tun. This, during the fermentation, is closed with a number of loose boards fitting into grooves, which serve to retain the yeast.

Squares are constructed of planks some two and a half to three inches in thickness; they are bolted together with iron bolts let into the thicknesses of the wood. They are generally made of fir, sometimes of oak, and in recent times American cedar has been much employed. The sides of the square, like those of the round, are several feet higher than the level of the wort to be fermented in it; one of them is similarly provided with an opening which can be closed with loose boards. Vessels of oak or fir require seasoning before use. This, with oak vessels, is a very simple matter; they only require to be scalded out several times with boiling water. As fir-wood contains a considerable amount of resin, this must be partially removed before the vat is brought into use. Southby recommends that such vessels be first filled with boiling water, which is run off the next day. The vat is then to be whitewashed out with a mixture of $2\frac{1}{2}$ lbs. of chloride of lime to each gallon of water.

After this coating has remained on for twenty-four hours, the vat is washed out with a mixture of one part of hydrochloric acid (spirit of salt) and four parts of water. It is then washed out several times with boiling water, and finally painted out with an ordinary solution of bisulphite of lime to remove the last traces of chlorine. American cedar has the advantage that it requires no seasoning, but may be used straight away after being well scrubbed out.

In those parts of the country where the stone square system of fermentation is almost exclusively employed, the fermenting vats are constructed for the most part of large slabs of hard, impervious stone; but, as material of the right quality is somewhat difficult to obtain, a softer stone is often employed. This is objectionable, since by use its surface becomes, in course of time, rough, and cannot be properly cleansed. Slate is now beginning to take the place of stone in the construction of the Yorkshire squares; it preserves a much better surface during wear.

The stone square (Fig. 70) differs materially from the wood square in having a second square built round it through which cold or warm water may be passed; it acts as an attemperator. The sides and bottom consist of single slabs of stone, which are fitted together by means of grooves, the joints being rendered water-tight with cement, while a number of iron bolts serve to retain the slabs in position. Since the weight of a stone square is very great, it is supported on a pillar of masonry. The water-jacket, C, is constructed in a similar manner by four slabs of stone of such a size that a space of about 2 inches is left between its internal surface and the outside of the square proper, and its sides are 3 or 4 inches lower in height than those of the square itself, so as to allow the attemperating water to run away. The square proper, A, is covered over with another slab having a circular aperture (the "man-hole") of 18 inches in diameter, which is surrounded with a stone ring some 5 or 6 inches high, on which a stone lid provided with a handle fits. In one of the corners of the covering slab is another opening situated a few inches from each of the sides and about 3 inches in diameter, provided with a brass valve, E, to which a chain is attached. From the under side of the valve a tube, D, extends to within a few inches of the bottom of the square; this is technically known as the "organ-pipe." Upon the upper side of the covering slab is placed the yeast-trough, B, constructed of four stone slabs. It has the same superficial area as the square, and a depth of from 24 to 30 inches. A pump is one of the necessary adjuncts to a stone square. It is of the simplest possible construction, made of tinned copper and of one uniform size (3 inches) in diameter, and with a stroke of 6 inches. When the square has been put together, it is carefully pointed out with cement, so that no crannies are left where yeast or other organic matter may be retained; and the pointing should be inspected every few months, to see that it remains intact, for, if it becomes defective, troubles are certain to make their appearance in the brewery.

Slate is now beginning to replace stone to a certain extent in the construction of this particular form of square, which, when made of this material, is generally not provided with a water-jacket, an ordinary attemperator being fitted inside instead. Slate possesses the advantage of having a smoother surface than stone,

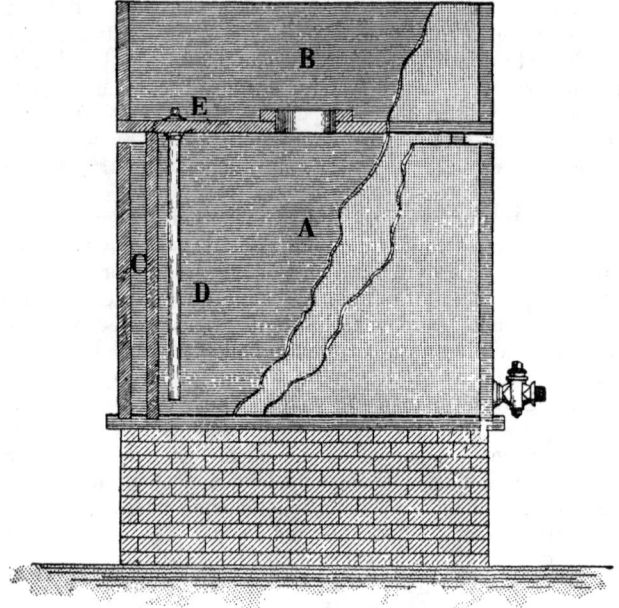

FIG. 70.—Yorkshire Stone Square.

A, the square; *B*, yeast-trough; *C*, water-jacket; *D*, organ-pipe; *E*, valve.

but it has the disadvantage of being a better conductor of heat, and consequently the wort in such a square is much more subject to the influence of changes of temperature in the external atmosphere. Slate will not bear the application of boiling water or steam without cracking or splitting, and it cannot, consequently, be sterilised by these agents. It is also readily attacked by sulphurous acid and the acid sulphites, and the only germicide of this nature which can be employed is the neutral sulphite of lime, with which the inside of the square may be whitewashed. A deposit of calcium oxalate is liable to form on its surface, which is somewhat difficult to remove; rubbing with a soft brick or the application of a strong solution of caustic potash or soda form the best methods for removing this.

Fermenting squares constructed of iron plates are not unknown; they are only employed in the production of black beers.

Outlets.—In fermenting vats worked on the skimming system, the outlet or outlets must be so arranged that the beer can be drawn off without disturbing the yeast, which invariably deposits to a certain extent on the bottom of the tun. This may be effected by having two cocks, one at the side some 3 or 4 inches above the floor of the tun, from which the clear beer is raked off into the trade casks, the other fixed into the bottom of the vat, and by this the remainder of the beer is drawn off into vessels where it may settle or into filter-backs. A single outlet may be provided at the bottom of the tun, through which a tube, having the cock at its lower extremity, slides through a stuffing-box. As soon as the beer has been drawn off to the level of the top of the tube, this is gradually lowered little by little as long as the beer runs bright. When yeast begins to appear, the tube is drawn down to its fullest extent, and the thick bottoms treated in the manner just described.

Loose Pieces.—In breweries where the cleansing system of fermentation is employed, the wort, after having been fermented to a certain stage in the square or round, is run off into a series of casks or puncheons which hold from three to four barrels each. These are provided with the usual bung and tap holes, and are placed on long troughs called "stillions," which catch the yeast that escapes from the bung-hole and runs down the side of the cask. The loose pieces are inclined a little to one side, so that the yeast may run down one side of the piece. In some cases the trade casks—that is, the casks in which the ale is sent out to the customers—are used for this purpose. Under this system the yeast has to run down the side of the cask, and this is obviously objectionable on the score of cleanliness. Various methods have been tried to overcome this faulty arrangement, such, for instance, as the employment of short, conical tinned pipes, which can be inserted water-tight into the bung-hole of each loose piece, and the upper end of which can be closed by a wooden plug. To the side of each conical tube there is fixed at right angles a pipe of about 2 inches in diameter, long enough to project over the end of the piece, and bent at its extreme end to a right angle. In this case the yeast trough is placed in front of the pieces, and may be made to serve for two adjacent and parallel sets; the pieces are filled up through the conical tube. When emptied, the pieces can be removed to the cask-washing department and treated there; this avoids the introduction of the large quantities of boiling water into the fermenting-room which is necessary in the system about to be described.

Burton Unions.—These are considered by many to be the highest development of the cleansing system. In it, as represented in Fig. 71, the casks, which are of about four barrels' capacity, are permanently mounted on tall wooden stands, to which they are slung by means of two axles, one attached to each head. These work in bearings, and permit the cask to be rotated on its axis, for which purpose one of the heads of the trunnions is made square, so that a handle may be attached to it when necessary. The bung-hole of each cask is provided with a conical brass socket, into which fits a hollow brass plug, carrying a pipe to convey away the yeast. This is called the "swan neck," and consists of a tinned copper tube, which, after being carried up vertically a foot and a half or two feet, makes a turn of half a circle, and curves over into a long wooden trough, which extends between the two rows of adjacent casks, and is called the "yeast trough." At one end of this another vessel is fixed, called the "feed trough," which has a capacity of five or six barrels. A tap is fixed into the bottom of this, from which a pipe of about two inches in diameter proceeds, extending in front of each row of unions, and giving off a short branch to each cask, with which it can be connected by means of a union joint to a tap permanently fixed in the head of the cask. Another cock is fixed in each cask exactly opposite the bung-hole, and is provided with a short tube which projects some little distance inside the cask, and which can be raised or lowered by means of a screw. This serves for the removal of the fermented beer, and, as the tube communicating with the tap is some little height above the bottom, it serves to hold back the bottoms. When a set of unions are cleansed, the swan-necks are first removed and the feed-pipe communications unscrewed; the handle is then attached to each cask in turn, boiling water poured into it, and the cask rotated on its axis; this is an objectionable feature in the system, for the introduction of large quantities of hot water into the fermenting-room necessarily raises its temperature.

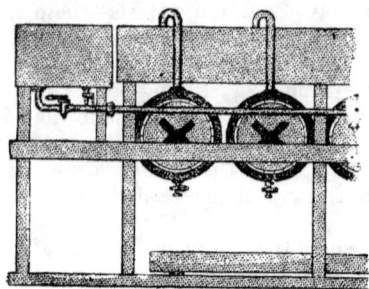

FIG. 71.—Burton Union.

Attemperators.—These are of two varieties, fixed and loose. The former consists of a coil of tinned copper which runs round the inside of the fermenting-vat, to the sides of which it is fixed by gun-metal brackets, which support it at such a distance from the sides of the vat as to afford sufficient room for cleaning purposes. The first round of the coil runs just under the surface of the wort; each succeeding turn is about six or eight inches lower than the preceding one, and, being supported on a longer bracket, stands out an inch or two farther into the tun. Cold water enters through the regulating tap at the upper extremity of the coil. The other extremity turns upwards on its exit from the vat, and is carried up to a level a few inches higher than the uppermost round of the coil, so that the whole of the attemperator is kept completely filled when in action. The end of the pipe then turns over into the funnel communicating with the waste-pipe, and this portion of the apparatus is placed in such a position that the person regulating the supply can readily see the quantity of water running away.

The movable attemperator is a flat coil of pipe, either fixed in a wooden frame, or, what is better, held together by metal connections. It is suspended by chains, by means of which it can be lowered or raised. The communication with the water supply and waste-pipe are made with indiarubber tubes. Sometimes this kind of attemperator, especially when it is to be used in long, narrow tuns, takes the form of a gridiron, the pipe being bent backwards and forwards. In any case, the apparatus should be easily removable from its attachments, so that it can be taken down and thoroughly cleaned. According to Southby, fifty-five square inches of cooling surface are required to each barrel of wort when working on the skimming system. This will be, where a one-inch pipe is employed, one and a half feet, or one foot where the pipe is one and a half inches in diameter.

Arrangements for the Removal of Yeast.—Small squares or rounds are often skimmed by hand with an appliance something like that used for skimming milk, but when the fermenting vessel is of any size the operation becomes tedious and troublesome, and some form of mechanical skimming is necessary. In rounds, the removal of the yeast is effected by one of the two forms of "parachute" and skimming-board. The first of these consists of a funnel attached to a pipe which passes through a packed gland fixed in the bottom of the tun. By an arrangement of rack-and-pinion work the mouth of the parachute can be moved upwards or downwards; it is placed about half an inch higher than the level of the wort. A valve is fixed in the throat of the funnel, to which a chain is attached. The parachute is placed in the centre of the round, and the skimming-board, which in this case is a strip of sheet-metal about six inches broad, takes the form of a helix mounted on an axle situated in the axis of the tun. The skimming-board, like the parachute itself, is provided with an arrangement by which it can be raised or lowered, and also another for rotating it. The board is placed in such a position that its lower edge just touches the surface of the wort, and in the course of a revolution in one direction it brings the yeast from the sides into the centre; the yeast then falls into the mouth of the parachute. In the second and more usual form, illustrated in Fig. 72, the funnel has the shape of a small sector of a circle, and extends from the centre to the circumference of the tun, the pipe from it having an eccentric position. The skimming-board in this case is straight, a little shorter than the radius of the tun, and is attached by one of its extremities to an axle rotating in the axis of the tun. In a single sweep round it pushes the yeast before it into the mouth of the parachute. Arrangements for raising and lowering the parachute

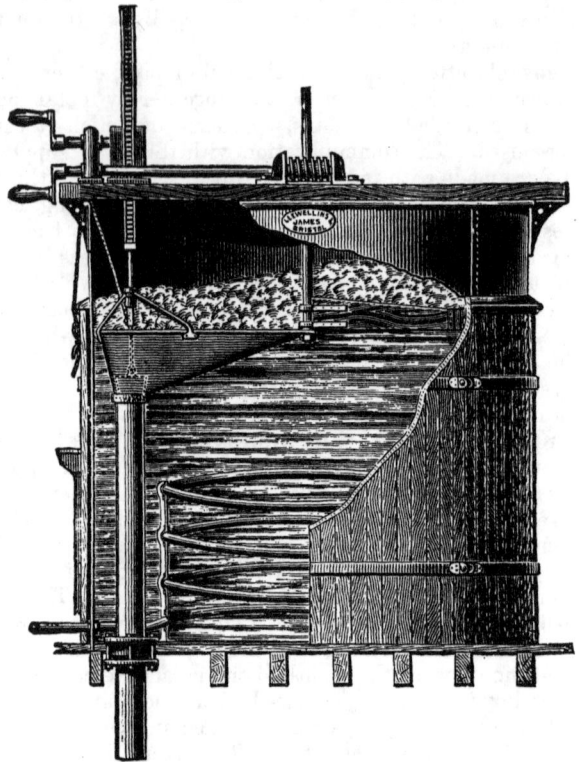

FIG. 72.—Fermenting Round with Parachute.

and skimming-board are provided; also a valve in the throat of the former. The parachute arrangement is sometimes applied to squares. In this case the funnel has the shape of a long thin parallelogram, situated at one end of the vessel. The skimming-board is straight, and moves on a travelling carriage, and in its passage from the end of the tun farthest from the parachute pushes the yeast before it into this apparatus. Squares are sometimes provided with a yeast-sluice, which consists of a flat plate of wood or metal having a projecting lip. This is fixed against an opening cut out at the top of one of the sides of the square, arrangements being provided by which it can be raised and lowered. A strip of indiarubber between the sluice-board and the side of the square renders the whole water-tight.

Rousers and Aërators.—The yeast is mixed into the wort by hand-rousers, each of which consists of a piece of flat board with a hole in its centre by means of which it is attached to the end of a long wooden handle. A simple contrivance for rousing and aërating at the same time consists of a small cask of about three gallons capacity, with both its ends removed, and having a number of holes bored through its sides. It is weighted with lead so as to sink readily, and suspended by a rope passing over a pulley. The cask is let down into the fermenting wort and pulled up suddenly to a short distance above its surface; by repeating this several times, a very efficient rousing and aëration is secured.

There are numerous forms of rousing apparatus worked by power, such as an arrangement similar to the screw propeller of a ship, a current of air forced in by a pump or bellows, or a centrifugal pump, which draws the wort from the bottom of the tun and delivers it at some height above the surface of the wort. The extremity of the delivery pipe may be made to terminate in a rose with fairly large holes; this divides the wort into a number of small streams, and secures an abundant aëration. By causing the pump to turn in the contrary direction, air may be forced into the wort.

Racking Squares.—Beers brewed on the skimming or cleansing system are often not racked off directly from the fermenting vessels, but run into a vessel which may be large enough to hold two brewings. These may be made of wood or of slate, the latter being much the better material for the purpose. They should not be above four feet deep, and should be fixed at such a height that the largest casks can be filled from them conveniently. A number of taps are inserted in the front of the racking square a few inches above the bottom; and to these, lengths of india-rubber hose are attached long enough to reach to the bottoms of the casks to be filled. To the end of each piece of hose a racking tap is often attached, the nozzle of which has a length of metal tube attached, sufficiently long to reach the bottom of the cask. When a number of small casks are being filled, a certain amount of succussion takes place in the hose each time the racking tap is closed, which being transmitted to the beer in the tank, disturbs the yeast. This may be avoided by placing an intermediary vessel between the racking tank and the hose, the wort flowing into this vessel from the tank through a ball-cock.

Vats and Casks.—The large vats for storing purposes are made of the best English oak, hooped with iron. Since they are subject to the attacks of a fungus which causes dry rot, they should be examined periodically, and measures for the eradication of the fungus taken as soon as its presence is detected.

The trade casks are generally made of foreign oak, occasionally of English. The best quality of foreign oak comes from Memel, the next best from Dantzic, other kinds from Odessa, Blumeza, and Riga. Ale casks have been constructed from time immemorial of wood, which is a substance of a porous nature, into the interstices of which bacterial and other minute organisms are extremely liable to effect an entrance. When certain species of these organisms have, in this way, invaded the substance of the wood, it is next to impossible to effect their removal. A cask so attacked is known as a "stinker," in which condition it is completely useless for the carriage of beer. On the Continent the casks are coated internally with a peculiar kind of pitch, so that the beer never comes into actual contact with the wood. Probably in the future some material which does not take up impurities so readily as wood will be employed in the construction of brewery casks, such as steel lined with tin, or wood with a lining of some indifferent metal.

CHAPTER IX.

ESTIMATION OF THE QUANTITIES OF MATERIALS FOR THE BREW.

BEFORE commencing the operation of brewing, it is necessary to calculate, or, in other words, work out the quantity of materials which will give the required quantity of beer of the desired gravity. This is a very simple matter, and it is here that the great convenience of the expression "brewers' pounds of extract" is seen. All that has to be done is to reckon up the total number of pounds of brewers' extract which will be contained in the number of barrels of beer, or, as it is generally called, "in the length of the brew," and then to reckon up the quantity of malt, or of malt and malt-substitutes, which will be required to furnish this amount of extract.

Suppose, for example, we wish to brew 150 barrels of beer at a specific gravity of 1050. Converting this into brewers' pounds ($1050 - 1000 \times 0.36 = 18$), we find that 18 lbs. per barrel corresponds to a specific gravity of 1050. The number of barrels in the length to be produced is then multiplied by the number of pounds of extract ($150 \times 18 = 2700$); thus we find the total number of pounds required will be 2700. If the brew is to be entirely from malt, the actual yield of extract per quarter of which we do not know, we take the average yield of malt, which is about 86 lbs. per quarter. Dividing the 2700 by 86, we obtain 31.395 quarters, and $0.395 \times 8 = 3.16$, or a trifle over three bushels. The quantity to be entered in the Excise-book will be therefore 31 quarters 3 bushels, if the malt weighs 42 lbs. to the bushel. If it does not, then its weight must be calculated into bushels of 42 lbs. weight. Suppose the malt only weighs 40 lbs., then, as $42 : 40 :: 31.395 : 29.9$, or it will be entered as 29 quarters 7 bushels. If a portion of the extract is to be derived from sugar, then this amount must be deducted from the total amount required for the brew. Suppose, in the foregoing brew, one-fifth of the extract is to be obtained from sugar, then, as $2700 \div 5 = 540$ lbs. of extract is to be derived from sugar, consequently $2700 - 540 = 2160$ will have to come from malt—that is, $2160 \div 86 = 25.12$, or 25 quarters 9½ bushels will be required. If the sugar gives 35 lbs. of extract per cwt., then $540 \div 35 = 15.42$ cwts. This figure, multiplied by 112, gives 1727 lbs. (which is the quantity to be entered in the Excise-book), or 15 cwts. 47 lbs.

When several beers of different gravities have to be produced from one mashing, or, in other words, the wort to be divided into "party gyles," the calculation is made in a similar manner. Supposing that three different beers have to be produced with one mash, say 80 barrels at 24 lbs., 60 at 20 lbs., and 50 at 14 lbs., then the total extract required will be 80 × 24 + 60 × 20 + 50 × 14 = 1920 + 1200 + 700 = 3820. From this amount the malt, or malt and sugar, required is calculated as before. When gelatinised materials, such as flaked rice or maize, are to be employed, since these yield extracts of from 100 to 105 lbs. per quarter of 336 lbs., a corresponding allowance must be made. Thus, suppose 520 lbs. of the extract of the brew are to be obtained from flaked maize, yielding an extract of 104 lbs. per quarter, 520 ÷ 104 = 5 quarters (1680 lbs.) will be required. This, under the new regulations, has to be entered in the Excise-book in the proper column for prepared grain. As, however, maize is bought by the cwt., and also used in the brewery by weight, the 5 quarters will have to be calculated into cwts., and as there are 3 cwts. to the quarter, the number is simply multiplied by 3.

Calculation of the Amount of Liquor required for the Brew.—As there are several sources of loss of liquor on its way from the hot-liquor back to its ultimate destination, the fermenting vat, much more liquor is needed for a brew than the actual quantity, or the length, of the wort which is to be collected. For instance, we commence in the mash-tun with dry malt and leave off with wet grains; these absorb on an average from 28 to 30 gallons of liquor for every quarter of malt used. In the subsequent boiling which ensues in the copper, a further quantity of liquor is lost by evaporation, and this differs considerably, according to the shape of the copper, the vigour of the ebullition, the state of humidity of the atmosphere, &c.; as an average, 8 per cent. per hour may be set down for this. The wort, when on the cooler, also loses water by evaporation, and to a slight extent also in its passage over the refrigerator; these two sources of loss may be taken collectively at about another 8 per cent. Some little, about 2 per cent., remains adherent to the various vessels it passes through. Finally, there is the liquor retained by the hops, which may be set down at about $1\frac{3}{4}$ barrels for each 100 lbs. of hops, or a little over half a gallon for each pound. The total amount of liquor lost by evaporation from copper, cooler, and refrigerator, and adhering to vessels, proportionate to the length of a brew, after being once ascertained, will be found to be tolerably constant for each brewery; as a rule, it averages about 30 per cent. The other two, the liquor absorbed by the malt and by the hops, will naturally vary with the respective quantities of each employed in different brews. Taking, as an example, a brewing in which 75 barrels of wort, of the gravity of 17 lbs. (sp. gr., 1047), are to be collected in the fermenting vessels, and in which hops are to be used at the rate of 8 lbs. per quarter of malt, taking malt which yields 85 lbs. extract per quarter, the number of quarters required will be 15, and 15 × 8 = 120, the number of pounds of hops required. The 15 quarters of malt will absorb 15 × 30 = 450 gallons, or 12½ barrels of liquor, and the hops will retain 120 × 0.0175 = 2.10 barrels nearly. The loss from these two sources amounts, therefore, to 14.6 barrels. Beyond this, there is the further loss of say 30 per cent. on the 75 barrels of wort required, which is found by multiplying 75 by 0.30, and gives 22½ barrels. This, with the 14.6 mentioned before = 47.1 barrels, and with another 6 barrels for contingencies, together with the 75 barrels of finished wort, will give a total of 128 barrels.

Heating the Mash Liquor.—In all properly arranged breweries, a hot-liquor back is provided in which water for the mash is heated by steam.[1] Although, theoretically, it is not necessary to raise the temperatune of the mash liquor beyond some 25° or 30° above that at which it is actually required to enter the mashing-machine, it has been found by practical experience always best to boil the liquor for a quarter or half an hour. The contents of the hot-liquor back are heated by steam, which may either pass through a coil, or, in breweries where the steam is perfectly free from contamination, it may be blown directly into the water; or, as it is termed, the liquor may be heated by "naked steam." The boiling of the mash liquor generally takes place on the evening before brewing, and during the night it will have cooled down to a lower temperature than that required for mashing; steam is therefore turned on for a short time in order to bring the liquor up to the proper temperature. In the boiling which the mash liquor is subjected to, obviously all living organisms and their germs will have been destroyed. Since the majority of waters contain acid calcium carbonate in solution, and this, by boiling, is decomposed into insoluble calcium carbonate (chalk), with liberation of carbon dioxide, a precipitate forms which has the effect of carrying down with it a large portion of the organic matter of the water. It is extremely undesirable that this precipitated calcium carbonate should obtain access to the mash-tun, because it would there partially neutralise the natural acidity of the malt; consequently, the aperture from which the mash liquor is drawn off should not be situated at the bottom of the back, but at its side some few inches above the bottom.

Addition of Hardening Materials.—If calcium sulphate or other salts are to be added to the mash liquor, this is best done when it is on the boil. The gypsum and other substances are mixed with water to the consistency of thin whitewash, and poured into the hot-liquor back when its contents are in a state of lively ebullition.

Many breweries are unprovided with a hot-liquor back, and

[1] In old-fashioned breweries many slipshod arrangements for heating the mash liquor are met with, such as the employment of the copper for this purpose.

in these the liquor is heated either in a separate copper or in the brewing copper itself. In such cases the heating and treatment of the liquor must be made to approximate to the method just described as nearly as the nature of the plant will permit.

Mashing.—This operation consists in intimately mixing the ground malt or "grist," as it is termed, with a definite proportion of water at the proper temperature. The mixture must be thorough, for any portions of malt which escape wetting are withdrawn from the action which ensues on mashing, and as their contents remain unaffected, they represent so much waste material. This can hardly apply now-a-days to places where mashing-machines are erected, for, unless these are worked in an extremely careless manner, they do their work most efficiently.

Temperature of the Mash.—From what has gone before it will be seen that the temperature at which the mash is conducted has an immense influence on the nature of those changes which the starch undergoes under the influence of the diastase, and these, as will be seen afterwards, exercise a profound influence on the subsequent fermentation and on the character of the finished beer. The most important temperature is that known as the "initial temperature" of the mash, for upon it chiefly depends what the nature of the products formed by the action of the diastase upon the starch constituents of the malt shall be. This temperature, which is taken as soon as the mashing process has been completed, is ascertained by means of a thermometer constructed especially for this purpose. The initial temperature is varied according to the quality of the malt and the character of the required beer. Beers intended for storage are, as a rule, brewed from pale malts. These are mashed at a somewhat higher degree of temperature than high-dried malts, for we require, in this class of beers, a fairly large quantity of those dextrins which ferment very slowly, and which, since they provide for a long, slow, continued fermentation, keep the ale (through the period of its storage) charged with gas. High-dried malts are mashed at a somewhat lower temperature, since the beers produced from them are quickly consumed, and here we require a wort which contains large quantities of maltose and of those dextrins which give fulness and sweetness to the beer. Between these two types of beer, which may be taken as extremes, there are many other intermediate ones that are brewed to suit the particular wants of different neighbourhoods. The mashing temperature will also vary according to the diastatic power of the malt. Malts in which this is high will require a higher mashing temperature than less diastatic ones. There is, excepting in very rare cases, an excess of diastatic power in all brewing malts, and this has to be checked or crippled to a certain extent in the mash-tun; hence heats have almost invariably to be employed, which destroy a certain portion of the diastase. If such were not the case, too much maltose would be yielded, and the resulting beer be deficient in fulness and body. The temperatures for the initial heat vary from about 145° F. to 155° F. Malt which is perfectly dry will require a somewhat lower mashing heat than malt which has become to a certain extent slack, because, when perfectly dry starch is mixed with water, a certain amount of heat is evolved owing to the chemical combination which takes place between the two. This slightly increases the heat of the mash, and naturally the elevation of temperature from this cause is greater the drier the malt. A delicate appreciation of the right initial heat to be employed with different malts to produce beers of the required character is only to be obtained by actual experience in working, and is one of those things for which hard and fast rules cannot be laid down.

The Operation of Mashing.—Before the actual mashing is commenced the mash-tun is thoroughly warmed, either by being partly filled with hot water which is afterwards run to waste, or it is heated by steam. In the actual mashing, where the tun is provided with a mashing-machine, water is run through this from the hot-liquor back, the contents of which have been brought to the proper mashing temperature, or, as it is often termed, the right "striking heat," until the plates are just covered. The slide of the grist-case is then opened, and the malt allowed to fall into the mashing-machine, where it is intimately mixed with the mash liquor. The consistency of the mixed malt and water as it falls into the tun is observed, and by a little practice the beginner is soon able to appreciate the stiffness of the mash which corresponds to a definite proportion of water to malt, and to regulate the flow of liquor and grist accordingly. The proportions generally used are from one and a half to two and a half barrels of water, or "mash-liquor," as it is generally termed, to each quarter of malt. The former would be considered a very stiff mash, and the latter a thin one. Obviously, the less liquor employed in the mash the more there remains for sparging, and the better chance there is for obtaining all the extract from the malt in the tun, or, as it is now termed, the "goods." Diastase exhibits a little more activity in a thin mash than in a stiff one. Mashing when performed in this way is carried out in the most perfect manner, for each portion of grist meets its own portion of water in the mashing-machine, and an absolutely uniform heating of the entire mash is secured.

In the case of the mash-tun which is not provided with a mashing-machine but with rakes, rather more than the total quantity of liquor required for the mash is run into the tun at a temperature a few degrees higher than that actually to be employed in the mashing. The object of this is to provide a sufficiency of heat to warm the mash-tun; the extra quantity necessary for this purpose will have to be determined by experiment for each particular tun. After the tun is sufficiently heated the taps are allowed to run for a few minutes so as to flush out the spend-pipes; they are then closed and the temperature of the mash liquor taken. This should now be a degree or two above the temperature actually required; a few revolutions of the rake machinery will quickly reduce it to the proper temperature. When this is reached the mash-tun is closed, the slide in the grist-case opened, and the grist allowed to fall rapidly into the mash liquor, the rakes being kept in motion at the rate of about one revolution round the tun in forty or fifty seconds. The rakes are kept on for about fifteen minutes after all the malt is in the tun, and then the temperature of the mash is taken.

In certain breweries the malt is run into the tun first and the hot liquor afterwards; this is an exceedingly faulty method, and is only mentioned to be condemned.

The latter two methods of mashing have this great disadvantage, that by neither of them is it possible to ensure that each portion of grist is raised to the same temperature in the mashing. Obviously, those portions which come in contact with the hot liquor first are raised to a much higher temperature than the last portions; consequently, the diastatic power of the former is considerably crippled, while that of the latter portions is scarcely, if at all, affected. Such a mash is therefore of a very irregular nature in this respect.

Some brewers who have no mashing-machine endeavour to get over this defect by running in the hot liquor and grist at the same time, keeping the rakes continuously in motion; this, no doubt, obviates to some extent the imperfection, yet it can never yield such a regular mash with regard to equable temperature as the mashing-machine affords.

Underlet.—It is customary after the mash has stood on for some little time, a quarter to half an hour, to admit a further portion of hot liquor, at a higher temperature than that used in mashing, under the goods by means of the underlet. Such an addition also goes by the name of "piece" liquor. The temperature of this varies with the temperature of the liquor employed in mashing, and is generally about 10° to 12° above it. In those cases where the mash-tun is unprovided with rakes, the piece liquor should be allowed to flow in slowly, so as to secure as even an admixture as possible under the circumstances. Where the mash tun is provided with rakes, after the piece liquor has been added, they are slowly revolved three or four times, in order to secure an even temperature throughout the goods.

Standing on.—The mash is then allowed to stand in the mash-tun until the conversion of the starch is completed. This is ascertained by taking out from time to time a small quantity of the liquid portion of the mash and testing it with iodine. In order to do this, a few drops of the mash are placed on a white tile or the bottom of an ordinary dinner-plate,[1] allowed to become cold, and a drop of a solution of iodine potassium iodide, about the colour of brown brandy, added. If this experiment is tried as soon as the mash is made, a deep blue colour will be produced. As the mashing process proceeds the colour gradually changes from purple to brownish-red, and at last no colour is yielded at all, the wort being simply coloured yellow by the iodine solution. As soon as this happens the taps may be opened and the wort run off.

Second Mash.—In some breweries, where the copper is not large enough to boil the whole mash at once, the mash, after being made in the usual manner, is sparged until one-half of the total length required is obtained. This is run or pumped up into the copper, and fresh mash liquor run on to the goods at the rate of about one and a half barrels per quarter, and allowed to stand on until the first half of the length in the copper has been sufficiently boiled. When this has been run off the taps are again set, and sparging carried on until the second half of the length is reached, which in its turn is brought into the copper and boiled. Very occasionally a third mashing is made in a similar manner. The second mash is a time-wasting arrangement, though it is strongly advocated by some brewers, who consider that it gives a better yield of extract. Opinions are, however, somewhat divided as to the value of the last portions of extract obtained in this way; some maintain that it is not only useless, but even prejudicial.

Use of Subsidiary Apparatus.—A number of these have been described, such as circulators, steam-coil in the mash-tun, &c., the object of all of them being to gradually raise the temperature of the mash. This may be performed for two very different purposes: either to increase the quantity of the less fermentable dextrins, or to deal with inferior or badly germinated malt.

In those cases where the circulator is used for raising the percentage of the higher dextrins, the apparatus is set in action ten minutes after the mashing is completed, and the operation continued for fifteen minutes, or for such longer period as is required to bring the mash up to the required temperature. The mash may be raised in this way to 158° F., or even higher, if it is necessary to restrict the diastase considerably. In this case it is usual to curtail the time of standing on to one hour, reckoning from the time the circulation is finished. One advantage of the use of a circulator is that the wort flows bright in a much shorter time after setting taps than it does under the ordinary conditions.

The same object may be effected by the use of a steam-coil under the plates, as first suggested by Dr. Charles Graham. The mash is made with a low initial temperature, 130° to 134° F., and directly this is completed, steam is turned on at such a rate as will

---

[1] Small slabs of plaster of Paris have been recently proposed for this purpose.

bring the temperature of the mash up to 160° F., or even to a few degrees higher, in from thirty to forty minutes. A stir round with the rakes is given every seven or eight minutes to distribute the heat. It must not be forgotten that under these conditions a very high temperature is required to cripple the diastase, since it is protected by the maltose, which is, in these circumstances, rapidly formed. The time of standing on may, with advantage, be considerably shortened when this method is employed.

When using any of these arrangements for dealing with badly vegetated malts, which are always more or less deficient in diastatic power, the mash should be made at a low initial temperature, such as 140° to 145° F., and, after remaining at this temperature for half an hour or so, the heat should be very slowly and gradually raised to 155° F., at which temperature the mash should be allowed to stand until it ceases to give a coloration with iodine. The object of this treatment is to dissolve out as much diastase as possible, and then assist its action by the subsequent heating up of the mash. Circulating, unless very high heats are employed, promotes diastatic action.

Hot Grist Mashing.—In this modification of the mashing process, which was introduced and patented by Mr. Charles Clinch, the grist is heated by means of hot air to a temperature slightly lower than that of the initial mash heat required, and the mash liquor is brought to the same temperature. A slight rise in temperature takes place in the mashing, owing to the heat developed by the combination of the starch with water, and this brings the mash up to the correct initial. The malt is said to be benefited by this heating process, its aroma being brought out. During the standing on, diastatic action proceeds with extreme vigour, and the mash must not be allowed to stand more than an hour. Brilliant dextrinous worts, which contain an abundance of food for the yeast, are said to be obtained by this method.

Limited Decoction.—This process seeks to combine, to a certain extent, the decoction processes employed on the Continent in the production of lager beer with our infusion system. A mash-tun provided with a steam-coil is employed. The mash is made in the usual manner with about two barrels to the quarter, at an initial temperature of 155° F. or a little higher. After standing for an hour, taps are set and wort run off at the rate of half a barrel per quarter, and this is kept in a suitable vessel until required. Steam is then turned on and the goods raised to a temperature closely approaching 212° F., at which it is kept for fifteen minutes. Cold liquor is now sparged on until the mash is reduced in temperature to about 160° F., with the rakes going all the time. The strong wort which was drawn off previously is now stirred in, the temperature of the mash brought up to 160°, if necessary. The rakes are then stopped and the mash allowed to stand for twenty minutes or half an hour, after which the taps are set and the remainder of the operation performed in the usual manner. A little more extract is obtained in this way from good malts, but the benefit is most marked when working with hard, steely, or ill-vegetated samples. Unless the plates are good and the perforations fine, there may be trouble with the drainage.

As in this method the starch is completely gelatinised, it is possible to use a portion, say 25 per cent., of raw grain.

American System of Using Large Amounts of Raw Grain.—The American brewers use as much as 50 per cent. of maize in the shape of grits, which are thoroughly gelatinised in a separate vessel, and cooled down to the mashing temperature. The mash is made with the malt in the usual manner, and after being allowed to stand on for from an hour to an hour and a half, the gelatinised maize is stirred in, and the whole allowed to stand on for a very short time, when the taps are set. As the starch of the maize is thoroughly gelatinised, the diastase acts upon it with extreme rapidity. When such large percentages of raw material as these are employed, a distinct kind of malt is necessary, made from the coarse-skinned varieties of barley, in order to facilitate the drainage in the mash-tun. The barley is fully germinated during the malting process, and the resulting malt dried at a low temperature, so that it possesses a maximum diastatic power.

This plan of using maize grits has been adopted to a limited extent in this country, and vessels called "converters" have been placed on the market. The maize grits (which should not contain more than one per cent. of oil), are introduced into the converter together with about five per cent. of ground malt; steam is then turned on, and the vessel and its contents brought up to 200° F., and kept at this temperature for half an hour. The gelatinised maize, which is partly converted into soluble starch, is mixed in with the goods in the mash-tun just before the taps are set.

Employment of Prepared Materials.—Within the last few years a number of materials, which are more or less gelatinised, have been introduced as partial substitutes for malt. One of the first of these was a grist in the form of a coarse powder, in which state it was apt to give rise to drainage troubles. Afterwards rice was gelatinised and rolled into flakes; and soon after this maize, after being deprived of its germ which contains a large quantity of oil, was similarly treated. Barley is also now flaked in a similar manner. Barley and maize are also torrefied by being passed through a hot revolving cylinder; the sudden application of heat ruptures the starch-cells and bursts the corn, the contents of which are left in a tender friable condition, and can be readily disintegrated by the malt-mill.

In using these materials no alteration is made in the method of mashing; they are simply mixed in with malt, and pass along with it through the malt-mill, or, in the case of the flaked preparations, they may be mixed in with the grist immediately after it is passed through the mill. As a rule, from ·15 to 25 per cent. of this class of material is employed.

Black Beers.—In the production of stout and porter it was formerly the custom to use a grist composed of pale, brown, and black or patent malts, but now the brown malt is generally omitted, and often a portion of the black malt is replaced by caramel. The relative proportions of these constituents vary considerably in different breweries, as is shown in the following table:—

	Pale.	Brown.	Amber.	Black.
	Per Cent.	Per Cent.	Per Cent.	Per Cent.
No. 1	48	24	24	4
No. 2	57	25	15	3
No. 3	58	30	10	2
No. 4	60	34	0	6
No. 5	91	0	0	9

It is also customary to use a certain amount of sugar in black beers; this is generally either invert or cane.

Return Wort.—In some breweries it is the rule to pump the last runnings from the mash-tun, or the liquor which has been used to sparge the hops, or both, into the hot-liquor back, and to use them for mashing in the next day's brew. This plan is of very doubtful value, for though the last runnings may contain a certain amount of extract, as shown by the hydrometer, yet it is extremely questionable if it is of much value. When hopped returned worts are used, they have a great tendency to cause what is known as a "dead" mash. Probably this is owing to the precipitation of the proteids of the malt by the tannin bodies of the hops; the diastase is also prejudiced from the same cause.

Setting Taps.—This is the designation given to the opening of the taps which are attached to the ends of the spend-pipes. The taps are first turned on very slightly, and as soon as the wort begins to run clear they are then gradually opened more fully. After the sparging is commenced they are adjusted so that the quantity of wort flowing from them is about equal to the amount of liquor being delivered by the sparge, but, as will be seen farther on, this method is subject to modifications.

Sparging.—In order to obtain the full amount of extract from the goods in the mash-tun, it is necessary that the sparging should be performed in an intelligent manner, and that the apparatus should be well arranged and in good order. Sparging should never be too rapid; it should, as a rule, occupy about four hours. Very often sparges are badly constructed, and do not deliver the liquor evenly on the surface of the goods; and the same thing arises if a number of the holes in the arms of the shape get blocked up with dirt. In order to ascertain if a sparger is working evenly, it should be tested with the trough described on page 439.

With reference to the heat of the sparge liquor, this should never be too high, or during the last part of the operation the diastase present in the mash may be completely destroyed, and soluble starch may make its appearance in the wort and give rise to subsequent difficulties. Since, during the time of standing on, the goods at the surface lose a certain amount of heat, it is customary to run on the first portions of the sparge liquor at a somewhat higher temperature than the latter portions. As a rule, the first half or three-quarter barrels of sparge liquor per quarter may be run on at a temperature of from 170° to 175° F. The temperature should then be allowed to decline gradually, until, at the final stage of the process, the temperature stands at 160°, or a few degrees lower.

Much difference exists as to the way the sparging is performed in relation to that in which the wort is drawn off. Some brewers sparge at the same speed as that at which the wort is running off, in which case the goods are kept at about the same level all through the operation. Others, after stopping the sparge, allow almost the whole of the wort to run off once or several times during the process. The former is probably the better plan for beers of low gravity, the latter for stronger ones, and this latter course becomes a necessity when a large quantity of concentrated wort is required. In this last case as much wort as possible is run off, the amount of wort drawn off being replaced by sparge liquor, or by underlet, or by both, two or three revolutions of the rakes given, and the whole allowed to stand for a short time before sparging is again commenced.

This plan of adding liquor by underlet, mixing with the rakes, and allowing a short stand, may be resorted to if the drainage becomes bad at any time from the goods getting down on the plates. Care should be taken to see that the sparge liquor does not form channels through the goods, a contingency which may occur, especially at the sides of the rake machinery. All the liquor passing through such channels escapes without contributing its share to the washing-out process.

The last runnings, if the sparging has been carefully and properly performed, should have a gravity of not more than 1004 sp. gr. for worts of medium strength, and a couple of degrees higher than this for strong worts; and this is perhaps as far as the washing should be carried, for, if carried farther, substances of doubtful value are then extracted. A quantity of water nearly equal in weight to the original weight of the malt used in mashing remains in the grains.

Dead Mash.—During the whole sparging process the goods should float on a substratum of wort, and their surface appear comparatively dry. When from any cause the goods fail to float in this way, the mash is said to be dead. In such cases, the goods lie obstinately on the plates, and the sparging liquor, instead of percolating through them, collects on their surface. This state of affairs may arise from several causes: too high a striking heat may have destroyed the diastase; the malt may have been too finely ground; malt-flour may have been separated from the husk by falling from too great a height into the grist-case, or by excessive knocking about in the elevator; the employment of too much huskless material; and excessive use of the rakes. Sometimes, when bad drainage depends upon a layer of starch becoming deposited on the plates, this after a time is dissolved by the action of the diastase, and the drainage then improves. Occasionally, a stir round once or twice with the rakes will mend matters, but in very bad cases such methods as draining off from the grain trap, or placing a bundle of straw in the goods and pumping from the middle of this have had to be resorted to. A dead mash is, however, only the result of carelessness.

Wort in Underback.—In some breweries this undesirable vessel is a necessity, because the copper is at a higher level than the taps, and the wort must have some receptacle to lodge in until it can be pumped into the copper; in others, where the mash is boiled off in two lengths, a large underback must be provided. As it is desirable to prevent the cooling of the wort as much as possible, the underback should always be provided with a steam-coil. In the former case, this vessel should be just sufficiently large to serve the purpose for which it is intended, and the wort should be allowed to remain in it as short a time as possible, for it must be remembered that diastatic action is continuously proceeding in the wort until it has reached a temperature of $180°$ or $190°$ F. in the copper.

Boiling and Hopping.—The mash being completed, the wort, which will have either been run into the copper, or pumped in to it as it left the mash-tun, or have been pumped in from the underback, is now boiled for a longer or shorter period.

Objects of Boiling.—The boiling is performed for several objects. The wort as it leaves the mash-tun is swarming with minute organisms and their germs, which, though to a certain extent paralysed by the heat to which they have been subjected in the mash-tun, are not killed. Consequently, if the mash were allowed to remain for any length of time in an unboiled condition, it would soon fall into a state of bacterial fermentation. The prolonged boiling to which it is subjected effectually destroys these organisms and renders the wort sterile. All further diastatic action is also put an end to. A considerable portion of the proteids, since they are coagulable by heat, are thrown out of solution and precipitated, and a further small portion are precipitated by the tannin bodies of the hops. During the boiling the hops are added, the functions of which are to give a pleasant aromatic flavour to the beer, to confer stability on it, and to assist in the precipitation of the proteid matters of the wort. The wort is also considerably concentrated during the boiling, and this permits a larger quantity of liquor to be used in the sparging and, consequently, a larger amount of extract to be obtained from the malt.

Addition of the Hops.—Much difference prevails as to the time when the hops should be added to the wort in the copper. Some brewers add them as soon as there is a fair quantity of wort in the pan, others just after the wort begins to boil. Probably the most rational method would be to add the first portion after the wort has been boiling for ten minutes or a quarter of an hour. By this time a large amount of proteids would have coagulated and been thrown out of solution by the mere action of the heat, and the tannin of the hops would then be able to expend its full action upon the portion still remaining in solution; in this way a larger total amount would be precipitated. As the fine delicate aroma of the hop is due to the presence of certain volatile oils, which, to a great extent, are dissipated and lost in the boiling process, it is now customary to add the coarser hops first and the finer ones about twenty minutes or half an hour before turning out, and this modification has much to recommend it. Hops contain various constituents, some of which are much more readily soluble than others, and it is necessary to have both of them in the finished beer. But, unfortunately, there are other constituents of a highly undesirable nature also present in hops; these are only brought into solution by prolonged boiling, and, as a consequence, are extracted along with some of the less soluble constituents, which are of a valuable nature. By adding the hops in two portions at different times, this difficulty is to a certain extent overcome. The first portion added is extracted to the fullest extent, and though the undesirable bodies are for the most part dissolved, yet this long-continued extraction is necessary to secure the solution of the hop resins on which the preservative power of the hops depends. The tannin and the essential oils are also dissolved, but the latter are almost completely lost during the prolonged boil. It is undesirable to have too large a quantity of the hop resins dissolved in the wort, because they are precipitated in the cooler or in the fermenting vat, and contaminate the yeast. But when a portion of the hops is added at a late stage of the boiling, only small quantities of hop-resin and of the objectionable matters are dissolved, while the bulk of the tannin and essential oils are perfectly extracted; and as the boiling is now short, these last very valuable constituents are only dissipated to a slight degree. It has even been proposed to add the hops in three portions, the first at the commencement, the second in the middle, and the third near the end of the boil. Naturally, the finest hops will always be added last, since it is the essential oils of these which it is the most desirable to retain. The preservative power of the hops is dependent on the soft resins they contain, and these, on excessive boiling, undergo a chemical change and are converted into bodies of a less soluble nature; hence hops should never be boiled a second time. It is highly probable that the long periods which hops are often boiled at the present time might be shortened with advantage.

When the wort is boiled at twice or three times, or when it is distributed over several coppers, each copper should have its due proportion of hops, calculated on the quantity and the respective gravity of the wort of each.

Boiling.—As soon as the wort is in the copper it should be brought to the boil as quickly as possible, for, until a certain temperature is reached, diastatic action slowly proceeds, and more and more maltose is formed. The boiling must be thorough; there is probably no more frequent source of unstable beer than insufficient or imperfect boiling.

As to the time of boiling there is much difference in practice. Some brewers only boil for an hour, some an hour and a half, others two hours or longer. Probably two hours may be taken as the happy medium, and this will give sufficient cooking. The unnecessary prolongation of the boil causes a useless expenditure of fuel. Whatever length of time is taken for the boil, during the whole of that period the contents of the copper should be kept in a state of vigorous ebullition.

The contents of the copper, unless constantly watched, are so liable to boil over that it is necessary for one person to give his whole attention to the process to see that this does not occur. In order to do away with this constant supervision, it is customary to employ a dome or fountain, which not only prevents boiling over, but also helps to secure an abundant aëration of the wort. Domes and fountains, especially the latter, are objected to by some, since they have a tendency to disintegrate the hops, in consequence of which they do not form such an efficient filter for the wort when it is run out of the hop-back as when they are unbroken. Much has been said on the comparative values of steam and fire coppers. Some maintain that a much more vigorous boil may be maintained in a copper heated by direct fire, but the balance of evidence shows that either form is equally good, provided the arrangements for the heating are properly constructed, and the steam-copper is undoubtedly much the more economical in the matter of fuel. Boiling by gas has been introduced into some breweries which are provided with the necessary apparatus for producing what is termed "water gas." It forms an exceedingly convenient source of heat for this purpose, since its intensity can be so easily regulated.

Boiling at Twice.—In what has gone before, it has been assumed that the whole length, that is, the whole quantity, of the wort is boiled in one copper; but often the copper is not large enough to contain the whole length, and in such a case

ACTION OF HOP-TANNIN IN BOILING.

it has to be boiled at twice, or in extreme cases at three times. Under these circumstances sparging is kept up until sufficient wort is obtained to fill the copper, the taps are then stopped, and the mash allowed to stand on until the contents of the first copper are boiled. It is customary in such cases to add a further amount of hot liquor to the goods in the mash-tun, either by sparging, or underlet, or both.

The next point to be determined is the respective proportions in which the hops are to be added to each copper. It is obvious that if we find the right proportion for the first copper, the remainder will be the proper quantity for the second. The calculation is based on the respective amounts of pounds of brewer's extract in each copper, and if we know the number of pounds in the first copper, that amount deducted from the total number of pounds required for the whole length will be the quantity contained in the second copper wort.

Taking, for example, a brew in which a hundred barrels of beer of a gravity of 21 lbs. are to be produced, and to which hops at the rate of 10 lbs. for each quarter of malt are to be added, the wort to be boiled in two equal lengths. The total number of pounds of brewer's extract required will be 2100, and if the malt yields 84 lbs. extract per quarter, twenty-five quarters of it will be the necessary quantity: 25×10 gives 250, the total quantity of hops required. On an average the total length of the brew of a hundred barrels, allowing for loss between the copper and the fermenting vessels, will amount to 134 barrels, half of which, or 67 barrels, have to be boiled in each copper length. Say this first length is found to have a gravity of 24 lbs., the total number of pounds of extract in the copper will be $67 \times 24 = 1608$ lbs. Then—

2,100	:	1,608	: :	250	:	191.4
Total lbs. of extract in whole length.		Lbs. of extract in first copper.		Total amount of hops.		Amount of hops for first copper.

The remainder, $250 - 192 = 58$ lbs., will go to the second copper.

Action of the Hop-Tannin Bodies during the Boiling Process.—As has been stated previously, both the hop-tannin and the phlobaphen combine with albumin; consequently, during the boiling process they will enter into combination with the proteid bodies dissolved in the wort. Hayduck found that hop-tannin forms with albumin a compound readily soluble in hot water and slightly so in cold, but that phlobaphen formed an absolutely insoluble compound. In a cold-water extract of barley, which had been boiled and filtered to remove the coagulated albumin, a solution of hop-tannin gave a precipitate which again dissolved on heating, but was reprecipitated as the solution became cold. When this precipitate had been removed by filtration, the presence of tannin could be detected in the clear cold filtrate by ferric chloride, and the presence of albumin could also be detected by tannin, showing that a portion of the tannin-proteid compound still remained in solution. When this solution, which was perfectly clear at the ordinary temperature, was further cooled, it again became turbid. The barley extract behaved in a similar manner without the addition of tannin, though to a less degree; if, after it had been boiled and filtered, it was boiled a second time, a fresh precipitate slowly formed, and this shows that a portion of the proteid matter in solution gradually became insoluble on boiling. When a small quantity of tannin was added to the barley extract, and the precipitate which formed filtered off, the solution, on subsequent boiling, behaved in a similar manner; it was determined experimentally that a small quantity of tannin was carried down with the precipitated matters. When, on the other hand, phlobaphen was added to the barley extract, a precipitate formed which was quite insoluble when the fluid was subsequently boiled. When an excess of phlobaphen was added to the barley extract, all the proteid matter was completely precipitated, for, neither on the addition of hop-tannin nor of ordinary tannin, was a precipitate produced.[1] From this it follows that the phlobaphen completely precipitates a certain amount of the proteid matters, whilst that portion which combines with the hop-tannin remains in solution during the boiling process; it is partially precipitated as the wort cools, and still further during the fermentation. It was found possible to remove the hop-resins from the scum which formed on the wort, by treating it with ether, and, by afterwards boiling the scum freed from the resins in this way with water, to detect the presence of hop-tannin in the solution. Hayduck concludes from these experiments that in the process of boiling, a certain portion of the soluble proteid matters are gradually converted into insoluble modifications, which probably carry a portion of the hop-tannin down with them. These matters cannot, however, be completely separated by boiling, for their presence could be invariably detected in beer by tannin, hop-tannin, or phlobaphen, though beers which were heavily hopped contained smaller quantities of albuminoid matter uncombined with tannin. He considers that the hop-tannin and phlobaphen, since they remove a portion of the proteid matter from beer, act indirectly as preservative agents, but that they have no direct influence. He had previously shown that hop-tannin exercised no action in arresting the growth of the lactic bacteria, and that an aqueous extract of hops, from which the resins had been removed by treatment with ether, had just as little effect.

From an experiment made by Briant and Meacham,[2] the quantity of albuminous matter removed by boiling with hops is very

---

[1] Probably the beneficial effect of a prolonged vigorous boil may be due to the conversion of the hop-tannin into phlobaphen.

[2] *Transactions of the Institute of Brewing*, vol. vi. p. 160.

small. Two brews were conducted under similar conditions, the sole difference being that in one (A) hops were used at the rate of 8 lbs. per quarter, in the other (B) at 16 lbs. The result was as follows:—

Soluble Albuminoids in Dry Extract.

	Copper Boiling.	Copper Out.	Albuminoids Precipitated.
Experiment A	Per Cent. 5.93	Per Cent. 5.75	Per Cent. 0.18
Experiment B	6.01	5.83	0.18

Turning Out.—After the wort has been boiled for a sufficient length of time it is "turned out"—that is, its contents are run out into the hop-back. The hops gradually settle down to the perforated bottom of this vessel, where they form a layer of filtering material, and as the wort runs off, hold back the various matters which have been precipitated during the boiling process; the wort consequently passes away in a bright and clear condition. Some breweries are not provided with a hop-back, but have a strainer fixed in the copper which retains the hops; this is a faulty arrangement, since under these conditions the wort only flows away very slowly and much waste of time ensues.

Cooling.—After a sufficient length of time has been given for the hops to settle in the back, the wort is either run or pumped on to the cooler, and as it is there exposed to the air in a thin layer, it rapidly loses heat. The cooling effect of the air is sometimes increased by a fan placed in the middle of the cooler, which, when set in motion, induces powerful currents of air, which are wafted over the surface of the wort. The wort, while on the cooler, takes up a quantity of oxygen, which, as was shown by Pasteur, enters into chemical combination with some of its constituents. This combined oxygen exercises a profound influence on the readiness with which the wort undergoes subsequent clarification. In a wort which has been well aërated the precipitated particles agglomerate into coherent masses, which subside readily, whilst in an unaërated wort they remain suspended in a state of fine division. The wort also becomes darker in colour under the influence of the combined oxygen; the darkening due to this cause, however, entirely disappears during the fermentation. The wort, during its passage through the cooler, deposits much matter, which is known as "cooler sludge," and this should not be allowed to be carried away with the wort. Wort should not be allowed to stay on the cooler for a longer period than is necessary to bring its temperature down to 140° F., for, if allowed to remain for any length of time at a temperature a few degrees below this, it is liable to take on changes of an objectionable character, the result of which is the production of a beer of a less stable character. As soon as, or a little before, the wort has arrived at 140° F., it should be passed on to the refrigerator.

Refrigerating.—In passing over the refrigerator, the wort takes up a further quantity of oxygen; this does not, however, chemically combine with it, but simply enters into a state of solution. This dissolved oxygen is of great importance in stimulating the yeast and starting a vigorous fermentation. Since vertical refrigerators afford the fullest exposure of the wort to the air, they are to be preferred to the other forms which do not possess this important advantage. The rate at which the wort passes over the refrigerator is so regulated that it leaves the apparatus at the temperature required for commencing the fermentation, which, as a rule, is 60° F., or two or three degrees below.

Employment of Germ-Free Air.—In recent times it has been proposed to carry out the operations of cooling and refrigerating in such a way that the wort is kept entirely out of contact with the outside germ-laden atmosphere, and only allowed to come in contact with air which has been deprived of its germs by filtration. Consequently, wort treated in this manner reaches the fermentation vessels in as sterile a state as it left the copper, and this is undoubtedly a step in the right direction. Though aëration with germ-free air in this way has been adopted to some extent on the Continent, it has only been applied in this country to the air which reaches the refrigerator. The wort is most liable to be contaminated by living germs during its passage over the refrigerator, for its temperature when on the cooler is so much higher, that any germs which fall in are either destroyed or have their vital energy considerably lowered.

Collection of the Wort.—After refrigeration the wort passes either into the collecting vessel or, more generally, it is run directly into the fermenting vat or vats. In either case it is here that the Excise officer takes a note of its quantity and gravity, and upon the calculation made from these data the amount of duty payable is based.

Excise Charges.—The duty on beer is levied at the rate of 6s. 9d. for each barrel at a specific gravity of 1055, an allowance of 6 per cent. being made for waste and loss during fermentation. This allowance is not reckoned off each day's brew, but is deducted at the end of the month, when the book is made up for payment. Consequently beer, at whatever gravity it is brewed, has to be reduced to its equivalent in gallons at the standard gravity of 1055. The calculation is effected in the following manner. The number of gallons in a brew is ascertained from the dip,[1]

---

[1] All fermenting vessels are gauged by the Excise officers, *i.e.* the number of gallons corresponding to each inch of depth ascertained. The number of gallons of wort actually contained in the vat is found by the dipping-rod, which is a kind of boxwood rule divided into inches.

EXTRACT ACTUALLY YIELDED.

and the figure thus obtained multiplied by the excess of the gravity over 1000. Suppose we have a brew of 3600 gallons of sp. gr. 1060; then $3600 \times 1060 - 1000 = 216,000$. This figure divided by 55, the excess of the standard gravity over 1000, gives $216,000 \div 55 = 3927.27$. Therefore, excluding the decimals, 3927 gallons of wort at 1055 sp. gr. are equivalent to the 3600 gallons at 1060 sp. gr., and this former figure (3927) is entered into the day-book. At the end of the month the amounts for each day are added together, the allowance of 6 per cent. for waste deducted, and the remainder brought to barrels by dividing by 36. Thus, suppose the amount brewed in one month was equivalent to 125,640 gallons at standard gravity (1055); this multiplied by 0.06 would give the 6 per cent. to be deducted for waste, which would be 7538.4, and $125,640 - 7538 = 118,102$, which is the gross number of gallons for which duty would be charged. The number thus obtained is then converted into barrels by dividing by 36; thus $118,102 \div 36 = 3280$ barrels and 22 gallons. The duty is not collected on the 22 odd gallons, but they are carried over as "odds" to the next monthly account. The duty per barrel, 6s. 9d., multiplied by 3280 gives £1107. From this it will be seen that the brewer, when not paying on goods, pays duty on the weight of extract yielded at the rate of 6s. 4.14d. for each 19.8 lbs. of brewer's extract, since each barrel of beer at a specific gravity of 1055 contains this amount of brewer's extract, and 6s. 9d. less the allowance of 6 per cent. equals 6s. 4.14d.

If the brewer does not obtain four barrels of wort of specific gravity of 1055°, less 4 per cent., from each quarter, or 336 lbs. of malt or its equivalent, the charge, instead of being levied on the wort, is levied on the materials used, and in this case a higher rate is paid. From the amount of the charge obtained in this way the 6 per cent. for waste is deducted as when the charge is made on the wort. 224 lbs. of cane sugar are deemed by the Excise authorities to be equivalent to a quarter of malt weighing 336 lbs., and similarly 256 lbs. of glucose or of invert sugar, or of flaked maize or rice, 272 lbs. of No. 1 syrup weighing 14 lbs. per gallon, or 328 lbs. of No. 2 syrup weighing 13 lbs. 2 oz. per gallon. It is only, however, with very badly arranged plant, or with the use of very indifferent materials or want of skill in working, that this condition of things can arise.

Extract per Quarter Actually Yielded.—The brewer should never omit to ascertain from day to day the actual amount of extract which his goods are yielding, as this at once affords valuable information as to the manner in which work is being carried on. To do this, the number of barrels in each brew is multiplied by the gravity expressed in brewer's pounds, and from this is deducted the amount of brewer's pounds obtained from sugar or syrup. The figure thus obtained is then divided by the number of quarters of malt used. For instance, if in a brew in which 25 quarters of malt were used, and 14 cwts. invert sugar yielding 35 lbs. per cwt., 125 barrels of wort were produced of a specific gravity of 1055°, the total number of pounds of brewer's extract contained in this would be 125 multiplied by 19.8 (this latter figure being the equivalent of sp. gr. 1055°, $i.e.$ 1055 − 1000 × 0.36 = 19.8). The sum obtained would be 2475 lbs. of brewer's extract, and of this 490 were obtained from sugar; deducting this, 2475 − 490 = 1985 lbs. were obtained from the malt; since 25 quarters of malt were used, then 1985 ÷ 25 = 79.4 lbs. per quarter.

Percentage of Extract upon the Malt.—As there is some doubt as to what the actual amount of solid matter in solution which corresponds to 1 lb. of brewer's extract is, or to a degree of specific gravity, it is impossible to give an exact method of finding the actual amount of solid matter dissolved from the malt. The multiplying factor generally employed is 2.59, which is based on the assumption that when 1 lb. of malt-extract is dissolved in water and the amount made up to 10 gallons, the specific gravity is raised 3.86 degrees (water = 1000). Though this is true for cane sugar in a 13 per cent. solution, it is not so with regard to malt-extract, 1 per cent. of which, according to O'Sullivan, raises the specific gravity 3.95°, in which case the factor would be 2.532.

For instance, if a malt yields 85 lbs. extract, and if the former factor is employed, the actual matter dissolved from a quarter of malt will be 85 × 2.59 = 220.15 lbs.; if the latter, 85 × 2.532 = 215.220.

From this the percentage yield of malt is obtained by the proportion sum 336 : 100 :: 220.15 : 65.5, which is the percentage when the factor 2.59 is employed.

FERMENTATION.

During the process of fermentation the wort is submitted to the action of yeast, through the influence of which considerable portions of its saccharine constituents are converted into alcohol and carbon dioxide; it is here that the fluid first assumes its true character of an alcoholic and fermented beverage. The methods in which this part of the operation of brewing is conducted vary considerably, though all have one common object. They may be roughly classified under the following three heads—the *cleansing* system, the *skimming* system, and the *stone square* system.

Addition of Yeast to the Wort.—This is technically known as "pitching." In many breweries the usual custom is to add the pitching yeast, the proper quantity having been previously ascertained by weighing, to the whole of the wort after it has been collected in the fermenting-tun. This plan cannot be recommended, since it delays the active reproduction of the yeast, and it should be the brewer's object to induce a rapid growth of yeast in the wort as soon as possible, for the sooner this is brought about the less chance foreign organisms have of obtaining a foothold. It is therefore much better first to run down a small portion of the wort at a temperature of from 65° to 75° F., and to mix the yeast with this. In this way a rapid and vigorous growth of yeast is secured from the onset, and the reproduction of any bacterial organisms, should these happen to be present, effectually held in check. The remainder of the wort is then run in at a slightly lower temperature than that which the whole bulk is to have when collected, so that at the finish the gyle may be at the proper heat. The wort, while being collected, is roused at frequent intervals in order that the yeast may be evenly diffused through it.

Choice of Yeast for Pitching.—The fitness of yeast for this purpose is determined by its general and microscopical characteristics. With regard to the former, it should possess an agreeable taste and odour, and, if derived from ale wort, should be of a pale yellow colour. The best yeast for pitching is obtained from medium gravity worts (18 to 20 lbs. gravity), which are not too heavily hopped. The first heads of yeast thrown off during fermentation are always very dirty, since they bring up with them much of the substances precipitated during cooling, such as the compounds of hop-tannin with the albuminoid matters of the malt, hop-resin, &c., and as they are quite unfit for pitching purposes, they are thrown away. The middle skimmings, which yield the best pitching yeast, are preserved. The last skimmings are also rejected, because the yeast finally thrown off is not nearly so strong and vigorous as that of the middle skimmings. Much stress has been laid upon the proper consistency of the pitching yeast, and it has been the rule only to use yeast from which the beer has been allowed to drain to such an extent that the yeast is left in a firm semi-solid pasty condition, without any tendency to sloppiness. Young yeast, which is generally thin because sufficient time has not elapsed for the beer to drain away from it, should not, if good in other respects, be rejected on account of its thinness; naturally a little more must be added to allow for the beer it contains. But thick yeast, which has become sloppy through age, should never be employed, for sloppiness in this case is an indication that putrefaction has already commenced. That these external characteristics may be at times delusive is shown by the No. 1 Carlsberg bottom yeast, which, though it has yielded excellent results during a long series of years, is said to be wanting in all those external characteristics which are usually regarded as of importance. As to the age of yeast, in the opinion of Southby and other reliable authorities, the younger it is the better; in extreme cases it should never be employed when more than a week old, and this during the coldest months in the year; during the hottest it cannot be safely used when more than four days old.

The yeast used for pitching should be frequently submitted to microscopic examination. When placed under the instrument, the cells should appear uniform in shape and size, detached from one another, plump, and with sharp, well-defined borders. The protoplasmic contents should be transparent, with cloudy markings here and there, the vacuoles clear and of moderate size. Very young cells have no vacuoles, and their contents are quite clear and transparent; as the cells mature the vacuoles increase in size, and the cell-contents become granular. The cells shrink in size as their age progresses, their walls become thickened and wrinkled, and their protoplasmic contents exceedingly granular. When this stage is reached, the cells are either quite dead, or so deteriorated as to be useless; consequently, a sample which contains even a trace of such cells should never be used for pitching purposes. The number of cells actually dead may be easily ascertained by running under the cover-glass a weak solution of one of the aniline dyes, such as methylene blue, which immediately stains the dead cells a deep blue colour.

Bacterial organisms should be carefully looked for during the microscopic examination, several fields of the same yeast being systematically gone over, and the number of bacteria present in each noted. Good and reliable yeast ought to be practically free from these organisms.

Hansen strongly advises the employment, for pitching purposes, of that portion of the yeast which is produced in the earlier stages of fermentation, since it is the least likely to be contaminated with wild yeasts.

Quantities of Yeast required for Pitching.—The quantity of yeast to be used for this purpose is generally estimated by weighing, and, when speaking of pounds of yeast, it must be understood that this refers to good thick stiff yeast. Should the yeast be insufficiently drained and contain beer, an allowance must be made for this, and proportionately more yeast added. The quantity of yeast required for pitching a tun is usually calculated at so many pounds of yeast per barrel of wort, and its amount varies with the different classes of beers, proportionately more being necessary for high gravity and for heavily hopped worts. From $1\frac{1}{2}$ lbs. to 2 lbs. per barrel will be necessary for a beer of 18 lbs. to 20 lbs. gravity, brewed with 5 lbs. to 6 lbs. of hops per quarter of malt. A 24-lb. wort, with 14 lbs. to 16 lbs. of hops per quarter, will require from $2\frac{1}{2}$ lbs. to $3\frac{1}{2}$ lbs. of yeast per barrel, and worts stronger than these some $\frac{1}{2}$ lb. per barrel more. The salts in the brewing water, when their amount is not inordinately large, seem to have little or no effect upon the quantity of yeast required; but when antiseptics are added to the mash or to the wort, they exercise a retarding action on the yeast, and hence necessitate its addition in slightly larger quantities. This is especially true of salicylic acid and the bisulphites, though it is questionable if the monosulphites have much action in this respect. In those systems of fermentation in which a considerable amount of rousing and aëration takes place, the conditions are especially favourable to the growth and multiplication of the yeast, and consequently less pitching yeast is required. The most vigorously active yeasts are those derived from beers of medium gravity which are not too heavily hopped, and such yeasts are therefore best adapted for general pitching purposes. Yeast, when continuously passed through several gyles of strong wort, gradually becomes more and more sluggish in its action, probably from being surfeited with nitrogenous matter. When frequently cultivated in heavily hopped worts, its surface acquires a coating of hop-resin, which naturally interferes with the fulfilment of its proper functions.

Changes of Yeast.—In some breweries the yeast, after being in use for a longer or shorter period, commences to evince signs of degeneration; it gradually becomes slower and slower in its action, and performs its work generally in an unsatisfactory manner. When this happens a change of yeast has to be obtained from some other brewery. The degeneration is in many cases brought about by bacterial contamination, and in this case a microscopic examination at once reveals the cause of the trouble. Yeast degeneration of this kind can only occur where brewing operations are carried on without a proper regard to cleanliness; it is often associated with a dirty condition of the wort pipes and other communications. In other cases, the cause of the degeneration is much more obscure, and probably depends upon the yeast not finding a sufficiency of some substance or substances in the wort which are necessary for its perfect nutrition.[1] Thausing lays the blame chiefly on inferior malt or faults in the mashing process. While some brewers find it necessary to have a change of yeast every few weeks or months, there are others who are able to work with the same yeast for almost unlimited periods; and, granting that the yeast invariably receives a proper supply of all the substances necessary for its nutrition, and that the brewing operations are conducted with a due regard to cleanliness, it is difficult to understand why this should not be more universally the case. The Carlsberg bottom yeast, No. 1, which has been used at

[1] A remarkable instance of this occurred at the Kalinkin Brewery, at St. Petersburgh, where Hansen's system of single-cell yeast is employed. A yeast which had yielded eminently satisfactory results for two years began to degenerate, and became at last unusable. The cause of the mischief was found to be the absence of sufficient lime in the worts, the water employed at the brewery, both for malting and brewing purposes, only containing 0.91 grain of lime (CaO) per gallon. The addition of gypsum to the steeping water and mash liquor brought back the yeast to its former pristine vigour, and restored the proper amount of lime to its ash, which was at first 4.48 per cent., but which had fallen as low as 0.54 per cent.

It was found experimentally that a vigorous yeast introduced into a wort from which the lime had been almost completely removed by dialysis caused a "bladdery" fermentation.

that brewery continuously since 1883, is an excellent instance of this.

In making a change of yeast, the fresh supply should be, if possible, secured from a brewery working on different lines, and, if possible, from one where the worts were entirely produced from malt. Too great dissimilarities in the type of yeast should be avoided; for instance, the slow-fermentation yeast used in the stone square system would be unsuitable for a brewer working on the skimming system. The new supply should be rigorously examined before being taken into actual use; and, when a change is being effected, it is well to keep some of the old stock in hand, as there is always a possible chance that the new may turn out worse than the old.

Making a change of yeast is always a matter of great uncertainty; matters may be improved by the new supply, or the change may prove for the worse, and it is here that the extreme utility of Hansen's reformed system shows itself. In a brewery supplied with a pure yeast propagating apparatus, the brewer, when his stock of yeast begins to show signs of degeneration, instead of having to seek a change of yeast from another brewery, with all its consequent uncertainties, simply turns to his own apparatus, with the full assurance that he will there find a supply of yeast of exactly the same nature and character as the one he has been using.

Temperature at which Fermentation is Conducted.—This varies somewhat with the different classes of beer. The weaker qualities, such as those of a gravity of from 18 to 20 lbs., are generally started at about 58° to 60° F., and are not allowed to rise beyond 70°. But the better plan is not to permit them to rise beyond 66°, since a temperature above this is liable to encourage the growth of bacterial organisms. With stronger beers greater latitude is often allowed; they are started at from 56° to 58° F., and allowed to run up to 75°, since the larger amount of alcohol formed, coupled with the more vigorous fermentation in a strong wort, serves to keep bacteria in check, but even these are all the better for being kept down to 66°. Low fermentation temperatures conduce to the production of beers having a fine flavour; moreover, they contain larger quantities of carbon dioxide in solution, and are more stable. Slightly higher temperatures are admissible in the case of black beers; the brown and black malts, to which these particular types of beer owe their distinctive characteristics, appear to exert a certain antiseptic influence on the worts. Where the plant is provided with powerful attemperators the fermentation may be commenced at a higher temperature and confined within narrower limits, say between 62° and 65° F., with good results.

Dressing.—When the fermentation is sluggish and the yeast crop does not develop with its usual vigour, the heads being persistently discoloured with hop-resin which ought to have been brought up by the first heads, it is usual to dress the gyle. The old-fashioned way was to take 1 lb. of wheat-flour and 4 ounces of salt for each ten barrels of wort in the gyle, and to mix these ingredients well together on the surface of the fermented wort, after which a thorough rousing was given. It was formerly supposed that the stimulative action of the dressing on the yeast was due to the albuminoids of the wheat, which were dissolved by the salt. Now it is known that the effect is caused by the diastase present in the wheat-flour, and that in those cases where dressing is beneficial, the cause of the trouble is too high a percentage of the more unfermentable carbohydrates in the wort, which, after being degraded by the diastase of the flour used in dressing, are rendered amenable to the action of the yeast. It is now more customary to use the more diastatic malt-flour instead of wheat-flour, and this without the addition of salt. Dressing is always objectionable, because a number of bacterial organisms and their germs are introduced into the wort; consequently, it should only be resorted to in very exceptional circumstances.

Appearance of the Heads of Yeast during Fermentation.—In all normal fermentations the yeast thrown up on the surface of the fermenting wort, technically called the head, undergoes certain characteristic changes, and considerable information may be obtained as to the way in which the yeast is fulfilling its functions by carefully observing the heads. When the wort has been pitched at the usual temperatures (58° to 60° F.), the commencement of fermentation may be detected in two or three hours by the appearance of small bubbles of carbon dioxide which rise to the surface. In another two or three hours froth begins to form round the sides of the vessel, and this gradually extends over the whole surface and increases in volume, until what is termed the "cauliflower" stage is reached; this then gradually passes into the "rocky head" stage. The heads go on steadily increasing for a time, and often attain a height of three or four feet above the surface of the wort. The more or less frothy head now commences to fall and the "yeasty head" commences to form. This is in a constant state of motion from the continual formation and bursting of the large bubbles of gas that is now being rapidly disengaged. With the commencement of the formation of the yeasty head, what is known as "skimming-point" is reached, the normal time for this being about forty-eight hours from the time of pitching. The gravity of the wort will by this time, according to circumstances, have been reduced to from one-half to two-thirds of its original gravity. It is at this point that the separation of the yeast from the beer begins in the cleansing and skimming systems, and it is also the point at which the treatment of the wort on the different systems diverges.

The Cleansing System.—Probably this is the oldest system of all, and is the survival of the old home-brewing, where the same tub did duty for mash-tun and fermenting vessel, the beer being cleansed in the casks in which it was eventually stored. This primitive method of cleansing is still employed in some breweries, the wort, after being fermented for about forty-eight hours, being run into the trade casks and there cleansed. Later on larger casks, holding from two to three barrels and known by the name of "loose pieces," butts, and puncheons, came into use. These pieces were mounted on a large wooden trough called a stillion, and from this arrangement doubtless arose the Burton union system. In the cleansing system the wort is pitched at from $56°$ to $60°$ F., and the fermentation allowed to proceed until its gravity has been reduced to about one-half, by which time its temperature will have risen to about $70°$, a point generally reached in thirty-six to forty hours from the time of pitching. The half-fermented wort is then run off into the cleansing casks. This distribution of the wort into much smaller quantities serves to keep its temperature down; consequently, when the temperature of the fermenting-room does not exceed $45°$ or $50°$ F., the contents of the cleansing casks never rise higher than $70°$ with worts of medium strength, or a few degrees higher with very strong worts. In the warmer months of the year, where the cleansing is performed in loose pieces, there is a danger of their temperature rising too high; consequently it is customary during the hotter months of the year to run off the wort from the fermenting vat at a somewhat earlier stage of fermentation before it has reached a temperature of $70°$ F. The Burton unions are provided with attemperators, and these are used in the summer months for keeping the wort at the proper temperature.

The yeast which is removed during the process of cleansing carries away with it a certain amount of beer; and unless some automatic arrangement, such as the Burton unions are provided with, is employed, the casks have to be replenished by hand, or, as it is technically termed, "topped up," at frequent intervals. If the casks are not kept completely full, a portion of the yeast, instead of being ejected through the bunghole, sinks to the bottom. The first topping up takes place in about four hours after the beer has been turned into the cleansing vessels, and is repeated every three hours while the fermentation is vigorous; when this slackens the filling up is performed at less frequent intervals. The beer used for this purpose should be as free from yeast as possible.

The Skimming System.—In this the fermentation is started in the same way as in the cleansing system, but when skimming-point is reached (p. 477), the wort, instead of being run off into cleansing casks, is well roused. As soon as the head begins to assume a distinct yeasty character it is skimmed off every six hours; in some breweries oftener. This may be effected by hand or by any form of skimming apparatus with which the tun is provided. The wort which passes off with the yeast should be freed from the latter as soon as possible and returned to the vat. As in this system no subdivision of the wort takes place, the regulation of its temperature is entirely dependent upon the attemperators; consequently these should be of ample power. When the temperature of the fermenting wort has risen to about 59° or 59½° F., the attemperator is started slowly, and the flow of water through it is so regulated that the heat is allowed to rise half a degree every three hours. When the temperature has reached 65° or 66° the attemperator is then set more vigorously into action, and any further rise in temperature prevented. When the fermentation on approaching completion begins to slacken, the flow of water in the attemperator is still further increased, and the temperature of the beer brought down, in the hotter months to 60° F., in the colder to a somewhat lower degree.[1] This cooling may, contrary to the opinion commonly held, be rapidly effected without any danger of the beer acquiring a yeasty flavour. The skimming is kept on until it is judged that the beer will be able to throw up just one more head of sufficient thickness to protect it from aërial contamination. The right point to stop skimming is found by pushing a small portion of yeast on one side and examining the surface of the beer thus exposed; when this appears black and clear, denoting that there is scarcely any more yeast in suspension, the skimming is stopped, and the head which subsequently forms is allowed to remain untouched. Should the beer appear brown and turbid, it is a sign that a considerable quantity of yeast is still in suspension, and skimming must still be kept on. It is better to err on the side of leaving too thick a final head of yeast than the reverse.

The Dropping System.—In this, which is a modification of the skimming system, the beer is collected in the upper vat and pitched as in the skimming system. After the fermentation has nearly reached skimming-point (p. 477), it is run down into the lower vessel, where it is treated exactly as in the skimming system. During the process of dropping, the wort is thoroughly roused and aërated, and leaves behind in the upper vessel much of those precipitated matters which are collectively termed "slummage."

The Stone Square System.—This system, which is exceedingly popular in Yorkshire and the North of England, yields beers which drink very full for their gravity, and which, since they retain large quantities of carbon dioxide, are full of life. The yeast employed is of a peculiar type, is exceedingly slow in its action, and requires the frequent rousings and aëration which are peculiar to the system to compel it to fulfil its functions properly. The wort is pitched at a temperature of 58° to 59° F., with a smaller proportion of yeast than under the other systems, from 1 to 1½ lbs. per barrel being sufficient. The yeast is

[1] The practice as to the temperature to which the wort is permitted to rise during fermentation, as also the time at which the attemperator is started, differs in nearly every brewery. Some brewers allow even their low gravity beers to rise to 72° F.

generally added to the wort when the whole of it is in the square; but a much better plan is to add it to the first portion of wort, running this down at a temperature slightly higher than the remainder, for the reasons already stated (p. 472). The square is pitched in the following manner :—The valve at the upper end of the organ-pipe is closed and a portion of wort run into the upper chamber; the yeast is then thoroughly roused in with this, the valve opened, and the mixed yeast and wort allowed to flow into the lower chamber. When the whole of the wort has been collected and pitched, it is left undisturbed for thirty-six hours, at the end of which it should have risen to about 62° F. It is now roused for the first time, the rousing being repeated every two hours during the next twelve hours, at the expiration of which (forty-eight hours after pitching) pumping commences. Before starting the pump the valve of the organ-pipe is shut, and as much wort pumped into the upper compartment as is delivered by fifteen strokes of the pump; it is then well roused, so as to mix in the yeast which has risen through the manhole, after which the valve is opened and the wort allowed to flow back into the lower chamber. Pumping and rousing are repeated every two hours, the number of strokes of the pump being increased at each repetition of the operations, beginning with fifteen strokes for the first pumping, and increasing by ten at each repetition. This is continued until the wort has reached a gravity some 1 or $1\frac{1}{2}$ degrees higher than that required at the finish. During the whole period of fermentation the temperature of the wort is kept within the necessary limits by means of the attemperating jacket. When the pumping and rousing stage is passed, the organ-pipe valve is closed, and the yeast which rises through the manhole removed every four hours. After each removal the valve is opened for a short time to allow the beer which has drained from the yeast to run down into the lower compartment. When, by observing the surface of the beer as in the skimming system, the yeast appears fully removed, the temperature of the beer is gradually brought down by attemperating to 60° F., or a little lower. The manhole is then covered with the cap, and the contents of the square allowed to remain undisturbed for two days, by which time the beer is ready for racking into the trade casks.

Settling.—When the primary fermentation in the fermenting vat is completed, the beer is allowed to remain at rest for twenty-four hours or longer, in order that it may deposit the bulk of the yeast which still remains in suspension. This part of the process may take place in the fermenting vat itself, or in the settling back, where this vessel is employed. The latter vessel is by no means so frequently used as it was formerly, for it is found that the less the beer is moved about and exposed to the air at this particular stage the better.

Racking.—As soon as the beer has become sufficiently bright it is run off, or, as it is termed, "racked off," either into the store or trade casks, each of which, before being filled, is carefully examined to ascertain that it is perfectly dry, clean, and sweet. Care is taken that only the clear beer passes off, which object is secured by having an arrangement of taps at different levels, or by a tap with an internal sleeve. For conveying the beer from the vat to the cask a leather or india-rubber hose is employed, one of its ends being attached to the tap in the vat, and the other provided with a metal thimble or a piece of metal pipe. The object of the latter is to keep the end of the racking-tube close to the bottom of the cask, so that violent agitation of the beer, which causes the excessive frothing known as "fobbing," may be avoided. Each cask, after being filled, is temporarily closed with a bung and removed to some convenient place, where, after a period of rest, it is filled up by hand, the shive inserted, and the cask taken to the store cellar.

Dry Hopping.—It is usual to add a small quantity of fresh hops to beer at the time of racking; this is known as "dry hopping." Hops of the best quality only are used for this purpose, such as Goldings or Worcesters, or the finest varieties of foreign hops, the former being added preferentially for their fine flavour and aroma, the latter for their preservative powers. The hops used should be absolutely free from mould, mildew, or other disease, and should be as little disintegrated as possible. The quantities employed vary from half a pound to one pound per barrel, or more for very strong ales. The amount required for each cask is weighed out and introduced into the empty cask by means of a wide funnel through which the hops are pushed with a short wooden rod, care being taken that the hops are simply loosened, and not broken into fragments. Hops added in this way confer additional flavour and aroma on the beer, since the whole of the essential oil which they contain is retained. They also assist in promoting condition, which is so highly conducive to stability. As was shown by Brown and Morris, hops contain a certain amount of diastase, which degrades some of the lower dextrins, and brings them into a form in which they can be fermented by the yeast which is always present in traces in every cask of beer.[1] Hops also assist the clarification since they form points of attraction for fine particles of matter in suspension, just as the wood chips or aluminium strips used by the German brewers do.

Secondary Fermentation.—Beers, when stored in vat or cask for any length of time, pass, under normal conditions, into a very slow fermentation, which, to distinguish it from that which takes place in the fermenting vat, is termed "secondary." Often the secondary fermentation becomes unduly excited; the beer is then said to "fret" or "kick up." Such a state of matters points

[1] Wahl (*Der Braumeister*, 1889, p. 307) found 295,000 yeast cells in each cubic inch of a bright finished beer, and in a brilliant sample 82,000.

either to faulty goods or defective manipulation, or both, in the previous operations. During secondary fermentation the more resistant dextrins are gradually and slowly degraded by the invertive action of the yeast, assisted by that of the diastase of the dry hops when these are added at the time of racking, and probably also to some extent by the carbonic acid existing under pressure. The dextrins, when broken down in this way, become amenable to the fermentative power of the yeast, and by their fermentation keep the beer constantly charged with carbon dioxide, in which condition it is much better able to resist the attack of disease ferments, if such organisms should happen to be present. As carbon dioxide holds the majority of disease ferments in check, saturation of the beer with this gas is an important factor in promoting its stability. The carbon dioxide produced under these circumstances is evolved in the nascent state and under pressure. Consequently, a portion of it combines with water to form the carbonic acid (H_2CO_3), which gives the beer that peculiarly sharp, stinging sensation on the palate, especially noticeable in bottled beers which have been stored for some time. Certain compound ethers or esters are also formed on prolonged storage, and these are much concerned in the fine flavour characteristic of fully ripened and matured ales. But the demand for beers of this class seems to be gradually passing away; the tendency now-a-days is towards less ripe and mature ales, and consequently the period of storage is often reduced to weeks, or is even omitted altogether. In the former case, the secondary fermentation is hurried on as rapidly as possible by frequently rolling the casks, and thus keeping the yeast in suspension. In the latter case, the beers are sent out immediately after racking, dry hopping, and fining; consequently, what little secondary fermentation there is takes place in the publican's or customer's cellar. The worts for such beers contain an abundance of readily fermentable matters, which disappear during the primary fermentation, and also some of the very low dextrins, which, though slowly fermentable bodies, pass off with sufficient rapidity to keep the beer charged with gas.

Priming.—Of late years it has become customary to "prime" beers which are to be quickly consumed immediately before being sent out, and in this way to induce a sort of artificial secondary fermentation in them. This is effected by the addition to the beer of a strong solution of one of the readily fermentable sugars, such as cane or invert sugar. The effect is transitory, being rather more permanent when invert sugar is employed, since, paradoxical as it may appear, the lævulose half of the sugar ferments with much greater rapidity when cane sugar is added than after the addition of inverted cane sugar. As lævulose is possessed of considerable sweetening power, it also serves to temporarily increase the palate-fulness of the beer.

The syrups used for priming are made of a specific gravity

approaching 1150°, but must not exceed this, since it is the highest gravity that the instrument used by the Excise officers is capable of verifying. But in order to secure the best results the syrup should be as strong as is permissible. It must be kept in a special vessel properly gauged and marked.

The Yeast of the Secondary Fermentation.—It will be re membered that Brown and Morris (p. 148) found that when a wort was pitched with the *Saccharomyces cerevisiæ* only a certain portion of its extract, which, under their theory, they assume to be free maltose, was fermented, but that on the addition of one of the wild yeasts, such as *S. ellipsoideus*, a further portion of the constituents of the wort was fermented away. From this they concluded that the secondary fermentation could only be effected through the agency of wild yeasts. They assumed that these obtained access to the wort during the time it is exposed to the air when on the cooler, refrigerator, &c.; and this view has met with many adherents. It has been found, however, that worts which have not been allowed to come in contact with anything but germ-free air after leaving the copper, and which have been seeded with a pure culture of *S. cerevisiæ*, pass through the secondary fermentation just as readily as those brewed in the ordinary way. If the lightly-hopped beer from the propagating cylinder of a pure yeast cultivation apparatus is racked off into a clean sterilised cask, it passes through a normal secondary fermentation, and here the chances of contamination with wild yeasts are completely excluded. The only conclusion which can be drawn from these facts is that the presence of wild yeasts is not absolutely necessary for securing the secondary fermentation. The *S. cerevisiæ* is quite able to effect this without the assistance of any other yeast. Thus, one of the principal objections which were urged against Dr. Hansen's proposal to employ a pure culture of the *S. cerevisiæ*, and which is still maintained in some quarters, falls to the ground.

Antiseptics.—These substances, which are used as a sort of corrective for faulty materials or processes, are occasionally added to the mash liquor or to the contents of the mash-tun, but more generally to the beer at the time of racking. All antiseptics have a prejudicial effect on the flavour of the beer, and some also, since they retard the action of the yeast during the secondary fermentation, on its conditioning. The antiseptics used in brewing are either the sulphites of calcium or sodium, or salicylic acid; and to these may be added the newly-introduced formalin, which is a strong aqueous solution of formaldehyde. Of these, bisulphite of lime is the one most frequently employed. It is prepared by the manufacturers in a solution of a gravity of about sp. gr. 1072 [1]

---

[1] The value of bisulphite of lime when estimated from its specific gravity is often delusive; it should always be determined from the amount of sulphurous acid which it contains.

degrees, and is added from the rate of a quarter to a half pint to each thirty-six gallons of beer. Salicylic acid powerfully arrests the growth of disease ferments, but exerts a strong inhibitory action on the yeast also; it may considerably retard, or even completely arrest, the secondary fermentation; consequently, beers treated with this antiseptic are liable to drink flat and insipid. When employed, it is added at the rate of a quarter to half an ounce per barrel of thirty-six gallons. Formalin has been successfully used in brewing, one fluid ounce being added to each 10 barrels of the goods in the mash-tun, and a similar proportion to the wort in the fermenting vat.[1] The fluorides have been proposed as bactericides, and, undoubtedly, they are very powerful ones; but they should never be employed in brewing, because they are highly poisonous.

Fining.—Beer fit for consumption should be absolutely bright and clear; in other words, all the matters which were in suspension should be either precipitated or removed. Owing to the taste now prevailing for an absolutely bright article, ales which do not fulfil these conditions are hardly saleable.

The tendency of all beers properly brewed from good materials is to fall bright spontaneously if stored for a sufficient length of time, but owing to the exigencies of the present day, in which quick consumption is the rule, sufficient time cannot generally be allowed for spontaneous clarification to take place; consequently it becomes necessary to employ some method which will rapidly remove suspended matters from the beer. Finings made by dissolving isinglass in sour beer or in a dilute solution of an acid are almost universally employed for this purpose.

Isinglass.—Though other substances have been occasionally employed for the manufacture of finings, isinglass alone has been found to yield a really satisfactory product. Isinglass, the best qualities of which are made from the swimming-bladder of the sturgeon and other allied species, comes into commerce in the form of thin shreds, leaves, pipes, and lumps; it varies in colour from white to deep yellow. The lighter coloured varieties command the best prices, but the latter are said to yield the most powerful finings. The article should be perfectly sound and show no signs of decomposition, such as are indicated by an unpleasant odour and flavour. It should dissolve without difficulty in slightly acid solutions, and the less insoluble matter, or "skeg," as it is termed, which it leaves behind the better.

Manufacture of Finings.—For this several large casks will be required; these are unheaded and provided with closely-fitting covers; also two sieves, one coarse, the other fine. The quantity

---

[1] Formalin added at the rate of half an ounce per barrel was found by Rideal and Slater to destroy the disease ferments without prejudicially affecting the yeast; when added in larger quantities it arrested the action of the yeast as well.

of isinglass required is about 2½ lbs. of the best qualities, or 3½ lbs. of the inferior ones, for each 36 gallons of finings. The amount of isinglass requisite for a batch, after having been weighed out, is placed in one of the casks, and as much water, to which the acid or acids to be employed have been added, poured in as will cover it. The cask is then closed, and the mixture well roused several times a day; enough water is added from day to day to keep the isinglass, which swells out considerably, submerged. When the "glass" is thoroughly softened, or, as it is termed, "cut," the mixture is passed through the coarser sieve, after which it is allowed to stand a few days longer. Finally, it is passed through the finer sieve, made up to the required bulk with water, and thoroughly roused, after which the finings are ready for use.

The Acids Used in Making Finings.—In former times sour beer was almost universally employed for the manufacture of finings, and it is considered by many that the most effective finings are obtained in this way. As sour beer is generally highly contaminated with acidifying bacteria, it is obviously a most unscientific procedure, and cannot be too strongly condemned. This objection, however, would not apply to sour beer which has been sterilised.

The acids which have been found most suitable for the manufacture of finings are acetic, tartaric, and sulphurous. The first of these cuts well, but yields an article which communicates an unpleasant flavour to the beer. Finings made with tartaric acid alone are apt to decompose and turn mouldy; hence, when this acid is employed, it is usual to add a small quantity of sulphurous acid, which acts as an antiseptic. These acids may be used in the proportion of 1 lb. of tartaric acid and 1 gallon of commercial (7.5 per cent. H_2SO_3) sulphurous acid for each 7 lbs. of isinglass. It was supposed that acetic or tartaric acid cut better than sulphurous, but it has been shown by Matthews and Lott that this is not the case.[1] In a series of comparative experiments they found that sulphurous cut more quickly than the two former acids, which were themselves about equal in cutting power. The finings prepared with sulphurous acid were sound at the end of a month; those made with acetic acid were also sound, but somewhat mouldy; those made with tartaric acid were both mouldy and decomposed. It was found best to commence the cutting with a sulphurous acid solution of 1 per cent. strength, sufficient water being added during the process to reduce this down to 0.2 per cent. at the finish. These proportions are approximately obtained by diluting one gallon of the ordinary commercial sulphurous acid (containing 7.4 per cent. H_2SO_3) with six gallons of water, and afterwards gradually adding water until the total quantity of finings amounts to 36 gallons. Lactic acid was also found to be a more

[1] *Trans. Inst. Brewing*, vol. iv. p. 201.

effective cutting agent than acetic acid, and to give somewhat clearer finings than sulphurous acid. Lactic acid is the chief cutting agent when sour beer is employed.

Condition of the Isinglass in Finings.—Matthews and Lott consider that the "glass," when converted into finings, is held by the acid in a state of true solution. They could dilute the liquid portion of finings, strained off through a cloth, indefinitely, without any precipitate being thrown down. A dilute solution of finings could also be filtered through filter-paper, but, when such a solution is made slightly alkaline, precipitation of the isinglass takes place.

Conditions of Fining.—Though the isinglass in finings is in a state of true solution, yet this state is one of an extremely unstable nature, and is affected by comparatively slight influences. Matters in suspension, such as cooler or beer grounds, inert substances, such as powdered glass, asbestos, yeast in suspension, &c., were found to cause coagulation and precipitation of the isinglass. Beer turbidity, when caused by bacteria, cannot be removed by fining, nor will beers in a state of fret take finings. The peculiar haziness known as "greyness," which is supposed to be due to the presence of nitrogenous matter in an extremely minute state of division, cannot be removed by fining. None of the ordinary sugars or dextrins appear to affect finings, but caramel causes precipitation, a large quantity of its colouring matter being simultaneously carried down. Hop-resin in suspension also precipitates finings, and is readily removable by them. Alcohol appears to favour the action of finings. The presence of 0.4 per cent. of free acetic acid, in addition to the normal amount of acid contained in the beer (0.1 per cent.), did not prevent the action of finings, but rather accelerated it. As little as 0.2 per cent. of free lactic acid considerably impairs their efficiency, and it is questionable if beer containing 0.5 per cent. or more of this acid would take finings at all. Carbon dioxide under pressure can dissolve isinglass in water, but is unable to effect its re-solution when it has been precipitated in beer. Hop-tannin, which has always been supposed to be the principal agent in causing the coagulation of finings, is present in far too minute traces in beer to exert any action of consequence, though in the case of dry-hopped beers its action may be greater. Some brewers add catechu or cutch, which contains a large quantity of catechu-tannic acid, and this exerts a marked influence on fining. Fining, to be effective, must never be conducted at a temperature below 50° F.; and finings should never, either during their manufacture or afterwards, be heated, or they completely lose their efficiency.

Methods of Using Finings.—Occasionally finings are added to beer in bulk, as, for instance, when it is in the settling or racking tank, but this plan cannot be recommended, although occasionally successful results have been obtained by it. The proper place to add finings to the beer is when it is in the trade casks, and then immediately before they are sent out. There are two distinct modes of fining, the one termed "fining in," where the finings are permanently retained in the cask; the other, called "fining out," in which the finings are ejected after a short time from the bung-hole of the cask. The chief merit of the latter mode, which is usually performed in the publican's cellar, is its quickness of action. Beers treated in this way become, as a rule, brilliant in a few hours.

In "fining in," the requisite quantity of finings is added to about three or four times its bulk of beer, the mixture well stirred or whisked together and poured into the cask, which is afterwards shived up and well rolled to distribute the finings. In "fining out," the finings are similarly mixed with some of the beer, the whole returned to the cask, and then well stirred in with a stick. The bung-hole is left open, and in the course of a short time the finings are ejected, or, as it is termed, "spurged out." The ejection is caused by the gas disengaged by the removal of pressure and by the agitation, which, becoming entangled with the coagulated finings, so lowers their specific gravity that they rise to the surface. In the former mode, no doubt, the finings float for a time from the same cause, but afterwards, as the entangled gas becomes absorbed owing to the increased pressure inside the cask, they sink.

Quantity of Finings Required.—For "fining in," about one pint of good finings, containing the eighth of an ounce of isinglass per barrel, is, as a rule, sufficient. Stubborn ales may require more, but the quantity used should not exceed two pints per barrel. Any excess beyond the quantity of finings absolutely necessary should be avoided, since they only increase the quantity of "bottoms" or precipitated insoluble matter in the cask. In "fining out," a larger proportion may be used without any prejudicial result, since nearly the whole of the added finings are ejected.

Bottling.—Ales produced for bottling are of two entirely distinct types. To the former belong the older varieties of strong, bitter, and pale ales, which are allowed a considerable length of storage in cask previous to being bottled. To the latter belong the lighter ales of more modern origin, in which the storage is shortened to periods varying from six to two months, according as a more or less ripe and mature article is demanded in the locality in which they are consumed. It includes also the very light bottled beers, which have come so much into vogue lately, and also those in which the cellaring previous to, and storage after, bottling are reduced to a minimum, condition being brought about artificially by impregnation with carbon dioxide gas.

Strong Bottled Ales.—In the production of beers of this class, only the best materials are employed, and these are brewed in the most careful manner with water containing calcium sulphate in considerable amount. They are often produced from malt alone, but more usually with 25 per cent. of glucose, and occasionally a small proportion of unmalted material as well. Hops are used at the rate of from 14 to 15 lbs. per quarter. The ale undergoes a protracted period of storage in cask, so that secondary fermentation is almost completed. Ales of this class should drop bright spontaneously. Immediately before being bottled any excess of gas is allowed to escape by means of porous spiling, care being taken that the beer is not flattened too much or it may absorb oxygen from the air, and this would be fatal to its subsequent brilliancy in bottle. The beer is run into the bottles by means of a tap, the nozzle of which is long enough to reach to the bottom of the bottle, or by means of a bottling machine. The bottles are corked immediately they are filled, and, as recommended by Pasteur, placed on their sides for twelve or twenty-four hours, in order that the oxygen in the small amount of included air may be absorbed, after which they are stood upright in the store cellar, which should be maintained at an even temperature of about 55° F. Just before consumption they are stored for a week or a fortnight at a temperature about 10° higher in order to bring them into condition. All bottles employed for bottling ales should be cleansed with the greatest care, and afterwards thoroughly dried, best by means of hot air; moisture left in the bottles simply acts as a trap for germs.

Light Bottled Beers.—In the production of these a blend of good sound English and foreign malts, with a small proportion of prepared material, such as flaked maize or rice, together with in some cases a portion of glucose or invert sugar, are used. Mr. Chapman, in a paper read before the Institute of Brewing,[1] urges with considerable force the employment of some form of limited decoction process of mashing, such as that described on p. 459, in conjunction with a brewing liquor containing a fair amount of calcium sulphate and alkaline chlorides. In this way considerable body and palate fulness are secured. Hops may be used at the rate of 6 to 8 lbs. per quarter of malt. The period of storage in cask for this class of beers varies in different localities from six months to a few weeks; where short periods of storage are adopted, the secondary fermentation has to be hurried along by frequent cask-rolling, the porous spile being frequently used, and great care is taken that the cask changes do not proceed too far before the beer is bottled off. The ale under these circumstances rarely drops bright of itself; consequently, fining has to be resorted to, and the smallest quantity of finings which will effect the object in view is used. The finings are added to the beer, previously freed from any excess of gas, a week or a fortnight before bottling takes place. It is then bottled off under all the pre-

[1] *Journal of the Federated Institutes of Brewing*, vol. ii. p. 274.

cautions described above in connection with the bottling of strong ales, after which the beer is stored in a room kept at a constant temperature of about 65° F., so as to bring it rapidly into condition.

Heating of Conditioning-Rooms for Bottled Beers.—The storerooms for this purpose are often heated during the colder months of the year with coke fires contained in braziers, but this is a very rough and ready way, since the heat is never equally distributed throughout the room. A much better plan, as suggested by Mr. Fletcher, is to have a number of ordinary gas jets placed at the level of the floor, which have their supply of gas regulated by one tap on the main. This will need adjusting several times in the twenty-four hours, so that the temperature, as indicated by a thermometer suspended in the room, may be kept as equable as possible. Larger storerooms may be economically and efficiently heated by means of four-inch cast-iron pipes, into which are inserted at intervals small steam jets, fed from the steam-boiler and regulated by one tap on the steam main. The large pipes are conveniently sunk under the floor and covered with gratings. A steam-trap, or some other means of getting rid of the condensed water, is a necessary adjunct. A small boiler, heated by gas, in connection with a coil of pipes, arranged in the same way as in a greenhouse, also forms an economical and efficient apparatus for keeping up the temperature of small conditioning-rooms.

Carbonated Bottled Beers.—As beers of this class are intended for rapid consumption, the materials used in their production need not be of the choice quality necessary in the two former cases. The mashing liquor should contain a smaller quantity of calcium sulphate, and a fairly large amount of the alkaline chlorides, and the form of limited decoction method of mashing just mentioned may be advantageously employed. The beer is all the better for being stored for six weeks or two months before being bottled; after storage it is fined, and, as soon as bright, carbonated and bottled off in one of the numerous machines constructed for this purpose, when it is ready for immediate consumption. It must be remembered that the flavour due to carbon dioxide in artificially carbonated beers is at first distinctly different to that of those in which the gas is generated by slow fermentation in bottle, but if such beers be kept for a time it improves. In the former case, the gas is, to a great extent, merely in a state of solution; in the latter, it is combined with water, and exists as carbonic acid, which is formed in the following manner: $CO_2 + H_2O = H_2CO_3$. Probably, as suggested by Professor Liebrich, traces of ethyl carbonate are formed during the secondary fermentation, and this ether confers its agreeable flavour on beers of the latter class.

CHAPTER X.

BEER AND ITS DISEASES.

BEER fit for consumption should possess an agreeable flavour and aroma, be well charged with carbonic acid gas, should possess a certain amount of palate-fulness according to its class, and, on being poured into a glass, form a persistent foamy head. Beer must, above all things, in accordance with the present taste, be perfectly clear and bright; it should also possess a slight, though hardly perceptible, degree of acidity.

Flavour and Aroma.—These depend materially upon the employment of good sound materials, for it is obviously impossible to brew a good-flavoured beer from fusty malt, and it is equally impossible to secure the proper malty flavour from raw and under-cured material. Manipulation has also much to do with flavour, so also has the quality of the water employed. The aroma will depend materially upon the quality of the hops used; it cannot be expected that an article with a fine aroma will be produced if only coarse-flavoured hops are employed in its manufacture.

The species of yeast employed has, as was first pointed out by Hansen, much influence on the flavour of beer, each different species possessing different properties in this respect. Hence the value of employing yeast derived from one single cell, and consisting absolutely of one species, by which one factor in the flavour of the beer is secured. It was first pointed out by this distinguished observer that the yeast which he had isolated and named *S. Pastorianus No. I.* was able to develop an intensely nauseous flavour in beer. Grönlund states that *S. Pastorianus No. II.* is also able to communicate a bitter twang to beer, and various other yeasts possessing similar properties have since been discovered by other observers. Another yeast, Hansen's *S. anomalus*, readily ferments wort, and communicates to the beer a fruity flavour.

The *Torulæ*, which are frequently met with in wort, do not, as a rule, cause disease in beer; one species which Grönlund met with, and named *T. novæ Carlsbergiæ*, ferments maltose, cane sugar, and glucose, and develops in beer an unpleasant bitter taste. Lafar has also isolated a mycoderma-like fungus from beer which produces acetic acid.

The bacteria met with in wort and beer principally cause trouble by the development of acidity. The longest known of these are the lactic ferments described on p. 246. They are all rod-shaped organisms, with the exception of Lindner's *Pediococcus acidi lactici*. The butyric ferments (p. 246) are rarely met with in beer. When they do intrude, they communicate to it the intensely nauseous odour and flavour of butyric acid. Acetic bacteria occasionally infect beer, and when permitted to develop there, give it a vinegar-like flavour, and turn it sour from the production of acetic acid. As these organisms cannot thrive without air, they are only met with in beer which has been allowed to become flat, and to which air has been allowed access. As a producer of bad flavour may also be mentioned the *Pediococcus cerevisiæ*, described by Balcke,[1] and a similar organism found by Reichard,[2] and termed *P. sarcinæformis*. It produces chiefly alcohol and lactic acid.

Condition.—This is maintained (except where the beer is artificially impregnated with gas) by the constant slow evolution of carbon dioxide during the secondary fermentation, which is secured by leaving a certain amount of the slowly fermentable dextrins in the beer at the close of the primary fermentation. The nature and quantity of these will depend materially upon the particular class of beer; ales which are stored for long periods will require fairly large quantities of very slowly fermentable bodies, those for quicker consumption less of these and more of the readier fermentable bodies, which should be exactly fitted to the type of the beer. If, in the case of store beers, too much readily fermentable matter is left, fretting ensues, and in the end too much extract is fermented away; such beers consequently, at the time they should be ripe and fit for consumption, drink thin and poor. In beers for quicker consumption, if these matters are too large in quantity or of too fermentable a nature, a tumultuous secondary fermentation is induced. This keeps the yeast in suspension, and, as a consequence, such beers will not take finings. Should, on the other hand, these slowly fermentable bodies be present in too small a quantity, then the secondary fermentation either entirely ceases or is so small that the beer is not kept in condition. It becomes flat in the cask, and in this state is especially amenable to the attacks of any disease ferments which it may happen to contain.

Palate-Fulness.—The amount of palate-fulness required in a beer will depend materially on its type, and on the taste prevalent in the neighbourhood where it is consumed.

Beers should never taste thin; an amount of extract should always be left after fermentation sufficient to give the necessary body. As to the nature of those constituents of the extract which essentially contribute to palate-fulness, little is known with certainty. This property does not appear to be altogether dependent on the quantity of the extract left, but rather on its quality; for of two beers brewed at the same gravity and fermented to the same degree of attenuation, the one may drink full

---

[1] *Woch. f. Brau.*, 1884, p. 185.
[2] *Zeit. f. g. Brau.*, 1894, p. 257.

and round, the other thin and poor. The more highly coloured malts, that is, those which have been exposed to a greater degree of heat on the kiln while containing a fairly high percentage of moisture than the pale varieties, produce fuller beers. Constituents which have a sweet taste also confer palate-fulness on beer, and this is heightened by the simultaneous presence of sodium chloride. The effect which priming has of giving fulness is well known, and probably depends to some extent on the sweetness of the sugar used for priming, and also on the increased amount of carbon dioxide evolved, for beers well charged with gas drink much fuller than flat samples. It is extremely probable that the nitrogenous constituents help in this direction, as also do the gummy and pectinous bodies of the malt. Prior[1] is inclined to attribute some of the palate-fulness of beer to the kind of yeast employed; thus it is often found, in two beers produced from the same materials and brewed on exactly the same lines, that the one fermented with one variety of yeast will be much fuller on the palate than that fermented with another variety. This he attributes to the gummy substance which some yeasts secrete in much greater abundance than others.

Thinness in Flavour.—This may be brought about by too great a reduction of the extract during the primary fermentation. Even when a sufficient amount of extract has been left, and this of the right quality, it may be removed by the action of the yeasts during the secondary fermentation, should these be of such a nature as to be able to hydrolyse and ferment the higher dextrins. Many bacteria also secrete diastatic enzymes, which, acting on the higher dextrins, render them amenable to the yeast. There seems to be no definite relation between the viscosity of a beer, as ascertained experimentally, and its palate-fulness. Beers which contain more than the normal amount of acidity also taste thin.

Head.—Good beer should, upon being poured out into a glass, form a good foamy head, capable of persisting for some time. The formation of this head depends upon two factors, the presence in the beer of a sufficient quantity of combined carbonic acid and of bodies which give it a certain amount of viscosity. This enables the beer to entangle and retain the liberated bubbles of carbon dioxide as a foam on its surface. What the particular constituents of the extract are which confer this property on beer is not known with certainty; it is supposed to be dependent on the gummy and pectinous substances of the malt, and perhaps to some extent to the higher dextrins and to the hop-resin. Wahl and also Windisch[2] attribute viscosity to the proteoses; these, when separated by salting out either with ammonium or zinc sulphate and redissolved in water, give a solution which readily forms a persistent foamy head. The latter considers that a sufficient

---

[1] *Chem. u. Physiol. d. Bier*, p. 502.
[2] *Woch. f. Brau.*, 1893, p. 1358; 1896, p. 1253; 1897, p. 21.

amount of the proteoses is already formed in the malt, and that any further proteolytic action on them in the mash-tun, such as that which can be brought about at temperatures under 150° F., is undesirable.

Brightness.—Beer ought to be perfectly clear and brilliant; it should show no sign of opalescence, dimness, or turbidity.

The causes of a reverse condition of things are manifold. A more or less turbid condition may be caused by unorganised suspended matters, such as starch, dextrin, proteids, galactoxylan, or hop-resin. Turbidity from starch and dextrin are rare, and can only arise from the employment of malt of extremely low diastatic power, or from some fault in the mashing process by which too much diastase is destroyed, or through the use of raw material in too large a quantity. In connection with this it must be borne in mind that Mittelmeier's erythrodextrin No. I. (see p. 163) is a comparatively insoluble body. Turbidity from this cause can be removed by the addition of cold-water malt extract. The proteids also at times develop turbidity; the cause for this may be, according to Prior,[1] under-germinated and incompletely modified malt; material made from barley too rich in nitrogen; too small an increase of yeast during the fermentation; or too low a degree of attenuation. The slightly soluble compounds which the proteids of the malt form with the tannin of the hops are probably the cause of that form of turbidity which appears in beers when their temperature is reduced, and which disappears as the temperature again rises. A rare form of haziness is said to be caused by the hop-resins being partially thrown out of solution; it probably never arises when good sound hops are employed, and is probably due to some prejudicial change having taken place in the hop-resins either during the harvesting or storage of the hops.

One of Hansen's most remarkable discoveries is that beer turbidity can be caused by certain species of yeast; this he first showed to be the case with reference to *S. Pastorianus I.* and *III.* and *S. ellipsoideus II.*, and several other yeasts have been found which behave in a similar manner. But it has been decided by direct experiment that these "wild yeasts" can only do harm when they are present beyond a certain quantity; when below this they are held in check by the ordinary yeast. It has also been shown by Hansen that two yeasts, which individually give bright beers, may, when employed in combination, yield a turbid beer; this was found to be the case with the *S. cerevisiæ No. I.* and *II.* of the Carlsberg Brewery.

It must be remembered that yeast turbidity is a normal condition during the primary fermentation, and it only becomes an abnormal one when prolonged beyond a certain stage. There are considerable differences in the way in which the various cultivated

[1] *Chem. u. Physiol. d. Bier*, p. 508.

yeasts behave at the close of the primary fermentation; some have a tendency to clot together in small aggregates and leave the beer clear, others appear to be of a more pulverulent nature, and tend to remain in suspension as clay does when mixed with water. The brewer who requires quick clarification will avoid the latter. Slow brightening, however, does not always denote a bad quality of yeast; for instance, the beer fermented with the Carlsberg *S. cerevisiæ No. I.* takes much longer to clarify than the beer brewed with *No. II.*, but it is more stable. Yeast turbidity, unless the beer is in a state of fret, is removable by fining.

Another set of organisms nearly allied to the *Saccharomycetes*, the *S. mycodermæ*, are frequently found as a film on beer, but, as a rule, they do not cause trouble. Lasche[1] has described four varieties of these organisms, of which Nos. 1, 3, and 4 were able to cause beer turbidity, while No. 2 did not appear to act in this way. The most serious kind of beer turbidity, however, is that induced by bacterial organisms, for when due to this cause it cannot be removed by finings, and it is generally associated with such profound alterations in the odour and the flavour of the beer as to render it unsaleable. All the bacteria which have been mentioned in connection with flavour, with the exception of Pasteur's lactic ferment, produce turbidity.

It has been a much-debated question as to whether the *Sarcinæ* are able to cause disease in beer or not. Balcke[2] found that the *P. cerevisiæ* caused turbidity and communicated an unpleasant flavour, and this was confirmed by Reincke[3] as well as by Lindner. On the other hand, Petersen and Hansen[4] regard this organism as a perfectly harmless one. Probably different species of *Sarcinæ* exist so much alike in appearance that it is impossible to differentiate them microscopically, some of which are innocuous, others not. The *P. sarcinæformis* of Reichard (p. 249) is often present in wort or beer without causing trouble; at other times it evinces an intense virulence. He showed that the noxiousness of the organism was much influenced by the temperature at which it was cultivated, low temperatures increasing, high temperatures diminishing its virulence. When pediococci, which had been cultivated in unhopped wort, were added to wort before the primary fermentation, they did not manifest signs of malignancy; but when the organisms had passed through a fermentation they became virulent, and those which were derived from an infected beer were the most baneful. The organism showed its influence most in weakly attenuated beers, in which the secondary fermentation had been violent. When growing in an anaërobic condition amongst the sedimentary yeast at the

---

[1] *Der Braumeister*, 1891, p. 200.
[2] *Woch. f. Brau.*, 1884, p. 105.
[3] *Ibid.*, 1885, p. 748.
[4] *Zeit. f. Brau.*, 1890, pp. 1 and 7.

bottom of the cask, it appears to be harmless, but when, as in a fretty beer, it is brought to the surface by the rapidly evolved gas, it there meets with small amounts of oxygen, and these appear to increase both its powers of reproduction and its virulence.

Ropiness.—This condition may be induced by the *B. viscosus*, Nos. *I.* and *II.*, described on p. 247. It is only when these organisms obtain access to the wort before the primary fermentation that they are able to exert their baneful influence; when added afterwards, they do not appear to exert an injurious action. According to Lindner, beers containing much assimilable nitrogenous matter, and also beers which have been attenuated to a considerable extent, are the most liable to be affected in this way. The ropiness is caused by the production of two mucilaginous substances by the bacteria, one of which contains nitrogen and is insoluble, the other is nitrogen-free and soluble in water. When the wort contains an amount of acid equal to 0.125 per cent. lactic acid, these organisms do not seem able to cause ropiness.

The *Pediococcus viscosus* (page 249), has been found to cause ropiness in German white beer. A series of intermittent outbreaks of this disease in beer was traced by Brown and Morris[1] to the aërial infection of the worts when on the cooler by this or some other closely allied organism. It formed small yellow, wax-like glistening colonies on meat-broth gelatin. Its habitat was traced to the soil of a yard adjoining the fermenting room, occupied by a pork-butcher, which was polluted with putrid animal refuse.

As bacteria are such dangerous foes, it should be one of the brewer's first cares, by the strictest attention to cleanliness in every portion of the brewery itself, in the plant, and especially in the wort mains, to eradicate as far as possible these obnoxious organisms. Even with all possible care they cannot be absolutely exterminated; yet, if kept in subjection, they are unable to do harm; it is only when allowed to get the upper hand that they are able to evince their mischievous propensities. Thanks to the teachings of Pasteur, their depredations are not now nearly so frequent as they formerly were; it is as rare now-a-days to meet with samples of yeast swarming with bacterial organisms as it was common twenty years ago.

[1] *Journal of the Federated Institutes of Brewing*, 1895, p. 14.

LIST OF
SOME OF THE PRINCIPAL BOOKS CONSULTED IN THE COMPILATION OF THIS WORK.

Natural Philosophy. **Deschanel.**
Heat. **Balfour Stewart.**
Miller's Chemistry. Part III. **Armstrong** and **Groves.**
Grundriss der Chemie. **Langer.**
Organic Chemistry. **Perkin** and **Kipping.**
Principles of Theoretical Chemistry. **Remsen.**
Lehrbuch der Agriculturchemie. **R. Sachsse.**
Handbuch der landwirtschaftlichen Gewerbe. **C. J. Lintner.**
Die Kohlenhydrate. Vol. ii. **Tollens.**
Modern Microscopy. **Cole** and **Cross.**
Einführung in das Studium der Bacteriologie. **C. Günther.**
Lectures on Plant Physiology. **Sachs.** Translation by Professor Marshall Ward.
Practical Botany. **Strasburger** and **Hillhouse.**
Articles on Mould Fungi, by Professor **Marshall Ward**, which appeared in the *Brewer's Journal.*
Micro-organisms of Fermentation. **Jörgensen.** Translated by Miller and Lennholm.
Text-Book of Biology. Vol. I. **J. R. A. Davis.**
Gährungschemie. Vol. I. **J. Bersch.**
Studies in Fermentation. **Pasteur.** Translated by Faulkner and Robb.
Fermentation. **Schützenberger.**
Practical Studies in Fermentation. **Hansen.**
Die Gärungschemie. **Adolph Meyer.**
Theorie der Gärung. **Nägeli.**
Essays on the Floating Matter in the Air. **Tyndall.**
Mikroskopische Betriebscontrolle in der Gärungsgewerbe. **Lindner.**
Parkes' Hygiene. **De Chaumont.**
Micro-organisms in Water. **P.** and **C. G. Frankland.**
Malt and Malting. **Stopes.**
Practical Brewing. **Southby.**
Malzbereitung and Bierfabrikation. **Thausing.**
A Text-Book of the Science of Brewing. **Moritz** and **Morris.**
A Handy Book for Brewers. **Wright.**
Chemie und Physiologie des Malzes und des Bieres. **Prior.**

APPENDICES

APPENDIX A.

Solution Weight and Solution Factor.—In a paper read before the Chemical Society,[1] Brown, Morris, and Millar state that, provided the solution weight of a substance for each degree of concentration is accurately known, they consider the estimation of the solids in solution from the density of the solution is capable of as great, or in most cases of greater, accuracy than by a determination effected in the usual manner by drying, unless special precautions, that are practically impossible in the ordinary routine of analytical work, are taken.

The authors have made a number of fresh determinations of the solution weights of a number of the sugars and of various starch transformation products. The last traces of moisture were removed from the substances dealt with by a process devised by Lobry de Bruyn and Van Leent. It consists in placing the sugar or other body in a small flask, which is placed in a water or oil bath, and connected with another small flask containing anhydrous phosphoric acid, a vacuum being maintained in the apparatus during the drying process.[2] From the results thus obtained, the solution factor was determined for various concentrations. They were found, as in the case of cane sugar, not to be directly proportional to the percentages present in solution, but might be expressed in the form of a series of curves. These are given in a table, and by consulting this the proper solution factor for any concentration of any of the sugars given can be found by inspection. The solution factors for solutions of several of the sugars and starch transformation products at a density of 1055 and at a temperature of 15.5 C., taken from the table, are as follows:—

Anhydrous glucose 3.825	Low starch conversion $[a]_D$ 149.7°	3.947
,, cane sugar 3.859	Medium ,, ,, 173.9°	3.985
,, invert sugar . . 3.883	High ,, ,, 188.6°	4.000
,, fructose 3.907	Amylin or dextrin	4.206
,, maltose 3.916		

They point out, in defence of their having used the 3.86 factor for starch conversion products, that, so far as ascertaining the percentages of the constituents is concerned, the factor employed is a matter of indifference, provided the specific rotatory powers and reducing

---

[1] *Jour. Chem. Soc.*, 1897, p. 72.

[2] Ost, in a recent contribution to the *Chem. Zeit.*, criticises the above work, and defends his own method for obtaining water-free maltose. He points out that the high temperatures employed by Brown, Morris, and Millar would lead to a slight decomposition of the substances, and that, consequently, all values obtained with such altered bodies must be incorrect.

values of the constituents corresponding with those of the particular divisor taken are used in the calculation. They can be readily calculated into the true amounts as soon as the true factor for the particular starch conversion is known. At the same time, they call attention in a footnote to the fact that this is only strictly correct when the solutions are of approximately the same density, and the constituents possess identically the same divisor; but they do not consider that the error thus introduced is sufficiently large to vitiate their former work, or the conclusions based on it.

When the curves of the divisors for the different grades of starch conversions are examined, it is found that for equal concentrations the divisor for high conversions is greater than that for low—in fact, there appears to be *some* inverse ratio between it and the amount of apparent maltose present.

It was found that if the mixed products of starch conversions were assumed to consist of dextrin and maltose, and that the maltose, which, according to the amyloïn theory, exists in combination with dextrin, was assumed to have the same solution density as free maltose, it became possible to obtain by calculation the divisor for the amylin or dextrin constituent. This was done for various conversions, and the curve thus found is given in the table. Though, as pointed out, it is somewhat improbable that the solution factor for combined and free maltose would be the same, yet it was found that when this dextrin curve was used in conjunction with the maltose curve it was possible to determine, within certain limits of concentration and with a fair amount of accuracy, the solution divisor for the mixed products of any starch conversion brought about by diastase, the apparent maltose percentage being either obtained from the opticity or cupric reducing power of the solution. A table calculated in this way is given showing the divisors to be employed for solutions of starch conversion products of various concentrations and of different opticities and reducing powers.

Cupric Reducing Power.—In a continuation of the paper the authors discuss the methods hitherto employed in estimating cupric reducing powers, and give a detailed description of the method now employed by themselves. O'Sullivan's method, where a filter-paper is employed, is abandoned, and a Soxhlet filter-tube used instead, the resulting cuprous oxide being weighed either after reduction to metallic copper or after oxidisation to cupric oxide. A table similar to that of Wein is appended, in which the number of milligrammes of maltose equivalent to the number of milligrammes of Cu or CuO for various concentrations is given when their method is employed. They found that within 150 to 300 milligrammes of Cu the κ 3.86 and the κ absolute of maltose were respectively 61.14 and 62.24.

They also find that an error has crept into Wein's table for maltose, the results there given being 0.5 per cent. too low.

APPENDIX B.

Specific Rotatory Power.—In the same paper the authors discuss at considerable length the methods for the exact determination of the opticity of bodies, and of the relations of the expressions $[a]_j$ and $[a]_v$.

They have compared the readings obtained with solutions of the various sugars and starch conversion products (some at different concentrations) in a Ventzke-Scheibler instrument with the readings obtained in another polarimeter used with the sodium light, and find that, with the exception of cane sugar, the two sets of readings are slightly disproportionate, showing that the dispersive powers of these are slightly higher than that of quartz or of cane sugar. Moreover, they find that the expression $[a]_j$ has been used for two rays of different refrangibility; in the one case, which they distinguish as $[a]_j$ Biot, 24°, and in the other, distinguished as $[a]_j$ Montgolfier, 24.5° are equivalent to 21.67° of instruments which use the sodium light. A table is appended showing these various relations.

APPENDIX C.

The Law of Definite Relation.—In another paper, Brown, Morris, and Millar [1] give a résumé of the former work by which this law was gradually established, and show that by the application of the refinements recently introduced, and which have just been described, that it still holds good. The authors now take for their factors soluble starch having a reducing power of $R=0$ and an opticity of $[a]_D=202°$, instead of the non-reducing dextrin, and maltose having a reducing power of $R=100$ and an opticity of $[a]_D=138°$; consequently, the specific rotatory power and cupric reducing powers of the products of starch hydrolysis are expressed by the equation $[a]_D=202-0.64\,R$.

The relations found in a large number of starch transformation products, and evaluated by means of this equation, are given in a table, and the differences between the observed and calculated values are exceedingly minute. The results are also plotted in a system of rectangular co-ordinates, and as the values fall practically on a straight line, this shows that the optical and reducing powers are proportional to one another. They also compare in the same way Lintner and Düll's figures, and find that (after making an allowance of 2 per cent. for water, which from the method of drying employed they consider the substances would probably contain) all, with the sole exception of isomaltose, are found to conform to the law. The same is the case with Ost's results, which, as they were taken on the water-free substances, required no modification.

The authors emphasise the fact that this law was deduced experimentally, and quite independently of any theoretical considerations. They are of opinion that when explained it will form the key to the question of starch hydrolysis.

Rolfe and Defren [2] have found that some similar law holds good among the products of starch hydrolysis by means of dilute mineral acids.

APPENDIX D.

Alcoholic Fermentation without Yeast-Cells.—The view propounded by Liebig (p. 291), Traube (*ibid.*), and Hoppe-Seyler (p. 334), that this fermentation is caused by an enzyme secreted by the yeast-

---

[1] *Jour. Chem. Soc.*, 1897, p. 123.
[2] *Jour. American Chem. Soc.*, 1896, p. 869.

cell, has lately received considerable confirmation from the recent investigations of Eduard Buchner,[1] who has found it possible to prepare a liquid from yeast which was capable of exciting the alcoholic fermentation. It was obtained by grinding together 1000 grammes of pressed brewery yeast (previously freed from adherent liquid by being submitted to a pressure of twenty-five atmospheres, or about 350 lbs. to the square inch) with an equal weight of quartz sand and 250 grammes of kieselguhr until the mass became moist and plastic. With this, 100 c.c. of water were intimately mixed, the whole placed in a press-cloth and submitted to a pressure of from 400 to 500 atmospheres (about three to four tons to the square inch) in a hydraulic press; about 300 c.c. of yeast-extract were obtained in this way. The cake was removed from the cloth, powdered, sifted, 100 c.c. of water mixed with it, and again similarly submitted to hydraulic pressure; in this way another 150 c.c. of yeast-extract were obtained. The secret of the operation appears to lie in the complete rupture of the cellulose envelopes of the yeast-cells, which can only be accomplished by long grinding. The yeast-extract thus obtained was a clear, slightly opalescent, yellow fluid, of a specific gravity of 1042, having a pleasant yeast-like odour. On being boiled, coagulation ensued to such an extent that the liquid became a semi-solid mass. The extract contained about 10 per cent. of solid matter in solution.

The most interesting property of the extract was its power of exciting the alcoholic fermentation when added to a solution of either glucose, fructose, cane sugar, or maltose. On equal volumes of the yeast-extract and of a concentrated solution of one of these sugars being mixed, a regular evolution of carbon dioxide commenced in from fifteen minutes to an hour, which lasted several days. A solution of this kind, kept in an ice-chamber, and in which fermentation had been proceeding for several days, gradually became turbid from the coagulation of proteid matters, but no trace of micro-organisms could be detected in it. The addition of chloroform did not arrest the fermentation, and a portion of the extract, which had been passed through a Berkefeld filter (which would certainly retain any micro-organisms), when mixed with a sterilised solution of cane sugar, fell into a state of fermentation in about a day's time, at the temperature of an ice-chamber. When a parchment-paper tube filled with the yeast-extract was suspended in a 37 per cent. solution of cane sugar, minute bubbles of gas made their appearance on its surface in a few hours, while the contents of the tube fell into a lively state of fermentation owing to the sugar which had diffused into the tube. The fermentative power of the extract gradually disappears; a quantity kept in an ice-chamber for five days was found to be unable to act upon a cane sugar solution; but, on the contrary, when a quantity of fresh extract was added to such a solution, and the mixture placed in an ice-chamber, fermentation continued for a fortnight.

In an attempt to ascertain the nature of the substance present in the yeast-extract which caused fermentation, it was found that when the liquid was heated for an hour to from 40° to 50° C., the filtrate from the coagulum in one experiment excited a feeble fermentation in a solution of cane sugar, but in another experiment it proved inactive. Evidently the active substance was coagulated at this comparatively low temperature.

[1] *Berichte*, xxx. p. 227.

In a second paper[1] the same author shows that the fermentative action is not brought about by minute fragments of yeast-protoplasm suspended in the liquid. He finds that the instability of the yeast-extract is not due to the action of atmospheric oxygen, but probably to that of a proteolytic enzyme; and this latter hypothesis also explains the preservative influence of a concentrated solution of cane sugar, which, it is considered, would destroy the activity of the proteolytic enzyme. Antiseptics, such as chloroform, benzine, &c., do not destroy the activity of the yeast-extract; but, as they possess a powerful inhibitory influence on the growth of living yeast-cells, it is only fair to presume that their influence on portions of protoplasm deprived of a protective cell-wall would be still greater. Buchner was able to evaporate the yeast-extract to dryness by exposing it in thin layers to a temperature of from 30° to 35° C. (86° to 95° F.) in a partial vacuum. When the solid matter obtained from the extract in this way was again dissolved in water and added to a 75 per cent. solution of cane sugar, fermentation commenced in from six to ten hours, and lasted for several days, the dry residue retaining its activity for twenty days. It was also found possible to isolate an active substance from the extract by precipitation with alcohol.

All these facts indicate that there is present in yeast an enzyme which causes the alcoholic fermentation, and this Buchner proposes to call "zymase." As a further proof of the validity of this theory, he found that yeast which had been dried at 37° C. (98.6° F.), and afterwards exposed for six hours to a temperature of 100° C. (212° F.), by which the yeast-cells were killed, when added to a sterilised solution of cane sugar was able to induce a vigorous fermentation. If, however, yeast dried in a similar manner was exposed to a temperature of 140° to 145° C. (284° to 293° F.), the active principle, and with it the fermentative power of the yeast, were destroyed.

APPENDIX E.

Fermentation in a Vacuum.—In discussing the question as to whether the differences in the fermentative powers of the Frohberg and Saatz yeasts are of a physiological nature or not, Prior[2] describes a method of fermentation by means of which the latter yeast may be made to ferment as much of the reducing portion (apparent maltose) of the extract of a wort as the former. In this the fermentation is conducted at a temperature of from 32° to 33° C. (89° to 91° F.). The liquid undergoing fermentation is kept in motion by means of air bubbling through it, and the process is conducted in a partial vacuum, the conditions being so arranged that half the liquid, which is removed by evaporation every twelve hours, is replaced twice a day with sterilised water. The achroodextrin III. (p. 164) is, under ordinary circumstances, only partially fermented by the Frohberg and Saatz yeasts, more being fermented by the former than by the latter; but, in the vacuum apparatus, it is completely fermented by both. Such being the case, Prior considers that any attempt to

[1] *Berichte*, xxx. p. 1110.
[2] *Bayerisches Brauerjournal*, 1895, pp. 193 and 326.

determine the various constituents of a wort by the fermentation powers of various yeasts is futile. In seeking to give a chemico-physical explanation of the phenomena observed, he states that although it is true that a definite carbohydrate is or is not fermentable by a certain yeast, yet it has been shown that there are readily and difficultly fermentable bodies, and that their complete fermentation is in a great measure dependent on the conditions under which the fermentation is conducted, as in the case of the ordinary fermentation of a wort, where a portion of the readily fermentable sugar maltose is incompletely fermented in the presence of dextrin. Prior attributes this to the arresting influence of the dextrins,[1] which, he considers, act in accordance with the laws governing the osmotic pressure exerted on the cell-walls of the yeast, by means of which the assimilation of nutritive matters by the yeast is controlled.

[1] *Bayerisches Brauerjournal*, 1896, p. 205.

INDEX

ACHROODEXTRIN III. of Prior, 164
Acids of the acetic series, 34
—— organic, Salts of, 36
—— to aldehydes and alcohols, Conversion of, 35
Aëration, Importance of, during yeast cultivation, 352
Aërators, 451
Air-filters, 357
Air, Investigation of, for living organisms, 222
—— of breweries, Hansen's investigations on the, 313
Albumins, 91
Alcoholimeter, Tralles', 23
Aldoses, The, 33
Amide constituents of plants, The, 99
Amides, The, 37
—— of malt, Nature of, 101
Amylan, α and β, 86
Amylodextrins, No. I. and II., 74
Anaërobic organisms, Cultivation of, 216
Antiseptics, 483
Apparatus for counting bacteria on plate cultivations, 217
—— Pure cultivation, for yeast, 355
Appert's method of preserving provisions, 287
Asparagine, 37, 99
Aspartic acid, 100
Aspergillus oryzæ, 264
Asymmetric carbon atoms, 41
Atoms, 28
Attemperators, 448
Autoclave, The, 208

BACILLUS, the hay, Pure culture of, 201
—— viscosus I. and II., 247
Backs, Hot and cold liquor, 430
Bacteria, The, 239
—— The acetic acid, 248
—— The butyric acid, 246
—— the cause of acidity in beer, 490
—— Excretion in, 243

Bacteria, The lactic acid, 247
—— Locomotion in, 244
—— met with in brewing operations, The principal, 245
—— Methods of obtaining, 200
—— Microscopic examination of, 200
—— Nutrition of, 242
—— Observation of the individuals forming colonies of, 214
—— Reproduction by fission in, 240
—— Reproduction of, by sporulation, 241
—— Respiration in, 242
—— Secretion in, 243
—— Structure of, 239
—— which can exist in wort, 245
—— which can survive the alcoholic fermentation, 247
Bacteriological methods, 205
Bacteriology, Methods of sterilising in, 205
Bacterium termo, 245
Barley, Average composition of the proteids of, 95
—— Chemical examination of, 417
—— Choice of, for malting, 395
—— Coldness, Maturity, Mouldiness, Damaged corns, 398
—— Colour, Skin, 400
—— Condition of the endosperm, 397
—— Germination of, 271, 275
—— Growth of, on the malting floor, 408
—— Influence of age on, 397
—— Odour, Size, Weight, Uniformity, 399
—— The proteids of, 92
—— Screening of, 401
—— Steeping of, 401
—— Testing for transparency, 398
—— Tests for the vitality of, 395
—— Varieties of, 394
—— Vegetative energy and vegetative capacity of, 397
—— Vitality of, 395
Barleycorn, Structure of the, 274

Barley starch, 65
Bates' saccharometer, Principles of the construction of, 23
Beer and its diseases, 490
—— Brightness of, 493
—— Condition in, 491
—— Capacity for producing and maintaining a head, 492
—— Flavour and aroma of, 490
—— Palate-fulness of, 491
—— Thinness of flavour in, 492
—— turbidity, caused by certain wild yeasts, 493
Beet and cane sugar, Difference between, 84
Betaine, 101
Bibasic acids, 36
Biology, Vegetable, 224
Black beers, Grists for producing, 461
Boiling, Action of the hop-tannin bodies during, 467
—— and hopping, 464
—— Objects of, 464
—— Precautions to be observed in, 465
—— at twice, 466
Boiling points of some liquids used in freezing machines, 14
Botrytis cinerea, 262
Bottled ales, Carbonated, 488
—— Light, 488
—— Strong, 487
Bottling of ales, The, 487
Brew, Amount of liquor required for, 454
—— Estimation of the quantity of the various materials for the, 453
Brewery, General arrangement of the plant of, 428
Burton unions, 448

CANE sugar, Properties of, 83
Carlsberg flasks, Sterilisation of, 352
Casks, 451
Cavendish estimates the products of fermentation, 285
Cellulose, 62
Cellulose walls of the endosperm, Signs of dissolution in, 275
—— of the endosperm, Microscopic tests for dissolution in, 277
Chalara mycoderma, 264
Chamber, The moist, 213
Chemical constitution, 27
—— union, 27
—— symbols, 25
Chemistry, with special reference to the materials used in brewing, 25
Chlorhydrins, The, 34

Choline, 101
Cold malt-extract without action on starch, 128
Colloids and crystalloids, 230
Colonies, Observation of, under the microscope, 214
Combining proportions, 25
Competition among the organisms in a fermenting fluid, 303
Compounds, Saturated, 28
—— Unsaturated, 28
Conditioning rooms for bottled ales, Temperature of, 488
Configuration of the glucoses, 44
Cooler, The, 442
Cooling, 469
—— Employment of germ-free air in, 469
Copper, The, 439
Couching, 407
Crystalloids and colloids, 230
Cultivation plate, Preparation of, 211
Culture medium, Preparation of solid, 209
—— Pure, Methods of obtaining, 304
Cupric oxide reducing power of a sugar, 53

DEMATIUM pullulans, 266
Density, 15
Dextrins, Attempts of Brown and Morris to purify, 145
—— Ling and Baker on, 165
—— Lintner and Düll on the, 157
—— Molecular weight of, 149
—— Ost on the, 157
—— Scheibler and Mittelmeier on the, 155
—— Unfermentability of, by the Sacc. cerevisiæ, 147
—— Yield of maltose from, 144
Dextrose, glucose, or grape sugar, 78
Diastase, Action of, on starch, 122
—— Barley, 111
—— Compound nature of, 155
—— Increase of, during germination, 404
—— of malt, Isolation of, 106
—— The multiple nature of malt, 110
—— Nature of, 109
—— Osborne's investigations on, 107
—— Osborne's preparation of, 109
—— of secretion, 276
—— Test for, 110
—— of translocation, 277
—— without action on maltose, 142
Diastases or amylolytic enzymes, The, 105

INDEX.

Dilution method of obtaining pure cultures, 342
Disaccharide sugars, 49
Dressing the gyle, 476
Dry hopping, 480
Drying kiln, The, 413
Drying of malt, The, 414
Drum pneumatic malting, 411

ELEMENTS, Table of the more common, 26
—— The, 25
Enzymes or hydrolysts, The, 101
—— Action of, similar to that of acids, 102
—— derived from the albuminoids, 103
—— Purpose of, in nature, 102
—— Various inverting, 114
—— The various groups of the, 104
Enzymes, proteolytic, Action of, 165
—— Action of, in germination, 279
Equation, The number eight, 141
Equations, Transformation from higher to lower, 140
Ethers, Compound, or esters, 36
Eurotium aspergillus glaucus, 260
Evaporation, 14
—— in a vacuum, 14
Examination of the air for germs (Pasteur), 305
Excise charges, 470
Excretion in plants, 228
—— in yeast, 234
Extract per quarter yielded in a brew, 471

FABRONI on fermentation, 286
Fatty constituents of barley, 88
Fehling's solution, Composition of, 51
Ferment organisms, Presence of, on the skins of ripe fruit, 307
—— seldom present in the atmosphere, 306
Fermentation, 280
—— Addition of the yeast to the wort, 472
—— Adrian Brown's experiments on, 335
—— Alcoholic, caused by the mucors, 255
—— Amounts of yeast formed in, 339, 340, 341
—— Antiquity of, 280
—— Brefeld's views on, 327
—— The butyric acid, 316
—— The cleansing system, 477
—— The dropping system, 479
—— Energy of, 365

Fermentation experiments conducted with different amounts of air (Pasteur), 319
—— experiments with inorganic nitrogenous substances, 315
—— Extension of Pasteur's theory on, to every living cell, 325
—— Hoppe-Seyler's theory of, 334
—— Importance of aëration in, 322
—— Lactic acid, 315
—— Meyer's theory on, 327
—— Nägeli's moleculo-physical theory of, 330
—— Pasteur's experiments on, 314
—— Pasteur's theory of, 323
—— is a phenomenon of nutrition, Pasteur's theory that, 317
—— The physiological or vital theory of, 292
—— and putrefaction, 243
—— Schützenberger's views on, 328
—— Secondary, 481
—— The skimming system, 478
—— The stone square system, 479
—— Temperatures at which it is conducted, 476
—— Views on, during the period of the Alchemists, 281
—— during the period of Iatrochemistry, 282
—— during the period of the foundation of modern chemistry, 285
—— during the age of phlogiston, 284
Fermentative energy of yeast, Hayduck's apparatus for determining, 365
Fermenting vessels, 444
Ferments, Many of the mould fungi are able to act as, 317
Filters, Air, 357
Fining, The conditions of, 485
—— of beer, The, 483
Finings, The acids used in making, 484
—— Manufacture of, 484
—— The methods of using, 486
—— The quantities of, required, 486
Fire-coppers, 440
Flask, The Carlsberg, 351
Floating matter in the air, General conclusions concerning, 312
—— Morris's investigations on the, 313
—— Frankland's method of investigating the, 311
—— Hansen's method, 311
—— Hueppe's method, 311
—— Koch's method, 310
—— Miquel's method, 311

INDEX.

Floating matter in the air, Miquel on the, 310
—— Petri's method of investigating the, 311
Flooring of malt, 407
Fructose, lævulose, or fruit sugar, 81
Fruit immersed in carbon dioxide produces alcohol, 325
Fungus formed and sugar decomposed, Relation of, in fermentation, 318
Furfural, 87

GAY-LUSSAC, on fermentation, 287
Gelatinisation of starch, 69
—— temperatures of the different starches, 70
Gelatin-meat broth, Objections to, in air analysis, 311
Germ, Parasitic nature of, 278
Germination of barley, 403
—— First signs of enzymic action in, 275
—— Increase of acidity during, 406
—— Proteids of barley become soluble during, 406
Germs present in the atmosphere, The origin and distribution of, 305
Gliadin, 91
Globulins, 91
Glucase, 112
Glucose, 47
—— characteristics of, 80
—— constitution of, 81
—— grape sugar, or dextrose, 78
—— Preparation of, 78
Glutamine, 38, 100
Glutaric acid, 100
Gluten group, 95
Glutenin, 92
Glycogen, 119
Granulose, Starch, 68
Grape sugar, glucose, or dextrose, 78
Griessmayer's dextrins, 123
Grist case, The, 434
Gum, 87

HÆMATIMETER, Method of using, for counting yeast cells, 335
Half-shadow polarimeter of Schmidt and Haensch, 59
Hanging-drop method, 199
Hansen's apparatus, method of working, 357
—— system of pure yeast culture, Advantages of, 360
Hardness of water, Clarke's test for, 372
Heaping up of malt, 416

Heat, 3
—— capacity or specific heat, 10
—— Conduction of, 3
—— Convection of, 4
—— Definition of, 3
—— Expansion by, 5
—— Radiation of, 5
—— of solidification, Latent, 12
—— of vaporisation, Latent, 13
—— unit of, 10
Hemi-organisation theory of Fremy, 307
Heptose, 50
Hop back, The, 442
—— plant, The, 171
Hopping, Dry, 480
Hops, 170
—— Addition of, to wort in copper, 464
—— Alkaloids of, 176
—— Bitter principles of, 174
—— Chemical constituents of, 173
—— Chemical analysis of, 177
—— Diastase of, 177
—— Drying and sulphuring of, 173
—— Essential oil of, 173
—— The resins of, 174
—— The tannin of, 175
—— Valuation of, 172
—— Varieties of, 172
Hordein, 91
Hydrazones, The, 39
Hydrometer, Principles of the construction of, 18
Hydrometers indicating directly degrees of specific gravity, 20
—— indicating percentage of substance in solution, 22
—— indicating pounds per barrel, 21

INACTIVE series into the L. and R. series, Resolution of, 47
Incubator, The, 213
Invert sugar, 84
—— Preparation of, 85
—— Properties of, 85
Isinglass, 484
—— The condition of the, in finings, 485
Isomaltose, Brown and Morris on, 160
—— Characters of, 153
—— Discovery of, 151
—— Investigations of Ling and Baker on, 158
—— Lintner's rejoinder to Messrs. Brown and Morris on, 161
—— Ost's investigations on, 161
—— Preparation of, 152

INDEX.

KAINIT, Composition of, 379
Ketones, The, 33
Ketoses, The, 34
Kilning process, Changes effected during the, 415

LACCASE, 112
Lactic acid, 35
Lactobionic acid, 50
Lactones, The, 35
Lævulose, fructose, or fruit sugar, 81
—— Preparation of, 81
—— Properties of, 82
Laurent polarimeter, 57
Lavoisier, on fermentation, 286
Law of definite relation, 143
Lechartier and Bellamy's experiments on self-fermentation in fruit, 325
Leucine, 100
Liebig, Views of, on fermentation, 289
Loose pieces, 447

MAIZE, The proteids of, 96
Maize starch, 66
Malt, Black or roasted, 419
—— Care in grinding, 433
—— Chemical examination of, 422
—— Diastatic power of, 424
—— Germinated barley, 419
—— The nitrogenous constituents of, 169
—— The proteids of, 96
—— Ready-formed sugars of, 426
—— Specific heat of, 10
—— Storage of, 416
—— substitutes, 418
—— Valuation of, 420
Malt-extract, Action of, on starch in the cold, 129; normal, 133
—— Temperatures at which the various starches dissolve in, 128
Malt-mill, The, 431
Malting, 400
—— Pneumatic, 409
Maltobionic acid, 50
Maltodextrin of Brown and Morris, 146
—— of Hertzfeld, 147
—— Unfermentability of, by the Sacc. cerevisiæ, 147
Maltol, 427
Maltosazone, 78
Maltose, 75
—— Action of yeast on, 77
—— Characteristics of, 76
—— Constitution of, 78
—— Discovery and re-discovery of, 126

Maltose, Hydrolysis of, 77
—— Preparation of, 75
Mannose group, 46
Mash, Dead, 463
—— Second, 459
—— Time of standing on, 458
Mash-liquor, Addition of hardening materials to, 455
—— Heating of the, 455
Mash-tun, The, 434
—— Appliances for raising the temperature of, 438
Mashing, Allowances for specific heat of malt, 11
—— Clinch's hot grist, 460
—— Employment of prepared materials in, 461
—— Limited decoction method of, 460
—— machines, External, 436
—— The operation of, 457
—— System of using large amounts of raw grain in, 461
—— Temperature of, 456
—— Use of subsidiary apparatus in, 459
Meat-broth gelatin, Preparation of, 209
Melibiase, 113
Mercury used in experiments may be a source of bacterial contamination, 309
Metamaltose, 164
Micro-chemical reagents, 191
Microscope, The, 179
—— Choice of, 189
—— The diaphragms of, 184
—— Drawing apparatus for the, 191
—— Eye-pieces for, 188
—— Illuminating apparatus for, 188
—— Illumination of, 190
—— Interpretation of objects seen under, 192
—— Nose-piece of, 183
—— Objectives for, 185
—— The stage of the, 183
—— The sub-stage of the, 184
Microscopes, Description of, 180
Microscopic cover-glasses, 191
—— glass slips, 191
Microscopical manipulation, 192
—— preparation, Conversion of temporary into permanent, 204
—— the stained cover-glass, 203
—— staining fluids, 192
Mineral food of plants, 229
Miquel's experiments on the floating matter in the air, 310
Modification of endosperm, 403

INDEX.

Molecular weight, determination of, 55
Molecules, 28
Monilia candida, 268
Mould fungi, Danger of, to brewing premises, 268
—— fungi, may act as ferments, 317
—— fungi, The, 250
Mucor circinelloides, 255
—— erectus, 255
—— mucedo, Growth of, 250
—— mucedo, Nutrition of, 253
—— mucedo, Sexual reproduction in, 252
—— mucedo, Sporulation of, 252
—— racemosus, 254
—— spinosus, 255
—— stoloniferus, 255
Multicellular organisms, Simple, 256
Musculus and Gruber's dextrins, 132
Mycoderms, The, 239

NASCENT state, 29
Nonose, 51
Nuclein, 121

OBJECTIVES, Microscopic, 185
Octose, 50
Oidium lactis, 265
—— lupuli, 265
Opticity or specific rotatory power, 60
Osazones, The, 39
Osmosis, 229
Osones, The, 41
O'Sullivan, on the dextrins, 124
O'Sullivan's equation A, 130
—— equation B, 131
—— equation C, 131
—— investigation on the action of diastase on starch, 126
Oven, The hot-air, 206
Oxalic acid, 35

PARACHUTE, The, 449
Pasteur flask, The, 298
Pasteur's investigations on spontaneous evolution, 297
Pectin, 88
Penicillium glaucum, 256
—— Growth of, 256
—— Respiration and nutrition in, 259
—— Sexual reproduction in, 258
Percentage of extract on the malt yielded in a brew, 471
Petri dishes, 215
Phenylhydrazine, 38
Piece liquor, 458

Piece liquor, Allowances for specific heat of, in mashing, 11
Plants, Substances produced in the different cells of, 271
—— The higher, 270
Plate-cultivation method, 343
—— Koch's, Hansen's modification of, 343
Pneumatic malting, 409
—— Advantages of, 412
Polarimeter, The, 57
Potato-starch, 66
Pouchet's experiments, 309
Pounds per barrel to specific gravity, Relation of, 21
Preparation, The impression, 215
Preservation of fermentable fluids in Pasteur flasks, 309
Previous sewage contamination of a water, 386
Priming of beer when in cask, 482
Proteid groups, The various members of the, 91
Proteids or albuminoids, The, 89
—— Action of pepsin on, 166
—— Action of the proteolytic enzymes on, 165
—— Action of trypsin on, 168
—— Effects of hydrolysis on, 91
—— Formation of, in plants, 227
—— General reactions of, 90
—— of barley, The isolation of, 92
—— of malt, Coagulation temperatures of, 98
—— Molecular constitution of, 90
Proteolysis during germination of barley, 168
Protococcus pluvialis, 224
Putrefaction, 243

RACKING off of the finished beer, 480
Racking squares, 451
Raffinose, 86
Rakes or internal mashing machine, The, 436
Raoult's method for determining molecular weights, 55
Reaction of medium determines the class of organism developing in it, 302
Refrigerating, 469
Refrigerators, 442
Relation, The law of definite, 143
Reproduction in the Protococcus, 231
—— of yeast by budding, 235
Respiration of barley during germination, 405
—— in plants, 228
—— in yeast, 234

INDEX.

Rice-starch, 66
Rootlets, methods of keeping root-production down, 405
—— Removal of, 416
Ropiness in beer, Causes of, 495
Rousers and aërators, 451

SACCHAROMETER, Bates', 23
Saccharomyces apiculatus, 237
Salts of organic acids, 36
Sarcinæ, The, 249
—— Influence of, in producing disease in beer, 494
Saturated compounds, 28
Schifferer's experiments on starch transformation, 153
Schizo-saccharomycetes, The, 238
Schröder and Dusch, Invention of cotton-wool air-filter by, 295
Schulze's experiments on spontaneous evolution, 295
Schwann, on fermentation, 293
Secondary fermentation, 481
Settling after fermentation, 480
Soluble starch, Preparation of, 72
—— Properties of, 73
Solution factor, 53
Sparging, 462
Specific gravity, 15
—— gravity bottle, 17
—— gravity to pounds per barrel, Relation of, 21
Specific gravities of gases, Methods of obtaining, 16
—— of liquids, Methods of obtaining, 17
—— of solids, Methods of obtaining, 16
Specific heat, 10
Specific rotatory power or opticity, 60
Spontaneous evolution, The doctrine of, 294
Sporulation of yeast, 235
Sprinkling of growing barley, 408
Stahl, on fermentation, 284
Starch, 64
—— Action of acids on, 70
—— Action of diastase on, 122
—— Action of malt-extract heated to 104° F., on, 136
—— Action of malt-extract heated to 122° F., on, 136
—— Action of malt-extract heated to 140° F., on, 137
—— Action of malt-extract heated to 151° F., on, 138
—— Action of malt-extract heated to temperatures higher than 151° F., 139

Starch, Action of malt-extract which had been heated to various temperatures, on, 135
—— Action of neutralised malt-extract on, 139
—— Behaviour of, with iodine, 68
—— bruised, Action of diastase on, 134
—— Characteristics of, 64
—— Composition of, 67
—— of the endosperm, Solution of the, 276
—— Effects of micro-chemical reagents on, 196
—— Formation of, 65
—— Gelatinisation of, 69
—— Molecular weight of, 149
—— Ost's conclusions on the action of diastase on, 162
—— soluble, 72
—— The molecular transformations of, 140
—— unruptured, Diastase has no action on, 134
Starch cellulose, 63, 67
—— conversions, Influence of temperature on, 129
—— Investigations of Brown and Heron on, 133
—— degradation, Mittelmeier's views on, 163
—— gelatinised, Action of malt extract in the cold on, 135
—— granules, Microscopic characters of, 65
—— molecule, Brown and Morris's hypothesis of the breaking down of the, 148
—— molecule, Brown and Morris's second hypothesis, 150
—— molecule, Constitution of, 70
—— transformation, Investigations of Brown and Morris on, 142
—— transformation, Views of Musculus and Gruber on, 132
—— transformation products, Criterion of purity of, 143
—— transformations, Table of equations for (Brown and Heron), 141
Starches, Microscopic examination of the, 195
Steam, Heating power of, 13
Steam-Copper, 441
Steep-water, Change of, 402
—— Quality of, 402
Steeping, Changes taking place during, 403
—— Signs of proper, 402
Stereoisomerides, Optical, 41

INDEX

Stereoisomerism, Cause of, 41
Steriliser, The steam, 208
Sterilisation, Influence of reaction of fluid in, 301
—— Organic fluids of different constitution require varying temperatures to effect their, 301
Sterilising by direct heat of flame, 206
Stone square, The, 445
Substances containing two or more asymmetric carbon atoms, 43
Succinic acid, 36
Sugar determinations, Tables for, 52
—— Estimation of, by Fehling's solution, 51
—— Formation of, in living plants, 49
—— group, Constitution of, 45
Sugars, Cupric reducing powers of, 53
Synthetic Sugars, 48

TAPS, setting, 462
Termobacteria, The, 245, 246
Thermometer, The, 6
—— Directions for using, 9
—— Graduation of, 6
—— Manufacture of, 6
—— Sources of error in, 8
—— Various scales of, 7
Thermometric scales, Conversion of the readings of, into each other, 8
Thermostat, The, 207
Torulæ, The, 239
Transformation of micro-organisms, 304
Traube, Views of, on fermentation, 291
Turning out of the copper wort, 468
Tyndall's experiments on the floating particles in the air, 299
Tyrosine, 100

UNDERBACK, wort in, 464
Underlet, 458

VALENCY, 26
Vats and casks, 451
Vernine, 101
Vessel, Collecting, 443
Vessels, Fermenting, 444
—— for primings, &c., 443
Vitality, Dormant, 231, 274
Vitellins, 91
Volumeters, 20

WATER, 367
—— Bacterial examination of, 216

Water, Hardness of, 372
—— Occurrence of, 367
Water analysis, process of Wanklyn, 387
—— analysis, Behaviour of water residues on ignition, 389
—— analysis, Chlorine viewed from an organic standpoint, 389
—— analysis, Comparison of the results by the Wanklyn and Frankland processes, 387
—— analysis, Determination of oxygen absorbed, 388
—— analysis, microscopic examination of the sediment, 392
—— analysis, Frankland's combustion process, 384
—— analysis, Heisch's sugar test, 391
—— analysis, Nitric acid and nitrates, 388
—— analysis, nitrogen existing as nitrites, 388
—— analysis, phosphoric acid, 389
—— analysis, Views of Matthews and Lott on Heisch's test, 392
—— analyses, Statements of the results of, 370
Water, Rain, 369
—— Well and spring, 369
—— General constituents of, 370
Waters adapted for producing pale ales, 375
—— considered from an organic point of view, 382
—— fitted for mild ales, 376
—— Hansen's method of the bacteriological examination of, 218
—— Influence of boiling on some, 382
—— Influence of filtration on, 383
—— suitable for black beer, 376
—— unfitted for brewing purposes, 381
—— with respect to the geological formation of the district, 372
Waters, brewing, Artificial treatment of, 376
—— considered inorganically, 373
—— Removal of iron from, 381
Waters, filtered, Biological examination of, 218
Waters, natural, Impossibility of artificially imitating, 371
Waters, polluted, Microscopic objects met with in, 200
Wheat-starch, 66
Wichmann's method of examining waters bacteriologically, 221
Wine manufacture in the Jura, 318
Wine yeast, Apparent origin of, 267

INDEX.

Withering during the malting process, 409
Wort, Collection of the, 470
Wort, Return, 462

YEAST, Amount of air absorbed by, during growth, 322
—— Amounts of nitrogenous matter removed from wort by, 365
—— for analysis, Time for taking, 362
—— Analysis of, by Hansen's method, 361
—— Appearance of the heads during fermentation, 477
—— Arrangements for the removal of, 449
—— Chemical composition of, 114
—— Changes of, 474
—— Choice of, for pitching, 472
—— Cultivation of, under free exposure to air, 321
—— Development of, by budding, 293
—— Differences in the behaviour of, with raffinose, 364
—— Differences in the fermenting power of, 362
—— Differences in the attenuative powers of, due to the enzymes secreted, 363
—— Differences in the action of the various species of, 362
—— Differences in fermenting power of, at various ages, 321
—— Differences in the properties of the various kinds of, 350
—— Differences in the reproductive powers of, 365
—— Different quantities of acid produced by, 364
—— Different enzymes secreted by, 364
—— Distinction of the different varieties of, 345
—— Elementary composition of, 115
—— The enzymes of, 112
—— The fat of, 119
—— grown under exclusion of oxygen, 319, 323
—— grown in presence of an unfermentable sugar, 324

Yeast, Hansen's six species and varieties, 346–348
—— Introduction of pure cultures of, into the brewery, 349
—— Investigations of Hansen on, 342
—— Microscopical appearances of, 198
—— Microscopic appearances of vigorous and worn-out, 321
—— Mineral constituents of, 116
—— Multiplication of a single cell of, to quantities sufficient for the brewery, 351
—— The nitrogenous constituents of, 120
—— Nutrition of, 233
—— The practical employment of pure cultures of, in the brewery, 350
—— Preparation of pure cultures of, 343
—— Products of the putrefaction of, 121
—— of the propagating apparatus, Analysis of the, 362
—— Pure cultures of, may work irregularly at first, 353
—— Pure cultivation apparatus for, 355
—— Quantity of, required for pitching, 474
—— of the secondary fermentation, 482
—— Specific gravity of, 121
—— Species and varieties of, 236
—— Variations in the quantities of alcohol produced by, 364
—— Vitality of, 314
Yeast cells, method of counting, by hæmatometer, 335
Yeast cellulose, 117
Yeast cultivation, apparatus of Berg and Jorgensen, 359
Yeast glucase, 113
Yeast gum, 118
Yeast mucilage, 118
Yeast plant, 232
Yeasts which cause disease in beer, 490
—— Wild, 237

Printed by BALLANTYNE, HANSON & CO.
Edinburgh & London